Lecture Notes in Artificial Intelligence   12109

Subseries of Lecture Notes in Computer Science

More information about this series at http://www.springer.com/series/1244

Cyril Goutte · Xiaodan Zhu (Eds.)

Advances in Artificial Intelligence

33rd Canadian Conference on Artificial Intelligence, Canadian AI 2020
Ottawa, ON, Canada, May 13–15, 2020
Proceedings

 Springer

Editors
Cyril Goutte 🔟
National Research Council Canada
Ottawa, ON, Canada

Xiaodan Zhu 🔟
Queen's University
Kingston, ON, Canada

ISSN 0302-9743   ISSN 1611-3349   (electronic)
Lecture Notes in Artificial Intelligence
ISBN 978-3-030-47357-0   ISBN 978-3-030-47358-7   (eBook)
https://doi.org/10.1007/978-3-030-47358-7

LNCS Sublibrary: SL7 – Artificial Intelligence

This Springer imprint is published by the registered company Springer Nature Switzerland AG
The registered company address is: Gewerbestrasse 11, 6330 Cham, Switzerland

Preface

We are particularly pleased to present this volume containing the collected work prepared for and presented at the Canadian AI 2020 conference, the 33rd Canadian Conference on Artificial Intelligence, held during May 13–15, 2020. The Canadian AI conference is one of the longest running AI conferences, running first biennially from 1976 to 2000, then annually ever since. This volume continues to show the thriving vitality and leadership of the Artificial Intelligence (AI) scene in Canada. The 2020 conference was supposed to take place in Ottawa, where Canadian AI had taken place roughly every 10 years (1990, 2001, and 2010). However, the spread of Covid-19 had a massive impact on conferences throughout the world in the spring and summer of 2020, and the Canadian AI conference was moved to a virtual, fully online format.

We received 145 submissions to the main conference, the highest number for a Canadian AI conference since at least 2008. Most submissions were reviewed by three Program Committee members, although a few got only two, or up to five reviews. Based on the recommendations of the Program Committee, 31 submissions were accepted as long papers (12 pages), and an additional 24 submissions were accepted as short papers (6 pages). The selected papers cover a wide range of topics, including machine learning, pattern recognition, natural language processing, knowledge representation, cognitive aspects of AI, ethics of AI, and other important aspects of AI research. They reflect some of the most recent and trending topics such as adversarial learning and reinforcement learning, as well as applications of AI to various problems in healthcare, social media and network analysis, affective computing, anomaly detection, or processing of sensor data. In addition, the Graduate Student Symposium, co-chaired by Pooya Moradian Zadeh and James Wright, ran its own selection process. From 30 submissions, they selected 8 for oral presentations, 4 of which are included at the end of this volume. In addition, David Nadeau from Innodata Labs organized an Industry Session on the topic "Industrial AI: A day in the life of an AI practitioner." We thank David, James, and Pooya for organizing these sessions and selecting the relevant contributions.

The contributions selected for this volume owe much to the work of the Program Committee and additional reviewers, who volunteered their time and worked dutifully to complete their reviews (mostly) on time, and provided additional feedback during the discussion to inform the final acceptance decisions. The conference would of course not exist without the contributions provided by the 385 authors and co-authors who submitted their work. We also thank the authors of the accepted papers for their efforts finalizing the camera-ready version of their papers and preparing their online presentations.

Three keynotes enriched the program of the conference, given by leading figures from the field: Giuseppe Carenini from the University of British Columbia, Csaba Szepesvári from the University of Alberta, and Pascal Poupart from the University of Waterloo. The conference also included a tutorial on "Reinforcement Learning" by

Pierre-Luc Bacon from Université de Montréal. We thank Pierre-Luc, Pascal, Csaba, and Giuseppe for volunteering their time and for their contribution to the program of the conference.

Canadian AI is sponsored by the Canadian Artificial Intelligence Association (CAIAC). We gratefully acknowledge the support of the Executive Committee of CAIAC: Leila Kosseim, Xin Wang, Richard Khoury, Denilson Barbosa, and Ziad Kobti. We also thank Fabrizio Gotti, who did a wonderful job designing and maintaining the conference website. Canadian AI was collocated with the 17th Conference on Computer and Robot Vision (CRV 2020). We wish to acknowledge the many constructive discussions we had with the CRV co-chairs Liam Paul and Michael S. Brown, as well as Michael Jenkin and Steven Waslander from the Canadian Information Processing and Pattern Recognition Society, while preparing the move to a virtual conference. Last but not least, we are extremely grateful to the general chairs of AI-CRV 2020, Marina Sokolova and Chris Drummond, for their help with planning and organization before and after the move to virtual conferences.

Finally we thank the sponsors who provided and maintained their financial support through the tribulations of this complicated season: Carleton University, The University of Ottawa (Office of Vice-President Research, Faculty of Medicine and Faculty of Engineering), Huawei, Innodata, Compusult, and Tenera Care.

March 2020   Cyril Goutte
                                                                Xiaodan Zhu

Organization

Program Committee

Esma Aimeur	University of Montreal, Canada
Xiangdong An	UT Martin, USA
Ebrahim Bagheri	Ryerson University, Canada
Caroline Barrière	University of Ottawa, Canada
Nabil Belacel	National Research Council Canada, Canada
Colin Bellinger	National Research Council Canada, Canada
Sabine Bergler	Concordia University, Canada
Gabriel Bernier-Colborne	University of Montreal, Canada
Virendra Bhavsar	University of New Brunswick, Canada
Nizar Bouguila	Concordia University, Canada
Scott Buffett	National Research Council Canada, Canada
Cory Butz	University of Regina, Canada
Laurence Capus	Laval University, Canada
Eric Charton	Yellow Pages, Canada
Colin Cherry	Google, Canada
Paul Cook	University of New Brunswick, Canada
Elnaz Davoodi	Google, Canada
Berry de Bruijn	National Research Council Canada, Canada
M. Ali Akber Dewan	Athabasca University, Canada
Chris Drummond	National Research Council Canada, Canada
Audrey Durand	McGill University, Canada
Ashkan Ebadi	National Research Council Canada, Canada
Ahmed Esmin	Federal University of Lavras, Brazil
Ali Etemad	Queen's University, Canada
Reza Faieghi	Toronto Rehabilitation Institute, Canada
Atefeh Farzindar	University of Southern California, USA
Yufei Feng	Queen's University, Canada
George Foster	Google, Canada
Michel Gagnon	Polytechnique Montreal, Canada
Sebastien Gambs	Université du Québec à Montréal, Canada
Alice Gao	University of Waterloo, Canada
Yong Gao	The University of British Columbia, Canada
Moojan Ghafurian	University of Waterloo, Canada
Ali Ghorbani	University of New Brunswick, Canada
Nizar Ghoula	National Bank of Canada, Canada
Cyril Goutte	National Research Council Canada, Canada
Yuri Grinberg	National Research Council Canada, Canada
Jiachen Gu	University of Science and Technology of China, China

Diego Guarín	Toronto Rehabilitation Institute, Canada
Michael Guerzhoy	Princeton University, USA
Jimmy Huang	York University, Canada
Diana Inkpen	University of Ottawa, Canada
Ilya Ioshikhes	University of Ottawa, Canada
Aminul Islam	University of Louisiana at Lafayette, USA
Nathalie Japkowicz	American University, USA
Dhanya Jothimani	Data Science Lab, Ryerson University, Canada
Fazel Keshtkar	St. John's University, USA
Shehroz Khan	Toronto Rehabilitation Institute, Canada
Kamyar Khodamoradi	University of Alberta, Canada
Richard Khoury	Laval University, Canada
Svetlana Kiritchenko	National Research Council Canada, Canada
Grzegorz Kondrak	University of Alberta, Canada
Leila Kosseim	Concordia University, Canada
Adam Krzyzak	Concordia University, Canada
Guillaume Lajoie	Université de Monatréal, Canada
Sébastien Lallé	The University of British Columbia, Canada
Luc Lamontagne	Laval University, Canada
Philippe Langlais	University of Montreal, Canada
Yves Lespérance	York University, Canada
Yifeng Li	Brock University, Canada
Fuhua Lin	Athabasca University, Canada
Jackie Lo	National Research Council Canada, Canada
Andrea Lodi	École Polytechnique de Montréal, Canada
Rongxing Lu	University of New Brunswick, Canada
Simone Ludwig	North Dakota State University, USA
Ali Mahdavi Amiri	Simon Fraser University, Canada
Rupam Mahmood	University of Alberta, Canada
Brad Malin	Vanderbilt University, USA
Gordon McCalla	University of Saskatchewan, Canada
Robert Mercer	University of Western Ontario, Canada
Marie-Jean Meurs	Université du Québec à Montréal, Canada
Malek Mouhoub	University of Regina, Canada
Isar Nejadgholi	National Research Council Canada, Canada
Feng Nie	Sun Yat-sen University, China
Jian-Yun Nie	University of Montreal, Canada
Roger Nkambou	Université du Québec à Montréal, Canada
Jian Pei	Simon Fraser University, Canada
Fred Popowich	Simon Fraser University, Canada
Sheela Ramanna	University of Winnipeg, Canada
Robert Reynolds	Wayne State University, USA
Mehdi Rezagholizadeh	Huawei Noah's Ark Lab, Canada
Samira Sadaoui	University of Regina, Canada
Fatiha Sadat	UQAM, Canada
Ehsan Sherkat	MALNIS, Dalhousie University, Canada

Zhan Shi	Queen's University, Canada
Daniel L. Silver	Acadia University, Canada
Michel Simard	National Research Council Canada, Canada
Marina Sokolova	University of Ottawa and Institute for Big Data Analytics, Canada
Bruce Spencer	University of New Brunswick, Canada
Sun Sun	National Research Council Canada, Canada
Stan Szpakowicz	University of Ottawa, Canada
Graham Taylor	University of Guelph, Canada
Alain Tchagang	National Research Council Canada, Canada
Thomas Tran	University of Ottawa, Canada
Chun Wang	Concordia University, Canada
Xin Wang	University of Calgary, Canada
Yunli Wang	National Research Council Canada, Canada
René Witte	Concordia University, Canada
James Wright	University of Alberta, Canada
Dan Wu	University of Windsor, Canada
Yang Xiang	University of Guelph, Canada
Jingtao Yao	University of Regina, Canada
Pooya Zadeh	University of Windsor, Canada
Harry Zhang	University of New Brunswick, Canada
Xiaodan Zhu	Queen's University, Canada
Nur Zincir-Heywood	Dalhousie University, Canada
Farhana Zulkernine	Queen's University, Canada

Additional Reviewers

Ahmad, Amir
Alhadidi, Dima
Ali, Samr
Aloise, Daniel
Arabzadeh, Negar
Azam, Muhammad
Chételat, Didier
De Blois, Sebastien
Etemadi, Roohollah
Fuggitti, Francesco
Gasmallah, Mohammed
Godbout, Mathieu
Hajizadeh, Yasin
Hamidi Rad, Radin
Hartford, Jason

Hasan, Mahedi
Hendijani Fard, Fatemeh
Hosseini, Hawre
Kargar, Mehdi
Lee, Greg
Lira, Wallace
Manouchehri, Narges
Mcintyre, Andy
Mirlohi Falavarjani, Seyed Amin
Najar, Fatma
Peet-Pare, Garnet Liam
Pieper, Michael
Qiu, Lingling
Queudot, Marc
Razavi-Far, Roozbeh

Rezvani, Reyhaneh
Samet, Saeed
Sharma, Deepak
Skiredj, Selma
Sobol, Magdalena

Vandewiele, Nick
Waroux, Renan
Xin, Weizhao
Yu, Fenggen
Zamzami, Nuha

Contents

Toward Adversarial Robustness
by Diversity in an Ensemble
of Specialized Deep Neural Networks

Mahdieh Abbasi[1]([✉]), Arezoo Rajabi[2], Christian Gagné[1,3],
and Rakesh B. Bobba[2]

[1] IID, Université Laval, Québec, Canada
mahdieh.abbasi.1@ulaval.ca
[2] Oregon State University, Corvallis, USA
[3] Mila, Canada CIFAR AI Chair, Quebec city, Canada

Abstract. We aim at demonstrating the influence of diversity in the
ensemble of CNNs on the detection of black-box adversarial instances
and hardening the generation of white-box adversarial attacks. To this
end, we propose an ensemble of diverse specialized CNNs along with
a simple voting mechanism. The diversity in this ensemble creates a
gap between the predictive confidences of adversaries and those of clean
samples, making adversaries detectable. We then analyze how diversity
in such an ensemble of specialists may mitigate the risk of the black-box
and white-box adversarial examples. Using MNIST and CIFAR-10, we
empirically verify the ability of our ensemble to detect a large portion
of well-known black-box adversarial examples, which leads to a significant reduction in the risk rate of adversaries, at the expense of a small
increase in the risk rate of clean samples. Moreover, we show that the
success rate of generating white-box attacks by our ensemble is remarkably decreased compared to a vanilla CNN and an ensemble of vanilla
CNNs, highlighting the beneficial role of diversity in the ensemble for
developing more robust models.

1 Introduction

Convolutional Neural Networks (CNNs) are now a common tool in many computer vision tasks with a great potential for deployement in real-world applications. Unfortunately, CNNs are strongly vulnerable to minor and imperceptible
adversarial modifications of input images a.k.a. adversarial examples or adversaries. In other words, generalization performance of CNNs can be significantly
dropped in the presence of adversaries. While identifying such benign-looking
adversaries from their appearance is not always possible for human observers,
distinguishing them from their predictive confidences by CNNs is also challenging since these networks, as uncalibrated learning models [1], misclassify them
with high confidence. Therefore, the lack of robustness of CNNs to adversaries
can lead to significant issues in many security-sensitive real-world applications
such as self-driving cars [2].

© Springer Nature Switzerland AG 2020
C. Goutte and X. Zhu (Eds.): Canadian AI 2020, LNAI 12109, pp. 1–14, 2020.
https://doi.org/10.1007/978-3-030-47358-7_1

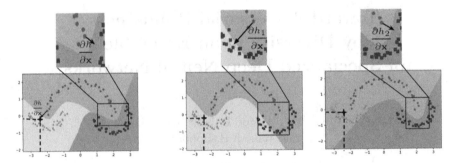

Fig. 1. A schematic explanation of ensemble of specialists for a 3-classes classification. On the left, a generalist ($h(.)$) trained on all 3 classes. In the middle and on the right, two specialist binary-classifiers $h_1(.)$ and $h_2(.)$ are trained on different subsets of classes, i.e. respectively (red,green) and (red, blue). A black-box attack, shown by **a black star**, which fools a generalist classifier (left), can be classified as different classes by the specialists, creating diversity in their predictions. Moreover, generation of a white-box adversarial example by the specialists can create two different fooling directions toward two unlike fooling classes. The fooling directions (in term of derivatives) are shown by black arrows in zoomed-in figures. Such different fooling directions by the specialists can harden the generation of high confidence white-box attacks (Sect. 3). *Thus, by leveraging diversity in an ensemble of specialists, without the need of adversarial training, we may mitigate the risk of adversarial examples.* (Color figure online)

To address this issue, one line of thought, known as *adversarial training*, aims at enabling CNNs to *correctly classify any type of adversarial examples* by augmenting a clean training set with a set of adversaries [3–7]. Another line of thought is to devise detectors to discriminate adversaries from their clean counterparts by training the detectors on a set of clean samples and their adversarials ones [4,8–10]. However, the performance of these approaches, by either increasing correct classification or detecting adversaries, is highly dependent on accessing a holistic set containing various types of adversarial examples. Not only generating such a large number of adversaries is computationally expensive and impossible to be made exhaustively, but adversarial training does not necessarily grant robustness to unknown or unseen adversaries [11,12].

In this paper, we aim at detecting adversarial examples by predicting them with high uncertainty (low confidence) through leveraging diversity in an ensemble of CNNs, without requiring a form of adversarial training. To build a diverse ensemble, we propose forming a *specialists ensemble*, where each specialist is responsible for classifying a different subset of classes. The specialists are defined so as to encourage divergent predictions in the presence of adversarial examples, while making consistent predictions for clean samples (Fig. 1). We also devise a simple voting mechanism to merge the specialists' predictions to efficiently compute the final predictions. As a result of our method, we are enforcing a gap between the predictive confidences of adversaries (i.e., low confidence predictions) and those of clean samples (i.e., high confidence predictions). By setting a threshold on the prediction confidences, we can expect to properly identify the adversaries. Interestingly, we provably show that the predictive confidence of our method in the presence of disagreement (high entropy) in the ensemble is upper-bounded by $0.5 + \epsilon'$, allowing us to have a global fixed threshold (i.e., $\tau = 0.5$) without requiring fine-tuning of the threshold. Moreover, we analyze our approach against the black-box and white-box attacks to demonstrate how, without adversarial training and only by diversity in the ensemble, one may design more robust CNN-based classification systems. The contributions of our paper are as follows:

- We propose an ensemble of diverse specialists along with a simple and computationally efficient voting mechanism in order to predict the adversarial examples with low confidence while keeping the predictive confidence of the clean samples high, without training on any adversarial examples.
- In the presence of high entropy (disagreement) in our ensemble, we show that the maximum predictive confidence can be upper-bounded by $0.5 + \epsilon'$, allowing us to use a fixed global detection threshold of $\tau = 0.5$.
- We empirically exhibit that several types of black-box attacks can be effectively detected with our proposal due to their low predictive confidence (i.e., ≤ 0.5). Also, we show that attack-success rate for generating white-box adversarial examples using the ensemble of specialists is considerably lower than those of a single generalist CNN and a ensemble of generalists (a.k.a pure ensemble).

2   Specialists Ensemble

Background: For a K-classification problem, let us consider training set of $\{(\mathbf{x}_i, \mathbf{y}_i)\}_{i=1}^{N}$ with $\mathbf{x}_i \in \mathcal{X}$ as an input sample along with its associated ground-truth class k, shown by a one-hot binary vector $\mathbf{y}_i \in [0,1]^K$ with a single 1 at its k-th element. A CNN, denoted by $h_{\mathcal{W}} : \mathcal{X} \to [0,1]^K$, maps a given input to its conditional probabilities over K classes. The classifier $h_{\mathcal{W}}(\cdot)^1$ is commonly trained through a cross-entropy loss function minimization as follows:

$$\min_{\mathcal{W}} \frac{1}{N} \sum_{i=1}^{N} \mathcal{L}(h(\mathbf{x}_i), \mathbf{y}_i; \mathcal{W}) = -\frac{1}{N} \sum_{i=1}^{N} \log h_{k^*}(\mathbf{x}_i), \tag{1}$$

where $h_{k^*}(\mathbf{x}_i)$ indicates the estimated probability of class k^* corresponding to the true class of given sample $\mathbf{x}_i$. At the inference time, the threshold-based approaches like our approach define a threshold τ in order to reject the instances with lower predictive confidence than τ as an extra class $K + 1$:

$$d(\mathbf{x}|\tau) = \begin{cases} \operatorname{argmax}_k h_k(\mathbf{x}), & \text{if } \max_k h_k(\mathbf{x}) > \tau \\ K + 1, & \text{otherwise} \end{cases}. \tag{2}$$

---

[1] For convenience, $\mathcal{W}$ is dropped from $h_{\mathcal{W}}(\cdot)$.

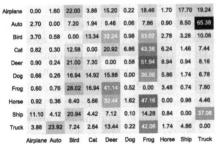

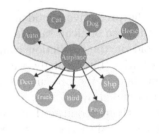

	Airplane	Auto	Bird	Cat	Deer	Dog	Frog	Horse	Ship	Truck
Airplane	0.00	1.60	22.00	3.88	15.20	0.22	18.46	1.70	17.70	19.24
Auto	2.70	0.00	7.20	1.94	5.46	0.06	7.86	0.90	8.50	65.38
Bird	3.70	0.58	0.00	13.34	32.24	0.98	33.02	2.78	3.28	10.08
Cat	0.82	0.30	12.58	0.00	20.92	6.86	43.38	6.24	1.46	7.44
Deer	0.90	0.24	21.00	7.30	0.00	0.58	51.94	8.94	0.94	8.16
Dog	0.66	0.26	16.94	14.92	15.88	0.00	36.96	5.86	1.74	6.78
Frog	0.60	0.76	28.02	16.94	41.14	0.52	0.00	3.48	0.74	7.80
Horse	0.92	0.36	6.40	5.66	32.44	1.62	47.16	0.00	0.98	4.46
Ship	11.10	4.12	20.94	4.42	7.12	0.10	14.28	0.84	0.00	37.08
Truck	3.88	23.92	7.24	2.64	13.44	0.22	42.06	1.74	4.86	0.00

(a) CIFAR-10 FGS fooling matrix   (b) The expertise domains of "Airplane" class

Fig. 2. (a) Fooling matrix of FGS adversaries for CIFAR-10, which is computed from 5000 randomly selected FGS adversaries (500 per class). Each row shows the fooling rates (in percentage) from a true class to other classes (rows and columns are true and fooling classes, respectively). (b) An example of forming expertise domains for class "Airplane": its high likely fooled classes (in yellow zone) and less likely fooled classes (in red zone) are forming two expertise domains. (Color figure online)

2.1   Ensemble Construction

We define the expertise domain of the specialists (i.e. the subsets of classes) by separating each class from its most likely fooled classes. We later show in Sect. 3 how separation of each class from its high likely fooling classes can promote entropy in the ensemble, which in turns leads to predicting adversaries with low confidence (high uncertainty).

To separate the most fooling classes from each other, we opt to use the fooling matrix of FGS adversarial examples $\mathbf{C} \in \mathbb{R}^{K \times K}$. This matrix reveals that the clean samples from each true class have a high tendency to being fooled toward a limited number of classes not uniformly toward all of them (Fig. 2(a)). The selection of FGS adversaries is two-fold; their generation is computationally inexpensive, and they are highly transferable to many other classifiers, meaning that different classifiers (e.g. with different structures) behave in similar manner in their presence, i.e. fooled to the same classes [13–15].

Using each row of the fooling matrix (i.e. $\mathbf{c}_i$), we define two expertise domains for i-th true class so as to split its high likely fooling classes from its less likely fooling classes as follows (Fig. 2(b)):

- Subset of high likely fooling classes of i: $\mathbb{U}_i = \cup\{j\}$ if $c_{ij} > \mu_i, \ j \in \{1, \ldots, K\}$
- Subset of less likely fooling classes of i: $\mathbb{U}_{i+K} = \{1, \ldots, K\} \setminus \mathbb{U}_i,$

where $\mu_i = \sum_{j=1}^{K} c_{ij}$ (average of fooling rates of i-th true class). Repeating the above procedure for all K classes makes $2K$ subsets (expertise domains) for a K classification problem. Note that the duplicated expertise domains can be removed so as to avoid having multiple identical expertise domains (specialists).

Afterwards, for each expertise domain, one specialist is trained in order to form an ensemble of specialist CNNs. A generalist (vanilla) CNN, which trained on the samples belonging to all classes, is also added to this ensemble. The ensemble involving $M \leq 2K + 1$ members is represented by $\mathcal{H} = \{h^1, \ldots, h^M\}$, where $h^j(\cdot) \in [0, 1]^K$ is j-th individual CNN mapping a given input to conditional probability over its expert classes, i.e. the probability of the classes out of its expertise domain is fixed to zero.

2.2   Voting Mechanism

To compute the final prediction out of our ensemble for a given sample, we need to activate relevant specialists, then averaging their prediction along with that of the generalist CNN. Note that we cannot simply use the generalist CNN to activate specialists since in the presence of adversaries it can be fooled, then causing selection (activation) of the wrong specialists. In Algorithm 1, we devise a simple and computationally efficient voting mechanism to activate those relevant specialists, then averaging their predictions.

Algorithm 1. Voting Mechanism

Input: Ensemble $\mathcal{H} = \{h^1, \ldots, h^M\}$, expertise domains $\mathbb{U} = \{\mathbb{U}_1, \ldots, \mathbb{U}_M\}$, input $\mathbf{x}$
Output: Final prediction $\bar{h}(\mathbf{x}) \in [0, 1]^K$

1: $v_k(\mathbf{x}) \leftarrow \sum_{j=1}^{M} \mathbb{I}\left(k = \mathrm{argmax}_{i=1}^{K} h_i^j(\mathbf{x})\right), \; k = 1, \ldots, K$
2: $k^* \leftarrow \mathrm{argmax}_{k=1}^{K} v_k(\mathbf{x})$
3: **if** $v_{k^*}(\mathbf{x}) = \lceil \frac{M}{2} \rceil$
4:     $\mathcal{H}_{k^*} \leftarrow \{h^i \in \mathcal{H} \,|\, k^* \in \mathbb{U}_i\}$
5:     $\bar{h}(\mathbf{x}) \leftarrow \frac{1}{|\mathcal{H}_{k^*}|} \sum_{h^i \in \mathcal{H}_{k^*}} h^i(\mathbf{x})$
6: **else**
7:     $\bar{h}(\mathbf{x}) \leftarrow \frac{1}{M} \sum_{h^i \in \mathcal{H}} h^i(\mathbf{x})$
8: **return** $\bar{h}(\mathbf{x})$

Let us first introduce the following elements for each class i:

– The actual number of votes for i-th class by the ensemble for a given sample $\mathbf{x}$: $v_i(\mathbf{x}) = \sum_{j=1}^{M} \mathbb{I}\left(i = \mathrm{argmax}_{\{1, \ldots K\}} h^j(\mathbf{x})\right)$, i.e. it shows the number of the members that classify $\mathbf{x}$ to i-th class.
– The maximum possible number of votes for i-th class is $\lceil \frac{M}{2} \rceil \leq K + 1$. Recall that for each row, we split all K classes into two expertise domains, where class i is included in one of them. Considering all K rows and the generalist, we end up having at maximum $K + 1$ subsets that involve class i.

As described in Algorithm 1, for a given sample $\mathbf{x}$, if there is a class with its actual number of votes equal to its expected number of votes, i.e. $v_i(\mathbf{x}) = \lceil \frac{M}{2} \rceil$, then it means all of the specialists, which are trained on i-th class, are simultaneously voting (classifying) for it. We call such a class *a winner class*. Then, the specialists CNNs voting to the winner class are activated to compute the final prediction (lines 3–5 of Algorithm 1), producing a certain prediction (with high confidence). Note that in the presence of clean samples, the relevant specialists in the ensemble are expected to do agree on the true classes since they, as strong classifiers, have high generalization performance on their expertise domains.

If no class obtains its maximum expected number of votes (i.e. $\nexists i$, $v_i(\mathbf{x}) = \lceil \frac{M}{2} \rceil$), it means that the input $\mathbf{x}$ leads the specialists to disagree on a winner class. In this situation, when no agreement exists in the ensemble, all the members should be activated to compute the final prediction (line 7 of Algorithm 1). Averaging of the predictions by all the members leads to a final prediction with high entropy (i.e. low confidence). Indeed, a given sample that creates a disagreement (entropy) in the ensemble is either a hard-to-classify sample or an abnormal sample (e.g. adversarial examples).

Using the voting mechanism for this specialists ensemble, we can create a gap between the predictive confidences of clean samples (having high confidence) and those of adversaries (having low confidence). Finally, using a threshold τ on these predictive confidences, the unusual samples are identified and rejected. In the following, we argue that our voting mechanism enables us to set a global fixed threshold $\tau = 0.5$ to perform identification of adversaries. This is unlike some threshold-based approaches [10,16] that need to tune different thresholds for various datasets and their types of adversaries.

Corollary 1. *In a disagreement situation, the proposed voting mechanism makes the highest predictive confidence to be upper-bounded by $0.5 + \epsilon'$ with $\epsilon' = \frac{1}{2M}$.*

Proof. Consider a disagreement situation in the ensemble for a given $\mathbf{x}$, where all the members are averaged to create $\bar{h}(\mathbf{x}) = \frac{1}{M} \sum_{h^j \in \mathcal{H}} h^j(\mathbf{x})$. The highest predictive confidence of $\bar{h}(\mathbf{x})$ belongs to the class that has the largest number of votes, i.e. $m = \max[v_1(\mathbf{x}), \dots, v_K(\mathbf{x})]$. Let us represent these m members that are voting to this class (k-th class) as $\mathcal{H}_k = \{h^j \in \mathcal{H} \mid k \in \mathbb{U}_j\}$. Since each individual CNNs in the ensemble are basically uncalibrated learners (having very high confident prediction for a class and near to zero for the remaining classes), the confidence probability of k-th class of those excluded members from $\mathcal{H}_k$ (those that do not vote for k-th class) can be negligible. Thus, their prediction can be simplified as $\bar{h}_k(\mathbf{x}) = \frac{1}{M} \sum_{h^j \in \mathcal{H}_k} h_k^j(\mathbf{x}) + \frac{\epsilon}{M} \approx \frac{1}{M} \sum_{h^j \in \mathcal{H}_k} h_k^j(\mathbf{x})$ (the small term $\frac{\epsilon}{M}$ is discarded). Then, from the following inequality $\sum_{h^j \in \mathcal{H}_k} h_k^j(\mathbf{x}) \leq m$, we have $\frac{1}{M} \sum_{h^j \in \mathcal{H}_k} h_k^j(\mathbf{x}) \leq \frac{m}{M}$ (I).

On the other hand, due to having no winner class, we know that $m < \lceil \frac{M}{2} \rceil$ (or $m < \frac{M}{2} + \frac{1}{2}$), such that by multiplying it by $\frac{1}{M}$ we obtain $\frac{m}{M} < \frac{1}{2} + \frac{1}{2M}$ (II).

Finally considering (I) and (II) together, it derives $\frac{1}{M} \sum_{h^j \in \mathcal{H}_k} h_k^j(\mathbf{x}) < 0.5 + \frac{1}{2M}$. For the ensemble with a large size, e.g. likewise our ensemble, the term $\epsilon' = \frac{1}{2M}$ is small. Therefore, it shows the class with the maximum probability (having the maximum votes) can be upper-bounded by $0.5 + \epsilon'$. ∎

3 Analysis of Specialists Ensemble

Here, we first explain how adversarial examples give rise to entropy in our ensemble, leading to their low predictive confidence (with maximum confidence of $0.5 + \epsilon'$). As well, we examine the role of diversity in our ensemble, which harden the generation of white-box adversaries.

In a **black-box** attack, we assume that the attacker is not aware of our ensemble of specialists, thus generates some adversaries from a pre-trained vanilla CNN $g(\cdot)$ to mislead our underlying ensemble. Taking a pair of an input sample with its true label, i.e. $(\mathbf{x}, k)$, an adversary $\mathbf{x}' = \mathbf{x} + \delta$ fools the model g such that $k = \arg\max g(\mathbf{x})$ while $k' = \arg\max g(\mathbf{x}')$ with $k' \neq k$, where k' is one of those most-likely fooling classes for class k (i.e. $k' \in \mathbb{U}_k$). Among the specialists that are expert on k, at least one of them does not have k' in their expertise domains since we intentionally separated k-th class from its most-likely fooling classes when defining its expertise domains (Sect. 2.1). Formally speaking, denote those expertise domains comprising class k as follows $\mathcal{U}^k = \{\mathbb{U}_j \mid k \in \mathbb{U}_j\}$ where (I) $\mathbb{U}_j \neq \mathbb{U}_i \; \forall \mathbb{U}_i, \; \mathbb{U}_j \in \mathcal{U}^k$ and (II) $k' \notin \cap \, \mathcal{U}^k$. Therefore, regarding the fact that (I) the expertise domains comprising k are different and (II) their shared classes do not contain k', it is not possible that all of their corresponding specialists models are fooled simultaneously toward k'. In fact, these specialists may vote (classify) differently, leading to a disagreement on the fooling class k'. So, due to this disagreement in the ensemble with no winner class, all the ensemble's members are activated, resulting in prediction with high uncertainty (low confidence) according to Corollary 1. Generally speaking, if $\{\cap \, \mathcal{U}^k\} \setminus k$ is a small or an empty set, harmoniously fooling the specialist models, which are expert on k, is harder.

In a **white-box attack**, an attacker attempts to generate adversaries to *confidently* fool the ensemble, meaning the adversaries should simultaneously activate *all* of the specialists that comprise the fooling class in their expertise domain. Otherwise, if at least one of these specialists is not fooled, then our voting mechanism results in adversaries with low confidence, which can then be automatically rejected using the threshold ($\tau = 0.5$). In the rest we bring some justifications on the hardness of generating high confidence gradient-based attacks from the specialists ensemble.

Instead of dealing with the gradient of one network, i.e. $\frac{\partial h(\mathbf{x})}{\partial \mathbf{x}}$, the attacker should deal with the gradient of the ensemble, i.e. $\frac{\partial \bar{h}(\mathbf{x})}{\partial \mathbf{x}}$, where $\bar{h}(\mathbf{x})$ computed by line 5 or line 7 of Algorithm. 1. Formally, to generate a gradient-based adversary from the ensemble for a given labeled clean input sample $(\mathbf{x}, \mathbf{y} = k)$, the derivative of the ensemble's loss, i.e. $\mathcal{L}(\bar{h}(\mathbf{x}), \mathbf{y}) = -\log \bar{h}_k(\mathbf{x})$, w.r.t. $\mathbf{x}$ is as follows:

$$\frac{\partial \mathcal{L}(\bar{h}(\mathbf{x}), \mathbf{y})}{\partial \mathbf{x}} = \frac{\partial \mathcal{L}}{\partial \bar{h}_k(\mathbf{x})} \frac{\partial \bar{h}_k(\mathbf{x})}{\partial \mathbf{x}} = \underbrace{-\frac{1}{\bar{h}_k(\mathbf{x})}}_{\beta} \frac{\partial \bar{h}_k(\mathbf{x})}{\partial \mathbf{x}} = \beta \frac{1}{|\mathcal{H}_k|} \sum_{h^i \in \mathcal{H}_k} \frac{\partial h_k^i(\mathbf{x})}{\partial \mathbf{x}}. \quad (3)$$

Initially $\mathcal{H}_k$ indicates the set of activated specialists voting for class k (true label) plus the generalist for the given input $\mathbf{x}$. Since the expertise domains of the activated specialists are different ($\mathcal{U}^k = \{\mathbb{U}_j \mid k \in \mathbb{U}_j\}$), most likely their derivative are diverse, i.e. fooling toward different classes, which in turn creates perturbations in various fooling directions (Fig. 1). Adding such diverse perturbation to a clean sample may promote disagreement in the ensemble, where no winner class can be agreed upon. In this situation, when all of the members are activated, the generated adversarial sample is predicted with a low confidence, thus can be identified. For the iterative attack algorithms, e.g. I-FGS, the process of generating adversaries may continue using the derivative of all of the members, adding even more diverse perturbations, which in turn makes reaching to an agreement in the ensemble on a winner fooling class even more difficult.

4 Experimentation

Evaluation Setting: Using MNIST and CIFAR-10, we investigate the performance of our method for reducing the risk rate of black-box attacks (Eq. 5) due to of their detection, and reducing the success rate of creating white-box adversaries. Two distinct CNN configurations are considered in our experimentation: for *MNIST*, a basic CNN with three convolution layers of respectively 32, 32, and 64 filters of 5×5, and a final fully connected (FC) layer with 10 output neurons. Each of these convolution layers is followed by a ReLU and 3×3 pooling filter with stride 2. For *CIFAR-10*, a VGG-style CNN (details in [17]) is used. For both CNNs, we use SGD with a Nesterov momentum of 0.9, L2 regularization with its hyper-parameter set to 10^{-4}, and dropout ($p = 0.5$) for the FC layers. For the evaluation purposes, we compare our ensemble of specialists with a vanilla (naive) CNN, and a pure ensemble, which involves 5 vanilla CNNs being different by random initialization of their parameters.

Evaluation Metrics: To evaluate a predictor $h(\cdot)$ that includes a rejection option, we report a risk rate $E_D|\tau$ on a clean test set $\mathcal{D} = \{(\mathbf{x}_i, \mathbf{y}_i)\}_{i=1}^{N}$ at a given threshold τ, which computes the ratio of the (clean) samples that are *correctly classified but rejected* due to their confidence less than τ and those that are *misclassified but not rejected* due to a confidence value above τ:

$$E_D|\tau = \frac{1}{N} \sum_{i=1}^{N} \Big((\mathbb{I}[d(\mathbf{x}_i|\tau) \neq K+1] \times \mathbb{I}[\arg\max h(\mathbf{x}_i) \neq \mathbf{y}_i])$$
$$+ (\mathbb{I}[d(\mathbf{x}_i|\tau) = K+1] \times \mathbb{I}[\arg\max h(\mathbf{x}_i) = \mathbf{y}_i]) \Big). \tag{4}$$

In addition, we report the risk rate $E_A|\tau$ on each adversaries set, i.e. $\mathcal{A} = \{(\mathbf{x}'_i, \mathbf{y}_i)\}_{i=1}^{N'}$ including pairs of an adversarial example $\mathbf{x}'_i$ associated by its true label, to show the percentage of misclassified adversaries that are not rejected due to their confidence value above τ:

$$E_A|\tau = \frac{1}{N'} \sum_{i=1}^{N'} (\mathbb{I}[d(\mathbf{x}'_i|\tau) \neq K+1] \times \mathbb{I}[\arg\max h(\mathbf{x}'_i) \neq \mathbf{y}_i]). \tag{5}$$

4.1   Empirical Results

Black-box Attacks: To assess our method on different types of adversaries, we use various attack algorithms, namely FGS [5], TFGS [7], DeepFool (DF) [18], and CW [19]. To generate the black-box adversaries, we use another vanilla CNN, which is different from all its counterparts involved in the pure ensemble– by using different random initialization of its parameters. For FGS and T-FGS algorithms we generate 2000 adversaries with $\epsilon = 0.2$ and $\epsilon = 0.03$, respectively, for randomly selected clean test samples from MNIST and CIFAR-10. For CW attack, due to the high computational burden required, we generated 200 adversaries with $\kappa = 40$, where larger κ ensures generation of high confidence and highly transferable CW adversaries.

Figure 3 presents risk rates $(E_D|\tau)$ of different methods on clean test samples of MNIST (first row) and those of CIFAR-10 (second row), as well as their corresponding adversaries $E_A|\tau$, as functions of threshold (τ). As it can be seen from Fig. 3, by increasing the threshold, more adversaries can be detected (decreasing E_A) at the cost of increasing E_D, meaning an increase in the rejection of the clean samples that are correctly classified.

To appropriately compare the methods, we find an optimum threshold that creates small E_D and E_A collectively, i.e. $\operatorname{argmin}_\tau E_D|\tau + E_A|\tau$. Recall that, as corollary 1 states, in our ensemble of specialists, we can fix the threshold of our ensemble to $\tau^* = 0.5$. In Table 1, we compare the risk rates of our ensemble with those of pure ensemble and vanilla CNN at their corresponding optimum thresholds. For MNIST, our ensemble outperforms naive CNN and pure ensemble as it detects a larger portion of MNIST adversaries while its risk rate on the clean samples is only marginally increased. Similarly, for CIFAR-10, our approach can detect a significant portion of adversaries at $\tau^* = 0.5$, reducing the risk rates

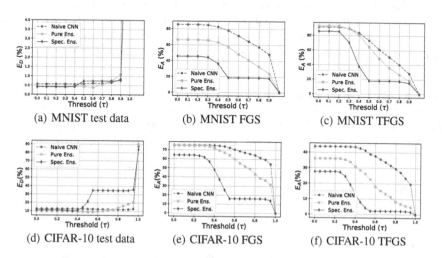

(a) MNIST test data   (b) MNIST FGS   (c) MNIST TFGS

(d) CIFAR-10 test data   (e) CIFAR-10 FGS   (f) CIFAR-10 TFGS

Fig. 3. The risk rates on the clean test samples and their black-box adversaries as the function of threshold (τ) on the predictive confidence.

Table 1. The risk rate of the clean test set ($E_D|\tau*$) along with that of black-box adversarial examples sets ($E_A|\tau*$) are shown in percentage at the optimum threshold of each method. The methods with the lowest collective risk rate (i.e. $E_A + E_D$) is underlined, while the best results for the two types of risk considered independently are in bold.

Task	Methods	Adversaries			
		FGS	TFGS	CW	DeepFool
		E_A / E_D	E_A / E_D	E_A / E_D	E_A / E_D
MNIST	Naive CNN	48.21 / 0.84	28.15 / 0.84	41.5 / 0.84	88.68 / 0.84
	Pure ensemble	24.02 / 1.1	18.35 / 1.1	28.5 / 1.1	72.73 / 1.1
	Specialists ensemble	**18.58** / **0.73**	**18.05** / **0.73**	**24** / **0.73**	**54.24** / **0.73**
CIFAR-10	Naive CNN	59.37 / 12.11	23.47 / 12.11	51.5 / 12.11	28.81 / 12.11
	Pure ensemble	36.59 / 18.5	8.37 / 13.79	4.0 / 13.79	7.7 / 18.5
	Specialists ensemble	**25.66** / 21.25	**4.21** / 21.25	**3.5** / 21.25	6.02 / 21.25

on adversaries. However, at this threshold, our approach has higher risk rate on the clean samples than that of two other methods.

White-box Attacks: In the white-box setting, we assume that the attacker has full access to a victim model. Using each method (i.e. naive CNN, pure ensemble, and specialists ensemble) as a victim model, we generate different sets of adversaries (i.e. FGS, Iterative FGS (I-FGS), and T-FGS). A successful adversarial attack $\mathbf{x}'$ is achieved once the underlying model misclassifies it with a confidence higher than its optimum threshold τ^*. When the confidence for an adversarial example is lower than $\tau*$, it can be easily detected (rejected), thus it is not considered as a successful attack.

We evaluate the methods by their *white-box attacks success rates*, indicating the number of successful adversaries that satisfies the aforementioned conditions (i.e. a misclassification with a confidence higher than τ^*) during t iterations of the attack algorithm. Table 2 exhibits the success rates of white-box adversaries (along with their used hyper-parameters) generated by naive CNN ($\tau^* = 0.9$),

Table 2. Success rate of white-box adversarial examples (lower is better) generated by naive CNN, pure ensemble (5 generalists), and specialists ensemble at their corresponding optimum threshold. An successful white-box adversarial attack should fool the underlying model with a confidence higher than its optimum τ^*.

Methods	MNIST			CIFAR-10		
	Adversaries					
	FGS	T-FGS	I-FGS	FGS	T-FGS	I-FGS
	$\epsilon=0.2$	$\epsilon=0.2$	$\epsilon=2\times10^{-2}$	$\epsilon=3\times10^{-2}$	$\epsilon=3\times10^{-2}$	$\epsilon=3\times10^{-3}$
Naive CNN	89.94	66.16	66.84	86.16	81.38	93.93
Pure ensemble	71.58	50.64	48.62	42.65	13.96	45.78
Specialists ensemble	**45.15**	**27.43**	**13.63**	**34.1**	**7.43**	**34.20**

pure ensemble ($\tau^* = 0.9$), and specialists ensemble ($\tau^* = 0.5$). For the benchmark datasets, the number of iterations of FGS and T-FGS is 2 while that of iterative FGS is 10. As it can be seen in Table 2, the success rates of adversarial attacks using ensemble-based methods are smaller than those of naive CNN since diversity in these ensembles hinders generation of adversaries with high confidence.

Gray-box CW Attack: In the gray-box setting, it is often assumed that the attacker is aware of the underlying defense mechanism (e.g. specialists ensemble in our case) but has no access to its parameters and hyper-parameters. Following [20], we evaluate our ensemble on CW adversaries generated by another specialists ensemble, composed of 20 specialists and 1 generalist for 100 randomly selected MNIST samples. Evaluation of our specialists ensemble on these targeted gray-box adversaries (called "gray-box CW") reveals that our ensemble provides low confidence predictions (i.e. lower than 0.5) for 74% of them (thus able to reject them) while 26% have confidence more than 0.5 (i.e. non-rejected adversaries). Looking closely at

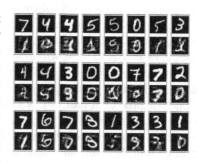

Fig. 4. Gray-box CW adversaries that confidently fool our specialists ensemble. According to the definition of adversarial example, however, some of them are not actually adversaries due to the significant visual perturbations.

those non-rejected adversaries in Fig. 4, it can be observed that some of them can even mislead a human observer due to adding very visible perturbation, where the appearance of digits are significantly distorted.

5   Related Works

To address the issue of robustness of deep neural networks, one can either *enhance classification accuracy of neural networks to adversaries*, or devise *detectors to identify adversaries* in order to reject to process them. The former class of approaches, known as adversarial training, usually train a model on the training set, which is augmented by adversarial examples. The main difference between many adversarial training approaches lies in the way that the adversaries are created. For example, some [3,5,7,21] have trained the models with adversaries generated on-the-fly, while others conduct adversarial training with a pre-generated set of adversaries, either produced from an ensemble [22] or from a single model [18,23]. With the aim detecting adversaries to avoid making wrong decisions over the hostile samples, the second category of approaches propose the detectors, which are usually trained by a training set of adversaries [4,8–10,24,25].

Notwithstanding the achievement of some favorable results by both categories of approaches, the main concern is that their performances on all types of adversaries are extremely dependent on the capacity of generating an exhaustive set of adversaries, which comprises different types of adversaries. While making such a complete set of adversaries can be computationally expensive, it has been shown that adversely training a model on a specific type of adversaries does not necessarily confer a CNN robustness to other types of adversaries [11,12].

Some ensemble-based approaches [26,27] were shown to be effective for mitigating the risk of adversarial examples. Strauss et al. [26] demonstrated some ensembles of CNNs that are created by bagging and different random initializations are less fooled (misclassify adversaries), compared to a single model. Recently, Kariyappa et al. [27] have proposed an ensemble of CNNs, where they explicitly force each pair of CNNs to have dissimilar fooling directions, in order to promoting diversity in the presence of adversaries. However, computing similarity between the fooling directions by each pair of members for every given training sample is computationally expensive, results in increasing training time.

6   Conclusion

In this paper, we propose an ensemble of specialists, where each of the specialist classifiers is trained on a different subset of classes. We also devise a simple voting mechanism to efficiently merge the predictions of the ensemble's classifiers. Given the assumption that CNNs are strong classifiers and by leveraging diversity in this ensemble, a gap between predictive confidences of clean samples and those of black-box adversaries is created. Then, using a global fixed threshold, the adversaries predicted with low confidence are rejected (detected). We empirically demonstrate that our ensemble of specialists approach can detect a large portion of black-box adversaries as well as makes the generation of white-box attacks harder. This illustrates the beneficial role of diversity for the creation of ensembles in order to reduce the vulnerability to black-box and white-box adversarial examples.

Acknowledgements. This work was funded by NSERC-Canada, Mitacs, and Prompt-Québec. We thank Annette Schwerdtfeger for proofreading the paper.

References

1. Guo, C., Pleiss, G., Sun, Y., Weinberger, K.Q.: On calibration of modern neural networks. In: Proceedings of the 34th International Conference on Machine Learning-Volume 70, JMLR. org (2017), pp. 1321–1330 (2017)
2. Eykholt, K., et al.: Robust physical-world attacks on deep learning models. arXiv preprint arXiv:1707.08945 (2017)
3. Madry, A., Makelov, A., Schmidt, L., Tsipras, D., Vladu, A.: Towards deep learning models resistant to adversarial attacks. arXiv preprint arXiv:1706.06083 (2017)

4. Metzen, J.H., Genewein, T., Fischer, V., Bischoff, B.: On detecting adversarial perturbations. In: Proceedings of the 5th International Conference on Learning Representations (ICLR) (2017)
5. Goodfellow, I.J., Shlens, J., Szegedy, C.: Explaining and harnessing adversarial examples. arXiv preprint arXiv:1412.6572 (2014)
6. Liao, F., Liang, M., Dong, Y., Pang, T., Zhu, J., Hu, X.: Defense against adversarial attacks using high-level representation guided denoiser. arXiv preprint arXiv:1712.02976 (2017)
7. Kurakin, A., Goodfellow, I., Bengio, S.: Adversarial examples in the physical world. arXiv preprint arXiv:1607.02533 (2016)
8. Feinman, R., Curtin, R.R., Shintre, S., Gardner, A.B.: Detecting adversarial samples from artifacts. arXiv preprint arXiv:1703.00410 (2017)
9. Grosse, K., Manoharan, P., Papernot, N., Backes, M., McDaniel, P.: On the (statistical) detection of adversarial examples. arXiv preprint arXiv:1702.06280 (2017)
10. Lee, K., Lee, K., Lee, H., Shin, J.: A simple unified framework for detecting out-of-distribution samples and adversarial attacks. In: Advances in Neural Information Processing Systems, pp. 7167–7177 (2018)
11. Zhang, H., Chen, H., Song, Z., Boning, D., Dhillon, I.S., Hsieh, C.J.: The limitations of adversarial training and the blind-spot attack. arXiv preprint arXiv:1901.04684 (2019)
12. Tramèr, F., Boneh, D.: Adversarial training and robustness for multiple perturbations. In: Proceedings of the Neural Information Processing Systems (NeurIPS) (2019)
13. Liu, Y., Chen, X., Liu, C., Song, D.: Delving into transferable adversarial examples and black-box attacks. arXiv preprint arXiv:1611.02770 (2016)
14. Szegedy, C., et al.: Intriguing properties of neural networks. arXiv preprint arXiv:1312.6199 (2013)
15. Charles, Z., Rosenberg, H., Papailiopoulos, D.: A geometric perspective on the transferability of adversarial directions. arXiv preprint arXiv:1811.03531 (2018)
16. Bendale, A., Boult, T.E.: Towards open set deep networks. In: Proceedings of the IEEE Conference on Computer Vision and Pattern Recognition, pp.1563–1572 (2016)
17. Simonyan, K., Zisserman, A.: Very deep convolutional networks for large-scale image recognition. arXiv preprint arXiv:1409.1556 (2014)
18. Moosavi-Dezfooli, S.M., Fawzi, A., Frossard, P.: Deepfool: a simple and accurate method to fool deep neural networks. In: Proceedings of the IEEE Conference on Computer Vision and Pattern Recognition (CVPR) (2016)
19. Carlini, N., Wagner, D.: Towards evaluating the robustness of neural networks. In: 2017 IEEE Symposium on Security and Privacy (SP), pp. 39–57. IEEE (2017)
20. He, W., Wei, J., Chen, X., Carlini, N., Song, D.: Adversarial example defenses: ensembles of weak defenses are not strong. arXiv preprint arXiv:1706.04701 (2017)
21. Huang, R., Xu, B., Schuurmans, D., Szepesvári, C.: Learning with a strong adversary. arXiv preprint arXiv:1511.03034 (2015)
22. Tramèr, F., Kurakin, A., Papernot, N., Boneh, D., McDaniel, P.: Ensemble adversarial training: attacks and defenses. arXiv preprint arXiv:1705.07204 (2017)
23. Rozsa, A., Rudd, E.M., Boult, T.E.: Adversarial diversity and hard positive generation. In: Proceedings of the IEEE Conference on Computer Vision and Pattern Recognition Workshops, pp. 25–32 (2016)
24. Lu, J., Issaranon, T., Forsyth, D.: Safetynet: detecting and rejecting adversarial examples robustly. In: The IEEE International Conference on Computer Vision (ICCV), October 2017

25. Meng, D., Chen, H.: Magnet: a two-pronged defense against adversarial examples (2017)
26. Strauss, T., Hanselmann, M., Junginger, A., Ulmer, H.: Ensemble methods as a defense to adversarial perturbations against deep neural networks. arXiv preprint arXiv:1709.03423 (2017)
27. Kariyappa, S., Qureshi, M.K.: Improving adversarial robustness of ensembles with diversity training. arXiv preprint arXiv:1901.09981 (2019)

Locating Influential Agents in Social Networks: Budget-Constrained Seed Set Selection

Rishav Raj Agarwal[1]([✉]), Robin Cohen[1], Lukasz Golab[1], and Alan Tsang[2]

[1] University of Waterloo, Waterloo, Canada
{rragarwal,rcohen,lgolab}uwaterloo.ca
[2] National University of Singapore, Singapore, Singapore
akhtsang@gmail.com

Abstract. The study of information spread in social networks has applications in viral marketing, rumour modelling, and opinion dynamics. Often, it is crucial to identify a small set of influential agents that maximize the spread of information (cases which we refer to as being budget-constrained). These nodes are believed to have special topological properties and reside in the core of a network. We introduce the concept of nucleus decomposition, a clique based extension of core decomposition of graphs, as a new method to locate influential nodes. Our analysis shows that influential nodes lie in the k-nucleus subgraphs and that these nodes outperform lower-order decomposition techniques such as truss and core, while simultaneously focusing on a smaller set of seed nodes. Examining different diffusion models on real-world networks, we provide insights as well into the value of the degree centrality heuristic.

1 Introduction

With the rise of big data tools and platforms, it has become easier to mine social networks. One topic of particular interest is the study of information spread through a network. Finding influential agents is often key, either to stem the spread of harmful content or to facilitate influence maximization for such positive aims as spreading HIV awareness among homeless youths [24] or increasing revenue with viral marketing [4].

Many approaches have been developed to locate influencers and track the spread of their communications to peers. The NP-hard optimization problem known as *influence maximization* [8] looks to find a set of n nodes that, when "activated", can spread information maximally throughout a given network under a given information diffusion model (see Li et al. for a survey of approximate algorithms for influence maximization [12]). Other heuristics consider properties of a given node such as its degree, as well as information about its local graph structure (for example, avoid nodes at the fringes of a graph that have a high degree but weakly connected neighbours [14]).

Two classes of topology-based heuristics to locate influential nodes are centrality based methods and subgraph decomposition methods. Centrality methods

© Springer Nature Switzerland AG 2020
C. Goutte and X. Zhu (Eds.): Canadian AI 2020, LNAI 12109, pp. 15–28, 2020.
https://doi.org/10.1007/978-3-030-47358-7_2

consider the degree of a node (degree centrality), the length of shortest paths from a node to all other nodes (closeness centrality), or the number of times a node occurs in the shortest paths (betweenness centrality). By contrast, [10] and [13] argue that less connected but strategically placed nodes may be better candidates for disseminating information. They turn to k-core [20] and k-truss decompositions [5], which identify subgraphs having high degree or many triangles, respectively (details in the next section). Simulation studies have found that k-core methods outperform some centrality based measures [10] and that k-truss methods, in turn, outperform k-core methods [13].

In this paper, we focus on scenarios where organizations have a limited budget to expend when engaging with potential influencers, and thus locating a small seed set of agents is paramount. Our main contribution is the evaluation of a new method for budget-constrained seed set selection based on nucleus decomposition [19]. A k-nucleus is a generalization of graph decomposition methods, and it has been observed that k-nuclei often overlap with the densest parts of k-cores and k-trusses. Using four real datasets, we compare the effectiveness of topology based methods – k-nucleus decomposition, k-truss, k-core and degree centrality – under three popular information diffusion models: Independent Cascade [8], Linear Threshold [7], and Susceptible-Infectious-Recovered (SIR) [15]. We further show that degree centrality, an often ignored heuristic, can perform as well as the nucleus in some cases, as long as sufficiently many high-degree nodes (e.g., as many as there are in a maximal k-nucleus) are selected. This observation is in contrast to prior work that only used the nodes with the *highest* degree in the network as influencers, which was not as effective as using core or truss decomposition [13]. Finally, we show that topology based methods often perform on par with an approximation algorithm that solves the underlying influence maximization problem (IMM [22]). Our analysis enables practitioners to better choose heuristics according to their choice of information diffusion model and to consider k-nucleus decomposition and degree centrality as important algorithms in their arsenal.

2   Methods

We start by describing the methods included in our study, followed by a discussion of the diffusion models. Let $G(V, E)$ be an undirected graph that models the underlying social network with $|V|$ nodes and $|E|$ edges. Let $v \in V$ be a node in G and let $e \in E$ be an edge in G. Finally, let k be a positive integer.

2.1   Graph Decomposition Methods

k-core Decomposition [20]: A $k-$core is a largest connected subgraph of G where each node has degree at least k. Each node $v \in V$ can be assigned a core value $c(v)$ that equals k if v belongs to a $k-$core but not a $(k+1)-$core. Using this concept, the influential nodes are those with the largest value of $c(v)$. To find a k-core subgraph, we repeatedly remove nodes with a degree of less than

k and their adjacent edges. Since removing edges reduces the degree of some of the remaining nodes, whenever a node is removed, we decrement the degree of the affected nodes, and we continue until all the remaining nodes have a degree at least k. The time complexity of this method is $O(|V| + |E|)$ since a node or an edge can be removed at most once.

k-truss Decomposition [5]: This method expands on k-core decomposition by considering triangles, i.e., cycles of length 3. A k-truss is a largest subgraph of G where each edge is contained in at least $k - 2$ triangles within the subgraph. Each edge e can be assigned a truss number t_e that equals k if e belongs to a k-truss but not a $(k+1)$ truss. Furthermore, the truss number t_v for a node v is equal to the maximum edge truss of the edges adjacent to v. Using this concept, the influential nodes are those with the largest value of t_v. To find a k-truss, we follow a similar methodology as that for k-core decomposition. However, instead of removing nodes directly, we repeatedly remove edges that are not part of at least $k - 2$ triangles, and we output the connected components that remain at the end (time complexity $O(|E|^{1.5})$).

k-nucleus Decomposition [19]: This method generalizes k-truss and k-core decomposition by finding subgraphs of *cliques*. Let r and s be two positive integers such that $r < s$. Let K_r be an r-clique, i.e., a clique with r nodes. Intuitively, a k-(r, s)-nucleus is a maximal subset of smaller r-cliques, each of which is part of many larger s-cliques. Formally, let χ be a set of s-cliques K_s in G. Let $K_r(\chi)$ be a set of smaller r-cliques K_r in some $S \in \chi$. The χ-**degree** of an r-clique $u \in K_r(\chi)$ is the number of larger s-cliques in χ that contain u. χ-**connected:** Two K_r, call them u and u', are χ- connected if there exists a sequence of r-cliques $u = u_1, u_2, ..., u_k = u'$ in $K_r(\chi)$ such that for each i, some s-clique $S \in \chi$ contains $u_i \cup u_{i+1}$. Finally, we define a k-(r, s) nucleus as a maximal union χ of s-cliques K_s such that the χ-degree of any r-clique $u \in K_r(\chi)$ is at least k and any r-clique pair $u, u' \in K_r(\chi)$ is χ-connected.

Setting $r = 1$ and $s = 2$ allows us to recover the definition of k-core from k-(r, s) nucleus. To see this, observe that any node is a 1-clique, and any edge is a 2-clique. Thus, the χ degree of a 1-clique is the degree of the node, and, by the χ-connected property, we simply get a set of edges connecting nodes of degree at least k. Similarly, setting $r = 2$ and $s = 3$ reduces to k-truss decomposition. Triangles are 3-cliques, and we get a set χ that is part of at least k triangles.

Let $RT(K_r)$ and $RT(K_s)$ be the time complexity of enumerating all $K_r \in G$ and all $K_s \in G$, respectively. The complexity of nucleus decomposition was shown to be bounded by $O(RT(K_s) + RT(K_r))$ [19]. In this paper, we consider k-$(3, 4)$ nucleus decomposition as complexity grows rapidly for K_4 and above. From now on, we refer to a k-$(3, 4)$ nucleus as k-nucleus for simplicity[1]. As in k-core and k-truss decomposition, each node can be assigned a nucleus value $n(v)$ that equals k if v belongs to a k-nucleus but not a $(k + 1)$-nucleus.

---

[1] [19] showed that $(3, 4)$-nucleus provides high-quality outputs in terms of density and network hierarchy; e.g., it finds both small sets of high density and large sets of low density.

The influential nodes are those with the largest value of $n(v)$. Some nodes of a $k + 1$ core are part of a k-truss and some nodes of a $k + 1$-truss are part of a nucleus. Figure 1 illustrates a graph and the corresponding 3-core, 2-trusses and 1-nuclei. The entire graph is a maximal 3-core (as each node has at least three edges). In the 3-core, there are two 2-trusses and two 1-nuclei. Note that the nuclei and trusses are smaller and identify denser subgraphs than the core.

2.2 Information Diffusion Models

Independent Cascade (IC) Model [8]: In this model, nodes that are *activated* can influence their neighbours. Activation proceeds one step at a time. Each *directed* edge $(v, v') : v \rightarrow v'$ in the underlying graph has a threshold value $p_{v,v'} \in [0, 1]$ denoting the propagation probability of information from v to v'. We begin with a set of nodes that are initially assumed to be active. The information then flows as follows. At time t, any active node $v \in V$ has a chance to activate an inactive child node v' with probability $p_{v,v'}$. If v succeeds then v' becomes active in step $t + 1$. If multiple parents of v' are active at the same time, their activation attempts are arbitrarily sequenced at time t. v only gets one chance to activate v' and cannot activate v' in subsequent rounds. The process terminates when no more activations are possible.

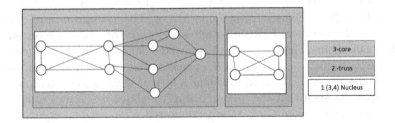

Fig. 1. Comparison of subgraph decompositions

Linear Threshold Model (LT) [7]: In this model, a node is influenced by every incoming neighbour v' with a weight $b_{v',v} \in [0, 1]$. Each $v \in V$ also has a threshold $\theta_v \in [0, 1]$, which represents the minimum pressure that has to be exerted on v to activate it. v is activated iff the sum of the weights of the active neighbours of v is greater than a threshold θ_v: $\sum_{v' \rightarrow v, v' \text{active}} b_{v',v} \geq \theta_v$. The information flow proceeds in discrete steps (from $t = 0$), with a seed set of active nodes S. For each neighbour v of $v' \in S \subseteq V$, we check the threshold condition. If a node satisfies its condition, it is activated in the next step. The algorithm continues until no more activations occur.

Susceptible Infected Recovered (SIR) Model [9]: In this model, a node can be in one of three states: Susceptible (S): not yet infected; Infected (I): can spread information to the rest of the population; Recovered (R): after a node has been infected for some period of time, it is considered to be immune and cannot further spread the information. To examine the spreading power of a set of nodes, we initially set these nodes as infected, and we set all other nodes as susceptible. Then, at each time step t of the process, every infected node can infect its susceptible neighbours with probability β (called infection rate), and afterwards, it can recover with probability γ (called recovery rate). A node cannot directly pass from state I to state R during the same time step. The process ends when no more nodes can be infected.

3   Results

We now explore the performance of k-core, k-truss, k-nucleus and degree centrality in locating influential nodes using four real-world social networks. We start by showing that k-nucleus decomposition identifies fewer nodes as being influential. We then show that despite being lower in number, the nucleus nodes have similar or better information spreading power than those identified by other methods. We also show that the nodes selected by nucleus decomposition are robust to low diffusion rates under the IC and SIR models. Finally, we show that choosing a sufficient number of high degree nodes can work well as the nucleus, and these topology based methods often perform on par with an approximation algorithm that solves the underlying influence maximization problem.

Table 1. Properties of datasets and their subgraph decompositions.

Dataset	Nodes	Edges	τ	k_{max}			v_{max}		
				Nucleus	Truss	Core	Nucleus	Truss	Core
WikiVote	7,115	88,750	0.00720	15	19	49	37	80	332
Slashdot	81,871	545,671	0.00074	26	34	54	65	77	118
Epinions	131,828	841,372	0.00540	97	105	121	112	135	149
EuEmail	265,214	420,045	0.00970	13	18	37	56	62	292

3.1   Datasets

We use four publicly available datasets [11]. **Slashdot** is a technology news website whose users form a signed social network as they can tag each other as "friend" or "foe". **Epinions** is a trust-based (who-trusts-whom) network between members of the Epinions.com product review website. **WikiVote** is a dataset showing who voted for whom in the Wikipedia election for membership. **EuEmail** is a "who talked to whom" network with an edge between two nodes, A and B, meaning A sent an email to B.

To compute subgraph decompositions, we use the code from [19], available at http://sariyuce.com/nucleus-master.zip. Table 1 summarizes the datasets and the properties of the corresponding k-core, truss and nucleus decompositions (v_{max} is the number of nodes in the maximal subgraph of order k_{max}).

The dataset statistics we report are the number of nodes and edges, and the inverse of the largest eigenvalue of the corresponding adjacency matrix, τ. It is known that epidemic spreading can be achieved by setting the propagation probabilities to be at least τ [3]. Below this threshold, the number of affected nodes decreases exponentially.

3.2 Subgraph Decomposition Properties

For the subgraph decompositions, Table 1 shows the largest values of k, denoted as k_{max}, that gave a non-empty core, truss and nucleus, and the number of nodes that were identified as influential, i.e., the number of nodes belonging to the k_{max}-cores, trusses and nuclei, denoted by v_{max}. As shown in Table 1, the number of maximal nucleus nodes (v_{max}) is smaller than the number of maximal truss nodes, which is smaller than the number of maximal core nodes. This is expected as truss decomposition relies on triangles and nucleus decomposition relies on cliques, which are increasingly stricter criteria. Furthermore, Table 2 reports the overlap between influential nodes identified by the different decompositions. Many nodes are common among the three decompositions. In fact, the entire k_{max}-nucleus is often a subset of a k_{max}-truss or k_{max}-core subgraph. This was also seen in our illustrative example in Fig. 1.

3.3 Analyzing Trust

Two datasets, Epinions and Slashdot, contain ground truth about who trusts whom in the network. This allows us to explore the contextual properties of our subgraphs. These two graphs have directed edges with binary edge weights: An edge from A to B has a weight of one if A marks B as a "friend" and zero if A marks B as a "foe". Only 15% of edges in Epinions are foe edges, and 23% of edges in Slashdot are foe edges. We assume that individuals trust their friends but not their foes. This is important in the context of influence maximization because, in practice, influential people are generally those who are trusted by others.

Table 2. Overlap among various selected sets: N is the maximal set of Nucleus nodes, T is the maximal set of Truss nodes, C is the maximal set of Core nodes, D is the set of top 100 Degree centrality nodes, and $I_{IC,\tau}$ is the set of nodes (number of nodes equal to the size of N) found by IMM under the IC model at τ. The left table is the percentage overlap with N and the right is the percentage overlap with D.

Dataset	$N \cap T$	$N \cap C$	$N \cap T \cap C$	Dataset	$C \cap D$	$T \cap D$	$N \cap D$	$I_{IC,\tau} \cap D$
WikiVote	100%	94.6%	94.6%	Wikivote	19%	15%	38.8%	5.4%
Slashdot	96.9%	96.9%	96.9%	Slashdot	5.9%	3.9%	24.6%	23.1%
Epinions	100%	100%	100%	Epinions	0%	0%	0%	29.5%
EuEmail	71.4%	100%	71.4%	Euemail	6.8%	3.2%	33.9%	53.57%

Table 3. Average trust metrics for Slashdot and Epinions.

Dataset	Subgraph	Trusted by	In degree	Out degree	Reputability (in %)
Slashdot	Whole	5.159	6.665	6.665	77.4
	Core	176.0	186.4	180.9	95.0
	Truss	180.7	191.2	185.5	96.2
	Nucleus	183.3	191.1	194.1	96.8
Epinions	Whole	5.444	6.382	6.392	49.0
	Core	177.4	183.1	239.3	96.9
	Truss	182.3	188.5	245.3	96.8
	nucleus	191.5	197.4	254.3	96.9

Table 3 presents the following statistics for the entire graphs and for their respective maximal core, truss and nuclei: the average number of nodes that trust a given node, the average node in and out degrees, and the average node *reputability*, defined as the percentage of nodes who trust the given node v and the node v's in-degree. We see that higher-order decompositions are more densely connected and have higher reputability (on average), reinforcing our belief that subgraph decomposition identifies topologically and contextually essential nodes. As users often rate things they like or not rate at all [17], being connected to more people makes one more likely to be positively rated and may have a cascading effect on reputability. The influential nodes identified by nucleus decomposition have the highest reputability in Slashdot, whereas in Epinions, all three tested methods have similar reputability scores.

3.4   Evaluating Spreading Performance

We now evaluate spreading effectiveness using the three information diffusion models. We test k-core, truss and nucleus decomposition as well as the *Degree Centrality* method for selecting the seed set. For degree centrality, we take the top-n highest degree centrality nodes, where n is the number of nodes in a maximal nucleus. While [13] use only the nodes having the highest degree as the seed set, we found that the number of highest-degree nodes can often be too small to be of any practical significance.

Furthermore, we compare our methods to the IMM algorithm [22], which is an approximation algorithm to solve the underlying NP-hard influence maximization problem.

Given the desired number of seed nodes, IMM identifies the (approximately) best such nodes given the underlying diffusion model. As we did in the degree centrality method, we set the desired number of seed nodes to be the number of nodes in a maximal nucleus.

We also note that previous work often used undirected versions of datasets. However, an undirected edge means that "if A trusts B, B also trusts A." This reciprocal behaviour may not always be true, which may affect the efficacy of the diffusion process. Thus, we use directed graphs in our simulations. The experimental setup for the three diffusion models is given below:

- **Independent Cascade Model:** We draw propagation probabilities from a uniform distribution. We set the propagation probability, or **activation rate**, of an edge (v, v') to $p_{v,v'} = u(0, t)$, where $u(0, t)$ is a uniform function between 0 and t for some $t \in (0, 1]$. We choose the uniform distribution as we do not have complete knowledge about users' propagation probabilities [4]. We limit the propagation probabilities to t and use this as a parameter for our experiments. For instance, a low value of t means that nodes are not easily influenced and thus can be thought of as low-trust networks. We start with $t = \tau$ as per Table 1.
- **Linear Threshold Model:** We set the activation thresholds based on a uniform distribution as we do not know the real thresholds for the nodes. Thus, $\theta_v = u(0, 1)$ where $u(0, 1)$ is a uniform function.
- **SIR Model:** We set the infection rate to be the threshold τ (see Table 1) and the recovery rate to be 0.08 as suggested in [13].

Table 4. Average Spreading performance (number of nodes activated per seed set node). Note: for SIR and IC, we use threshold τ.

Dataset	Subgraph	LT	SIR	IC	Dataset	Subgraph	LT	SIR	IC
WikiVote	Core	7.85	1.15	0.16	Epinions	Core	8.23	6.01	0.32
	Truss	14.99	1.15	0.30		Truss	9.79	6.50	0.36
	Nucleus	17.65	2.00	0.35		Nucleus	11.01	8.47	0.43
	DC	28.44	0.24	0.25		DC	38.2	14.41	1.13
	IMM	28.66	3.34	0.72		IMM	68.39	7.16	1.76
Slashdot	Core	47.6	0.68	0.04	EuEmail	Core	9.45	2.82	0.79
	Truss	71	0.24	0.05		Truss	58.21	3.15	1.39
	Nucleus	87.46	1.12	0.05		Nucleus	78.27	6.96	2.48
	DC	79.42	1.08	0.05		DC	74.71	4.02	2.46
	IMM	85.2	1.27	0.06		IMM	76.67	9.48	2.52

For each trial, we activate all the nodes that were reported by a given method as being influential. We then run the diffusion process until no new nodes can be activated. We repeat the simulation 1,000 times and report the average number of nodes activated divided by the seed set size. This metric is our **spreading efficiency**.

Table 4 shows the results. We present the spreading performance for IC and SIR at the minimum threshold chosen (τ). We see that the absolute value of the final spreading efficiency is low for SIR and IC. This is because the reported values are at the lowest threshold we tested. Furthermore, we see that both

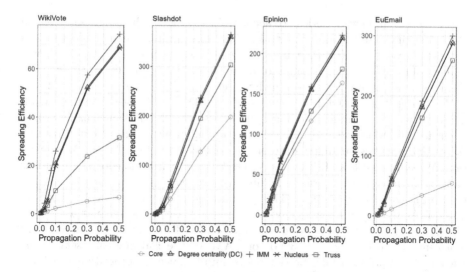

Fig. 2. Impact of propagation probabilities on spreading performance (number of nodes activated per seed set node) for IC model. Note: the standard errors were too small to report (standard error < 0.15).

Degree Centrality (DC) and nucleus decomposition performs better than core and truss decompositions, even at low thresholds. In some cases, DC outperforms the tested graph decomposition methods.

Impact of Propagation Probabilities: We now illustrate the impact of propagation probabilities in the Independent Cascade (IC) Model. We focus on the IC model as SIR has been discussed in depth in [13], and SIR reduces to IC when the propagation probability is the same for all nodes (called infection rate for SIR).

For each method and dataset, Fig. 2 shows the number of activated nodes for the following propagation probabilities, shown from left to right: τ, 0.01, 0.03, 0.05, 0.1 and 0.5 (recall that τ is the inverse of the largest eigenvalue of the corresponding adjacency matrix and gives a reasonable lower bound for the propagation probability threshold). We stop at 0.5 as information spread tends to saturate at some propagation probability threshold, typically ≈ 0.5. As expected, the number of activated nodes increases for all methods as the propagation probability increases. Second, as nucleus decomposition starts with a smaller seed set, nucleus nodes have better per-node spreading efficiency on average. The degree centrality nodes also perform on par with the nucleus nodes.

Comparison with IMM Algorithm: The methods we considered so far select influential nodes based on graph properties. In contrast, IMM selects influential nodes by solving the underlying influence maximization problem approximately to $1 - 1/e - \epsilon$, where ϵ controls the approximation (a higher value trades solution quality for runtime). Since we randomly assign propagation probabilities in our simulations, we found that different experimental runs of IMM on the same graph gave different influential nodes. To account for this, for each experiment, we run IMM 100 times and output the nodes that were most frequently identified as influential over the 100 runs. (Again, as mentioned earlier, the total number of influential nodes we select is equal to the number of nodes in a maximal nucleus.) IMM requires the user to set ϵ, which we set to 0.1 following prior work [22].

From Fig. 2 and Table 4, we conclude that IMM has the best average spreading performance in many situations, and is nearly as good as nucleus decomposition for Slashdot and EuEmail. Our results align with those from [1], which shows that there is no single state of the art technique in Influence Maximization. We also note that IMM took about 1500 s per iteration for the Epinions dataset (IC model, 0.5 threshold), while nucleus decomposition took 126 s, truss took 5.2 s, and core took 0.2 s. The runtimes on all the datasets are shown in Table 5. As we explained above, we ran IMM 100 times to arrive at a "stable" set of seed nodes, giving a total runtime of over 41 h. Moreover, the memory footprint of IMM is high (> 30GB of RAM for the EuEmail dataset). Thus, nucleus decomposition may be the algorithm of choice for practitioners willing to sacrifice some effectiveness for much faster runtime.

Comparison with Degree Centrality: According to Fig. 2 and Table 4, DC performs better than nucleus decomposition in some cases. In Table 2, we see that the top 100 nodes ranked by degree centrality contain $\sim$ 20% of nucleus nodes. Interestingly, there is no overlap between the top degree centrality nodes and subgraphs for Epinions. One possible reason for this could be the sparse nature of the Epinions graph, as seen in Table 1. We also see that DC has a high overlap with the nodes found by IMM, indicating that the optimal nodes chosen by IMM often have high degree as well.

Table 5. Runtimes in seconds. For IMM, we report the time at threshold 0.5 for IC

Dataset	Core	Truss	Nucleus	IMM
WikiVote	0.16	0.35	4.1	83.13
Slashdot	0.10	0.81	5.2	373.8
Epinions	0.19	4.00	126.2	1275.2
EuEmail	0.11	0.38	2.0	786.7

4   Discussion

Identifying influential nodes that can disseminate information to a large part of a network is of particular interest in social network research. k-core, a subgraph decomposition based on maximal node degrees, has mainly been studied in this context and found to be effective [10,15]. However, even k-core can overlook critical features in the graph, motivating the use of a higher-order decomposition called k-truss [13]. The promising results using k-truss decomposition motivated us to consider even more dense substructures, and we arrived at a generalized notion of nucleus decomposition [19]. By imposing more restrictions on nodes in terms of topological and positional factors, our experiments reveal that nucleus decomposition significantly reduces the number of candidates for influential spreaders. This means that in practice, fewer nodes need to be engaged to obtain similar spreading performance. Nucleus decomposition can be used in conjunction with other influence maximization algorithms to reduce their search space for even better results. For example, one may start with a set of nucleus nodes and choose a subset of them with the highest degree. However, as seen in the experimental results using the Epinions dataset, nucleus decomposition may not always produce the most influential nodes.

We explored three models for information spread in order to gauge the performance of topology-based IM methods. The Linear Threshold (LT) model is sensitive to the number of neighbours that can influence a node: a large number of neighbours make it more likely for information to spread. Thus, Degree Centrality works well in the LT model. The SIR and Independent Cascade (IC) models use propagation probabilities that may be different for different node pairs. In these cases, nucleus decomposition performs better as it tends to identify strategically placed nodes. We notice that Degree Centrality often performs on par with nucleus decomposition, especially for higher propagation probabilities. This may be due to the high overlap between the nucleus and high degree-centrality nodes, as seen in Table 2. Furthermore, we found that in many situations, DC and nucleus decomposition performs similar to IMM in terms of average spreading performance. However, they take much less time to be computed. This suggests the benefits of topology-based methods such as nucleus decomposition compared to approaches that solve the underlying influence maximization problem.

Modeling and optimizing influence spread has garnered much interest from multiagent systems researchers. Some researchers in this field contend that graph properties may be prone to an error in predictions of information spread, and that actual behaviour in certain networks, especially ones of more modest sizes (e.g., for homeless youth HIV prevention [23]) may play out differently, integrating more connection to those currently outside the network. These authors also advocate considering the set of seed nodes as a multiagent team, with interconnections. We view the work in this paper as complementary to these research threads in the multiagent systems domain. For one, if it is indeed critical to be examining relationships between the nodes in the seed set, this can be done all the more effectively if operating with a smaller set of nodes, the behaviour of which can be examined in detail. It is also possible to use nucleus decomposition together with other models of diffusion, which are more generous to the integration of external nodes.

5   Conclusions and Future Extensions

This work provides vital new insights into how to track influence within social networks when operating with constrained resources, revealing the effectiveness of smaller seed sets, of use for a host of applications. Calibrating the value of k-nucleus is an important part of our effort. Recall that truss and core can be thought of as $(2,3)$ and $(1,2)$-nuclei, respectively. A $(3,4)$-nucleus is an even denser subgraph with fewer nodes that have the potential to exhibit good spreading power. While there is a marked improvement, there will be a diminishing return on computation time investment on successively mining denser subgraphs (as noted by [19]). This opens several avenues for future work.

A limitation of subgraph mining methods is that they usually consider undirected graphs, and thus some information may be lost. A potential solution is to identify d-cores [6], which separately consider the in-degree and out-degree of nodes and thus may be more suitable for directed graphs. Furthermore, other graph decomposition based approaches have been proposed, such as k-meanoid [25] and modified k-shell [2]. A comparison of k-nucleus against these methods would be an ideal next step.

Various empirical studies to date have provided insights into the theoretical advantages of different algorithms (e.g. [8]); for future work, it would be valuable to expand these kinds of discussions to nucleus decomposition. Additionally, since nucleus decomposition gives a verifiably smaller set of nodes with better-spreading properties than other methods such as core and truss, it can also be used as a preprocessing step for optimal algorithms. Moreover, there are now parallelized algorithms available for nucleus decomposition, which can improve the efficiency of our approach [18].

Interestingly, the work of [15] on tracking real-world information flow found that the most influential nodes lie in the k-core subgraph, and it would be valuable to show that they lie in the nucleus or truss subgraphs as well. It would also be interesting to empirically compare subgraph based methods with the greedy algorithm of [8]. [16] assigns a "Klout Score" of influence to 750 M users by extracting features from user interactions in multiple social networks and then aggregating them into a hierarchical scoring structure. Combining these content-based methods with k-nucleus decomposition is another potential direction for future work.

One final avenue for future work is to experiment with something other than a uniform propagation probability distribution, determining which scenarios benefit from lifting that assumption. Examining other methods for measuring influence would also be valuable (for example, [21] uses social media posts to create a content based metric).

References

1. Arora, A., Galhotra, S., Ranu, S.: Debunking the myths of influence maximization: an in-depth benchmarking study. In: ACM International Conference on Management of Data, pp. 651–666 (2017)

2. Brown, P.E., Feng, J.: Measuring user influence on twitter using modified k-shell decomposition. In: Fifth international AAAI conference on weblogs and social media (2011)
3. Chakrabarti, D., Wang, Y., Wang, C., Leskovec, J., Faloutsos, C.: Epidemic thresholds in real networks. ACM Trans. Inf. Syst. Secur. **10**(4), 1 (2008)
4. Chen, W., Wang, C., Wang, Y.: Scalable influence maximization for prevalent viral marketing in large-scale social networks. In: Proceedings of the 16th ACM SIGKDD international conference on Knowledge discovery and data mining, pp. 1029–1038. ACM (2010)
5. Cohen, J.: Trusses: Cohesive subgraphs for social network analysis. National Security Agency Technical Report **16** (2008)
6. Giatsidis, C., Thilikos, D.M., Vazirgiannis, M.: D-cores: measuring collaboration of directed graphs based on degeneracy. In: 2011 IEEE 11th International Conference on Data Mining (ICDM), pp. 201–210. IEEE (2011)
7. Granovetter, M.: Threshold models of collective behavior. Am. J. Sociol. **83**(6), 1420–1443 (1978)
8. Kempe, D., Kleinberg, J., Tardos, É.: Maximizing the spread of influence through a social network. In: Proceedings of the ninth ACM SIGKDD international conference on Knowledge discovery and data mining, pp. 137–146. ACM (2003)
9. Kermack, W.O., McKendrick, A.G.: A contribution to the mathematical theory of epidemics. Proceedings of the royal society of london. Series A, Containing papers of a mathematical and physical character **115**(772), 700–721 (1927)
10. Kitsak, M., et al.: Identification of influential spreaders in complex networks. Nat. Phys. **6**(11), 888 (2010)
11. Leskovec, J., Krevl, A.: SNAP Datasets: stanford large network dataset collection. http://snap.stanford.edu/data (2014)
12. Li, Y., Fan, J., Wang, Y., Tan, K.L.: Influence maximization on social graphs: a survey. IEEE Trans. Knowl. Data Eng. **30**(10), 1852–1872 (2018)
13. Malliaros, F.D., Rossi, M.E.G., Vazirgiannis, M.: Locating influential nodes in complex networks. Sci. Rep. **6**, 19307 (2016)
14. Mislove, A., Marcon, M., Gummadi, K.P., Druschel, P., Bhattacharjee, B.: Measurement and analysis of online social networks. In: Proceedings of the 7th ACM SIGCOMM conference on Internet measurement, pp. 29–42. ACM (2007)
15. Pei, S., Muchnik, L., Andrade Jr., J.S., Zheng, Z., Makse, H.A.: Searching for superspreaders of information in real-world social media. Sci. Rep. **4**, 5547 (2014)
16. Rao, A., Spasojevic, N., Li, Z., Dsouza, T.: Klout score: measuring influence across multiple social networks. In: IEEE International Conference on Big Data, pp. 2282–2289 (2015)
17. Sardana, N., Cohen, R., Zhang, J., Chen, S.: A bayesian multiagent trust model for social networks. IEEE Trans. Comput. Soc. Syst. **5**(4), 995–1008 (2018)
18. Sariyüce, A.E., Seshadhri, C., Pinar, A.: Local algorithms for hierarchical dense subgraph discovery. Proc. VLDB Endow. **12**(1), 43–56 (2018)
19. Sariyüce, A.E., Seshadhri, C., Pinar, A., Çatalyürek, Ü.V.: Nucleus decompositions for identifying hierarchy of dense subgraphs. ACM Trans. Web **11**(3), 16 (2017)
20. Seidman, S.B.: Network structure and minimum degree. Soc. Netw. **5**(3), 269–287 (1983)
21. Sun, B., Ng, V.T.Y.: Identifying influential users by their postings in social networks. In: Atzmueller, M., Chin, A., Helic, D., Hotho, A. (eds.) MSM/MUSE -2012. LNCS (LNAI), vol. 8329, pp. 128–151. Springer, Heidelberg (2013). https://doi.org/10.1007/978-3-642-45392-2_7

22. Tang, Y., Shi, Y., Xiao, X.: Influence maximization in near-linear time: a martingale approach. In: Proceedings of the 2015 ACM SIGMOD International Conference on Management of Data, pp. 1539–1554. ACM (2015)
23. Wilder, B., et al.: End-to-end influence maximization in the field. In: 17th International Conference on Autonomous Agents and MultiAgent Systems, pp. 1414–1422 (2018)
24. Yadav, A., et al.: Bridging the gap between theory and practice in influence maximization: raising awareness about HIV among homeless youth. In: IJCAI, pp. 5399–5403 (2018)
25. Zhang, X., Zhu, J., Wang, Q., Zhao, H.: Identifying influential nodes in complex networks with community structure. Knowl. Based Syst. **42**, 74–84 (2013)

Investigating Relational Recurrent Neural Networks with Variable Length Memory Pointer

Mahtab Ahmed$^{(\boxtimes)}$ [ID] and Robert E. Mercer [ID]

Department of Computer Science, University of Western Ontario, London, Canada
mahme255@uwo.ca, mercer@csd.uwo.ca

Abstract. Memory based neural networks can remember information longer while modelling temporal data. To improve LSTM's memory, we encode a novel Relational Memory Core (RMC) as the cell state inside an LSTM cell using the standard multi-head self attention mechanism with variable length memory pointer and call it LSTM$_{RMC}$. Two improvements are claimed: The area on which the RMC operates is expanded to create the new memory as more data is seen with each time step, and the expanded area is treated as a fixed size kernel with shared weights in the form of query, key, and value projection matrices. We design a novel sentence encoder using LSTM$_{RMC}$ and test our hypotheses on four NLP tasks showing improvements over the standard LSTM and the Transformer encoder as well as state-of-the-art general sentence encoders.

Keywords: Relational Memory Core · Long short term memory · Sentence encoder · Attention · Semantic similarity

1 Introduction

Humans use their memory system to retrieve important information irrespective of when they are perceived [8,12,15]. Recent neural network based research has successfully encoded memory inside the LSTM cell [7,12] and even designed core memory augmented neural networks [6,13]. These memory based networks have efficient information storing and retrieval capabilities reinforced by bounded computational cost, augmented memory capacities and overcoming the vanishing gradient problem. In NLP, it is necessary to analyze and compare phrases as well as sentence constituents seen at different time steps to extract the meaning. Giving an LSTM's common hidden memory access to more of the previous representations using a relational memory core (RMC) gives a performance boost in tasks demanding particular types of temporal relational reasoning [12]. This RMC is designed using the standard multi-head self attention framework [18].

In this paper, we design an improved relational memory core (RMC) having access to previously seen representations through a variable length memory pointer. Our idea is that the memory created at each time step should reflect the previously created representations whereas the LSTM gates should be updated

© Springer Nature Switzerland AG 2020
C. Goutte and X. Zhu (Eds.): Canadian AI 2020, LNAI 12109, pp. 29–35, 2020.
https://doi.org/10.1007/978-3-030-47358-7_3

only with the last one. We encode this new RMC as the cell state inside an LSTM cell and design a sentence encoder model to encode a pair of sentences for the sentence pair modelling task. Evaluation done on textual entailment (TE), semantic relatedness (SR), paraphrase identification (PI), and question answer pairing (QAP) improves on standard LSTM encoders on all tasks and is best on the textual inference and second best on question-answer pair tasks.

2   Related Work

Memory based models based on attention have been used to modify standard and tree LSTMs. Sukhbaatar et al. [17] train a memory based neural network in an end-to-end fashion to solve a compartmentalization problem with a slot-based memory matrix. Santoro et al. [14] propose a plug-and-play neural network to perform relational reasoning task which involves two MLPs as the composition functions. Ahmed et al. [1] encode a single head dot product attention block inside a tree-LSTM cell. General purpose sentence encoders have been state-of-the-art on the four tasks of interest in this paper. Cer et al. [3] propose a sentence encoder model based on the encoder portion of Transformer [18] and perform an element-wise sum of the encoded representations at each word position to get a fixed length sentence representation. Conneau et al. [4] also propose a universal sentence representation model based on LSTMs where they first train it on the Stanford Natural Language Inference dataset and then uses transfer learning to evaluate on a range of tasks including QAP, PI, and SR. Zhou et al. [20] propose a sentence pair ranking model where they encode attention in the tree structure of the hypothesis based on the sequential representation of the premise and vice versa. Zhao et al. [19] propose a self-adaptive hierarchical model which first extracts an intermediate representation of all possible phrases and finally takes the convex combination of them through gating. Socher et al. [16] propose a recursive auto-encoder-based paraphrase identification model that first reconstructs each of the phrases from the tree representation and then extracts the sentence representation with a min-pooling block.

3   The Model

To improve the design principle of the current RMC [12], we extend the scope of the memory pointer in RMC by giving the self attention module more to explore. In the classic setting [12], the network maintains a fixed length memory pointer which begins with a random memory $M_{1 \times b \times d}$ and is updated at each time step. It creates a new tensor $\widetilde{M}$ by concatenating the previous memory and a linear projection of the current input as follows,

$$\widetilde{M}^t_{2 \times b \times d} = [M^{t-1}_{1 \times b \times d}; \mathbf{W} x^t_{1 \times b \times d}] \tag{1}$$

Following this, the `multi-head self attention` (MSA) block is applied to see how much information to take from the current input and the previous memory to create the next memory. This MSA block first creates query,

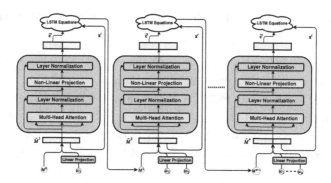

Fig. 1. Sample encoding of a sentence using RMC with variable length memory pointer (from left to right direction).

$Q(= M\mathbf{W}^q)$, and key, $K(= M\mathbf{W}^k)$, and then applies the `dot attention` A $(= softmax(\frac{QK^T}{\sqrt{d_k}}))$ on top of them to get the scalar weights. Finally, the value tensor, $V(= M\mathbf{W}^v)$, is scaled with these weights giving the final memory $\widetilde{M}$.

A residual connection is then added to the resultant tensor $\widetilde{M}$ followed by a `Layer-Normalization` (LN) block to speed training. Unlike the separate versions of mean and variance projection matrices in the original RMC work, we maintain the same parameter over all dimensions. Next, n non-linear projections of h^t are applied followed by a residual connection. Eq. 2 shows this step for $n = 2$.

$$X = \mathbf{f}(\mathbf{W}^{(1)}\mathbf{f}(h^t\mathbf{W}^{(2)})) + h^t \tag{2}$$

Here, we use `ReLU` as `f`. The new memory is extracted from the resultant tensor X from Eq. 2 (has shape $2 \times b \times d$) by splitting it at the cardinal dimension $M^t_{1 \times b \times d} = X^1_{1 \times b \times d}$ and used as the candidate cell state in LSTM equations. The x^t's in the LSTM equations are replaced with $\mathbf{W}x^t$ from Eq. 1.

Unlike Eq. 1 we continually expand the area on which the MSA operates. At each time step t, we do this by concatenating all the projected inputs from time step t to 1 with the previous memory as follows,

$$\widetilde{M}^t_{(t+1) \times b \times d} = [M^{t-1}_{1 \times b \times d}; \mathbf{W}x^t_{1 \times b \times d}; \mathbf{W}x^{t-1}_{1 \times b \times d}; \cdots ; \mathbf{W}x^1_{1 \times b \times d}] \tag{3}$$

We keep the weight ($\mathbf{W}$) shared over all the words. Next, we apply MSA, LN and a series of projections to get a new transformed representation X as in Eq. 2. We can always keep track of the memory as it is fixed at the first index. Finally, as done earlier, we use $\mathbf{W}x^t$ as x^t in the LSTM equations. We limit the past view to a maximum window size meaning that we cannot use a fixed set of mean and variance projection matrices in the LN block as it will make the gradient accumulation of few parameters unstable. Finally, as the sentence pair modelling architecture, we choose Infersent [4] as is except its LSTM block is replaced by LSTM$_{RMC}$.

Table 1. Hyper-parameters used for the experiments (in boldface) and the ranges that were searched during tuning.

Config	Value	Config	Value
Initial learning rate	**0.1**/0.05/0.001	maxNorm	**5**
Batch size	10/**16**/25	Learning rate decay	**0.99**
No. of Attention layers	1/**2**/3	Dropout FC	**0.0375−0.5**
Hidden dimension	256, 512, **1024**	No. of Heads	**8**
Word embedding	Glove 300D	W^q, W^k, W^v dimension	**128**

Table 2. Model performance compared to top performing models. † indicates models that we implemented. **FLMP/VLMP** = Fixed/Variable length memory pointer.

Model	MSRP	AI2-8grade	SICK-E	SICK-R
	Acc.	Acc.	Acc.	r/MSE
LSTM$_{RMC}$ + FLMP	74.67	74.72	85.38	0.8107/0.3452
LSTM$_{RMC}$ + VLMP (WINDOW SIZE = 5)	75.89	74.72	84.28	0.8440/0.2925
INFERSENT [4] †	74.46	74.10	84.62	0.8563/0.2732
LSTM [4] †	70.74	74.24	76.80	0.8291/0.3244
BILSTM PROJECTION LAYER [4] †	74.24	75.15	85.20	0.8037/0.3667
INNER ATTENTION [9] †	69.74	74.32	72.01	0.7863/0.3944
CONVNET ENCODER [19] †	73.96	75.15	83.82	0.8520/0.2806
TRANSFORMER ENCODER [3]	74.96	–	81.15	–/0.5241
SEQ-LSTMs [20]	71.70	63.30	–	0.8528/0.2831
TREE LSTM [20]	73.50	69.10	–	0.8664/0.2610
TREE LSTM + ATTN. [20]	75.80	72.50	–	0.8730/0.2426
TREE GRU [20]	73.96	70.60	–	0.8672/0.2573
TREE GRU + ATTN. [20]	74.80	72.10	–	0.8701/0.2524
RAE [16]	76.80	–	–	–

4   Experimental Setup

Datasets: Model evaluation uses three datasets. **MSRP** (paraphrase identification) [5], **SICK** (two tasks: (1) classify sentence pairs into three classes: Entailment, Neutral and Contradiction, and (2) assign a score between 1 and 5 for a sentence pair based on their semantic relatedness) [10], and **AI2-8grade** (pair a question with its correct answer given multiple answers per question) [2].

Experimental Setup: Word vectors are initialized with the 300D GloVe embeddings [11] and are frozen during training. To smooth the update, we normalize the gradients by batch size and adopt the decaying learning rate paradigm. Table 1 shows the hyper-parameter settings used during the experiments. Training is done on a GeForce GTX 1080Ti GPU with the 'SGD' optimizer. Models are implemented in PyTorch 0.4.1 under the Linux environment.

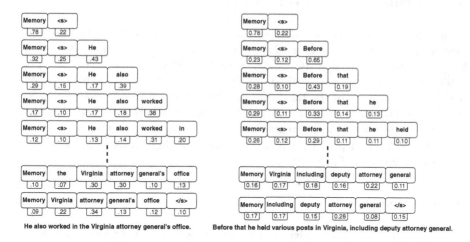

Fig. 2. Change in attention weights as the attention window shifts over time (memory pointer length/window size = 5). Sentences shown are paraphrases.

5   Experimental Results and Analysis

Table 2 displays our model's performance in terms of task specific evaluation metrics. The first group contains the results of $LSTM_{RMC}$ in the classic fixed pointer (FLMP) and our variable pointer (VLMP) configurations. Notably, on the MSRP task, the VLMP model is behind only RAE [16] which leverages additional grammatical information. The VLMP model achieves better accuracy than the FLMP model indicating the importance of checking word overlapping in a local context for this task. On the AI2-8grade dataset our model is behind only BiLSTM Projection Layer [4] and ConvNet Encoder [19]. Our application of classic FLMP on the SICK-E task achieves the best accuracy compared to the other sequential models. On the SICK-R task, the 0.8440 Pearson's r and 0.2925 MSE puts us in the top three sequential models on this task. These scores clearly tell that $LSTM_{RMC}$ is always in the top three general purpose sentence encoders on all of the four different corpora.

Figure 2 visualizes attention weights on different segments (5-gram in this case) of the sentence pair when doing the comparison. Over stop-words, it puts less attention and distributes the probability on other content words inside the window. In the left sentence, as "Virginia attorney general's office" is a content phrase, attention on memory is less significant compared to attention on the 5-grams at the last time step. In the right sentence, the memory gets comparatively higher weight as the content phrase appears at the time step before.

6   Conclusion and Future Work

In this paper, we modify the classical RMC with the concept of a variable length memory pointer allowing it to use local context for computing enhanced memory. Encoding this improved RMC inside an LSTM cell we design a sentence pair modelling architecture and evaluate on four different tasks. We achieve on par performance on most of the tasks and best performance on one of them. Being equipped with this new memory, our model interprets the attention shifting very well. Our limited experiments show that the memory pointer length does not follow a uniform pattern across all datasets, making it an interesting area to investigate for our future studies.

References

1. Ahmed, M., Samee, M.R., Mercer, R.E.: Improving Tree-LSTM with tree attention. In: 13th International Conference on Semantic Computing (ICSC), pp. 247–254 (2019)
2. Baudiš, P., Stanko, S., Šedivý, J.: Joint learning of sentence embeddings for relevance and entailment. In: Proceedings of the 1st Workshop on Representation Learning for NLP, pp. 8–17 (2016)
3. Cer, D., et al.: Universal sentence encoder. arXiv preprint arXiv:1803.11175 (2018)
4. Conneau, A., et al.: Supervised learning of universal sentence representations from natural language inference data. arXiv preprint arXiv:1705.02364 (2017)
5. Dolan, B., Quirk, C., Brockett, C.: Unsupervised construction of large paraphrase corpora: exploiting massively parallel news sources. In: Proceedings of the 20th International Conference on Computational Linguistics, p. 350 (2004)
6. Graves, A., Wayne, G., Danihelka, I.: Neural turing machines. arXiv preprint arXiv:1410.5401 (2014)
7. Hochreiter, S., Schmidhuber, J.: Long short-term memory. Neural Comput. 9(8), 1735–1780 (1997)
8. Knowlton, B.J., et al.: A neurocomputational system for relational reasoning. Trends Cogn. Sci. 16(7), 373–381 (2012)
9. Lin, Z., et al.: A structured self-attentive sentence embedding. arXiv preprint arXiv:1703.03130 (2017)
10. Marelli, M., et al.: SemEval-2014 Task 1: Evaluation of compositional distributional semantic models on full sentences through semantic relatedness and textual entailment. In: Proceedings of the 8th International Workshop on Semantic Evaluation, pp. 1–8 (2014)
11. Pennington, J., Socher, R., Manning, C.: Glove: global vectors for word representation. In: Proceedings of the 2014 Conference on Empirical Methods in Natural Language Processing (EMNLP), pp. 1532–1543 (2014)
12. Santoro, A., et al.: Relational recurrent neural networks. In: Advances in Neural Information Processing Systems, pp. 7299–7310 (2018)
13. Santoro, A., Bartunov, S., Botvinick, M., Wierstra, D., Lillicrap, T.: Meta-learning with memory-augmented neural networks. In: International Conference on Machine Learning, pp. 1842–1850 (2016)
14. Santoro, A., et al.: A simple neural network module for relational reasoning. CoRR abs/1706.01427 (2017). http://arxiv.org/abs/1706.01427

15. Schacter, D.L., Tulving, E.: Memory Systems. MIT Press, Cambridge (1994)
16. Socher, R., et al.: Dynamic pooling and unfolding recursive autoencoders for paraphrase detection. In: Advances in Neural Information Processing Systems, pp. 801–809 (2011)
17. Sukhbaatar, S., Szlam, A., Weston, J., Fergus, R.: Weakly supervised memory networks. CoRR abs/1503.08895 (2015). http://arxiv.org/abs/1503.08895
18. Vaswani, A., et al.: Attention is all you need. In: Advances in Neural Information Processing Systems, pp. 5998–6008 (2017)
19. Zhao, H., Lu, Z., Poupart, P.: Self-adaptive hierarchical sentence model. In: IJCAI, pp. 4069–4076 (2015)
20. Zhou, Y., Liu, C., Pan, Y.: Modelling sentence pairs with tree-structured attentive encoder. arXiv preprint arXiv:1610.02806 (2016)

Unsupervised Monocular Depth Estimation CNN Robust to Training Data Diversity

Valery Anisimovskiy$^{(\boxtimes)}$ (iD), Andrey Shcherbinin, Sergey Turko, and Ilya Kurilin

Samsung R&D Institute Russia, Moscow, Russia
{valery.a,s.andrey,s.turko,i.kurilin}@samsung.com

Abstract. We present an unsupervised learning method for the task of monocular depth estimation. In common with many recent works, we leverage convolutional neural network (CNN) training on stereo pair images with view reconstruction as a self-supervisory signal. In contrast to the previous work, we employ a stereo camera parameters estimation network to make our model robust to training data diversity. Another of our contributions is the introduction of self-supervision correction. With it we address one of the serious drawbacks of the stereo pair self-supervision in the unsupervised monocular depth estimation approach: at later training stages, self-supervision by view reconstruction fails to improve predicted depth map due to various ambiguities in the input images. We mitigate this problem by making depth estimation CNN produce both depth map and correction map used to modify the input stereo pair images in the areas of ambiguity. Our contributions allow us to achieve state-of-the-art results on the KITTI driving dataset (among unsupervised methods) by training our model on hybrid city driving dataset.

Keywords: Depth estimation · Deep learning · Unsupervised learning

1 Introduction

Monocular depth estimation is one of the tasks which is well-known to be easily performed by humans, while performing the same task by computer vision systems has proven to be a particularly difficult problem. One of the reasons for such a good depth perception by humans is their ability to utilize a rich set of depth-related cues available in the image, such as perspective, relative sizes of familiar objects, occlusions, lighting and shading of object surfaces. Recently, convolutional neural networks (CNNs) have become famous for being capable of leveraging similar rich sets of visual cues for solving various computer vision tasks, like image classification, content-based image retrieval, semantic segmentation, etc. So, unsurprisingly, several latest works successfully applied CNNs for monocular depth estimation task.

V. Anisimovskiy and A. Shcherbinin—Equal contribution.

© Springer Nature Switzerland AG 2020
C. Goutte and X. Zhu (Eds.): Canadian AI 2020, LNAI 12109, pp. 36–48, 2020.
https://doi.org/10.1007/978-3-030-47358-7_4

However, most of those works used supervised learning approach, which limits the applicability of the methods to scene categories where large training datasets containing dense ground truth depth data are available. This is a serious restriction, since ground truth depth values are typically obtained by depth sensor hardware which often either suffers from short range and high noise levels (IR sensors) or provides sparse non-instantaneous point cloud measurements at very high cost (LiDAR sensors).

To overcome the limitations of supervised methods, there has lately been a surge in the number of works trying to solve the task of monocular depth estimation via unsupervised learning. Most of these methods are based on view reconstruction used as self-supervision. Not only do such methods provide much wider coverage of scene categories due to learning on readily available and abundant unlabelled datasets, but they also achieved remarkable levels of depth estimation accuracy.

Our approach builds upon the latest success of the end-to-end unsupervised monocular depth estimation approaches, but adds several crucial contributions:

- **Stereo camera parameters estimation network**, aimed at predicting gain and bias parameters which are used to transform inverse depth map produced by depth estimation network to obtain disparity map subsequently used for view reconstruction.
- **Multi-task learning for the depth estimation network**: this network produces input image correction maps in addition to inverse depth maps. These correction maps are added to the input images which are further used for view reconstruction and self-supervising loss computation.

2   Related Work

Supervised Approaches. In a pioneering work [8], depth estimation was formulated as a regression task with deep CNN used as regression model. Fu et al. [10] combine regression and classification CNNs in the cascaded network to achieve very good depth estimation accuracy at the cost of high model complexity. Kundu et al. [19] and Zheng et al. [39], while claiming their approaches being unsupervised, actually make use of synthetic depth ground truth data in their variants of Generative Adversarial Network (GAN) based domain adaptation. Similarly, Jiang et al. [17], though also claiming their model being self-supervised, use sparse LiDAR depth values to fine-tune their network. Still, such methods heavily rely on high quality, pixel-aligned ground truth depth data for training.

Unsupervised Approaches. To overcome the limitations of supervised approaches, unsupervised methods have been attracting increasing attention over the last few years. Garg et al. [11] proposed training on stereo pairs using opposite view reconstruction loss as self-supervision signal for learning monocular depth estimation. Several works have built on the success of this stereo pair-based approach, including: Godard et al. [14] use bilinear sampler for opposite view reconstruction and left-right consistency regularization loss, Repala and Dubey [29] leverage Siamese CNN processing for left and right views, Poggi et al. [26] trade accuracy for efficiency in their real-time inference-capable low complexity model.

Another unsupervised approach removes the requirement of training on stereo pairs by utilizing neighbouring frame reconstruction loss for self-supervision in monocular video sequences: Zhou et al. [40] designed a model relying on estimation of both the depth map of a target image and the relative camera poses in a short image sequence captured by a moving camera. Wang et al. [31] augmented the framework of [40] with perceptual loss.

3   Method

This section describes our method for monocular depth estimation. The task of the method is generation of dense depth map given a single input image. We build our method basing upon the framework of the state-of-the-art unsupervised monocular method of [14]. Our modifications to this framework allow us to set a new state-of-the-art in the unsupervised monocular depth estimation and are described in the following subsections.

3.1   Depth Estimation as Image Reconstruction

Similar to [14], we formulate a depth estimation task as a subtask in the image reconstruction framework (see Fig. 1). Given a calibrated stereo pair consisting of two RGB images I_L and I_R at training time, our model first predicts per-pixel inverse depth maps $\hat{z}_L$ and $\hat{z}_R$ using two Siamese UNet-style CNNs (the weights of which are kept tied during training). Alongside inverse depth maps, correction maps ΔI_L and ΔI_R are estimated by the same CNNs. Note that, in contrast to [14], a single inverse depth map is predicted from each image (rather than a pair of disparity maps, as in [14]), while each correction map is a three-channel image which is added to the corresponding input image to form the corrected left and right images $\hat{I}_L$ and $\hat{I}_R$.

Two small subsets of high-level feature maps produced in the bottleneck of each depth estimation CNN are concatenated and subsequently fed into camera parameters estimation CNN which outputs two scalar values: gain G and bias B. These parameters are used to derive the primary disparity maps d_L and d_R from the corresponding inverse depth maps $\hat{z}_L$ and $\hat{z}_R$ using per-pixel affine transform:

$$d_L = G(\hat{z}_L + B)$$
$$d_R = -G(\hat{z}_R + B), \tag{1}$$

where multiplications and additions are element-wise. We draw inspiration for this particular transform from Eq. 3 of [34], where similar affine transform was used to translate depths to disparities.

The primary disparity maps are used to reconstruct the left image from the right one and vice versa using the same bilinear sampler as in [14]: $I'_L = \hat{I}_R(d_L)$ and $I'_R = \hat{I}_L(d_R)$. The secondary disparity maps d'_L and d'_R are derived from the

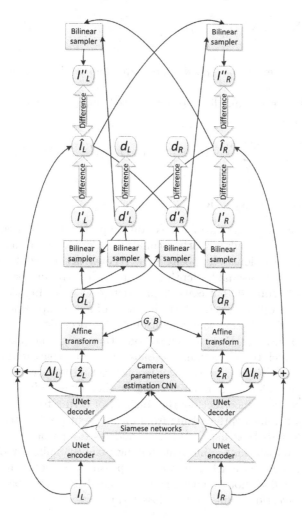

Fig. 1. The block diagram of our model.

primary ones using bilinear sampler, too: $d'_L = -d_R(d_L)$ and $d'_R = -d_L(d_R)$. Furthermore, the secondary disparity maps are also used for image reconstruction, producing auxiliary reconstructed images: $I''_L = \hat{I}_R(d'_L)$ and $I''_R = \hat{I}_L(d'_R)$. Note that for all aforementioned images and maps, predictions at four different scales are produced and used in subsequent operations (except for correction maps, for which two scales are used), with each scale doubling in spatial resolution with respect to preceding scale.

The reconstructed images provide primary supervisory signal for training our model via L_1 reconstruction loss:

$$\mathcal{L}_{rec} = \sum_s \frac{1}{N_s} \sum_{i,j} \left| I_L^{\prime(s)}(i,j) - \hat{I}_L^{(s)}(i,j) \right| + \left| I_R^{\prime(s)}(i,j) - \hat{I}_R^{(s)}(i,j) \right|, \qquad (2)$$

where s is scale index, i, j are pixel coordinates and N_s is the number of pixels at scale s.

Additionally, we use auxiliary reconstructed images to define auxiliary L_1 reconstruction loss:

$$\mathcal{L}_{aux} = \sum_s \frac{1}{N_s} \sum_{i,j} \left| I_L^{\prime\prime(s)}(i,j) - \hat{I}_L^{(s)}(i,j) \right| + \left| I_R^{\prime\prime(s)}(i,j) - \hat{I}_R^{(s)}(i,j) \right|. \qquad (3)$$

3.2 Input Image Correction

We motivate the introduction of correction maps by the observation that in the unsupervised monocular depth estimation approaches training on stereo pairs, the self-supervision by opposite view reconstruction, while being highly efficient in learning reasonable depth map at earlier training stages, at later stages becomes to some extent counter-productive when trying to find geometric explanation for differences in left/right views of a stereo pair actually arising from non-geometric reasons: camera sensor noises, occlusions, reflections from non-Lambertian surfaces. Due to such reasons, there might be pixels in left and right views for which there's actually no matching counterpart in the same scanline of the opposite view. However, the reconstruction loss drives the model to seek matchings even for such unmatched pixels, thereby introducing false disparities and, respectively, incorrect depths. Instead of trying to smooth the noises in disparities or inverse depths arising from inconsistent pixel matching by using disparity smoothness loss (like in [14]), we address the problem by introducing input image correction maps aimed at making pixel matching by disparities in the corrected images as much consistent as possible. To constrain image correction so that it does not remove geometric cues from stereo pair images, we introduce L_1 correction regularization loss term:

$$\mathcal{L}_{cor} = \sum_s \frac{1}{N_s} \sum_{i,j} \left| \Delta I_L^{(s)}(i,j) \right| + \left| \Delta I_R^{(s)}(i,j) \right|. \qquad (4)$$

3.3 Camera Parameters Estimation

Another issue of the state-of-the-art unsupervised monocular depth estimation methods trained on stereo pairs is their intrinsic assumption of stereo rig parameters constancy across the training dataset. Taking as a baseline the method of [14], where the CNN processes the left view and outputs two disparity maps for reconstructing left and right views, such a setting assumes that the baseline distance between the cameras is the same for all stereo pairs. Indeed, given the left view image, for different baseline distances disparity values needed to reconstruct the right view from the left one would be different: longer baseline distance implies larger disparity values. So, we argue that in general setting, where the baseline distance for training stereo pairs is variable, the left view image alone is insufficient for predicting disparity maps.

To remove the requirement for constant (or known from external sources) baseline distance, we introduce stereo camera parameters estimation network. We rely on the intuition that a pair of left and right views contains visual cues sufficient for estimation of the baseline distance between the monocular cameras of the stereo rig used to obtain the stereo pair (at least, up to a scale factor). So, we concatenate two small equally-sized subsets of the high-level feature maps produced in the bottleneck of the depth estimation UNet for the left and right views and use the resulting combined feature maps as the input to the stereo camera parameters estimation network. It allows our model to adapt during training to stereo pairs originating from stereo set-ups with varying stereo baseline and focal length. We argue that such adaptivity makes our model robust to diversity and variability of training dataset in terms of stereo set-up settings, thereby allowing us to efficiently train our model on hybrid dataset containing data from datasets produced by differently configured stereo set-ups.

3.4   Total Training Loss

Similar to [14], we use L_1 left-right disparity consistency penalty:

$$\mathcal{L}_{con} = \sum_s \frac{1}{N_s} \sum_{i,j} \left| d_L'^{(s)}(i,j) - d_L^{(s)}(i,j) \right| + \left| d_R'^{(s)}(i,j) - d_R^{(s)}(i,j) \right| \quad (5)$$

Note that the auxiliary reconstruction loss given in Eq. 3 provides additional left-right consistency indirectly by using secondary disparity maps d_L' and d_R' for image reconstruction.

So, the total training loss for our model is composed of all the above loss terms:

$$\mathcal{L}_{tot} = \mathcal{L}_{rec} + \alpha_{cor}\mathcal{L}_{cor} + \alpha_{con}\mathcal{L}_{con} + \alpha_{aux}\mathcal{L}_{aux}, \quad (6)$$

where α_{cor}, α_{con} and α_{aux} are weighting coefficients for corresponding loss terms.

4   Results

In this section we provide details of the implementation of our method, describe the training datasets and procedure and compare the performance of our method with prior monocular depth estimation methods (both unsupervised and supervised).

4.1   Implementation Details

The architecture of depth estimation CNN is similar to the one of [14], though we modified it to enlarge network's receptive field and to improve output depth map's detailization.

For camera parameters estimation CNN we use relatively shallow network since it accepts as input the high-level feature maps from the bottleneck of the depth estimation CNN.

Once the model is trained, the depth prediction is performed by running depth estimation CNN on an input image and inverting its output using the following formula:

$$z = \frac{1}{P_{99.99}(\hat{z}) - \hat{z}} \tag{7}$$

where $P_{99.99}$ is 99.99th percentile which is introduced here since the inverse depth maps are produced using tanh activation, so they may take both positive and negative values and thus require proper shifting before inverting.

The model was implemented in MXNet framework [2] and trained from scratch for 50 epochs on Nvidia GeForce GTX 1080 Ti GPU with a batch size of 8 using Adam [18] with the learning rate of $\lambda = 10^{-4}$ and $\beta_1 = 0.9$, $\beta_2 = 0.999$, $\epsilon = 10^{-8}$.

4.2   Datasets

We use two datasets to train our model: hybrid city driving (CS+K) dataset and Stereo Movies (SM) dataset.

Hybrid city driving dataset is a mixture of stereo pairs from KITTI [12] (using split of [8]) and Cityscapes [5] datasets, rescaled to 768 × 320. To the best of our knowledge, our method is the first monocular depth estimation method trained on the mixture of those datasets, while other methods typically employ pre-training on Cityscapes and fine-tuning on KITTI (e.g. [14,26,27,37,40]). Training on CS+K dataset poses a serious challenge for unsupervised methods based on stereo pair training, since KITTI and Cityscapes datasets have been collected using different stereo rigs having different baseline distances and focal lengths. On the other hand, training on such diverse dataset forces the model to seek for more general geometric cues in the images. The weights for the loss terms when training on CS+K dataset were set as follows: $\alpha_{cor} = 1.7$, $\alpha_{con} = 0.2$, $\alpha_{aux} = 0.5$.

Stereo Movies dataset has been collected by us using 28 commercially available stereo movies, totalling to over 4 million stereo pairs rescaled to 768 × 384. The movies used for collecting dataset were selected to be very diverse in genre and scenery. This dataset is quite challenging for monocular depth estimation algorithms: supervised methods cannot use it since there're no ground truth depths for stereo movies, while for the unsupervised methods based on monocular video sequence training (like [40]) there's no camera intrinsics information required by such methods. On the other hand, this dataset provides vast diversity in content and high variability in stereo camera parameters. The weights for the loss terms when training on SM dataset were set as follows: $\alpha_{con} = 0.01$, $\alpha_{aux} = 0.5$ and no correction maps were used for this dataset (because of much higher quality of its images).

For quantitative evaluation of the trained models we utilized the widely-used test subset of KITTI (Eigen split) following the evaluation procedure described in [14] as well as Depth In the Wild (DIW) dataset [4].

Table 1. Results on KITTI 2015 [12] using the split of Eigen et al. [8]. Best results in each section are in bold font, our results are in red color. Datasets: **K** is the KITTI dataset [12], **CS+K** is the mixture of Cityscapes [5] and KITTI, **CS→K** means pre-training on Cityscapes [5] and fine-tuning on KITTI, **MD→K** means pre-training on MegaDepth dataset [23] and fine-tuning on KITTI.

Method	Output resolution	Training data			Training dataset	Error metric (lower is better)					Accuracy metric (higher is better)			#Weights[c], ×10^6
		#Input images	Depth	Camera params		ARD	SRD	RMSE	RMSE (log)	RMSE(sc.inv.)	$\delta < 1.25$	$\delta < 1.25^2$	$\delta < 1.25^3$	
Unsupervised approaches (cap at 80 m)														
Zhou et al. [40]	416 × 128	3		✓	CS→K	0.198	1.836	6.565	0.275	0.270	0.718	0.901	0.960	31.6
Poggi et al. [26]	512 × 256	2[b]			CS→K	0.146	1.291	5.907	0.245	–	0.801	0.926	0.967	1.9
Yang et al. [36]	830 × 254	2 × 2[a]		✓	CS→K	0.114	1.074	5.836	0.208	–	0.856	0.939	0.976	45.1
Chen et al. [1]	512 × 256	2[b]			K	0.145	1.267	5.786	0.244	–	0.811	0.925	0.965	31.7
Zhan et al. [38]	608 × 160	2 × 2[a]	✓		K	0.135	1.132	5.585	0.229	–	0.820	0.933	0.971	5.94
Wang et al. [32]	416 × 128	3		✓	CS→K	0.148	1.187	5.496	0.226	–	0.812	0.938	0.975	31.6
Ranjan et al. [28]	832 × 256	5		✓	K	0.148	1.149	5.464	0.226	–	0.815	0.935	0.973	23.6
Godard et al. [13]	640 × 192	3 × 2[a]		✓	K	0.114	0.991	5.029	0.203	–	0.864	0.951	0.978	14.1
Godard et al. [14]	512 × 256	2[b]			CS→K	0.118	0.923	5.015	0.210	0.205	0.854	0.947	0.976	31.6
Godard et al. [14] ResNet	512 × 256	2[b]			CS→K	0.114	0.898	4.935	0.206	0.204	0.861	0.949	0.976	58.5
Poggi et al. [27]	512 × 256	2[b]			CS→K	0.111	0.849	4.822	0.202	–	0.865	0.952	0.978	78.9
Ours (w/o correction)	768 × 320	2[b]			CS+K	0.118	0.883	4.776	0.197	–	0.869	0.953	0.979	25.8
Ours	768 × 320	2[b]			CS+K	0.118	0.809	4.692	0.196	0.192	0.865	0.954	0.979	25.8
(Semi-)supervised approaches (cap at 80 m)														
Eigen et al. [8]	142 × 27	1	✓		K	0.203	1.548	6.307	0.282	–	0.702	0.890	0.958	54.2
Li and Snavely [23]	512 × 384	1	✓		MD→K	0.141	1.328	5.90	0.241	–	–	–	–	5.4
Li et al. [22]	310 × 94	1	✓		K	0.113	–	4.687	–	–	0.856	0.962	0.988	65.2
Kuznietsov et al. [20][d]	320 × 96	2[b]	✓		K	0.113	0.741	4.621	0.189	–	0.862	0.960	0.986	63.5
Hu et al. [16]	378 × 76	1	✓		K	0.107	–	4.604	–	–	0.875	0.970	0.990	41.5
Yang et al. [35]	512 × 256	2[b]	✓	✓	K	0.097	0.734	4.442	0.187	–	0.888	0.958	0.980	95.3
Dharmasiri et al. [6]	320 × 240	2 × 2[a]	✓	✓	K	0.096	–	4.301	0.173	–	0.895	0.968	0.987	117.0
Luo et al. [25]	1280 × 384	2[b]	✓		K	0.094	0.626	4.252	0.177	–	0.891	0.965	0.984	412.8
He et al. [15]	320 × 224	1	✓	✓	K	0.086	–	4.014	–	–	0.893	0.975	0.994	121.4
Tang and Tan [30]	–	2	✓		K	0.083	0.025	3.640	0.134	–	–	–	–	63.6
Fu et al. [10]	192 × 96	1	✓		K	0.084	0.386	3.072	0.136	–	0.911	0.979	0.993	166.2
Fu et al. [9]	513 × 385	1	✓		K	0.072	0.307	2.727	0.120	–	0.932	0.984	0.994	–

[a] Sequence of stereo pairs

[b] Stereo pair

[c] Weights required for depth estimation at test time only

[d] Semi-supervised variant

4.3   Depth Estimation Performance Evaluation

Table 1 shows the results of our method (trained on CS+K dataset) in comparison with the previous works (both unsupervised and supervised). Since the depth predicted by our method is defined up to a scale factor, we use the same median normalization approach for depth maps as in [40]. We compare various methods using depth estimation accuracy metrics widely used in the previous works [8]: absolute relative difference (ARD), squared relative difference (SRD), root-mean-square error (RMSE), logarithmic RMSE, scale-invariant RMSE and three thresholds $(\delta < 1.25^{\{1,2,3\}})$. In contrast to many previous works, we also compare inference model complexities using the number of inference model weights. Finally, to complete the comparison, we also added information on the output depth map resolution, training data requirements (number of input images, depth supervision, camera intrinsics) and training dataset.

From Table 1 one can see that our method outperforms previous unsupervised results in terms of depth estimation accuracy across most of metrics and output depth map resolution, while being one of the least complex in terms of inference model weights number. Our method is also one of the least demanding in terms

Table 2. Results on DIW for our model trained on SM dataset. Best results are in bold font, our results are in red color.

Method	Supervised	WHDR, %
Laina et al. [21]	✓	31.65
Zhou et al. [40]		31.24
Godard et al. [14]		30.52
Liu et al. [24]	✓	28.27
Eigen and Fergus [7]	✓	25.70
Ours (SM dataset)		23.92
Li and Snavely [23]	✓	22.97
Xian et al. [33]	✓	11.37
Chen and Deng [3]	✓	**10.59**

of training data requirements. Even when compared to the supervised results, our method outperforms many methods, while being much less complex than the supervised methods having better accuracy.

To quantitatively evaluate our model trained on SM dataset, we have to take into account the fact that during training it was presented movie images having quite shallow depth of field (DOF) with most of the background blurry and with detailed objects in focus being quite shallow in their depth range. Such peculiarities restrict the depth-related visual cues our model can learn from, thus we only expect it to provide relative depth predictions rather than metric depths. So, for quantitative evaluation we use DIW dataset, containing very diverse set of sceneries annotated with relative depth labels, and Weighted Human Disagreement Rate (WHDR) [4] as depth estimation accuracy metric. Table 2 shows the results of our method (trained on SM dataset) in comparison with the previous works. Our method outperforms both unsupervised methods of [40] and [14] and even some of the supervised methods. Remarkably, the result of our unsupervised method is close to that of [23], even though they used supervised learning on densely-labelled MegaDepth dataset.

Ablation Study. In Table 1 we provide the results of our method obtained without image correction, using different consistency loss term weight $\alpha_{con} = 0.1$. The results are very close to those with correction maps, indicating that most of the improvement in quantitative results comes from the other contributions. We do not perform ablation for camera parameters estimation network, since removing it would disable conversion of inverse depths to disparities and break the subsequent processing. However, to verify the efficiency of our contributions in achieving robustness to training dataset diversity, we performed additional set of experiments: we trained our model separately on KITTI and Cityscapes datasets and then fine-tuned on KITTI the model pre-trained on Cityscapes. The results were compared to our baseline model of [14] trained on the same

Table 3. Results of test of robustness to training dataset diversity. Best results for each method are in bold font.

Method	Training dataset	Error metric (lower is better)					Accuracy metric (higher is better)		
		ARD	SRD	RMSE	RMSE (log)	RMSE (sc.inv.)	$\delta < 1.25$	$\delta < 1.25^2$	$\delta < 1.25^3$
Godard et al. [14]	CS	0.652	8.960	13.544	0.518	0.233	0.078	0.443	0.866
	K	0.141	1.186	5.677	0.238	0.229	0.809	0.928	0.969
	CS→K	**0.118**	**0.923**	**5.015**	**0.210**	**0.205**	**0.854**	**0.947**	**0.976**
	CS+K	0.152	1.328	5.748	0.232	0.227	0.805	0.934	0.972
Ours	CS	0.314	2.158	7.309	0.366	0.353	0.509	0.769	0.928
	K	0.130	0.973	5.127	0.210	0.206	0.842	0.946	0.976
	CS→K	**0.114**	**0.793**	**4.676**	**0.193**	**0.190**	**0.870**	**0.955**	**0.980**
	CS+K	0.118	0.809	4.692	0.196	0.192	0.865	0.954	0.979

datasets (using their published code[1]). This comparison is shown in Table 3. From this table one can see that the method of [14], when trained on CS+K dataset, drastically deteriorates its performance, which is not the case for our method. Moreover, our method significantly outperforms the method of [14] even when trained on KITTI or Cityscapes alone.

The model of [14] failed to converge on SM dataset at all. We hypothesize that the reason for that is that stereo pairs from different movies have different stereo baselines and varying focus lengths, making prediction of disparity maps from the left view, suitable for reconstructing the right view, infeasible. Our model does not suffer from this problem, since our depth estimation network only predicts inverse depth maps (which are independent of stereo rig parameters) and we derive disparity maps using those inverse depth maps along with stereo camera parameters estimated from the data extracted from both left and right views.

5   Conclusion

We have presented an unsupervised deep neural network for single image depth estimation. While incorporating some of the advantageous ideas from the previous state-of-the-art methods in the field, such as training on stereo pairs with opposite view reconstruction error as a supervisory signal and left-right consistency loss, we propose significant novel improvements to the previous unsupervised design: camera parameters estimation network and regularized input image correction. These improvements allow us to outperform the previous unsupervised results and make our model capable of training on very diverse and challenging datasets, such as hybrid city driving dataset and our own Stereo Movies dataset, resulting in highly generalizable depth estimation model.

---

[1] https://github.com/mrharicot/monodepth.

References

1. Chen, L., Tang, W., John, N.: Self-supervised monocular image depth learning and confidence estimation. arXiv preprint arXiv:1803.05530 (2018)
2. Chen, T., et al.: Mxnet: A flexible and efficient machine learning library for heterogeneous distributed systems. arXiv preprint arXiv:1512.01274 (2015)
3. Chen, W., Deng, J.: Learning single-image depth from videos using quality assessment networks. arXiv preprint arXiv:1806.09573 (2018)
4. Chen, W., Fu, Z., Yang, D., Deng, J.: Single-image depth perception in the wild. In: Advances in Neural Information Processing Systems, pp. 730–738 (2016)
5. Cordts, M., et al.: The cityscapes dataset for semantic urban scene understanding. In: 2016 IEEE Conference on Computer Vision and Pattern Recognition (CVPR), pp. 3213–3223 (2016)
6. Dharmasiri, T., Spek, A., Drummond, T.: Eng: end-to-end neural geometry for robust depth and pose estimation using CNNs. arXiv preprint arXiv:1807.05705 (2018)
7. Eigen, D., Fergus, R.: Predicting depth, surface normals and semantic labels with a common multi-scale convolutional architecture. In: Proceedings of the IEEE international conference on computer vision, pp. 2650–2658 (2015)
8. Eigen, D., Puhrsch, C., Fergus, R.: Depth map prediction from a single image using a multi-scale deep network. In: Advances in neural information processing systems, pp. 2366–2374 (2014)
9. Fu, H., Gong, M., Wang, C., Batmanghelich, K., Tao, D.: Deep ordinal regression network for monocular depth estimation. In: Proceedings of the IEEE Conference on Computer Vision and Pattern Recognition, pp. 2002–2011 (2018)
10. Fu, H., Gong, M., Wang, C., Tao, D.: A compromise principle in deep monocular depth estimation. arXiv preprint arXiv:1708.08267 (2017)
11. Garg, R., B.G., V.K., Carneiro, G., Reid, I.: Unsupervised CNN for single view depth estimation: geometry to the rescue. In: Leibe, B., Matas, J., Sebe, N., Welling, M. (eds.) ECCV 2016. LNCS, vol. 9912, pp. 740–756. Springer, Cham (2016). https://doi.org/10.1007/978-3-319-46484-8_45
12. Geiger, A., Lenz, P., Urtasun, R.: Are we ready for autonomous driving? the kitti vision benchmark suite. In: 2012 IEEE Conference on Computer Vision and Pattern Recognition, pp. 3354–3361 (2012)
13. Godard, C., Mac Aodha, O., Brostow, G.: Digging into self-supervised monocular depth estimation. arXiv preprint arXiv:1806.01260 (2018)
14. Godard, C., Mac Aodha, O., Brostow, G.J.: Unsupervised monocular depth estimation with left-right consistency. In: 2017 IEEE Conference on Computer Vision and Pattern Recognition (CVPR), pp. 6602–6611. IEEE (2017)
15. He, L., Wang, G., Hu, Z.: Learning depth from single images with deep neural network embedding focal length. IEEE Trans. Image Process. **27**(9), 4676–4689 (2018)
16. Hu, J., Ozay, M., Zhang, Y., Okatani, T.: Revisiting single image depth estimation: toward higher resolution maps with accurate object boundaries. arXiv preprint arXiv:1803.08673 (2018)
17. Jiang, H., Learned-Miller, E., Larsson, G., Maire, M., Shakhnarovich, G.: Self-supervised depth learning for urban scene understanding. arXiv preprint arXiv:1712.04850 (2017)
18. Kingma, D.P., Ba, J.: Adam: a method for stochastic optimization. arXiv preprint arXiv:1412.6980 (2014)

19. Kundu, J.N., Uppala, P.K., Pahuja, A., Babu, R.V.: Adadepth: unsupervised content congruent adaptation for depth estimation. arXiv preprint arXiv:1803.01599 (2018)
20. Kuznietsov, Y., Stückler, J., Leibe, B.: Semi-supervised deep learning for monocular depth map prediction. In: Proceedings of the IEEE Conference on Computer Vision and Pattern Recognition, pp. 6647–6655 (2017)
21. Laina, I., Rupprecht, C., Belagiannis, V., Tombari, F., Navab, N.: Deeper depth prediction with fully convolutional residual networks. In: 2016 Fourth International Conference on 3D Vision (3DV), pp. 239–248. IEEE (2016)
22. Li, B., Dai, Y., He, M.: Monocular depth estimation with hierarchical fusion of dilated cnns and soft-weighted-sum inference. Pattern Recogn. **83**, 328–339 (2018)
23. Li, Z., Snavely, N.: Megadepth: learning single-view depth prediction from internet photos. In: Proceedings of the IEEE Conference on Computer Vision and Pattern Recognition, pp. 2041–2050 (2018)
24. Liu, F., Shen, C., Lin, G.: Deep convolutional neural fields for depth estimation from a single image. In: Proceedings of the IEEE Conference on Computer Vision and Pattern Recognition, pp. 5162–5170 (2015)
25. Luo, Y., et al.: Single view stereo matching. In: Proceedings of the IEEE Conference on Computer Vision and Pattern Recognition, pp. 155–163 (2018)
26. Poggi, M., Aleotti, F., Tosi, F., Mattoccia, S.: Towards real-time unsupervised monocular depth estimation on CPU. arXiv preprint arXiv:1806.11430 (2018)
27. Poggi, M., Tosi, F., Mattoccia, S.: Learning monocular depth estimation with unsupervised trinocular assumptions. arXiv preprint arXiv:1808.01606 (2018)
28. Ranjan, A., Jampani, V., Kim, K., Sun, D., Wulff, J., Black, M.J.: Adversarial collaboration: joint unsupervised learning of depth, camera motion, optical flow and motion segmentation. arXiv preprint arXiv:1805.09806 (2018)
29. Repala, V.K., Dubey, S.R.: Dual CNN models for unsupervised monocular depth estimation. arXiv preprint arXiv:1804.06324 (2018)
30. Tang, C., Tan, P.: Ba-net: Dense bundle adjustment network. arXiv preprint arXiv:1806.04807 (2018)
31. Wang, A., Fang, Z., Gao, Y., Jiang, X., Ma, S.: Depth estimation of video sequences with perceptual losses. IEEE Access **6**, 30536–30546 (2018)
32. Wang, C., Buenaposada, J.M., Zhu, R., Lucey, S.: Learning depth from monocular videos using direct methods. In: Proceedings of the IEEE Conference on Computer Vision and Pattern Recognition, pp. 2022–2030 (2018)
33. Xian, K., et al.: Monocular relative depth perception with web stereo data supervision. In: Proceedings of the IEEE Conference on Computer Vision and Pattern Recognition, pp. 311–320 (2018)
34. Xie, J., Girshick, R., Farhadi, A.: Deep3D: fully automatic 2D-to-3D video conversion with deep convolutional neural networks. In: Leibe, B., Matas, J., Sebe, N., Welling, M. (eds.) ECCV 2016. LNCS, vol. 9908, pp. 842–857. Springer, Cham (2016). https://doi.org/10.1007/978-3-319-46493-0_51
35. Yang, N., Wang, R., Stückler, J., Cremers, D.: Deep virtual stereo odometry: leveraging deep depth prediction for monocular direct sparse odometry. arXiv preprint arXiv:1807.02570 (2018)
36. Yang, Z., Wang, P., Wang, Y., Xu, W., Nevatia, R.: Every pixel counts: unsupervised geometry learning with holistic 3D motion understanding. arXiv preprint arXiv:1806.10556 (2018)
37. Yin, Z., Shi, J.: Geonet: unsupervised learning of dense depth, optical flow and camera pose. In: Proceedings of the IEEE Conference on Computer Vision and Pattern Recognition (CVPR), vol. 2 (2018)

38. Zhan, H., Garg, R., Weerasekera, C.S., Li, K., Agarwal, H., Reid, I.: Unsupervised learning of monocular depth estimation and visual odometry with deep feature reconstruction. In: Proceedings of the IEEE Conference on Computer Vision and Pattern Recognition, pp. 340–349 (2018)
39. Zheng, C., Cham, T.J., Cai, J.: T2net: synthetic-to-realistic translation for solving single-image depth estimation tasks. arXiv preprint arXiv:1808.01454 (2018)
40. Zhou, T., Brown, M., Snavely, N., Lowe, D.G.: Unsupervised learning of depth and ego-motion from video. In: 2017 IEEE Conference on Computer Vision and Pattern Recognition (CVPR), pp. 6612–6619. IEEE (2017)

The *K-Closest* Resemblance Classifier for Remote Sensing Data

Nabil Belacel[1]([⊠]), Cheng Duan[2], and Diana Inkpen[2][iD]

[1] Digital Technologies, National Research Council, Ottawa, ON, Canada
`nabil.belacel@nrc.gc.ca`
[2] School of Electrical Engineering and Computer Science, University of Ottawa,
Ottawa, Canada
{`cduan092,Diana.Inkpen`}`@uottawa.ca`

Abstract. The supervised learning algorithms provide a powerful tool to classify and process the remotely sensed imagery data sets. They have the strengths to handle high-dimensional data and to map classes with very complex characteristics. However, the usual supervised machine learning algorithms face issues that limit their applicability especially in dealing with the knowledge interpretation and with imbalanced labeled data sets. To address these issues, the prototype based classifier *K-Closest Resemblance K-CR* was proposed. *K-CR* is inspired by the social choice theory and preference modeling, which argues that the classifiers based on preference modeling are simple, do not need to normalize the features, and do not have loss of information during learning. The effectiveness of the proposed classifier is evaluated by comparing with the other well-known existing classifiers for remote sensing data set. The obtained results indicate that our proposed classifier is an effective tool for land cover classification from remote sensing data.

Keywords: Machine learning · *K-CR* classifier · Land cover classification

1 Introduction

Over the last few years machine learning algorithms have become highly successful and very popular techniques in remote sensing image classification. The Urban Land Cover (ULC) classification is one of the widest used applications in the field of remote sensing. A wide range of machine learning methods for ULC classification continues to be proposed and assessed. They show that supervised machine learning leads to higher accuracy compared to traditional statistical parametric methods, especially for complex data with high-dimensional feature spaces [9]. Many supervised classification algorithms, such as K-Nearest Neighbor (K-NN), Artificial Neural Network (ANN), Random Forests (RF) and Support Vector Machines (SVM), have been applied for classifying the ULC [9]. However, they do not deal with the imprecision of the attributes' scores and

C. Goutte and X. Zhu (Eds.): Canadian AI 2020, LNAI 12109, pp. 49–54, 2020.
https://doi.org/10.1007/978-3-030-47358-7_5

they allow the compensation between features, which will lead to the low classification accuracy [8]. In order to improve the ULC classification accuracy, we have introduced the *K-CR* classifier. The proposed method is inspired by the social choice theory through the application of the outranking approaches to calculate the similarity "the closeness" of unlabeled sample to the prototype of the classes [2]. Comparing to the well-known classifiers, *K-CR* possesses several advantages, such as:

- *K-CR* does not require any normalization during the pre-processing steps compared with other classifiers such as *ANN* and *SVM*. This is because *K-CR* is based on the aggregation of the partial pairwise comparison of each feature. It uses the outranking comparison approach from the *ELECTRE* methods [5,10]. Therefore, the *K-CR* method can deal with all types of data without sensitivity to noise [2].
- *K-CR* results are automatically explained which provides the possibility to access to more detailed information concerning the classification results.
- *K-CR* is based on each feature's weight that reflects the intrinsic relative importance of the feature to different classes. This property is important especially for the multiclass imbalanced data classification problems and for feature selection.

We show that *K-CR* is an appropriate technique for classifying the ULC from remote sensing data. In this paper, the *K-CR* classifier is presented and compared with other well-known classifiers using the case of the ULC data set.

2   The K-Closest Resemblance Classifier

The *K-CR* classifier is an exemplar-based generalization learning models, where the knowledge representation are formed by generalized exemplars like in the case of hyper-rectangular models [7]. It was successfully applied to bioinformatics, clinical decision support system and Content-based recommendation system [1,4]. *K-CR* is considered as a weighted voting classifier in which each feature votesfor the class membership of an unlabeled sample according to which class's prototype is the closest [1,2]. The partial comparison used by *K-CR* to calculate the resemblance between the sample and prototypes avoids resorting to conventional metric and non-metric distances that aggregate the score of all features in the same value unit. Hence, it helps to find the correct pre-processing and normalization data techniques without losing information. The *k-CR* classifier proceeds in two phases:

1. *Learning phase*: For each class C^h *K-CR* determines a set of prototypes from the training set. More precisely, for each prototype and each attribute, an interval is defined. To define these intervals we follow the same approach from the *PROAFTN* classifier as described in [3].
2. *Class prediction*: To classify an unlabeled sample s, *K-CR* applies the following steps:

1. *Outranking relation between the prototypes*: If b^h and b^l represent two prototypes of classes C^h and C^l respectively, then the outranking relation P^s expresses the degree with which the resemblance between the sample s and prototype b^h is stronger than the resemblance between the sample s and prototype b^l. *K-CR* takes as input the partial outranking indices induced by the set of attributes and aggregates them into global outranking relation $P^s(b^h, b^l)$.

2. *Scoring function*: Based on the outranking relations P^s between the prototypes of all classes, *K-CR* selects the best prototypes in terms of their resemblance to the unlabeled sample. The scoring function is used to select a subset of prototypes that are more closely to the sample s. Different scoring functions can be used as shown in [1,2]. In our method we used the flux scoring function as used in *PROMETHEE* methods [5].

3. *Class prediction*: Once a subset of K closest resemblance prototypes to s is determined, a majority voting rule is applied to assign the unlabeled sample to the closest class.

3   Experiments and Evaluation

The data set presented in this paper is obtained from the University of California, Irvine Machine Learning Repository [6]. The data set was created based on high-resolution aerial geographical images which have been classified into 9 types of ULC including Asphalt, Building, Car, Concrete, Grass, Pool, Shadow, Soil, and Tree. The class target Y is specified from 0 to 8 based on these 9 types of ULC. The training set contains 168 instances in total. In addition, each instance is described by 147 features which represent information about spectral properties, size, shape, and texture. One of the main challenge of this data set is the imbalanced distribution of the classes. Accordingly, the objective of this paper is to apply the proposed model on this data set, and to explore its ability of handling imbalanced data sets. The experiments involve comparison between the *K-CR* and other classifiers including RF, K-NN, SVM and MLP for ANN. The classifiers RF, K-NN and SVM were implemented using packages from 'sklearn' and MLP was implemented using TensorFlow. The parameters of *K-CR* are set to 0.1 for the threshold and to 2 for the number of intervals for each attribute. Apart from that, RF constructs 500 trees. The linear function is set as SVM's kernel function, while the number of neighborhood K in K-NN is set to 3. Furthermore, the MLP with 2 hidden layers is trained for 200 epochs with a learning rate of 0.0001 and using the Adam optimizer. In our experiments, the 10-fold cross-validation (10-CV) technique is used to randomly partition the training data set into 10 equal sized subsets. A single subset is used for testing and the remaining 9 folds are used for the training. The cross-validation process is then repeated 10 times and we calculate the average score from 10 results. Performance evaluation is conducted using the generalization of F1-measure and AUC (Area Under The Curve) ROC (Receiver Operating Characteristics) curve since the standard classification accuracy (the % of correct classification) is not suitable for multi-class imbalanced data sets. Therefore, the F1-measure which seeks a balance between precision and recall is taken into consideration. More precisely, a weighted average F1-measure is used in these experiments. The contributions of all classes are aggregated so as to obtain a 'fair' average among unevenly-distributed classes. ROC analysis is a popular metric which gives a standard for selecting possibly optimal models while discarding the impact of the class distribution. ROC is a probability curve and AUC represents degree or measure of separability. It tells how much model is capable of distinguishing between classes. The ROC curve is plotted with True Positive Rate (TPR) against the False Positive Rate (FPR) where TPR is on y-axis and FPR is on the x-axis. Theoretically, in a ROC curve, a perfect one goes straight up along the y-axis and straight right along the x-axis.

4 Results and Discussion

Table 1 shows the weighted average F1-measure on training file using 10-CV. The first column represents the 9 classes within the data-set. The rest of columns show the F1-measure of each class, followed by the average score for each classifier. Among all 5 weighted average F1-measure scores, K-CR obtained 0.91 which is the highest score. Comparing with the second highest score 0.82 from RF, K-CR significantly improves the performance by 9%. By looking at the scores of each individual class, the improvements vary from 3% to 21%. The class Soil with 14 training instances improved with 21%. Another class, Asphalt, which also has only 14 training instances improved with 5%.

Table 1. Weighted average F1-measure on the training file using 10-CV

	K-CR	RF	SVM	KNN	MLP
Asphalt	0.89	0.84	0.61	0.22	0.52
Building	0.91	0.78	0.62	0.57	0.60
Car	0.90	0.87	0.87	0.87	0.61
Concrete	0.90	0.75	0.47	0.38	0.23
Grass	0.94	0.87	0.58	0.44	0.33
Pool	0.93	0.86	0.73	0.77	0.48
Shadow	0.88	0.84	0.64	0.32	0.29
Soil	0.88	0.67	0.40	0.09	0.21
Tree	0.94	0.87	0.57	0.41	0.38
Weighted average F1-measure	0.91	0.82	0.60	0.46	0.40

From another perspective, the AUC-ROC curve shown in Fig. 1 also visually demonstrate such improvements. In Fig. 1, K-CR has a micro-average area score of 0.95, while in figure of RF has 0.9. The lines of the Fig. 1 obviously centralize more to top-left corner compared to the figure of RF.

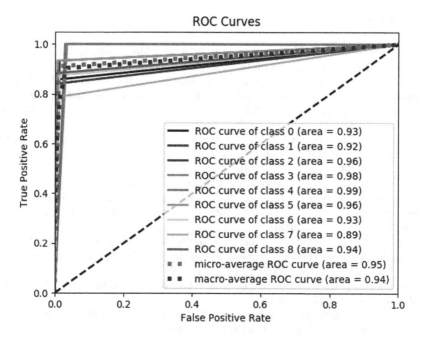

Fig. 1. ROC curve for *K-CR* using 10-CV

In Maxwell et al. [9] the same data set was tested using SVM, DT, RF, Boosted decision tree, ANN and k-NN and the results showed that RF has the highest accuracy of 81.5%. Therefore their results are in concordance with ours, which show that RF is better than SVM, KNN and MLP.

5   Conclusion

In this work, we presented an experimental study using machine learning to classify the remotely sensed imagery data sets for ULC classification problem. Our results demonstrate the power of prototype based classifier that based on preference modeling and simple discretization technique for ULC classification problems. In addition, the proposed *K-CR* classifier can face the challenges of sensing data sets in multi-class imbalanced data sets, and the method presented here provides an improvement when compared with well-known classifiers in ULC classification problem such RF and SVM. In future work, we hope to apply the developed *K-CR* classifier on other types of ULC classification problems.

References

1. Belacel, N.: Méthodes de classification multicritère: Méthodologie et application à l'aide au diagnostic médical. Ph.D. thesis, Univ. Libre de Bruxelles (1999)
2. Belacel, N.: The k-closest resemblance approach for multiple criteria classification problems. In: Hoai, L., Tao, P. (eds.) Modelling, Computation and Optimization Information and Management Sciences, pp. 525–534. Hermes Sciences Publishing, London (2004)
3. Belacel, N., Cuperlovic-Culf, M.: *PROAFTN* classifier for feature selection with application to alzheimer metabolomics data analysis. Int. J. Pattern Recognit. Artif. Intell. **33**(11), 1940013 (2019)
4. Belacel, N., Wei, G., Bouslimani, Y.: The k closest resemblance classifier for amazon products recommender system. In: Proceedings of the 12th International Conference on Agents and Artificial Intelligence - Volume 2: ICAART, pp. 873–880. INSTICC, SciTePress (2020)
5. Brans, J.-P., Mareschal, B.: Promethee methods. In: Multiple Criteria Decision Analysis: State of the Art Surveys. ISORMS, vol. 78, pp. 163–186. Springer, New York (2005). https://doi.org/10.1007/0-387-23081-5_5
6. Dua, D., Graff, C.: UCI machine learning repository (2017). http://archive.ics.uci.edu/ml. Accessed 17 Jan 2020
7. Guvenir, H.A., Şirin, İ.: Classification by feature partitioning. Mach. Learn. **23**(1), 47–67 (1996). https://doi.org/10.1023/A:1018090317210
8. Ma, L., Li, M., Ma, X., Cheng, L., Du, P., Liu, Y.: A review of supervised object-based land-cover image classification. ISPRS J. Photogramm. Remote Sens. **130**, 277–293 (2017)
9. Maxwell, A.E., Warner, T.A., Fang, F.: Implementation of machine-learning classification in remote sensing: an applied review. Int. J. Remote Sens. **39**(9), 2784–2817 (2018)
10. Roy, B.: Multicriteria Methodology for Decision Aiding. Nonconvex Optimization and Its Applications. Springer, New York (2013)

Reinforcement Learning in a Physics-Inspired Semi-Markov Environment

Colin Bellinger[1]([⊠]), Rory Coles[2], Mark Crowley[3], and Isaac Tamblyn[1,4]

[1] National Research Council of Canada, Ottawa, Canada
{Colin.Bellinger,Isaac.Tamblyn}@nrc-cnrc.gc.ca
[2] University of Victoria, Victoria, Canada
rfcoles@uvic.ca
[3] University of Waterloo, Waterloo, Canada
mark.crowley@uwaterloo.ca
[4] Vector Institute for Artificial Intelligence, Toronto, Canada

Abstract. Reinforcement learning (RL) has been demonstrated to have great potential in many applications of scientific discovery and design. Recent work includes, for example, the design of new structures and compositions of molecules for therapeutic drugs. Much of the existing work related to the application of RL to scientific domains, however, assumes that the available state representation obeys the Markov property. For reasons associated with time, cost, sensor accuracy, and gaps in scientific knowledge, many scientific design and discovery problems do not satisfy the Markov property. Thus, something other than a Markov decision process (MDP) should be used to plan/find the optimal policy. In this paper, we present a physics-inspired semi-Markov RL environment, namely the phase change environment. In addition, we evaluate the performance of value-based RL algorithms for both MDPs and partially observable MDPs (POMDPs) on the proposed environment. Our results demonstrate deep recurrent Q-networks (DRQN) significantly outperform deep Q-networks (DQN), and that DRQNs benefit from training with hindsight experience replay. Implications for the use of semi-Markovian RL and POMDPs for scientific laboratories are also discussed.

Keywords: Reinforcement learning · Semi-Markov decision processes · Materials science

1 Introduction

Developing new materials is seen as a key to advance in many areas of science and society [11]. Currently, state-of-the-art methods for developing new materials are slow, unpredictable, and have high associated costs. Artificial intelligence has the potential to make significant contributions to problems of this nature.

In recent years, deep reinforcement learning (RL) has achieved significant advancements, and produced human level performance on challenging video

© Springer Nature Switzerland AG 2020
C. Goutte and X. Zhu (Eds.): Canadian AI 2020, LNAI 12109, pp. 55–66, 2020.
https://doi.org/10.1007/978-3-030-47358-7_6

games, board games, and in robotics [4, 7, 12]. These results have garnered much attention across a wide variety of domains, including the fields of chemistry and physics. RL has, for example, been applied in quantum physics and chemistry [1, 15]. The latter is partially motivated by work with scientific laboratory robots [6, 9].

Our research focuses broadly on the application of RL to materials science. We hypothesise that RL has a great potential to speed up the materials design and discovery process. From an AI perspective, this application area embodies many interesting challenges. In materials, for example, evaluating prospective solutions can be costly, time consuming and destructive. Therefore, sample efficiency is a key requirement. On the other hand, an agent may have multiple goals, and/or new goals may be added overtime. Thus, multi-agent learning with shared experience and transfer learning are of interest. The rewards are often binary and significantly delayed, which motivates the need for strategies to handle rewards, and improve sample efficiency. Moreover, important information to the materials design process is often hidden due to costs and scientific limitations. Thus, the AI must be suitable for semi-Markov decision processes.

To date, there has not be a systematic investigation of the suitability of deep RL algorithms for applications in materials science involving semi-Markov decision processes. In this paper, we commence this exploration by presenting a new physics-inspired semi-Markov learn task; specifically the semi-Markov phase change environment. Subsequently, we conduct an initial evaluation of the potential for value-based deep RL algorithms in the environment, and discuss the challenges to be faced in future real-world applications.

1.1 Contributions

We make the following contributions in this paper:

- Introduce the semi-Markov phase change environment;
- Compare the performance of deep Q-networks (DQN) to deep recurrent Q-networks (DRQN) on the proposed environment;
- Evaluate the benefit of hindsight experience replay (HER) on DQN and DRQN; and,
- Discuss the performance gap between these methods and the optimal policy.

2   Related Work

Q-learning is an off-policy temporal difference control algorithm [14] where the objective is to learn an optimal action-value function, independent of the policy being followed. DQN is a recent variation of Q-learning that takes advantage of the generalizing capabilities of deep learning. DQNs have been shown to produce human-level performance on challenging games on Atari 2600 [8].

DQNs offer a solution approach for Markov decision processes (MDPs). Specifically, problems where the state observation emitted from the environment is sufficient to select the next action. Cases where the Markov property does not hold, require a *partially observable MDP (POMDP)*. In these cases, the representation of the current state alone is not sufficient to select the next action. This can occur due to unreliable observations, an incomplete model (i.e. latent variables), noisy state information or other reasons.

In [5], the authors propose the use of a recurrent neural network architecture in place of the feed-forward network in DQN. Leveraging recurrent neural networks, it is argued, enables the Q-network to better handle POMDPs. Specifically, with the recurrent neural network, the agent can build an implicit notion of its current state based on the recent sequence of state observation resulting from actions taken. The authors show that Deep Recurrent Q-Networks (DRQN) presented with a single frame at each time-step can successfully integrate information through time, and thereby replicate the performance of DQNs on standard Atari 2600 games. In this work, we extend the evaluation of DRQNs to the phase change environment in order to better understand the potential of DRQN on real-world POMDPs.

Many of the recent achievements of deep RL have been produced in simulated environments because RL agents must gather a large amount of experience. Deep Q-Networks, for example, famously required approximately 200 million frames of experience for trainings and approximately 39 days of real-time game playing, on the Atari 2600 [8]. Model-based RL methods, such as DYNA-Q [13], aim to replace a large portion of the agent's real-world experience with experience collected from a surrogate or other models of the environment. Model-based methods, however, have seen most of their successes in environments where the dynamics are simple and can easily and accurately be learned. This is decidedly *not* the case for most physics and chemistry environments.

Learning in many physics and chemistry environments is made more challenging by sparse, binary rewards. Andrychowicz *et al.* in [2], proposed Hindsight Experience Replay (HER), which extends the idea of training a universal policy [10]. Inspired by the benefit that humans garner by learning from their mistakes, HER simulates this by re-framing a small, user-defined, portion of the failed trajectories as successes. It is applicable to off-policy, model-free RL, and to domains in which multiple goals could be achieved. HER was shown to improve sample efficiency, and make learning possible in environments with sparse and binary reward signals.

Multi-goal learning environments with sparse, binary rewards, and the necessity for sample efficiency are key features of many physics and chemistry applications, such as materials design. As a result, HER is potentially of great value in these domains. To date, however, it has not been evaluated in semi-Markov decision processes nor has it been explored in conjunction with DRQN.

3   Semi-Markov Phase Change Environment

Our new semi-Markov phase change environment[1] is implemented based on the OpenAI Gym framework [3] and is depicted in Fig. 1. Within the physical

---

[1] The environment is available at http://clean.energyscience.ca/gyms.

sciences, Fig. 1 is known as a phase diagram - a convenient representation of a materials behaviour where, within a "phase", symmetry is preserved over a wide range of experimental conditions (in this case temperature, T, and pressure, P). In general, it is possible to alter the pressure or temperature of a material while remaining within the same phase (e.g. cold water and warm water are both liquids). Within a single phase, adding heat (Q^+) results in a positive change of temperature, while removing it, (Q^-) does the opposite. Similarly, within a single phase, doing positive work (W^+) increases the pressure, while negative work, (W^-) results in a pressure decrease.

Importantly, we note that the relationship between heat, work, temperature, and pressure is different at the boundary between some phases. Thus, the state transition dynamics are different at the boundary. Specifically, symmetries change when crossing a discontinuous phase boundary (e.g solid-liquid). This change is accompanied by the addition or removal of a latent heat. On a phase diagram, such a boundary is denoted with a solid line. Because of the latent heat, under equilibrium conditions, two or more phases can co-exist with one another in a stable state. As a result, when visualized on a phase diagram, a trajectory of constant heating will temporarily stall at a phase boundary. There is an apparent lack of progress at the boundary while this energy is used to convert the material from one phase into another at constant (P, T) (e.g. the size of an ice-cube decreases while the amount of liquid water increases).

In our environment, the agent's goal is to take a series of actions that add or remove energy in the system by two independent mechanisms (heat and work) to modify a material from its start state M_s to some goal state M_g. The result of the actions is measured in terms of the pressure and temperature of the material.

The environment has a discrete 4-action space, $\mathcal{A} \in \{a_0 = Q^-, a_1 = Q^+, a_2 = W^-, a_3 = W^+\}$. Thus, the agent must learn to navigate from some start position in the two-dimensional temperature-pressure space $M_s = (t_s, p_s)$ to some goal state $M_g = (t_g, p_g)$ in as few steps as possible. The episodes terminates immediately after the agent takes the action to transition in to M_g. The agent receives a reward of 1 when it reaches the goal, and zero elsewhere. The optimal policy in the environment is to apply the minimum number of actions (steps) to get to the goal. The environment emits a state observations in terms of T and P. The initial version of the environment has discretized pressure and temperature measurements, and a limit to the range. This results in a 2-dimensional grid state-space with vertical movements analogous to changes in pressure (resulting from $W^+/-$) and horizontal movements corresponding to changes in temperature (resulting from the $Q^+/-$).

The environment is designed to weakly approximate the process of adding small, fixed amounts of energy (in the form of heat or work) to an initial phase (e.g. a liquid) to convert it to another one (or for the case of the phase boundary, a mixture of different fractions solid, liquid, and gas). In order to make the problem extra challenging, we include the requirement that the agent invoke two different actions when it crosses through the boundary. While this would not strictly be required physically for equilibrium processes, it makes the learning task more difficult and relevant for real world examples which involve nucleation, activation barriers, etc.

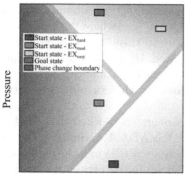

Dynamics Within a Phase

Next State	Physics	Action
(s_{x-1}, s_y)	Decrease temperature	$a = 0$
(s_{x+1}, s_y)	Increase temperature	$a = 1$
(s_x, s_{y-1})	Decrease pressure	$a = 2$
(s_x, s_{y+1})	Increase pressure	$a = 3$

Dynamics at a Boundary

Next State	Physics	Action
(s_{x-1}, s_y)	Decrease temperature	$\{a = 2, a = 0\}$
(s_{x+1}, s_y)	Increase temperature	$\{a = 3, a = 1\}$
(s_x, s_{y-1})	Decrease pressure	$\{a = 0, a = 2\}$
(s_x, s_{y+1})	Increase pressure	$\{a = 1, a = 3\}$

Fig. 1. (Left) Phase change environment. (Top right) State transition dynamics for within-phase states. (Bottom right) State transition dynamics for phase change boundary. This assumes that the agent is at some boundary states.

Unlike the traditional grid-world setup, the grid-based phase change environment does not have any barriers that might prevent an agent from moving in a certain direction. The challenge, as we have discussed from the scientific perspective above, is learning to efficiently navigate through partially observable phase change boundaries. The state transition dynamics are presented in the tables on the right in Fig. 1.

The dynamics for the phase change boundary are as such, when in some boundary state (s_x, s_y), the agent must apply a sequence of two actions to transition into the state on the other side of the boundary. In order for the agent to move in the direction of increasing pressure, for example, it must apply action $a = 1$ followed by action $a = 3$. This leads to the following state-action sequence:

$$...(s_x, s_y), a = 1, (s_x, s_y), a = 3, (s_x, s_{y+1})... \tag{1}$$

4 Experimental Setup

4.1 Reinforcement Learning Algorithms

In order to assess the suitability of value-based RL in a semi-Markov materials-inspired environment, we compare the performance of DQN to DRQN. We evaluate DRQN with a trace length of one (*i.e.*, a one state history) as this makes it directly comparable to DQN. This forces the network to rely solely on its internal architecture to remember the implicit state of the system. Finally, we explore the benefit of HER on DQN and DRQN in the semi-Markov environment.

Deep Q-Learning: In this work, DQN receives a state vector $s = [P, T]$ as the input and emits a value for each action $Q = (a_1, a_2, a_3, a_4)$ at the output layer. A greedy agent in state s will take $a = \arg\max Q$. The parameters of the network θ are updated as, $\theta_{i+1} = \theta_i + \alpha\Delta_\theta\mathcal{L}(\theta_i)$, to minimize the loss function, $\mathcal{L}(s, a|\theta) \approx (r + \gamma\max_a \mathcal{Q}(s', a|\theta) - \mathcal{Q}(s, a|\theta))^2$, where r is the reward, γ is a discount factor and α is the learning rate. In this case, a single network is generating the update target and being updated. Updating based on a single network has been shown to lead to instability in some cases, and can be improved upon by having a separate target network. However, this was not necessary in the phase change environment.

In the following experiments, we applied a neural network with a single 48 unit hidden layer with ReLU activation and the ADAM optimizer. The free parameters were set as follows, the discount factor $\gamma = 0.95$ and the exploration rate $\epsilon = 1$ with linear $\epsilon_\delta = 0.00001$ decay. After an initial period of experience gathering, the network was updated after every episode by sampling a batch of size 127 from the experience replay buffer.

Deep Recurrent Q-Learning: DRQN follows the same setup as that presented for DQN above. Specifically, the input, output and objective function, and optimizers are the same. The key difference is that the fully connected hidden layer in DQN is replaced by a recurrent network. In our experiments below, we use a 128 unit Gated Recurrent Unit.

Training with Hindsight Experience Replay: HER is a training framework that requires the current state and the goal state to jointly form the state space. Thus, all experiments related to HER have an expanded state space. We edited 5% of the tuples corresponding to failed actions (*i.e.*, action with zero reward) to be seen as successful. Specifically, we set the reward to 1 and the goal state to the current state, prior to adding the tuple to the experience replay buffer.

4.2   Evaluation Method

In order to thoroughly assess the impact of the non-Markov phase change boundaries on the RL algorithms, we evaluate each method from three deterministic starting locations. From each of these starting locations, the agents must learn to navigate to a single goal. In experiment 1 (EX_{hard}), the agents start off farthest from the goal and must cross two phase change boundaries. The agent starts marginally closer to the goal in experiment 2 (EX_{mod}). Here, the agent must cross a single non-Markov barrier. Finally, in experiment 3 (EX_{easy}) the agent starts close to the goal and is not required to cross any non-Markov barriers.

To further our analysis of the impact of the non-Markov phase change boundaries on the RL algorithms, we repeat each of the above experiments in a Markov version of the phase change environment. In the Markov version, the dynamics in the phase change boundaries are equivalent to the inner-phase dynamics, and all of the states are fully observable.

The performance of each agent is recorded on intervals of 50 episodes. Specifically, after each increment of 50 episodes of training, each agent is applied for one episode (or a maximum of 10,000 steps) of testing with an ϵ-greedy policy ($\epsilon = 0.2$). Thus, for $20,000$ episodes of training, in each of the 30 iterations, we collect 400 test results. These are averaged and reported in the plots below.

5 Results

5.1 DQN on the Semi-Markov Phase Change Environment

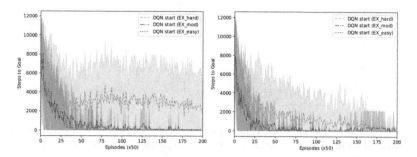

Fig. 2. Mean number of steps per episode for DQN on (left) the semi-Markov phase change environment, and (Right) the Markov phase change environment

The plot on the left in Fig. 2 shows the average number of steps per episode that an agent learning with DQN takes to the goal in the semi-Markov environment when starting at each of the three starting locations EX_{easy}, EX_{mod}, and EX_{hard}. For comparison, the results on the right show the performance when the phase change environment is made fully Markovian.

These results demonstrate that DQN is affected by both the distance between the starting state and goal (sparsity of reward) and semi-Markov decision process resulting from the phase change boundaries. From the left plot, it is clear that the agent in EX_{hard} learns much slower than the agents in EX_{mod} and EX_{easy}. The mean number of steps by episode for EX_{mod} and EX_{easy} are nearly indistinguishable, whereas the mean number of steps for EX_{hard} remains significantly higher throughout training. Two factors are contributing to this, the crossing of phase change boundaries and the distance from the goal state.

To understand which factor is impacting the performance in EX_{hard} more, we compare the corresponding plots on the left (semi-Markov) and the right (Markov). On the semi-Markov environment, initially the mean number of steps drops quickly, before plateauing at what is still a large mean number of steps to the goal. Alternatively, in the Markov environment, the agent starting from EX_{hard} consistently learns to take fewer steps to the goal. Here, it converges to a mean number of steps that is much closer to optimal. This suggests that while DQN is harmed by the reward sparsity, it is the non-Markov phase change boundaries that prevent it from converging to the optimal number of steps.

5.2 DRQN on the Semi-Markov Phase Change Environment

Given the significant effect caused by the non-Markov phase change boundaries, we now investigate the extent to which the hidden representation and sequential nature of recurrent neural networks enables agents learning with DRQN to better navigate the non-Markov phase change boundaries.

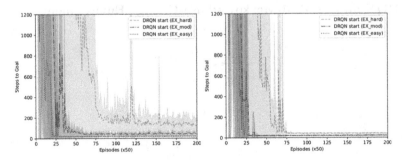

Fig. 3. (left) Mean number of steps per episode for DRQN on the semi-Markov phase change environment. (Right) Mean number of steps per episode for DRQN on the Markov phase change environment

The left plot in Fig. 3 shows the average number of steps per episode that the agent learning with DRQN takes on route to the goal in the semi-Markov environment. The results for the Markov version of the phase change environment are shown on the right.

For the most challenging case EX_{hard}, DRQN converges to approximately 530 steps after 10,000 episodes (200×50). By contrast, when the agent learns with DQN on EX_{hard}, it does not converge after 20,000 episodes of training. Thus, DRQN provides a good improvement in terms of the convergence speed and the average number of steps taken on route to the goal.

Comparing the semi-Markov results on the left and the Markov results on the right reveals that the DRQN agent on the semi-Markov problem is still not equivalent to the agent on the Markov problem. The gap, however, is closed significantly from what we found with DQN. In the Markov environment, the DRQN agent in EX_{hard} converges after approximately 3,750 episodes to 40 steps (which is optimal), versus approximately 530 steps after around 10,000 episodes for the semi-Markov environment.

5.3 Agents with Hindsight Experience Replay

The above results demonstrate that learning with DRQN can produce a significant reduction in the number of steps taken to the goal, and a significant speed up in the rate of learning in comparison to DQN. Nonetheless, the number of episodes DRQN requires to converge is more than double on the semi-Markov problem, and the converged agent takes on average over 10 times more steps.

In the following two subsections, we evaluate whether HER helps to improve the rate of convergence on the semi-Markov phase change environment, and assess how it compares to the Markov environment.

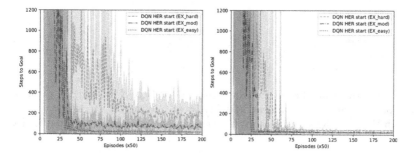

Fig. 4. (left) Mean number of steps per episode for DQN with HER on the semi-Markov phase change environment. (Right) Mean number of steps per episode for DQN with HER on the Markov phase change environment

DQN + HER: Figure 4 shows the mean performance for DQN with HER in the semi-Markov phase change environment on the left and in the Markov environment on the right. Once again, we will focus on the performance on EX_{hard} as it produces the most insightful results. The plot demonstrates that the DQN agent learns significantly faster with HER than without. This is consistent with previously published results. In particular, the agent on EX_{hard} converges to approximately 195 steps after 6,000 episodes of learning on average. Whereas, without HER, DQN had not converged after 20,000 episodes. Interestingly, this is faster convergence, and to fewer steps than the DRQN results reported in the previous section. This is likely due to improved efficiency within the phases, whilst the accuracy of the action selection in the non-Markov phase change boundaries remains less than optimal. The performance gap with the Markov environment is narrowed, but still wide. Specifically, in the Markov setup, DQN + HER converges to approximately 43 steps (approximately optimal) after an average of 3,150 episodes.

DRQN + HER: Finally, we evaluate the benefit of using HER with DRQN in the semi-Markov phase change environment. These results are presented in Fig. 5. The earlier results with DRQN on EX_{hard} amounted to 530 steps after approximately 10,000 episodes. With the addition of HER, the agent converges to approximately 145 steps on average after 4,000 episodes. This shows that DRQN receives a good performance boost from the addition of HER in terms of average number of steps and the rate convergence. For comparison, the agent learning with DQN + HER on EX_{hard} converges to approximately 195 steps after 6,000 episodes of learning on average. Thus, DRQN + HER is the better of the two methods on the semi-Markov phase change environment.

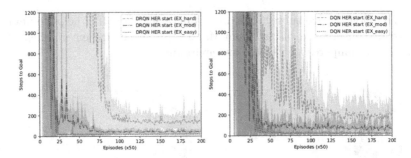

Fig. 5. (Left) Mean number of steps per episode for DRQN with HER on the semi-Markov phase change environment. (Right) Mean number of steps per episode for DQN with HER on the semi-Markov phase change environment (re-plotted for ease of analysis).

Despite its superiority, there is a noteworthy lag in the learning curve for DRQN + HER for EX_{hard} before the mean number of steps steeply drop off. Whereas, DQN + HER has a relatively consistent drop in the mean number of steps from the outset. This difference suggests that agents learning with recurrent neural network models may suffer from an initial lag in performance due to the added complexity of training the GRU.

6 Discussion

Our results have extended the previous analysis of DRQN as a method to solve POMDPs to problems beyond the standard Atari 2600 game suite. In particular, our results show that agents trained with DRQN learn significantly better value-functions for a physics-inspired semi-Markov phase change environment in comparison to DQN. Specifically, adding the recurrent architecture to the DQN enables the agent to takes fewer steps on route to the goal. Moreover, we show that DRQN is further improved in terms of the learning rate and the number of steps to the goal when HER is incorporated into the training process.

In spite of the significantly improved performance, DRQN does not learn a value-function that implements a perfect policy for the semi-Markov phase change environment. After convergence, DRQN + HER takes on average 3-times the optimal number of steps on route to the goal in EX_hard in the semi-Markov environment. Without HER, DRQN takes on average 13-times more steps than optimal. As can be seen in Fig. 5, the gap is significantly narrowed for EX_mod, and is completely closed for EX_easy. This suggests that the portion of non-Markov states has a non-linear impact of the learning difficulty.

A potential method to improve the performance of the DRQN is to use longer trace length. Longer trace lengths would provide more direct information about the state sequence, and potentially simplify the problem. Our current analysis does not reveal where the extra steps are taken. Nonetheless, it highly likely that the agent would still struggle with the semi-Markov phase change boundaries.

Our ongoing research aims to identify where the DRQN is failing to learn the optimal actions in order to propose improvements. There is clearly room for improvement; here we have established a strong baseline for future work.

From an experimental science perspective, these results suggest that RL has the potential to have a significant, positive impact of the advancement of materials, and other experimental science. We note, however, that application of RL in the laboratories will involve several more layers of complexities on top of partial observability. Each of these challenges needs to be clearly understood and analyzed from an RL perspective in order to leverage the right tools from the current state-of-the-art and to develop new RL theories and methodologies where necessary. In the list below, we outline a few characteristic of laboratory learning that we see as being pertinent.

- The existence of different classes of sensors, each of which provide different information content, costs, and data representation;
- The value and cost of each sensor depends on time and space;
- Because observations are costly, the agent should have the ability to make active decisions about when to take an observations and which observations to make;
- Sensors have intrinsic quantifiable uncertainties associated with them.
- In a significant number of experiments there is a simple phenomenological model which can roughly predict the outcome.

A straightforward extension of the results presented here would be to include simulated spectroscopic sensor input. This is closer to the conditions that human operators face. Additionally, in our simple model of material phases, the mapping between energy input to change in conditions (P,T) did not vary across the different phases. In general, this is not true and depends on the specific heat and compressibility of the material. Finally, throughout, we assumed equilibrium conditions - i.e. the timescale of internal relaxations was short compared to the observation time.

7   Conclusion

We introduced the phase change environment to evaluate RL algorithms on a semi-Markov problem inspired by physics and laboratory science. We compared DQN and DRQN with and without HER in the environment. Our results show that DRQN learns significantly faster and converges to a better solution than DQN in this domain. Moreover, we find that the number of episodes to convergence in DRQN is further improved by the incorporation of HER. Nonetheless, the hypothesis that the implicit state estimate maintained by the recurrent network in DRQN would enable it to learn to behave optimally in the phase change environment was not realized in these experiments. Specifically, DRQN + HER converges to approximately 3-times the optimal number of steps on EX_{hard}.

Our ongoing research is evaluating the benefit of longer trace lengths for DRQN and alternative algorithms for semi-Markov decisions processes. In addition, we are developing more materials-inspired RL environments to evaluate existing algorithms and promote the development of new, superior algorithms for materials design and discovery.

Acknowledgements. Work at NRC was performed under the auspices of the AI4D Program.

References

1. Andreasson, P., Johansson, J., Liljestrand, S., Granath, M.: Quantum error correction for the toric code using deep reinforcement learning. Quantum **3**, 183 (2019)
2. Andrychowicz, M., et al.: Hindsight experience replay. In: Advances in Neural Information Processing Systems, pp. 5048–5058 (2017)
3. Brockman, G., et al.: OpenAI gym (2016)
4. Gu, S., Holly, E., Lillicrap, T., Levine, S.: Deep reinforcement learning for robotic manipulation with asynchronous off-policy updates. In: 2017 IEEE International Conference on Robotics and Automation (ICRA), pp. 3389–3396. IEEE (2017)
5. Hausknecht, M., Stone, P.: Deep recurrent Q-learning for partially observable MDPs. In: 2015 AAAI Fall Symposium Series (2015)
6. MacLeod, B.P., et al.: Self-driving laboratory for accelerated discovery of thin-film materials. arXiv preprint arXiv:1906.05398 (2019)
7. Mnih, V., et al.: Playing Atari with deep reinforcement learning. In: NeurIPS: Deep Learning Workshop (2013)
8. Mnih, V., et al.: Human-level control through deep reinforcement learning. Nature **518**(7540), 529 (2015)
9. Roch, L.M., et al.: ChemOS: orchestrating autonomous experimentation. Sci. Robot. **3**(19), eaat5559 (2018)
10. Schaul, T., Horgan, D., Gregor, K., Silver, D.: Universal value function approximators. In: International Conference on Machine Learning, pp. 1312–1320 (2015)
11. National Academies of Sciences, Engineering, and Medicine: Frontiers of Materials Research: A Decadal Survey. The National Academies Press, Washington, DC (2019)
12. Silver, D., et al.: Mastering the game of Go with deep neural networks and tree search. Nature **529**(7587), 484 (2016)
13. Sutton, R.S.: Integrated architectures for learning, planning, and reacting based on approximating dynamic programming. In: Machine Learning Proceedings 1990, pp. 216–224. Elsevier (1990)
14. Watkins, C.J.C.H.: Learning from delayed rewards. Ph.D. thesis, King's College, Cambridge (1989)
15. Zhou, Z., Kearnes, S., Li, L., Zare, R.N., Riley, P.: Optimization of molecules via deep reinforcement learning. Sci. Rep. **9**(1), 1–10 (2019)

Deep Multi Agent Reinforcement Learning for Autonomous Driving

Sushrut Bhalla(✉)(iD), Sriram Ganapathi Subramanian(iD), and Mark Crowley(iD)

University of Waterloo, Waterloo, ON N2L 3G1, Canada
{sushrut.bhalla,s2ganapa,mcrowley}@uwaterloo.ca

Abstract. Deep Learning and back-propagation have been successfully used to perform centralized training with communication protocols among multiple agents in a cooperative environment. In this work, we present techniques for centralized training of Multi-Agent Deep Reinforcement Learning (MARL) using the model-free Deep Q-Network (DQN) as the baseline model and communication between agents. We present two novel, scalable and centralized MARL training techniques (MA-MeSN, MA-BoN), which achieve faster convergence and higher cumulative reward in complex domains like autonomous driving simulators. Subsequently, we present a memory module to achieve a decentralized cooperative policy for execution and thus addressing the challenges of noise and communication bottlenecks in real-time communication channels. This work theoretically and empirically compares our centralized and decentralized training algorithms to current research in the field of MARL. We also present and release a new OpenAI-Gym environment which can be used for multi-agent research as it simulates multiple autonomous cars driving on a highway. We compare the performance of our centralized algorithms to existing state-of-the-art algorithms, DIAL and IMS based on cumulative reward achieved per episode. MA-MeSN and MA-BoN achieve a cumulative reward of at-least 263% of the reward achieved by the DIAL and IMS. We also present an ablation study of the scalability of MA-BoN showing that it has a linear time and space complexity compared to quadratic for DIAL in the number of agents.

Keywords: Multi-agent reinforcement learning · Autonomous driving · Emergent communication

1 Introduction

Multi Agent Reinforcement Learning (MARL) is the problem of learning optimal policies for multiple interacting agents using RL. Current autonomous driving research focuses on modeling the road environment consisting of only human drivers. However, with more autonomous vehicles on the road, a shared cooperative policy among multiple cars is a necessary scenario to prepare for.

To overcome the problem of non-stationarity in the training of MARL agents, the current literature proposes the use of centralized training using message sharing between the agents [11]. The message shared between the agents is generated using the policy network and trained using policy gradients. This approach leads

© Springer Nature Switzerland AG 2020
C. Goutte and X. Zhu (Eds.): Canadian AI 2020, LNAI 12109, pp. 67–78, 2020.
https://doi.org/10.1007/978-3-030-47358-7_7

to sub-optimal messages being shared between agents as the message is inadvertently tied to the policy of the agent [3]. Current approaches thus show a poor performance in large-scale environments with sparse rewards and a long time to horizon as shown in our experiments section.

In this paper, we propose centralized training (MA-MeSN) algorithms for MARL environments which are a generalization of the MARL algorithms currently in literature. Our approach allows separation of policy and communication models and provides a stabilized method for training in an off-policy method. We also compare our centralized training algorithm against DIAL (Differentiable Inter-Agent Learning) [5] and IMS (Iterative Message Sharing) [14] on a large scale multi-agent highway driving simulator we developed as part of this work. We present techniques (MA-MeSN-MM) to derive a cooperative decentralized policy from the trained centralized policy (MA-MeSN). All algorithms are compared based on various metrics our treadmill driving simulator and OpenAI's multi-agent particle environments [13] for formal verification of our algorithms.

2   Related Work

MARL has a rich literature (particularly in the robotics domain [2]). Independent cooperative tabular Q-learning with multiple agents has been studied in [15]. The empirical evaluation shows that cooperative behavior policy can only be achieved by information sharing, for example, other agents' private observations, policies or episode information.

There is a vast literature on the emergence of communication between agents in the same environment [4,9,13,14]; which propose training messages shared between agents using backpropagation. The work in [5,14] extends the techniques of message sharing between agents to multi-agent reinforcement learning (MARL). The authors in [14] employ a message sharing protocol where an aggregated message is generated, by averaging the messages from all agents, and passing it back as an input to the agents along with their observation's hidden state representation to compute the final state-action values. This Iterative Message Sharing (**IMS**) is iterated P times before the final action for all agents is computed using ϵ-greedy method. Differentiable Inter-Agent Learning **DIAL** [5] also trains communication channels, through back-propagation, for sequential multi-agent environments. DIAL presents an on-policy training algorithm which uses the past history to generate messages for inter-agent communication. In this paper, we present a generalization of the MARL algorithms currently available in literature for centralized training. Our algorithm is able to outperform DIAL and IMS on large scale environments while achieving a better time and space complexity during training and execution.

Multi agent environments require a decentralized execution of policy by agents in the environment. Work in [7] has shown that the MARL agents could be executed with discrete communication channels by using a softmax operation on the message. This approach provides a partial decentralization of the trained centralized policy. The authors in [6] successfully train multiple independent agents by stabalizing the experience replay for multi-agent setting. The stabilization is done by prioritizing newer experiences in the experience buffer for training as they represent the current transition dynamics of the environment. We also compare the training of our decentralized policy against the independent agents trained using Stabilized Experience Replay (**SER**). Our algorithm allows the centralized trained cooperative policy to be easily extended to a decentralized setting while maintaining acceptable performance.

3 Background on Multi-Agent Reinforcement Learning

In this section we present a background on multi-agent reinforcement learning and the variables used in the paper. A short background on Deep Q-Networks [12] can be found in Appendix (https://uwaterloo.ca/scholar/sites/ca.scholar/ files/mcrowley/files/deep_multi_agent_reinforcement_learning_for_autonomous_ driving-full.pdf). In this work we consider a general sum multi-agent stochastic game G which is modeled by the tuple $G = (X, S, A, T, R, Z, O)$ with N agents, $x \in X$, in the game. The game environment presents states $s \in S$, and the agents observe an observation $z \in Z$. The observation is generated using the function $Z \equiv O(s, x)$ which maps the state of each agent to its private observation z. The game environment is modeled by the joint transition function $T(s, \mathbf{a_i}, s')$ where $\mathbf{a_i}$ represents the vector of actions for all agents $x \in X$. The dependence of the transition matrix on behavior policy of other agents gives it the non-stationary property in multi-agent environments. We use the subscript notation i to represent the properties of a single agent x, a bold subscript $\mathbf{i}$ to represent properties of all agents $x \in X$ and $-\mathbf{i}$ to represent the properties of all agents other than x_i. We use the superscript t to represent the discrete time-step. All agents share the same utility function R, which provides agents with an instantaneous reward for an action a_i. Our game environment represents a Decentralized Partially Observable Markov Decision Process (DEC-POMDP) [1]. The agents can send and receive discrete messages between each other, which are modeled based on speech act theory, represented as m_i^t. The game environment does not provide a utility function in response to the communication/message actions performed by an agent. The major challenges in the domain of multi-agent reinforcement learning include the problem of dimensionality, coordinated training, and training ambiguity. Having strong communication between agents can solve some of these problems.

4 Methods

4.1 Multi-Agent Message Sharing Network (MA-MeSN)

The DIAL and IMS methods demonstrated that emergent communication between multiple agents can be achieved by optimizing messages shared between agents using backpropagation. DIAL presents a model where the communicative actions (generated by the message policy) and non-communicative actions (generated by the behavior policy) are generated using the same model. This approach forces a strong correlation between the communicative and non communicative actions, but leads to sub-par results. Behavior policy of the agents might be similar in a cooperative environment, but their message policy is focused on achieving high information sharing between agents. Using the same model to predict the behavior and message policy would lead to conflicting updates to the neural network due to different objectives.

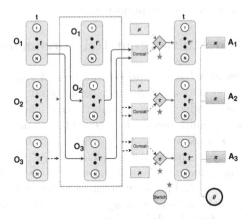

Fig. 1. Architecture for Multi-Agent Message Sharing Network (MA-MeSN)

We thus present a generalization of the communication based MARL algorithms in Fig. 1 where each agent uses a different model for message policy and behavior policy. The f'' neural network maps the message received from the other agents m_{-i} along with its partial observation of the environment z_i to a state-action-message value function $f'' = Q(z_i, a_i, m_i)$. We refer to this network as the (behavior) policy network. The message is generated by the other agents x_{-i} using the neural network approximator f'_i which maps the agent's private observation to a communication action m_{-i}. We refer to this network as the message (policy) network. The message passing interaction/negotiation can be extended to multiple iterations for faster convergence. During the training, we allow only allow a single pass of messages between agents. In contrast to previous work in DIAL, we train the message network using the cumulative gradients of all policy networks as shown in Algorithm 1. Optimizing the message network with cumulative gradients leads to messages which are generalizable to all agent policies.

Comparison to Previous Work. This approach has two advantages over DIAL. The messages $m_{-i}^t(z_{-i}^t, f(z_i^t))$ are conditioned on the entire observable state at time t, as opposed to DIAL, where messages $m_{-i}^t(z_{-i}^{t-1})$ are a function of the previous time-step observation of each agent z_{-i}^{t-1}. Generating a message based on the past introduces the message network's dependency on the transition dynamics; which as discussed exhibits a non-stationary property in multi-agent environments and thus lead to divergence. On the other hand, in MA-MeSN,

---

Algorithm 1. Multi-Agent Message Sharing Network (MA-MeSN)

---

for $i = 1, N$ **do**

    Initialize replay memory $\mathcal{D}_\rangle$; $where \, i \in \{1..N\}$ to capacity M

    Initialize the online and target, message and policy networks $f'_{i,\theta}, f''_{i,\theta}, f'_{i,\theta'}, f''_{i,\theta'}$

end for

for $episode = 1, E$ **do**

    for $t = 1, T_{convergence}$ **do**

        for $i = 1, N$ **do**

            Select a random action a_i^t with probability ε

            Otherwise, select $a_i^t = \arg\max_a Q_{f''_i}(o_i^t, m_{-i}^t, a; f''_\theta)$

            Execute action a_i^t, collect reward r_i^{t+1} and observe next state o_i^{t+1}

            Store the transition $(o_i^t, a_i^t, r_i^{t+1}, o_i^{t+1})$ in $\mathcal{D}_\rangle$

            Sample mini-batch of transitions $(o_i^j, a_i^j, r_i^{j+1}, o_i^{j+1})$ from $\mathcal{D}_\rangle$

            Generate the messages from other agents $m_{-i}^j = f'_{-i}(o_{-i}^j)$

            Set $y_i^j = \begin{cases} r_i^{j+1}, & \text{if } o_i^{j+1} \text{ is terminal} \\ r_i^{j+1} + \gamma\max_{a'} Q_{f''_i}(o_i^{j+1}, m_{-i}^{j+1}, a'; f''_{i,\theta'}), & \text{otherwise} \end{cases}$

            Compute gradients using target value y_i^j for policy network f''_θ

            $\Delta Q_{f''} = y_i^j - Q_{f''_i}(o_i^j, m_{-i}, a; f''_{i,\theta})$

            Apply gradients $\nabla\theta_{i,f''}$ to $f''_{i,\theta}$

            Collect gradients $\nabla\theta_{i,f'}$ from all policy networks

            Apply gradients $\nabla\theta_{i,f}$ to f'_i

        end for

        Every C steps, set $\theta'_{i,f''} \leftarrow \theta_{i,f''} \forall i$

        Every C steps, set $\theta'_{i,f'} \leftarrow \theta_{i,f'} \forall i$

    end for

end for

---

training the message network to generate messages m_{-i}^t based on the current observation reduces the dependence on the environment's transition dynamics. Secondly, this allows for our algorithm to train off-policy using a step based experience replay. Whereas DIAL requires on-policy training using recorded trajectories.

Fully Decentralized Cooperative Policy. The messages shared between agents are discrete of size 2 bytes. We generate discrete message by applying Gumbel-Softmax Sampling [7] on the prediction of the message network. To achieve fully decentralized execution without message sharing, we utilize a LSTM memory module μ in conjunction with each agent's policy network. The $LSTM_\mu$ learns a mapping from agent's private observation history to the message generated by the other agents in the environment. The model $LSTM_\mu$ mimics the message received from other agents. Thus the individual memory modules μ along with their policy network f'' can be independently used for fully decentralized execution of the learned cooperative policy (MA-MeSN-MM). The message memory module $LSTM_\mu$ is trained in a supervised fashion in parallel to the policy and message networks during centralized training.

4.2   Multi-Agent Broadcast Network (MA-BoN)

The generalization of communication based centralized MARL algorithms presented in the previous section allows us to develop communication models with distinct message types. We constraint our MA-MeSN model to a single message to rule them all approach and develop a broadcast model as shown in Fig. 2. The neural network f' (message network) maps the shared partial observation encoding from all agents to a broadcast message bm^t. We study the properties of MA-MeSN and MA-BoN in Sect. 6.3 and show that this network is feasible in multi-agent general sum games.

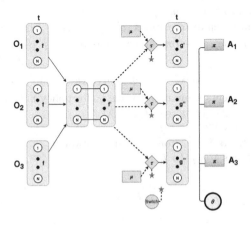

Fig. 2. Multi-Agent Broadcast Network (MA-BoN)

The NN f' learns a combined communication message as the broadcast message (bm^t). Each agent can now independently evaluate the action-value for their private observation using the function $g'(z_i^t, bm^t)$, which is a function of the complete observed state of the environment. This network also allows for parallel action-value evaluations with a single forward pass of the network and avoids the $|P|$ iterations required by IMS, and provides a linear space and runtime complexity as shown in Sect. 6.2. MA-BoN can also be decentralized by the use of a memory module $LSTM_\pi$ trained parallel to the policy network (MA-BoN-MM).

5   Experimental Methodology

In this paper, we compare our algorithms with MARL algorithms in the literature on three different MARL environments. We present the treadmill driving environment simulator in this section. The OpenAI particle environments are used to show the validity of our algorithms on public testbeds. The results can be found in Appendix (https://uwaterloo.ca/scholar/sites/ca.scholar/files/mcrowley/files/deep_multi_agent_reinforcement_learning_for_autonomous_driving-full.pdf).

5.1 Treadmill Driving Environment

The treadmill environment simulates an infinite highway with multiple cars driving in the presence of an adversary. The highway is simulated using a treadmill, which is always running and thus creates an infinite highway. The size of the treadmill is currently kept fixed at $[100, 100]$ steps. Agents can enter or exit the treadmill from the front and back. The treadmill contains a minimum of 2 cooperative autonomous agents and at least 1 adversary agent. These agents can be controlled using Deep RL methods and the adversary (aggressive) car is controlled with a stochastic behavior policy which can cause a crash with the closest autonomous car. The cooperative autonomous vehicles can sense the closest car as part of its partial private observation of the environment, but do not receive information to distinguish between their behavior (cooperative/adversary). The agents can send messages to other agents using a discrete communication broadcast channel, to which other agents subscribe. The private reward received by an autonomous car is the normalized distance from the closest observed car and a large negative reward for a crash. The agents' actions include 3 angles of steering in 8 directions and 3 discrete levels of acceleration/deceleration. The reward function does not provide explicit rewards for cooperation between the agents or for maintaining stable emergent communications between agents. The episode is terminated when the distance between any two agents is 0 (collision is encountered).

6 Results and Discussions

In this section, we present the results of training our algorithms in the treadmill driving environment. In all our algorithms (MA-MeSN, MA-BoN, DIAL, IMS, independent DQN, independent DQN with SER), we use a hierarchical neural network structure [8]. We provide an evaluation of hierarchical DQN on treadmill driving environment domain in Appendix (https://uwaterloo.ca/scholar/sites/ca.scholar/files/mcrowley/files/deep_multi_agent_reinforcement_learning_for_autonomous_driving-full.pdf). In this section, we focus on presenting performance results for our multi-agent algorithms on the treadmill driving simulator.

6.1 Centralized Training on Multi-Agent Driving Environment

All experiments are run for a minimum of $4K$ episodes ($0.8M$ steps). All neural networks consist of two layers with 4096 neural units in the first layer with 12 neurons in the second layer. DIAL network consists of two layers with 6144 units in the first layer to allow for fair evaluation to other algorithms. The maximum size of message shared between agents is 2 bytes. We use Adam optimizer with a learning rate of 5×10^{-4}. The batch-size for updates is 64 and the target network is updated after 200 steps, except DIAL's target network is updated after 40 episodes. For the IMS algorithm, we arrived at using $P = 5$ for communication iterations through cross-validation.

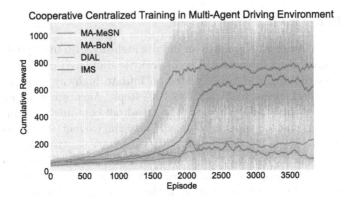

Fig. 3. Comparison of Cumulative Reward for Centralized Training Algorithms in Multi-Agent Driving Environment.

The cumulative reward achieved during centralized training of our MARL algorithms is shown in Fig. 3. All experiments are repeated 20 times and averaged to produce the learning curves. We achieve the highest cumulative reward with MA-MeSN followed by the MA-BoN algorithm. The IMS and DIAL algorithms are able to improve on the policy achieved by independent DQN, as they have the advantage of message sharing over independent DQN policy. IMS shows a slow learning curve compared to other algorithms with $P = 5$ communication iterations. IMS training also requires curriculum learning approach to train the network efficiently [14]. However, to maintain fairness, this was left out in our experiments. DIAL shows steady improvement in performance, however, the performance of the final policy is weak when compared to MA-MeSN.

As results show, our generalized MARL algorithm (MA-MeSN) is able to perform superior to DIAL and IMS. The benefit of having a separate model for message policy and behavior policy prediction. The separation of message policy model and behavior policy model leads to each neural network achieving a more optimal solution than competing approaches. Whereas, DIAL and IMS constraint the training of message and behavior policy to a single neural network which produces sub-optimal results. MA-BoN also constraints the inter-agent message sharing to a single broadcast message which also leads to a sub-par result in comparison to MA-MeSN.

6.2   Ablation Study of Scalability of MA-BoN

In this section, we demonstrate the scalability of the MA-BoN approach compared to DQN with stabilized experience replay, IMS and DIAL. We carry out an ablation study of our approach by varying the number of cars in the environment and present the results in Fig. 4. The Fig. 4 shows a comparison of the inference time it took to complete an episode and the average cumulative reward achieved per episode when the number of agents in the environment is increased. The results for cumulative reward comparison are computed by averaging results

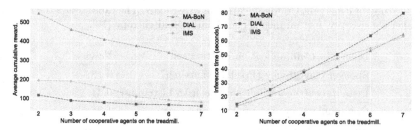

Avg. cumulative reward achieved at convergence with varying number of agents in the environment.

Avg. inference time of the algorithms with varying number of agents in the environment.

Fig. 4. Scalability comparison on the treadmill environment.

of 5 training runs for each algorithm with different seed values. The training of all algorithms was completed over $15,000$ episodes or $2.5M$ steps. We see that our approach MA-BoN is able to sustain better performance compared to other approaches when the complexity of the environment was increased. The inference time grows linearly for MA-BoN in comparison to the quadratic increase for DIAL. MA-BoN shows better scalability as the message generation network for each agent is optimized separately using the cumulative gradients from all agent's temporal difference loss. Thus the message is more generalizable in complex settings, while DIAL and IMS suffer from the problem of optimizing the joint objective for communicative and non-communicative policy; which leads to reduced robustness of the messages shared between agents.

6.3 Theoretical Study of Emergent Communication

In this section, we study the inter-agent emergent communication achieved during training of our MARL algorithm, MA-MeSN. Table 1 shows the results for MA-MeSN using common metrics [10] to measure the effect of these messages using our domain. *Speaker consistency (SC)* is used to measure positive signaling as it measures the mutual information between the communicative m_i and behavior a_i policy of an agent. We see a small positive value of 0.18 for SC; which suggests that the objective for message policy and behavior policy are indeed different. Thus our approach of generalizing MARL algorithms with separate message and policy networks is necessary. *Instantaneous Coordination (IC)* measures the positive listening between agents, which is measure of the mutual information between the speaker's communicative actions m_{-i} and the listener's behavior/locomotive actions a_i. We achieve a value of 0.41 for IC which indicates that the listener agent's policy are dependent on the messages of the speaker agent, which is necessary for emergent communication. We also study the *Communication Message Entropy* which measures if the listener receives the same message for a given input. We achieve a value of 1.27 for entropy, which shows that the speaker is not using different messages **for the same input** and is rather consistent in its signals.

Table 1. Study of Emergent Communication in MA-MeSN. The table shows the results for speaker consistency, instantaneous coordination and entropy [10].

Emergent communication metric used	Value
Speaker consistency (Positive signaling)	0.18
Instantaneous coordination (Positive Listening)	0.41
Communication message entropy	1.27
Message Input Norm (MIN)	63.75
Cumulative reward with white noise	319.07

To further study the effects of communication, we probe our MA-MeSN model, calculate the *L2-norm* of the fully connected weight matrix for message input for the listener agent, and report the results in Table 1. The weight matrix for the message input has an *L2-norm* much higher than 0.0, which suggests that the message indeed does get used by the listener agent's policy network. We extend our analysis of the MA-MeSN by replacing the messages received by the agents with white noise on a trained MA-MeSN model. We see a reduction in the mean cumulative reward achieved by the algorithm from 746.8 to 319.07 in the stochastic environment. The reduction in the cumulative reward shows that emergent communication did develop between agents and is an integral part of the final cooperative policy achieved.

6.4 Fully Decentralized Cooperative Policy in Driving Environment

We compare our method of using message memory models for decentralized execution (MA-MeSN-MM) with independent DQN, DQN with stabilized experience replay and distributed behavior cloning of centralized cooperative policy

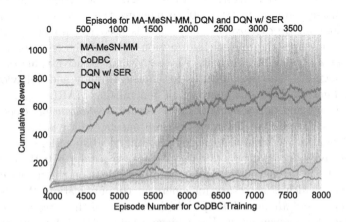

Fig. 5. Comparison of cumulative reward for decentralized training in Multi-Agent Driving Environment.

(CoDBC). CoDBC policy is trained using imitation learning of the (expert) centralized cooperative policy from MA-MeSN. All of the hyper-parameters and experimental setup are exactly the same as the experiments for the centralized training section. The learning curve for decentralized policies is shown in Fig. 5. As the treadmill environment does not explicitly reward agents for cooperation, we see poor performance from DQN and DQN with SER; however DQN with SER is more stable during training compared to DQN. DQN with SER applies a weight to each training sample's gradient. The weight is computed using a linearly decaying function based on the episodes elapsed since a sample was collected. Thus, DQN with SER is able to prioritize its training on the latest samples (which represent the latest policies of other agents) collected in the DQN's buffer and thus avoids divergence. However, the final policy achieved by DQN w/ SER is worse than MA-MeSN-MM and CoDBC.

While the CoDBC method outperforms DQN and DQN w/ SER, the number of episodes required to learn a cooperative policy is nearly 8000 episodes, as CoDBC needs to be run sequentially after MA-MeSN policy training has converged. Our method MA-MeSN-MM achieves decentralized cooperative policy by learning a function mapping from private observations to the messages received from other agents. The message module (MM) is trained in parallel to the policy network and thus does not require additional training after MA-MeSN has converged. This approach is ideal for real-time agents in MARL environments with a goal of cooperation as communication channels are unreliable and induce a time-latency.

7   Conclusion and Future Work

In this paper we present that generalization of the current work in MARL field leads to large improvements in the final multi-agent policy. Our approach allows for variability in the message format which is useful for various domains. MA-MeSN and MA-BoN both outperform the algorithms found in current literature based on learning curve results. Our algorithms also provide improvements in the time and space complexity over DIAL and IMS. MA-MeSN and MA-BoN are easier to train as they can be trained in an off-policy setting. We also present a decentralized model which achieves higher cumulative reward compared to some of the centralized techniques and all decentralized techniques. This paper also presents a new large scale multi-agent testing environment for further MARL research.

References

1. Bernstein, D.S., Givan, R., Immerman, N., Zilberstein, S.: The complexity of decentralized control of Markov decision processes. Math. Oper. Res. **27**(4), 819–840 (2002)
2. Busoniu, L., Babuska, R., De Schutter, B.: A comprehensive survey of multiagent reinforcement learning. IEEE Trans. Syst. Man Cybern.-Part C: Appl. Rev. **38**(2), 2008 (2008)

3. Das, A., Kottur, S., Moura, J.M.F., Lee, S., Batra, D.: Learning cooperative visual dialog agents with deep reinforcement learning, pp. 2970–2979, October 2017. https://doi.org/10.1109/ICCV.2017.321
4. Das, A., Kottur, S., Moura, J.M., Lee, S., Batra, D.: Learning cooperative visual dialog agents with deep reinforcement learning. arXiv preprint arXiv:1703.06585 (2017)
5. Foerster, J., Assael, I.A., de Freitas, N., Whiteson, S.: Learning to communicate with deep multi-agent reinforcement learning. In: Advances in Neural Information Processing Systems, pp. 2137–2145 (2016)
6. Foerster, J., Nardelli, N., Farquhar, G., Torr, P., Kohli, P., Whiteson, S.: Stabilising experience replay for deep multi-agent reinforcement learning. In: ICML 2017: Proceedings of the Thirty-Fourth International Conference on Machine Learning, June 2017. http://www.cs.ox.ac.uk/people/shimon.whiteson/pubs/foerstericml17.pdf
7. Jang, E., Gu, S., Poole, B.: Categorical reparameterization with gumbel-softmax. arXiv preprint arXiv:1611.01144 (2016)
8. Kulkarni, T.D., Narasimhan, K., Saeedi, A., Tenenbaum, J.: Hierarchical deep reinforcement learning: integrating temporal abstraction and intrinsic motivation. In: Advances in Neural Information Processing Systems, pp. 3675–3683 (2016)
9. Lazaridou, A., Peysakhovich, A., Baroni, M.: Multi-agent cooperation and the emergence of (natural) language. arXiv preprint arXiv:1612.07182 (2016)
10. Lowe, R., Foerster, J., Boureau, Y.L., Pineau, J., Dauphin, Y.: On the pitfalls of measuring emergent communication. arXiv preprint arXiv:1903.05168 (2019)
11. Lowe, R., Wu, Y., Tamar, A., Harb, J., Pieter Abbeel, O., Mordatch, I.: Multi-agent actor-critic for mixed cooperative-competitive environments, pp. 6379–6390 (2017). http://papers.nips.cc/paper/7217-multi-agent-actor-critic-for-mixed-cooperative-competitive-environments.pdf
12. Mnih, V., et al.: Human-level control through deep reinforcement learning. Nature 518(7540), 529 (2015)
13. Mordatch, I., Abbeel, P.: Emergence of grounded compositional language in multi-agent populations. arXiv preprint arXiv:1703.04908 (2017)
14. Sukhbaatar, S., Szlam, A., Fergus, R.: Learning multiagent communication with backpropagation. In: Lee, D.D., Sugiyama, M., Luxburg, U.V., Guyon, I., Garnett, R. (eds.) Advances in Neural Information Processing Systems 29, pp. 2244–2252. Curran Associates, Inc. (2016). http://papers.nips.cc/paper/6398-learning-multiagent-communication-with-backpropagation.pdf
15. Tan, M.: Multi-agent reinforcement learning: independent vs. cooperative agents. In: Proceedings of the Tenth International Conference on Machine Learning, pp. 330–337 (1993)

Incremental Sequential Rule Mining
with Streaming Input Traces

Andriy Drozdyuk[1,3], Scott Buffett[2(✉)], and Michael W. Fleming[3]

[1] National Research Council Canada, Ottawa, ON K1A 0R6, Canada
`Andriy.Drozdyuk@nrc.gc.ca`
[2] National Research Council Canada, Fredericton, NB E3B 9W4, Canada
`Scott.Buffett@nrc.gc.ca`
[3] University of New Brunswick, Fredericton, NB E3B 5A3, Canada
`mwf@unb.ca`

Abstract. Traditional static pattern mining techniques, such as association rule mining and sequential pattern mining, perform inefficiently when applied to streaming data when regular updates are required, since there is significant repetition in the computation. Incremental mining techniques instead reuse information that has been previously extracted, and apply newly received data to compute the updated set of patterns. This paper proposes a new algorithm for incrementally mining sequential rules with streaming data. An existing rule mining algorithm, *ERMiner* is presented, and an incremental extension, called *IERMiner* is proposed and demonstrated. Experiments show that *IERMiner* significantly decreases the run time required to update the set of patterns when compared to running *ERMiner* on the full dataset each time.

Keywords: Data mining · Incremental sequential rule mining · Streaming data

1 Introduction

Data mining involves discovering interesting information in large quantities of data. A popular subfield of data mining is tasked with finding patterns in *sequences* of data. Much work has been done in *itemset* and *association rule* mining [1]. However, these approaches do not consider time or the sequential order of the data. *Sequential pattern mining* and *sequential rule mining* solve this limitation.

Sequential pattern mining (SPM) [2,3] is a research discipline within the field of data mining that focuses on identifying frequently occurring sequences of objects or events in an input set of sequences. Here, patterns of interest are sequences themselves, and such patterns are declared frequent if they are contained (i.e. as a subsequence) by a sufficient number of the input sequences. An extension to sequential pattern mining is *sequential rule mining* [4], which aims to find rules that dictate under which circumstances (i.e. given the presence

C. Goutte and X. Zhu (Eds.): Canadian AI 2020, LNAI 12109, pp. 79–91, 2020.
https://doi.org/10.1007/978-3-030-47358-7_8

of certain items) that items or sequences appear with sufficient frequency, given a set of input sequences. These rules can then used to predict which items might appear in the future, given what has been observed in the past. Specifically, we attempt to identify *partially ordered sequential rules* (also referred to as *pos rules*), as proposed by Fournier-Viger et al. [5]. This type of rule dictates when a set of items are likely to appear in the future, conditionally on the presence of a second set of items. There exist various algorithms for pos rule mining [4,6], with the fastest today being *ERMiner* [6].

In this paper, we explore the problem of partially ordered sequential rule mining on *streaming* data, where there is never a static, complete dataset to mine, but rather data is being received continuously. Examples of this type of data are live customer transactions, up-to-date weather measurements, and ongoing patient registrations at a hospital. In such situations where there is a desire to update the set of identified patterns of interest with regularity, it can be wasteful to ignore information that was found in previous iterations and simply mine the set of patterns from scratch each time. To address this, *incremental algorithms* have been proposed that process new information efficiently and, by maintaining previously identified information effectively, are able to generate an updated set of patterns more quickly than static algorithms can process the entire dataset from scratch [7–12].

The key contribution of the work presented in this paper is an incremental algorithm for partially ordered rule mining with unbounded data. Specifically, we propose an extension of the ERMiner algorithm, called IERMiner, that incrementally mines rules from *streaming* input sequences. In this case, the number of sequences remains static; sequences grow over time, with newly arriving data appended to the ends of those sequences. We show that our solution is effective by comparing the performance of our algorithm with that of re-running the existing algorithm for bounded data on an ever-increasing set of data.

2   Background

2.1   Literature Review

Initial exploration into sequential pattern mining can be attributed to Srikant and Agrawal [13], who first proposed the *AprioriAll* algorithm, as well as its first key extension, the *Generalized Sequential Patterns (GSP)* algorithm. Further improvements were proposed by Zaki [14] with a vertical sequence representation for the *SPADE* algorithm, Ayres *et al.* [15], who introduced the use of bitmaps for the *SPAM* algorithm, and Pei *et al.* [16] with the introduction of the *pattern growth* mining method with the *PrefixSpan* algorithm.

Solutions for the concept of incremental sequential pattern mining were initially proposed by Parthasarathy *et al.* [10] who utilized a sequence lattice to contain all of the potential frequent sequential patterns with the *ISM* algorithm. Masseglia *et al.* [9] next provided an advancement with the *ISE* algorithm that utilized a candidate generate-and-test approach. Cheng *et al.* [8] provided further improvements to the efficiency of incremental search with the *IncSpan* algorithm.

As mentioned above, sequential rule mining is a relatively new area with a small number of algorithms having been proposed [4,6], with the current state of the art algorithm being *ERMiner*, which is described in detail in Sect. 3.

To the best of our knowledge, this paper is the first to propose an incremental algorithm for sequential rule mining on streaming data.

2.2 Sequential Pattern Mining

Sequential pattern mining is an extension of itemset mining applied to data that possess temporal order [17]. Let I be a set of *items*, and let S be a set of *input sequences*, where each $s \in S$ consists of an ordered list of *itemsets*, or sets of items from I, also known as *transactions*. A sequence $\langle a_1 a_2 \ldots a_n \rangle$ is said to be *contained* in another sequence $\langle b_1 b_2 \ldots b_m \rangle$ if there exist integers $i_1, i_2, \ldots, i_n$ with $i_1 < i_2 < \ldots < i_n$ such that $a_1 \subseteq b_{i_1}, a_2 \subseteq b_{i_2}, \ldots, a_n \subseteq b_{i_n}$. A sequence $s \in S$ *supports* a sequence s' if s' is contained in s. The support $sup(s')$ for a sequence s' given a set S of input sequences is the percentage of sequences in S that support s', and equals $sup(s') = |\{s \in S \mid s \text{ supports } s'\}| \,/\, |S|$. A sequence s' is deemed a *sequential pattern* if $sup(s')$ is greater than some minimum.

2.3 Sequential Rule Mining

Where a sequential pattern is essentially a smaller sequence that is contained within the set S of input sequences with sufficiently high frequency or support, a sequential rule $r \in R(S)$ dictates when the presence of one or more items implies the presence of others, where some sequence-based properties are satisfied with both sufficient support and confidence. In this paper, we specifically consider *partially ordered sequential rules*. Formally, a partially ordered sequential rule (henceforth referred to simply as a *sequential rule*) $X \rightarrow Y$ is a relationship between the nonempty sets of items $X \subseteq I$ and $Y \subseteq I$, such that $X \cap Y = \emptyset$. A rule defines a relationship between two sets of items, specifically that if the items in the set X occur within a sequence, then items in Y will occur afterwards in the same sequence. The left and right sides of the rule are called the *antecedent* and the *consequent*, respectively [4]. For example, a rule might state that customers who buy bread and butter are also likely to purchase milk later that week.

The problem of sequential rule mining is defined as follows: given a database of sequences, find all the sequential rules that satisfy some given constraints. Two commonly encountered constraints (also referred to as interestingness measures [4]) are *support* and *confidence*.

Let $X \rightarrow Y$ be a candidate sequential rule and let $s_i \in S$ be an input sequence containing items $U(s_i, m, n) \subseteq I$ in its itemsets from position m (inclusively) to position n (exclusively). The sequence s_i *contains* $X \rightarrow Y$ iff there exists an integer k such that $X \subseteq U(s_i, 0, k)$ and $Y \subseteq U(s_i, k, |s_i|)$. The overall *support* $\sigma(X \rightarrow Y)$ is then the fraction of input sequences that contain it. If the support of a candidate rule exceeds some prespecified threshold, it is said to be *frequent*.

The *confidence* $\omega(X \rightarrow Y)$ is then the fraction of sequences containing the rule's *antecedent* that also support the rule. Thus $\omega(X \rightarrow Y) = |\{s \in S | s$ supports $X \rightarrow Y\}| / |\{s \in S | X \subseteq U(s_i, 0, |s_i|)\}|$. If a candidate rule is frequent and its confidence is greater than or equal to some prespecified threshold, it is considered to be a *sequential rule*.

For example, consider the following sample database:

(1) a, c, b, k, e, f   (2) c, a, d, e   (3) k, g, f, p   (4) k, r, f, p

The set $\{a, c\}$ occurs in sequences 1 and 2. This gives rise to a partially ordered sequential rule $\{a, c\} \rightarrow \{e\}$, which occurs in two sequences out of the four, so the support is $\frac{2}{4} = 0.5$. The whole rule occurs in both of the sequences in which the rule's antecedent occurs, so the confidence is $\frac{2}{2} = 1$.

3   ERMiner

ERMiner [6] is an algorithm developed by Fournier-Viger et al. for mining partially ordered sequential rules. *ERMiner* uses a notion similar to equivalence classes to keep track of which rules need to be further inspected. It differs from a previous algorithm, RuleGrowth [4], in that it does not recursively expand the left or right sides of the rules, but instead proceeds by left and right merges.

To describe the *ERMiner* algorithm, we first demonstrate the concept of *rule equivalence classes*. Formally, a left equivalence class is:

$$LE_{W,i} = \left\{ W \rightarrow Y \mid (Y \subseteq I) \wedge (|Y| = i) \right\}$$

where $W \subseteq I$, i is an integer, and rule $W \rightarrow Y$ is frequent. In other words, given a particular itemset W and an integer i, a left equivalence class is defined to be the set of frequent rules $W \rightarrow Y$ for all possible itemsets Y of size i.

The right equivalence class is:

$$RE_{W,i} = \left\{ X \rightarrow W \mid (X \subseteq I) \wedge (|X| = i) \right\}$$

where $W \subseteq I$, i is an integer and rule $X \rightarrow W$ is frequent. In other words, given a particular itemset W and an integer i, a right equivalence class is defined to be the set of frequent rules $X \rightarrow W$ for all possible itemsets X of size i.

Now we are ready to introduce the concept of *merging*.

Left merge is a process of combining two rules $r : W \rightarrow X$ and $s : W \rightarrow Y$ from a left equivalence class $LE_{W,i}$ into a single rule $W \rightarrow X \cup Y$. We require that the itemsets differ by exactly one item, i.e. $|X \cap Y| = |X| - 1 = |Y| - 1$.

Right merge is a process of combining two rules $r : X \rightarrow W$ and $s : Y \rightarrow W$ from a right equivalence class $RE_{W,i}$ into a single rule $X \cup Y \rightarrow W$. We require that the itemsets differ by exactly one item, i.e. $|X \cap Y| = |X| - 1 = |Y| - 1$.

As the authors note, we can generate any left equivalence class $LE_{W,i+1}$ by performing all left merges on pairs of rules from $LE_{W,i}$, and we can generate right equivalence class $RE_{W,i+1}$ by right merging all pairs of rules from $RE_{W,i}$.

Both $LE_{W,i}$ and $RE_{W,i}$ are initialized by a set of 1×1 rules, which are easily computed by considering all pairs of individual frequent items. For example, if we know that the sequences in which the item a occurs are $[1, 2, 3]$ while the item b occurs in $[2, 3, 4]$, then we can calculate the intersection of the two to be $[2, 3]$. If a occurs before b in both, and this satisfies our minimum support requirement, we can form a simple rule $a \rightarrow b$, and we initialize $LE_{a,1} = \{a \rightarrow b\}$ and $RE_{b,1} = \{a \rightarrow b\}$. We proceed similarly for all the other pairs of items.

Finally, the algorithm performs left merges on the results of all the right merges. This is done by storing all equivalence classes generated by the right merge into a special structure called the *left store*. Left store is simply a way to turn right equivalence classes into left equivalence classes. For example, if we have two rules generated by the right merge like so: $RE_{\{e,f\},1} = \{\{a\} \rightarrow \{e, f\}, \{b\} \rightarrow \{e, f\}\}$ then the left store would contain $LE_{a,2} = \{\{a\} \rightarrow \{e, f\}\}$ and $LE_{b,2} = \{b\} \rightarrow \{e, f\}\}$. These two equivalence classes would then be merged by a left merge.

4   Incremental Sequential Rule Mining

4.1   Information Maintenance

It is crucial to the overall benefit of an incremental mining algorithm to retain knowledge about the dataset from state to state as new information is received, in order to alleviate the necessity to spend resources on re-analyzing previously processed data. To facilitate this information maintenance, the algorithm proposed in this paper takes advantage of the following characteristics. Please see [18] for developed proofs of each of the claims presented below.

Consider a database S of input sequences. Suppose a new event (i.e. itemset) is added to the end of a sequence in S, giving the new database S'. Then, for any r in the set of rules $R(S)$:

- The number of sequences that contain r can only increase, and thus support of r in S' is greater than or equal to that in S.
- If r is frequent in S then it cannot become infrequent in S'.
- The confidence of r can increase, decrease or remain unchanged in S'.
- If r is infrequent in S, then there exists a possible sequence of itemset additions, resulting in database S'' in which r would become frequent.

A key result of the above statements is that, as the database grows, equivalence classes (as defined by the *ERMiner* and adapted for the incremental version proposed here) cannot lose rules.

4.2   The IERMiner Algorithm

Here we will present the *IERMiner* algorithm. Unlike the bounded versions of Sequential Rule Mining algorithms, the *IERMiner* algorithm operates on events instead of sequences. Each new event triggers a complete run-through of the algorithm, which then computes a new set of rules.

We begin presentation of the *IERMiner* pseudocode with the main procedure called *mine*, which uses a procedure that generates simple 1×1 rules.

IERMiner Mine Procedure

1. Receive an event
2.   Generate 1×1 Rules from Event
3.   For each 1×1 Rule above
4.   Store the 1×1 Rule in the Aggregator
5.   Process the 1×1 Rule with Left Equivalence Class Processor
6.   Store the resulting rules in the Aggregator
7.   Process the 1×1 Rule with the Right Equivalence Class Processor
8.   Store the resulting rules in the Aggregator
9.   Process resulting rules with Left Equivalence Class Processor
10.   Store the resulting rules in the Aggregator

Generate 1×1 Rules Procedure

1. Given an Event and Sequence Id
2.   Find New Frequent Items in the Event
3.   Append Event to the Sequence with the given Id
4.   For each combination of New and Old frequent Item
5.   Generate a Rule of the form New Item $\rightarrow$ Old Item
6.   Return all Generated Rules

Next we detail the *Left Equivalence Class Generation* pseudocode.

Equivalence Class Generation

1. Given a Rule
2.   Generate Left Equivalence Class from the Rule
3.   Find Left Rules with Left Search on equivalence class above
4.   Remember all the Rules in Left Rules that are new or have changed
5.   Return the new and changed Rules to the caller or an empty set

Generate left equivalence class procedure:

1. Given a Rule
2.   Let $W =$ Rule's l.h.s.
3.   Let $Y =$ Rule's r.h.s.
4.   Let i be the number of items in Y
5.   The unique pair (W, i) will designate the Equivalence Class
6.   If (W, i) does not exist in our store
7.   initialize a map with the key (W, i) to be the equivalence class with just the rule $W \rightarrow Y$
8.   Else

9. Find the equivalence class (W, i)
10. Update all rules in it with new confidence information
11. If the rule $W \rightarrow Y$ is present in equivalence class:
12. update its support
13. Append the rule $W \rightarrow Y$ to the equivalence class with the determined support and confidence
14. Return the set of all rules in the (W, i) equivalence class.

As was the case for the left equivalence classes, there is a *Right Equivalence Class Processor* for the right equivalence classes, which utilizes a *Generate right equivalence class* procedure. These are developed in a manner analogous to that of the left equivalence, and are thus not outlined here.

Finally the developed rules are merged, as outlined in the *Left Search* and *Right Search* procedures, with *Right Search* being a procedure that makes use of the *Left Search*.

Left Search Procedure:

1. Given a Left Equivalence Class, Support and Confidence
2. Merge all the rules in the Equivalence Class that differ by 1 item in R.H.S.
3. Return only the Merged rules that have required Support and Confidence

Right Search Procedure:

1. Given a Right Equivalence Class, Support and Confidence
2. Merge all the rules in the Equivalence Class that differ by 1 item in L.H.S.
3. Keep only the Merged rules that have the required Support and Confidence
4. Recursively perform a Right Search on the resulting Merged rules
5. Return the union of Merged rules and rules from the recursive call

4.3 IERMiner Example

Here we provide a partial example of the *IERMiner* algorithm in action, where a new rule is identified as a result of a new item being added.

Each rule R in the example is of the form:

$$X \rightarrow Y \; [R_s] \, \langle R_c \rangle,$$

where R_s, called *supportive*, is a set of sequence identifiers denoting sequences that contain the rule. R_c, called *confident*, is a set of sequence identifiers denoting sequences that contain at least X. Notice that it is always true that $R_s \subseteq R_c$. We use this augmented notation to better illustrate the merging process.

We present the example in two parts. Part 1 will illustrate the operation of a conventional *ERMiner*, while Part 2 shows how an individual event is integrated into an existing structure to produce the new rules.

Part 1. Given minimum support and confidence of 0.5, and the following sequences:

1. (a, b), (c), (f), (g), (e)
2. (a, d), (c), (b), (a, b, e, f)
3. (a), (b), (f), (e)
4. (b), (f, g, h)

The first steps are to generate initial 1×1 rules and then to gather these into equivalence classes. The left equivalence classes, LE_1, of size one are shown in Fig. 1a.

From LE_1, we can produce left equivalence classes of size two, LE_2, by performing a left merge. The results are shown in Fig. 1b.

$LE_{a,1}$ $a \to b\, [2,3]\, \langle 1,2,3 \rangle,$
$\quad a \to c\, [1,2]\, \langle 1,2,3 \rangle,$
$\quad a \to e\, [1,2,3]\, \langle 1,2,3 \rangle,$
$\quad a \to f\, [1,2,3]\, \langle 1,2,3 \rangle$
$LE_{b,1}$ $b \to e\, [1,2,3]\, \langle 1,2,3,4 \rangle,$
$\quad b \to f\, [1,2,3,4]\, \langle 1,2,3,4 \rangle,$
$\quad b \to g\, [1,4]\, \langle 1,2,3,4 \rangle$
$LE_{c,1}$ $c \to e\, [1,2]\, \langle 1,2 \rangle,$
$\quad c \to f\, [1,2]\, \langle 1,2 \rangle$
$LE_{f,1}$ $f \to e\, [1,3]\, \langle 1,2,3,4 \rangle$

$LE_{a,2}$ $a \to (b,c)\, [2]\, \langle 1,2,3 \rangle,$
$\quad a \to (c,f)\, [1,2]\, \langle 1,2,3 \rangle,$
$\quad a \to (b,f)\, [2,3]\, \langle 1,2,3 \rangle,$
$\quad a \to (e,f)\, [1,2,3]\, \langle 1,2,3 \rangle,$
$\quad a \to (c,e)\, [1,2]\, \langle 1,2,3 \rangle,$
$\quad a \to (b,e)\, [2,3]\, \langle 1,2,3 \rangle$
$LE_{b,2}$ $b \to (e,f)\, [1,2,3]\, \langle 1,2,3,4 \rangle,$
$\quad b \to (e,g)\, [1]\, \langle 1,2,3,4 \rangle,$
$\quad b \to (g,f)\, [1,4]\, \langle 1,2,3,4 \rangle$
$LE_{c,2}$ $c \to (e,f)\, [1,2]\, \langle 1,2 \rangle$

(a) Left equivalence classes LE_1 (b) Left equivalence classes LE_2

Fig. 1. Example results of ERMiner algorithm

We then clean these up by throwing out all the rules that do not meet the minimum support requirement of 0.5, like $a \to (b,c)\, [2]\, \langle 1,2,3 \rangle$, which occurs only in one sequence #2 out of the total of four, making its support $\frac{1}{4} = 0.25$.

Execution continues in a similar manner to generate the remaining larger left equivalence classes, as well as the right equivalence classes.

Part 2. We now consider that an event (c) is added to the end of the sequence 3, so that the whole now looks as follows:

1. (a, b), (c), (f), (g), (e)
2. (a, d), (c), (b), (a, b, e, f)
3. $\boxed{\text{(a), (b), (f), (e), (c)}}$
4. (b), (f, g, h)

We next scan all the 1×1 rules with c in the consequent, and update their supportive and confident indices. In this case only one rule is affected:

$a \to b$ $[2,3]$ $\langle 1,2,3 \rangle$ $\qquad\qquad$ $b \to f$ $[1,2,3,4]$ $\langle 1,2,3,4 \rangle$

$\boxed{a \to c \ [1,2,3] \ \langle 1,2,3 \rangle}$ $\qquad\qquad$ $b \to g$ $[1,4]$ $\langle 1,2,3,4 \rangle$

$a \to e$ $[1,2,3]$ $\langle 1,2,3 \rangle$ $\qquad\qquad$ $c \to e$ $[1,2]$ $\langle 1,2 \rangle$

$a \to f$ $[1,2,3]$ $\langle 1,2,3 \rangle$ $\qquad\qquad$ $c \to f$ $[1,2]$ $\langle 1,2 \rangle$

$b \to e$ $[1,2,3]$ $\langle 1,2,3,4 \rangle$ $\qquad\qquad$ $f \to e$ $[1,3]$ $\langle 1,2,3,4 \rangle$

Next we proceed to update the equivalence classes that were produced as a result of combining the above rule with others. Here, too, only one is updated:

$LE_{a,1}$ $a \to b$ $[2,3]$ $\langle 1,2,3 \rangle$,

$\qquad\quad$ $\boxed{a \to c \ [1,2,3] \ \langle 1,2,3 \rangle,}$

$\qquad\quad$ $a \to e$ $[1,2,3]$ $\langle 1,2,3 \rangle$,

$\qquad\quad$ $a \to f$ $[1,2,3]$ $\langle 1,2,3 \rangle$

So we proceed to inspect and update $LE_{a,2}$ only:

$LE_{a,2}$ $\boxed{a \to (b,c) \ [2,3] \ \langle 1,2,3 \rangle,}$

$\qquad\quad$ $\boxed{a \to (c,f) \ [1,2,3] \ \langle 1,2,3 \rangle,}$

$\qquad\quad$ $a \to (b,f)$ $[2,3]$ $\langle 1,2,3 \rangle$,

$\qquad\quad$ $a \to (e,f)$ $[1,2,3]$ $\langle 1,2,3 \rangle$,

$\qquad\quad$ $\boxed{a \to (c,e) \ [1,2,3] \ \langle 1,2,3 \rangle,}$

$\qquad\quad$ $a \to (b,e)$ $[2,3]$ $\langle 1,2,3 \rangle$

Note that now the rule $a \to (b,c)$ $[2,3]$ $\langle 1,2,3 \rangle$ meets the minimum support requirement, thus demonstrating the identification of a new rule.

5 Experiments

We propose that updating an existing set of sequential rules with our incremental *IERMiner* algorithm works faster than mining the entire set of rules again using the *ERMiner* algorithm when a new event is added to a database. Each experiment starts by running each algorithm on an initial set of empty sequences. Events are then added to the sequences one at a time, running each algorithm each time. We test the methods on two publicly available datasets, *Kosarak*[1] and *MSNBC*[2]. We use 31790 sequences from the *MSNBC* dataset. We split the *Kosarak* dataset into one dataset with 5000 sequences, (and call it *Akosarak 5000*) and another with *25,000* sequences (denoted by *Akosarak 25000*). This splitting is done to facilitate experimenting with different values of minimum support and minimum confidence. Several experiments were run, using different combinations of values for the number of sequences, the minimum support level

---

[1] http://fimi.ua.ac.be/data/.

[2] http://archive.ics.uci.edu/ml/datasets/msnbc.com+anonymous+web+data.

and the minimum confidence level. The results presented in this section are for a representative sample of these experiments.

Experiments were run on a Macbook Pro with a 2.2 Ghz Intel Core i7 processor with 16 GB DDR3 memory and a 256 GB SSD hard drive. Algorithms were coded using Python, and tests were run using a Linux OS. Each experiment was conducted at least twice to ensure that accurate results were produced.

In all cases, the results showed that, in the long term, *IERMiner* significantly decreases the running time required to update the set of patterns when compared to running *ERMiner* on the full dataset each time.

5.1 Experiment Akosarak 5000

This experiment was defined by the following parameters:
Number of sequences: 5,000, Minimum Support: 0.1, Minimum Confidence: 0.1. To plot the data (Fig. 2), we took the average time for every set of 100 consecutive events.

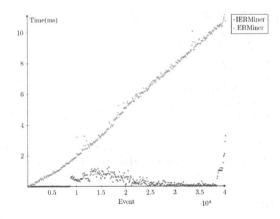

Fig. 2. Experiment Akosarak 5000 (times averaged in groups of 100 events)

5.2 Experiment Akosarak 25k

This experiment was defined by the following parameters:
Number of sequences: 25,000, Minimum Support: 0.05, Minimum Confidence: 0.1. To plot the data (Fig. 3), we took an average for every 1000 events.

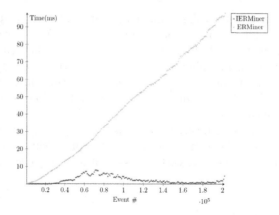

Fig. 3. Experiment Akosarak 25k (times averaged in groups of 1,000)

5.3 Experiment MSNBC 31790

This experiment was defined by the following parameters:
Number of sequences: 31,790, Minimum Support: 0.1, Minimum Confidence: 0.1.
To plot the data (Fig. 4), we took an average for every 2000 events.

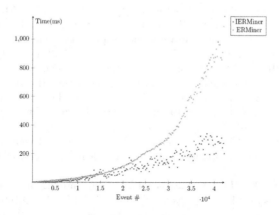

Fig. 4. Experiment MSNBC 31790 (times averaged in groups of 2,000)

6   Concluding Remarks

In this work, we aimed to show that it is possible to optimize data mining for
unbounded data. We created an algorithm, *IERMiner*, that can use previously
examined data to help it mine future data faster. *IERMiner* operates on more
granular data called events. This distinguishes it from other algorithms and
allows it to function on unbounded input streams. The result of the algorithm

is used to construct output, rules, similar to that of other algorithms that use bounded data. This allowed us to compare performance. We showed that, while the *IERMiner* algorithm initially performed more slowly than the state-of-the-art *ERMiner* algorithm for mining *Sequential Partially Ordered Rules*, in the long term *IERMiner* produced substantial savings in run-time. In the world of streaming data, such an algorithm provides value and can be used as a building block for further research.

Some potential topics for future work include using parallelization to improve performance, extending the algorithm to allow the processing of more than one new event at a time, and dealing with the issue of receiving events out of order.

References

1. Han, J., Cheng, H., Xin, D., Yan, X.: Frequent pattern mining: current status and future directions. Data Min. Knowl. Disc. **15**(1), 55–86 (2007)
2. Agrawal, R., Srikant, R.: Mining sequential patterns. In: Proceedings of the Eleventh International Conference on Data Engineering, ICDE 1995, pp. 3–14. IEEE Computer Society (1995)
3. Mooney, C.H., Roddick, J.F.: Sequential pattern mining - approaches and algorithms. ACM Comput. Surv. **45**(2), 19:1–19:39 (2013)
4. Fournier-Viger, P., Nkambou, R., Tseng, V.S.M.: RuleGrowth: mining sequential rules common to several sequences by pattern growth. In: Proceedings of the 2011 ACM Symposium on Applied Computing, SAC 2011, pp. 956–961. ACM (2011)
5. Fournier-Viger, P., Gueniche, T., Tseng, V.S.: Using partially-ordered sequential rules to generate more accurate sequence prediction. In: Zhou, S., Zhang, S., Karypis, G. (eds.) ADMA 2012. LNCS (LNAI), vol. 7713, pp. 431–442. Springer, Heidelberg (2012). https://doi.org/10.1007/978-3-642-35527-1_36
6. Fournier-Viger, P., Gueniche, T., Zida, S., Tseng, V.S.: ERMiner: sequential rule mining using equivalence classes. In: Blockeel, H., van Leeuwen, M., Vinciotti, V. (eds.) IDA 2014. LNCS, vol. 8819, pp. 108–119. Springer, Cham (2014). https://doi.org/10.1007/978-3-319-12571-8_10
7. Kao, B., Zhang, M., Yip, C.L., Cheung, D.W., Fayyad, U.: Efficient algorithms for mining and incremental update of maximal frequent sequences. Data Min. Knowl. Disc. **10**(2), 87–116 (2005)
8. Cheng, H., Yan, X., Han, J.: IncSpan: incremental mining of sequential patterns in large database. In: Proceedings of the Tenth ACM SIGKDD International Conference on Knowledge Discovery and Data Mining, pp. 527–532. ACM (2004)
9. Masseglia, F., Poncelet, P., Teisseire, M.: Incremental mining of sequential patterns in large databases. Data Knowl. Eng. **46**(1), 97–121 (2003)
10. Parthasarathy, S., Zaki, M.J., Ogihara, M., Dwarkadas, S.: Incremental and interactive sequence mining. In: Proceedings of the Eighth International Conference on Information and Knowledge Management, pp. 251–258. ACM (1999)
11. Wang, K., Tan, J.: Incremental discovery of sequential patterns. In: 1996 ACM SIGMOD Data Mining Workshop: Research Issues on Data Mining and Knowledge Discovery (SIGMOD 1996), pp. 95–102 (1996)
12. Zheng, Q., Xu, K., Ma, S., Lv, W.: The algorithms of updating sequential patterns. arXiv preprint cs/0203027 (2002)

13. Srikant, R., Agrawal, R.: Mining sequential patterns: generalizations and performance improvements. In: Apers, P., Bouzeghoub, M., Gardarin, G. (eds.) EDBT 1996. LNCS, vol. 1057, pp. 1–17. Springer, Heidelberg (1996). https://doi.org/10.1007/BFb0014140

14. Zaki, M.J.: Spade: an efficient algorithm for mining frequent sequences. Mach. Learn. **42**(1–2), 31–60 (2001)

15. Ayres, J., Flannick, J., Gehrke, J., Yiu, T.: Sequential pattern mining using a bitmap representation. In: Proceedings of the Eighth ACM SIGKDD International Conference on Knowledge Discovery and Data Mining, pp. 429–435. ACM (2002)

16. Pei, J., et al.: PrefixSpan: mining sequential patterns efficiently by prefix-projected pattern growth. In: 2013 IEEE 29th International Conference on Data Engineering (ICDE), p. 0215. IEEE Computer Society (2001)

17. Laxman, S., Sastry, P.S.: A survey of temporal data mining. In: SADHANA, Academy Proceedings in Engineering Sciences, vol. 31, pp. 173–198. The Indian Academy of Sciences (2006)

18. Drozdyuk, A.: Mining partially ordered sequential rules on unbounded data. Master's thesis, University of New Brunswick, Canada (2018)

FASTT: Team Formation Using Fair Division

Jeff Bulmer$^{(\boxtimes)}$, Matthew Fritter, Yong Gao, and Bowen Hui

University of British Columbia, Vancouver, Canada
bulmer.jeffk@gmail.com

Abstract. We consider the problem of multiple team formation within a project-based university course. Given several tasks with requirements and several students with skills, we investigate the problem of assigning teams of students to tasks as fairly as possible so that each task's requirements are maximally met. Instead of using traditional team formation techniques, we adapt the fair division formulation by considering tasks as agents and students as items. Furthermore, we present a novel framework that generalizes fair division to account for order within the assignment phase. Finally, we present an algorithm to address instances of team formation within this new setting. Our empirical experiments show that this new algorithm performs better than existing fair division algorithms in terms of speed and fairness, as defined by complete balance ordered and up to one individual.

Keywords: Team formation · Fair allocation · Social choice

1 Introduction

In many real-world situations, people must be grouped into teams to accomplish various tasks. Typically, a team must meet some requirements in order to be considered viable. For example, a task may require some minimum skill set for it to be accomplished. Finding teams that meet a set of requirements for a task is the domain of the well-explored Team Formation Problem (TFP) (e.g., [13,15,17]). Much of the research into TFP centres on creating teams that meet skill requirements of a single task, or that work well together by minimizing *communication cost* – prioritizing teams with members who get along. To date, most of this research addresses the creation of one team for a single task. However, many real-world scenarios require distinct teams for multiple tasks, without sacrificing the consideration of skill, communication cost, or other requirements that make TFP research so attractive.

Consider a project-based course in which students work in teams. Suppose further that each project is a task consisting of a set of requirements and each student has a set of skills, with skills and requirements drawn from the same set. Ideally, the instructor would like to assign students to teams that work

© Springer Nature Switzerland AG 2020
C. Goutte and X. Zhu (Eds.): Canadian AI 2020, LNAI 12109, pp. 92–104, 2020.
https://doi.org/10.1007/978-3-030-47358-7_9

well together, while ensuring that the teams can meet as many of the project requirements as possible.

Intuitively, this example can be viewed as multiple instances of TFP. However, with simple examples, it can be shown that TFP does not easily generalize to multiple tasks. In fact, it is impossible to guarantee that all requirements are met once a second task is introduced.

Instead, we propose that the above problem be tackled as an instance of *fair allocation*. Fair allocation is the problem of assigning a set of items among several agents, with each item possessing a *value* for each agent. In a typical fair division problem, either all items are desirable *goods* with positive value, or all items are undesirable *chores* with negative value. Once all items are assigned, the resulting allocation is evaluated on metrics such as *envy* to determine whether items were assigned fairly.

To cast the above example of assigning course projects as a problem of fair division, we switch the roles of projects and students by viewing each student as an *item* and each project as an *agent*. In this way, we arrive at a familiar setting of assigning items to agents (i.e., assigning students to projects). While somewhat counter intuitive, the values now represent how well-matched the project requirements are by each student. Thus, each item (i.e., student) has a positive or negative value with respect to an agent (i.e., project).

Although this formulation fits squarely into the realm of fair division, we note that at each step during which a student is assigned to a project (assignment step), an item's contribution to the total value of an agent's bundle may change. As a practical example, a project whose requirements have been completely fulfilled will gain nothing from an additional student – even if she has a valuable skill set. The potential for each item's value to change means that no item can be viewed as solely beneficial or detrimental to an agent. Since most existing research into fair division focuses on solely good or solely bad items [3,4,8], current frameworks are insufficient to tackle this problem. Instead, we propose a broader framework to capture this case of *dependent* values. Our framework is a generalization of fair division of indivisible items [1], which is capable of capturing all previous examples of fair division. Additionally, our setting allows several other types of problems to be cast as fair division, though we consider only team formation here.

For this framework, we present in Sect. 3 case studies to illustrate why current methods of fair division are insufficient in this more general setting. We additionally present a formulation of the team formation problem within this framework, and present an algorithm to solve specific cases in Sect. 4. Finally, Sect. 5 shows empirical results of our algorithm alongside other fair division algorithms to highlight its usefulness. We show that our algorithm gives satisfactory results when assigning teams in a multi-task setting, but that some existing fair division algorithms also perform surprisingly well in our setting after changing only the value functions.

2   Related Work

2.1   Team Formation

Within the multiagent systems community, extensive research has been done on team formation based on decision preferences [10,11,18]. Most notably, the Group Activity Selection Problem (GASP) [11] presents a setting in which many individuals are assigned to many groups based on their preference over both those groups and their potential teammates [14].

In contrast, the operations research and optimization communities have focused on the assignment of individuals to a single team for the purposes of completing one task [15,17,19]. Historically, the TFP in this space aims to form a group for some task such that the requirements of that task are met by the skills of the group members. Recently, consideration has also been given to the underlying social structure of the team [12,15]. Defining two different measures of communication cost, Lappas et al. show that the problem of finding a group of individuals for a task such that all requirements are met, while also minimizing their defined communication costs is NP-complete. When a solution exists, the researchers present algorithms which are able to find solutions efficiently.

While TFP is normally considered in the context of a single task, we are aware of at least one paper which attempts to apply team formation techniques and consideration of skill sets to a multi-task setting. Gutierrez et al. [13] propose the Multiple Team Formation Problem (MTFP) that formulates TFP in a multi-task setting using constraint programming and discrete optimization. In MTFP, each individual is considered to have only one skill, and requirements are given as the number of individuals possessing each skill required to complete a task. Additionally, individuals are allowed to split their time between multiple tasks.

The main limitation of team formation research is that there is no one-size-fits-all solution. While solutions to GASP create multiple groups, they do not consider requirements of tasks or skills of individuals. Solutions to TFP consider requirements and skills, but form only one group for a single task. Finally, solutions to MFTP consider requirements, skills, and create multiple groups, but require the problem space to be restricted to one skill per person, and may form non-distinct groups. While each is interesting and appropriate for certain scenarios, it is easy to identify real-world examples for which all three are insufficient.

2.2   Fair Division

Given a set of agents (typically people) and a set of items, the problem of fair division aims to assign subsets of items to agents such that the resulting allocation is *fair*. Research focuses heavily on identifying notions of fairness on which allocations can be evaluated [2,7]. Algorithms can then be developed to solve the allocation problem by assigning items within those constraints. There are several established fairness notions, often drawing inspiration from metrics from other fields [1,11], most notably *envy-freeness*.

An allocation is said to be *envy-free* if no agent would get greater value by switching their assigned items with a different agent. As this is often impossible to achieve, as will be shown in later sections, recent developments have explored the less restrictive notion of *envy-freeness up to one good* [7]. This fairness notion has become well-studied in its own right [3,7,9], though primarily for *goods allocation*, in which items have a positive value for all agents. *Chores allocation*, in which items having a negative value for all agents, is less well-studied. Recent approaches have attempted to address this by considering the mixed case, in which items may have positive or negative value [1,4,6]. Most notably, Aziz et al. [1] propose a setting in which items may have positive or negative value for any agent, including the case where an item may be a good for one agent, but a chore for another. They present this setting as a generalized version of goods allocation, including a generalization of the notion of envy-freeness up to one good as *envy-freeness up to one item*.

Brederck et al. [6] present another generalized setting allowing items to have positive or negative values depending on the agent, with the additional consideration of the multiplicity of objects. Interestingly, Brederck et al.'s motivating example is similar to ours, considering research groups to which items – including graduate students – must be assigned. The setting we propose differs in a few key ways from that of Brederck et al. In particular, we consider only students, while allowing these a number of features (i.e. skills). Additionally, our setting is powerful enough to consider an ordering of items, but does not account for multiplicity.

3   A Framework for Fair Division with Dependent Values

We formulate the problem of forming multiple teams from a set of individuals as an instance of fair division. To do this, we first need to generalize the fair division setting to allow values to change between assignment steps.

Let $N = \{1, ..., n\}$ denote a set of n tasks for which teams must be assigned, and M a set of individuals. Additionally, let $X_i \subseteq M$ denote the set of individuals assigned to task i. Each subset X_i is referred to as a *team*.

For each individual m, the *value* of m to a team X_i is defined by a two-parameter function $u_i(m, X_i)$. Because the value of $u_i(m, X_i)$ depends on the current members of team X_i, we say that task i has a *dependent value* for m.

Note under this definition that an individual's value may change at each assignment step. For this reason, the order in which individuals are assigned is important. Therefore, we define the value of a team iteratively by $U_i(X_i) = \sum_{k=0}^{|X_i|-1} u_i(X_i[k], X_i[1 : k-1])$, where $X_i[1 : k-1]$ is the ordered bundle consisting of the first k items of X_i, and $X_i[k]$ is the next item to be added.

An *allocation* π is a function assigning each project such that for each pair of distinct tasks $i, j \in N$, $\pi(i) \cap \pi(j) = \emptyset$. An allocation π is *complete* if $\bigcup_{i \in N} \pi(i) = M$. We take our definition of allocation from Aziz et al. [1].

Next, we define fairness concepts within our model, specifically envy-freeness (cf. Sect. 2), requiring that agents do not envy each other. In our setting, we rename this to *balance*. Specifically, given an allocation π, we say that i is *unbalanced* w.r.t j if $U_i(\pi(i)) < U_i(\pi(j))$. An allocation π is *completely balanced* (CB) if no task is unbalanced w.r.t. other tasks.

Consider the case of two tasks, and one individual with positive value for both. Since this individual can only be assigned to one task, this simple example illustrates the impossibility of achieving CB in general.

A relaxation of EF, introduced by Budish [7] and generalized by Aziz et al. [1], is *envy-freeness up to one item*. We rename this to *complete balance up to one individual* (CB1), and generalize CB1 to account for order by introducing the concept of *complete balance ordered and up to one individual* (CBO1).

Definition 1. *An allocation π is completely balanced up to order and one individual (CBO1) if for all $i, j \in N$ either i is balanced w.r.t. j, or if there is an individual $m \in \pi(i) \cup \pi(j)$ such that $u_i(\pi(i)\backslash\{m\}) \geq u_i(\pi(j)\backslash\{m\})$, where the ordering of $\pi(i)$ and $\pi(j)$ is retained except for the exclusion of m.*

Within this new setting, we can now formulate our problem as an instance of fair division. Let N and M be projects and students, respectively, in a project-based course. Further, suppose each project has a set of requirements, and each student a set of skills, with both requirements and skills drawn from the same set. Finally, suppose the communication cost is recorded in a social network graph modeling the student relationships. This results in the following instance:

- $N = \{1, ..., n\}$ is a set of projects
- $M = \{m_1, m_2, ..., m_p\}$ is a set of students
- $G(M, E)$ is a graph representing the underlying social network. A low weight on the edge between q, t signifies that m_q and m_t are friends and would work well together
- $R = \{r_1, ..., r_n\}$ is the sets of requirements for each project
- $S = \{s_1, ..., s_p\}$ is the sets of skills for each student, drawn from the same set as project requirements

We call the above formulation Fair Allocation of Several Teams to Tasks (FASTT). FASTT is adapted from the TFP instance defined by Lappas et al. [15]. The main difference is that N consists of several projects instead of only one. Additionally, we consider student skill sets and project requirements as separate from students and projects themselves.

Because TFP considers only one task, it is easily shown that allocations are not guaranteed when multiple tasks are involved. In comparison, FASTT chooses to instead to maximize task *coverage*. The coverage c of a task i is defined as the number of requirements of the task which are met by the students assigned to it, i.e. $|(r_i) \cap \bigcup_{s \in \pi(i)}|$.

Similarly, fairness concepts do not apply to TFP since only one task is involved. Overall, in most real-world scenarios, FASTT is more applicable because it can simultaneously assign all students within a course and find teams for multiple tasks. Moreover, our iterative definition for a student's value to a project generalizes to both offline and online scenarios. Specifically when teams already exist, we are able to add to an existing allocation. This is useful in examples where new students join a course after teams have been formed and need to be added to existing teams.

4   Algorithms

Given an instance of FASTT, we first ask how to define each value function. Motivated by the problem's similarities to TFP as described by Lappas et al. [15], we model the value of a student for a project as a function of the number of additional requirements met by that student and the communication cost of adding her to the team. Here, we measure cost as the *graph diameter*, i.e. the longest shortest path between any two students in the social network graph imposed by the resulting team. That is, $d_{X_i} = \max_{m,v \in X_i}(\delta(m,v))$, where $\delta(m,v)$ is the sum of edge lengths for the shortest path between students m and v.

Our value function is defined as:

$$u_i(m_{n+1}, X_i[1:n]) = ((r_i \setminus \cup_{x \in X_i[1:n]} s_x) \cap s_{m_{n+1}})$$
$$+ (d_{X_i[1:n]} - d_{X_i[n] \cup m_{n+1}})$$

where $U_i(X_i[1:n])$ is the value of the team, given by

$$U_i(X_i[1:n]) = |r_i \cap (s_1 \cup s_2 \cup \cdots s_n)| - d_{m_1 \cup m_2 \ldots \cup m_n}$$

where each student's value is given by the number of skills required by that project, exhibited by that student, and not currently present in the team assigned to that project, plus the change in diameter after adding the student. This value function has the useful property that a student's value may change at every assignment step, but the final value is independent of order. Note that, though our value functions are order-agnostic, it is possible to define order-aware value functions. However, when order-aware value functions are used, the number of possible allocations increases dramatically, heavily increasing the runtime for algorithms that must compare allocations. For this reason, we elected to define value functions in such a way that value to change at every step, but for which the final result does not depend on order.

To solve an instance of FASTT, we look to the mixed case of fair division explored by Aziz et al. [1]. Aziz et al. propose an algorithm called Double Round Robin (DRR) to solve mixed instances of fair division. As our values are not additive, this algorithm alone is insufficient in our setting. We note that there are several settings allowing for non-additive utility values [6,7,16], though these approaches tend toward proofs of more general results. Instead, we carefully adjust the DRR algorithm combined with elements of the Adjusted Winner (AW) rule [5] to produce an algorithm we call Multiple Round Robin with Adjusted Winners (MRR). We present this as Algorithm 1 and evaluate its performance in the next section.

MRR works by running several truncated selection rounds until all students have been assigned, each beginning with the project with the current lowest-valued team. At the beginning of each round, dummy students are created such that the total number of students and dummy students is a multiple of the

Input: An instance $I = (N, M, G(M, E), S, R, U)$
Output: An allocation π
while *There are students left to assign* **do**
    Order all projects in increasing order by the value of their teams
    if *There are any students with positive value for any project* **then**
        Create dummy students such that the total number of students plus
        dummy students is a multiple of the number of projects
    end
    for *i in projects* **do**
        if *There are no students with positive value, but there are still dummy*
        items **then**
            Select a dummy student and assign them to the project
        else
            Select the student with the highest value for this project and assign
            them to the project
            Reorder all projects as above, and set i to the project with the
            lowest team value
        end
    end
end
AdjustWinners(π)

Algorithm 1: Multiple Round Robin with Adjusted Winners (MRR)

Input: An allocation π
Output: An allocation π'
for *Each pair of projects (i, j)* **do**
    if *i is not CBO1 w.r.t. j* **then**
        for *Each student m assigned to j* **do**
            if *Moving m from j to i increases i's total value from X_i* **then**
                Moved m from j to i in π'
            end
        end
    end
end

Algorithm 2: AdjustWinners function

number of projects. During each round, students are assigned to a project in increasing order of the total value of its current team. As soon as a student is assigned, the values and order are recalculated and a new round begins. If at any point there are no students with strictly positive value for the current project, a dummy student may be assigned to that project, as long as one exists. The number of dummy students created in each round is low enough to guarantee that at least one student is assigned. However, because the number of dummy students is so low, these should be generated in each round to ensure that dummy students are always available at the start of the selection round.

Once all students have been assigned, the resulting allocation is checked for pairwise CBO1. If any pair is not CBO1, the AdjustWinners method moves items between them until the resulting allocation is pairwise CBO1. By running this correction function for each project pair i, j, we fix every CBO1 error, usually resulting in a complete CBO1 allocation. Note that proceeding in this way may, in rare cases, result in the introduction of additional CBO1 errors between pairs that have already been checked, which are therefore not fixed. For this reason, Algorithm 2 does not provably result in a CBO1 allocation in every instance. Regardless, we show in the next section that CBO1 allocations can be obtained in the majority of practical scenarios.

MRR runs in $O((|M| \cdot n) + (n \cdot (n-1) \cdot |M|^2))$ time. The while loop must iterate through all students, and then through all projects, achieving a complexity of $O(|M| \cdot n)$. The AdjustWinners method then checks all pairs of projects, and swaps students between them, while respecting the order of the projects, resulting in a complexity of $O(n \cdot (n-1) \cdot |M|^2)$. The verification step that an allocation is CBO1 happens pairwise after each swap, and has a worst case running time of $O(|M|^2)$. This is because verification of CBO1 must iteratively build up the total values of each team. The final complexity to that stated previously.

5 Empirical Evaluation

In this section, we evaluate our proposed algorithm for instances of FASTT using data extracted from the DBLP bibliography server. We used a snapshot of the DBLP database taken on July 1, 2019 to create a benchmark dataset. The dataset was then filtered in the same way as Lappas et al. [15]. We refer to the resulting dataset as the DBLP dataset.

Our students are a set of authors with at least three papers in the DBLP dataset. Each author has a list of skills defined by the keywords listed in at least two of their papers. Finally, a social network containing all chosen authors was created. Weights were assigned based on the frequency of co-authorships. Specifically, if two authors have never co-authored together, their edge receives a high weight, 5. On the other hand, if two authors co-authored with each other more than with anyone else, then their edge receives a low weight, 1. All other co-authors receive a weight of 3.

Test cases were generated by taking a random sample of authors as the set of students and their combined skill sets as the set of possible requirements for a number of projects. A project is generated by randomly choosing 4 requirements. Each test case consisted of 3 of these projects. Each trial in our simulation experiment consists of a test case with 3 projects (each having 4 requirements) and a variable number of students ranging between 3 and 25. Each trial was run 500 times (with different projects, project requirements, and students).

5.1 Results

We report on the performance of the fair division algorithms based on average running time and average team coverage over 500 trials, and the total number of times a CBO1 allocation was found in those trials. As a comparison, we use the

Double Round Robin (DRR) and *Generalized Envy Graph* (GEG) algorithms from [1] as a baseline algorithm. Each algorithm was run on the same test cases.

Our experiment was performed by running algorithms on a series of t2.micro instances with 1 GB of memory on Amazon's AWS platform.

Figure 1 shows the resulting runtimes on a logarithmic scale. The MRR and DRR algorithms are bounded by a quadratic runtime. We also note that the highest average runtime per trial is barely over 100 s and can therefore always be feasibly run when the number of students is low. Nonetheless, as the lowest average running times are all well below 1 ms per trial, this growth illustrates the algorithmic complexity quite well.

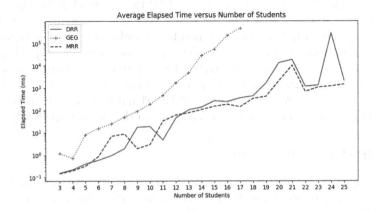

Fig. 1. Average runtimes of each algorithm.

Figure 2 reports the standard deviation for the runtimes of each algorithm. We note that DRR behaves erratically. This is likely due to DRR's selection round. Because of the dependent nature of values in our setting, it is not enough to determine the order of selection once at the beginning of each selection round, as values may change at any point. MRR's truncated selection round addresses this, and is reflected in the algorithm's standard deviation.

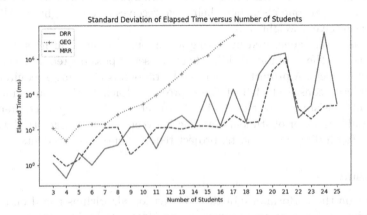

Fig. 2. Standard deviation of runtimes for each algorithm.

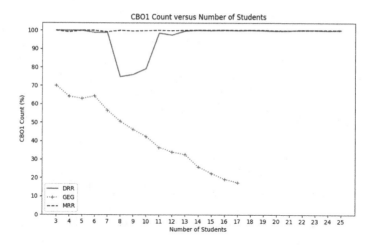

Fig. 3. Total number of times a CBO1 allocation was found by each algorithm.

Figures 3 and 4 show the number of times CBO1 allocations were found and the average project coverage achieved respectively. While MRR methods find CBO1 allocations in nearly all cases, several data points fall below 100%. This shows that MRR does not guarantee CBO1 even on a small number of projects. DRR methods perform surprisingly well in terms of average project coverage, approaching the average project coverage achieved by MRR methods with an increasing number of students.

Finally, we note that GEG could only be run up to 17 students. As the number of students increases, the runtime of GEG grows exponentially. Additionally, the cycle detection performed by GEG [1] results in GEG getting stuck on certain test cases when a cycle cannot be resolved. Finally, note that the percentage of CBO1 allocations found by the GEG algorithm decreases as the number of students increases, while the project coverage remains consistently lower than DRR or MRR. We conclude that GEG is unsuitable in our setting.

Figure 5 shows the results of running the DRR and MRR methods while varying the number of projects and students. We fixed the requirements per project at 4, and varied the number of students between 23 and 24. We note that the average project coverage decreases as the number of projects approaches the number of students. This is unsurprising, as more projects with the same number of students results in students being spread more thinly. We also note that MRR nearly always achieves as high or higher coverage than DRR.

Furthermore, it is worth noting that, in a realistic setting such as a class of 24 students with 6 projects, we can still achieve an average project coverage of over 50%. Therefore, in this setting, projects could be expected to be completed to an acceptable degree while still allowing for some independent learning (for the percentage that is not covered).

It should be noted at this point that our method of student selection and project generation resulted in a vast majority of cases in which student skill sets

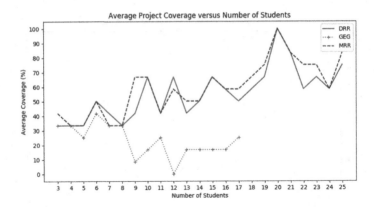

Fig. 4. Average project coverage of each algorithm.

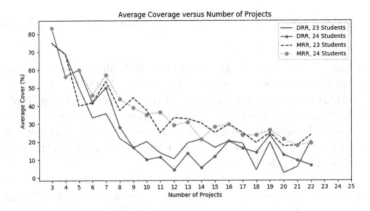

Fig. 5. Average project coverage of MRR as the number of projects increases.

were distinct, but many projects had overlapping requirements. For this reason, total project coverage is not possible in the majority of test cases.

6   Conclusions and Future Work

We presented a multi-project formulation of team formation utilizing fair division techniques. This formulation generalizes the fair division problem, allowing item values to change between allocation steps. Within this setting, we additionally generalized envy-freeness up to one item to CBO1 to account for order.

Our work paves the way for a more detailed examination of both fair division and multi-project team formation. In particular, it remains to be seen whether there are conditions which guarantee that an instance of FASTT has a CBO1 allocation. We also note that several other fairness concepts exist which we have not considered here, most notably *proportionality* and *envy-freeness up to the least valued item* (EFX) [9]. As well, recall that the work of Bredereck et al. [6]

introduces the idea of multiplicity. Future work could combine these settings, by introducing multiplicity to FASTT, or order to the setting presented by Bredereck et al. Additionally, Bredereck et al. provide proofs of the fixed-parameter tractability of several special cases of fair division, which may be applicable in our setting as well.

References

1. Aziz, H., Caragiannis, I., Igarashi, A., Walsh, T.: Fair allocation of indivisible goods and chores. In: IJCAI, pp. 53–59 (2019)
2. Aziz, H., Rauchecker, G., Schryen, G., Walsh, T.: Algorithms for max-min share fair allocation of indivisible chores. In: AAAI, pp. 335–341 (2017)
3. Barman, S., Krishnamurthy, S.K., Vaish, R.: Finding fair and efficient allocations. In: Proceedings of the 2018 ACM Conference on Economics and Computation, EC 2018, pp. 557–574. ACM, New York (2018)
4. Bogomolnaia, A., Moulin, H., Sandomirskiy, F., Yanovskaia, E.: Dividing bads under additive utilities. Soc. Choice Welfare **52**(3), 395–417 (2018). https://doi.org/10.1007/s00355-018-1157-x
5. Brams, S.J., Taylor, A.D.: Fair Division: From Cake-Cutting to Dispute Resolution. Cambridge University Press, Cambridge (1996)
6. Bredereck, R., Kaczmarczyk, A., Knop, D., Niedermeier, R.: High-multiplicity fair allocation: lenstra empowered by n-fold integer programming. In: Proceedings of the 2019 ACM Conference on Economics and Computation, EC 2019, pp. 505–523. ACM, New York (2019)
7. Budish, E.: The combinatorial assignment problem: approximate competitive equilibrium from equal incomes. J. Polit. Econ. **119**(6), 1061–1103 (2011)
8. Caragiannis, I., Kaklamanis, C., Kanellopoulos, P., Kyropoulou, M.: The efficiency of fair division. Theory Comput. Syst. **50**(4), 589–610 (2012)
9. Caragiannis, I., Kurokawa, D., Moulin, H., Procaccia, A.D., Shah, N., Wang, J.: The unreasonable fairness of maximum nash welfare. In: Proc. of the ACM Conference on Economics and Computation (EC). pp. 305–322. New York, USA (2016)
10. Darmann, A.: Group activity selection from ordinal preferences. In: Walsh, T. (ed.) ADT 2015. LNCS (LNAI), vol. 9346, pp. 35–51. Springer, Cham (2015). https://doi.org/10.1007/978-3-319-23114-3_3
11. Darmann, A., Elkind, E., Kurz, S., Lang, J., Schauer, J., Woeginger, G.: Group activity selection problem. In: Goldberg, P.W. (ed.) WINE 2012. LNCS, vol. 7695, pp. 156–169. Springer, Heidelberg (2012). https://doi.org/10.1007/978-3-642-35311-6_12
12. E Gaston, M., Simmons, J., desJardins, M.: Adapting network structure for efficient team formation, January 2004
13. Gutiérrez, J.H., Astudillo, C.A., Ballesteros-Pérez, P., Mora-Melià, D., Candia-Véjar, A.: The multiple team formation problem using sociometry. Comput. Oper. Res. **75**, 150–162 (2016)
14. Igarashi, A., Peters, D., Elkind, E.: Group activity selection on social networks. CoRR
15. Lappas, T., Liu, K., Terzi, E.: Finding a team of experts in social networks. In: Proceedings of the 15th ACM SIGKDD International Conference on Knowledge Discovery and Data Mining, pp. 467–476. ACM, New York (2009)

16. Lipton, R.J., Markakis, E., Mossel, E., Saberi, A.: On approximately fair allocations of indivisible goods. In: Proceedings of the 5th ACM Conference on Electronic Commerce, EC 2004, pp. 125–131. ACM, New York (2004)
17. Majumder, A., Datta, S., Naidu, K.: Capacitated team formation problem on social networks. In: Proceeding of the 18th ACM SIGKDD International Conference on Knowledge Discovery and Data Mining, pp. 1005–1013. ACM, New York (2012)
18. Marcolino, L.S., Jiang, A.X., Tambe, M.: Multi-agent team formation: diversity beats strength? In: IJCAI, pp. 279–285 (2013)
19. Zzkarian, A., Kusiak, A.: Forming teams: an analytical approach. IIE Trans. **31**(1), 85–97 (1999)

Empirical Confidence Models
for Supervised Machine Learning

Margarita P. Castro[1](✉) ⓘ, Meinolf Sellmann[2], Zhaoyuan Yang[2],
and Nurali Virani[2]

[1] University of Toronto, Toronto, ON, Canada
mpcastro@mie.utoronto.ca
[2] General Electric Gobal Research Center, Niskayuna, NY, USA
{meinolf,zhaoyuan.yang,nurali.virani}@ge.com

Abstract. We present a new methodology for assessing when data-based predictive models can be trusted. Particularly, we propose to learn a model from experimentation that determines, for a given labeled data set and a learning technique, when the model generated by the respective technique on the given data can be trusted to perform within specified accuracy limits. That is to say, we apply machine learning to machine learning: We repeatedly use a technique to generate models, referred as primary model, for a supervised regression problem. Based on the resulting model performance on a hold-out validation set, we then learn when the trained primary model can be expected to perform well and when there is a concern regarding the trustworthiness of that model.

Keywords: Supervised learning · Uncertainty quantification

1 Introduction

Artificial intelligence is increasingly being used to provide decision support in high stakes contexts, where errors can be very costly [1]. Examples for this trend range from operating machinery in transportation to diagnostics in healthcare. With this trend emerges a need for trusted and robust models. One way to gain a certain level of trust and acceptance is to convince with very good performance (e.g. as measured by accuracy, precision, and recall for predictive models, or the optimality gap or quality guarantees for prescriptive optimization models).

However, there always will be limits to practical performance, be it due to an imperfect assessment of the world or computational limitations. Despite the inherent performance limitations, we can potentially achieve a higher level of trust by embedding the model in a process that handles exceptions. That is, rather than requiring the model to make a call under all circumstances, we allow the model to declare its incompetence in a concrete situation.

This paper investigates how machine learning (ML) can be applied to ML to critically assess model competence. In particular, we develop techniques that learn when models generated by certain learning techniques on a particular data

© Springer Nature Switzerland AG 2020
C. Goutte and X. Zhu (Eds.): Canadian AI 2020, LNAI 12109, pp. 105–117, 2020.
https://doi.org/10.1007/978-3-030-47358-7_10

set can be expected to perform well, and when not. The core idea is to let the machine learn, from repeated experimentation, for which test inputs models generated by an ML technique will perform well, and for which vectors in the feature space it is prudent to exercise caution when considering the predictions of the generated primary model.

2   Related Works

Despite the high accuracy of modern ML models and their many applications, human end-users often distrust this models [7]. To increase trust in ML forecasts and recommendations, adversarial testing [17], black-box testing [9], and human-in-the-loop [6] approaches help identify and prevent model failures. Several works on trust focuses on interpretability, explainability [15,22], and failure perception [19].

We study an aspect of trust that is orthogonal to interpretability, namely the *self-assessment of the reliability of an ML model* for a concrete testing instance at run-time. A key results in the area was recently introduced by Jian et al. [11]. The authors theoretically and empirically show that classification confidence can be assessed effectively based on a *reliability measure* that consists in computing the ratio of the distance of the test feature vector to the nearest training example of the predicted class and the distance to the nearest training example that is outside the predicted class. The lower this ratio, the more we trust the primary classifier. The keys of this reliability measure are to carefully analyze the training set and to set the *threshold* for the distance ratio following the training data. In other words, the reliability metric is general but the thresholds used to decide when to trust the primary classifier is derived from the data at hand. Unfortunately, the method by Jian et al. [11] does not generalize to regression, which is the objective of this paper.

As in [11], the measures used to assess confidence in the primary model are not enough. At best, they give us a *relative* sense of trust. To make these measures prescriptive, we need to find thresholds or a mapping from these quantities to the actual performance of the primary model. The research on this transformation step for primary regression models is scarce. Bosnić et al. [5] propose a secondary regression model that predicts the *signed* error of a (primary) regression model. The authors perform a sensitivity analysis [3] over their primary model to generate their secondary model features. The output of the secondary model is then used to correct the predictions of the primary model. Hence, the approach focuses on boosting the primary model rather than assessing reliability.

Bosnić et al. [4] also propose an internal cross-validation and a meta-classifier that predicts the reliability for a given problem domain and regression model. Their decision-tree meta-learner uses features that describe the regression model and the training data to predict which of their nine reliability metrics works best.

Matsumoto et al. [14] use a meta-model to estimate when the primary regression model will err significantly. Their approach estimates a threshold value to create acceptable and critical error regions. Their secondary classifier uses the original training data and the predictions of the primary model as features to predict whether the prediction is acceptable or has a critical error.

Our technique belongs to the same class as Matsumoto et al. [14] work: We also propose to learn empirically when a primary regression model can be trusted. However, we propose a paradigm shift: Instead of assessing the competence of the primary model directly, we focus on the *learning technique* that is used to generate the primary model and to learn how models generated by this technique perform under different scenarios. The thresholds derived from our experimentation will still be specific to the training data, but the repeated experimentation with the learning technology gives us a much richer supervised training set for the secondary confidence assessor.

We use reliability measures derived from concepts introduced in the literature. We consider confidence and prediction intervals [10] which can be used to measure the reliability of regression tasks [18]. We use metrics based on *conformal predictions* which are used to create "regions of trust" for classification and regression tasks [20]. We also leverage *non-conformal measures* [16] to assess the performance of the learned model as a function of proximity to the training data.

3   Empirical Confidence Models

The goal is to build a *competence assessor pipeline* that accompanies a supervised regression model. We consider the following scenario: An ML model is built based on a set of labeled training examples. Once deployed, this primary model regresses the feature inputs it receives to predict a value for the dependent variable. The system does not receive the actual labels after making the predictions, i.e., we are not able to learn about any model drift [13] after deployment.

The top portion of Fig. 1a illustrates the process. Here and in the following, colored boxes mark data vectors and labels, gray boxes are algorithms programmed by human developers, and gray plaques with rounded corners denote machine-learned models. The *primary model* is generated by an ML technique (e.g., random forest) based on a given training set of labeled examples (see bottom Fig. 1a). At run-time, the primary model and competence assessor pipeline receive the same input feature vector, where the competence assessor provides a secondary *classification* on the primary model competence. The pipeline itself consists of (1) a meta-feature builder which relates the current input to the primary model and the training set, and (2) the competence assessor that attempts to classify the competence of the primary model for the respective input as either *trusted, cautioned,* or *not trusted.*

This work aims to overcome the assessment of supervised regression models purely by summary statistics. Summarized statics give an overall view of the quality of a model in the data set but can be misleading when it comes to asses the accuracy of a specific prediction. For example, while a primary model has a 95% accuracy, there is no guarantee that this performance will be achieved for each new test instance. In fact, it is quite likely to be regions of the input space where the model has similar or higher accuracy (i.e., trusted regions), and regions where the accuracy is low (i.e., cautioned regions).

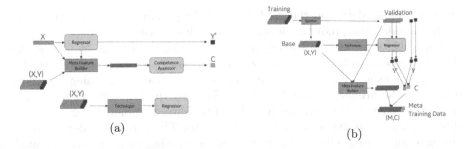

Fig. 1. Deployment of primary model and assessor pipeline (right) and training data generation for meta-model (left).

Therefore, we aim to learn when an ML model can or should not be trusted, conditioned by the concrete feature input given. Note that this is different from predicting local model accuracy, as considered by the sensitivity analysis in [5]. The label of a concrete example may be off due to the uncontrollable error introduced by label noise, but if the example falls into a region where our model generally performs well, it is still correct to trust the model. On average, we should see good accuracy of the primary model on inputs for which we trust our model. On the other hand, the given input may fall into a region where the model has no competence at all (e.g., a decision tree for a test instance where no training example is anywhere close). Even if the model has a low error, it is correct to distrust the model. Of course, on average we expect the accuracy of the model to be low on inputs where we consider the model to have low competence.

We now present the steps to build our empirical competence assessor pipeline. We first introduce the meta-features that relate the input with the primary model training data. Then, we show how to train the competence classifier. We explain how we generate training data for training the competence assessor, in particular, the base/validation splits and the techniques to train the assessor.

3.1 Reliability Measures as Meta-Features

Naturally, the assessor can consider all original features that are being fed to the primary model [14]. If the assessor were to learn only the primary model's behavior, these original features might even be enough to learn when the primary model is likely to perform within the desired quality bounds and when not.

Our approach goes one step further: We train the assessor for a *technique* rather than the concrete resulting model. Thus, we aim to learn how a model trained by a technique (e.g., random forest) performs based on the *relation of the given feature input and the data used to train the primary model*.

First, we need a distance measure between the run-time features and the training set. Given a distance function $d : F \times F \to \mathbb{R}_{\geq 0}$, where F denotes the primal feature space, we compute the k nearest neighbors to the input feature vector x (we use $k = 5$ in our experiments but this parameter could be tuned).

For any x, $N(x)$ denotes the set of the k nearest labeled training examples $(x_i, y_i)_i$. Then, the first meta-feature, M_1, is the average distance to points in $N(x)$, i.e., M_1 measures the of training experience in the vicinity of x.

$$M_1(x) := \sum_{(x',y') \in N(x)} \frac{d(x', x)}{k}. \tag{1}$$

Consider the regression forecasts by the primary model as $f : F \to \mathbb{R}$. Then, M_2 considers the average prediction difference between x and its k nearest neighbors. Analogously, M_3 is the deviation from the primary model's prediction and the weighted nearest neighbor average in the neighborhood of x. Thus, M_2 and M_3 represent to what extent the primary model's prediction coincides with the labels and forecasts in the immediate vicinity of the current input.

$$M_2(x) := \sum_{(x',y') \in N(x)} \frac{|f(x) - f(x')|}{k}, \tag{2}$$

$$M_3(x) := \left| f(x) - \sum_{(x',y') \in N(x)} y' \frac{s(x)}{d(x', x)} \right|, \quad s(x) = \frac{1}{\sum_{(x'',y'') \in N(x)} \frac{1}{d(x'', x)}}. \tag{3}$$

Next, we consider the average and variance of the distance-weighted training error on the neighborhood of the given input x. Thus, M_4 and M_5 assess the primary model's performance in the neighborhood of the current input.

$$M_4(x) := \sum_{(x',y') \in N(x)} |f(x') - y'| \frac{s(x)}{d(x', x)}, \quad M_5(x) := \sum_{(x',y') \in N(x)} \frac{(|f(x') - y'| - M_4(x))^2}{k - 1}. \tag{4}$$

Lastly, we consider the target value variability in the neighborhood of x. Thus, M_6 gives a measure of how much true label variance we expect in the immediate vicinity of the given input vector.

$$M_6(x) := \sum_{(x',y') \in N(x)} \frac{(y' - \bar{y})^2}{k - 1}, \quad \text{where} \quad \bar{y} = \sum_{(x',y') \in N(x)} \frac{y'}{k}. \tag{5}$$

The notion of a neighborhood is central to our features. We can employ the Euclidean distance in the feature space but this has all the drawbacks of k-nearest neighbor methods: The need to adjust for the respective influence of each feature, the distance-distorting influence of non-informative features, and the implicit assumption that feature importance is the same.

Consequently, we propose using technique-specific measures of distance, i.e., we assess similarity based on what the primary model considers similar. For example, random forest considers two feature vectors similar if they often end up in the same leaf node [2]. For a linear regression model with parameters $w \in \mathbb{R}^{m+1}$ and an m-feature regression problem, we can consider $d(x', x) \leftarrow \sum_{i=1}^{m} |w_i| |x_i' - x_i|$ to define the dissimilarity of feature vectors x and x'. In both examples, feature vectors that result in similar regression values are not necessarily considered similar by the model. The feature vectors would be similar only if the primary model's reasoning underlying the regression is similar.

3.2   Meta Training Data

Given the meta-features relating the run-time feature vectors to the primary model training data, we aim to learn when a model trained by a particular learning technique can be trusted to perform well, and when not. To this end, we split the training data several times into base and validation sets. The goal is to provide the meta-model with information on possible regions where the primary model might have a poor performance.

Then, for each split and validation instance, we assess whether the model trained performs with sufficient quality or not (see Fig. 1b). Using our meta-features, we can create a new training data set for the meta-model, where each validation instance in a split of the training data creates one new example instance for the meta-model: Its features are the original features augmented by the six features introduced above to characterize how the validation instance relates to the base data that was used to build the primary model.

The classification labels for the meta-model are determined by the true error of the learned model on the base data set. To this end, we first order the absolute values of the residuals for each base instance, and then determine the two maximum errors such that 80% or 95% of all base instances have a prediction error below that respective error rate. Then, we use three different classes to label our validation instances: *not trusted* if the forecast is above the 95% error, *cautioned* if the prediction is off by more than the 80% error but less than the 95% error and *trusted* otherwise. Note that the values 80% and 95% can be adjusted naturally to increase or decrease the desired levels of trust in the primary model. For the given used case, if regression errors are costly, a lower trust percentage can be used. If frequent escalation resulting from an incompetent regressor puts an undue burden on operations, a higher trust percentage may be chosen.

The remaining question is how to split the training data. The first method is cross-validation, where each training example is randomly assigned to one of h buckets. We consider h splits (for $h = 3$, $h = 5$, and $h = 10$) where, in each split, one bucket is used as a validation set and the others as a base set. Thus, we hope to give the meta-model enough training data to learn, for various levels of training density surrounding the current input, when the learned model can be expected to perform well or not.

We also wish the meta-model to learn to assess when the i.i.d. assumptions underlying the learning of the primary model are broken. To this end, we construct splits that create "interpolation" and "extrapolation" scenarios: We project the training data onto the first (and later the second) principal component (PC) dimension. Then, we partition the respective line in h intervals (the left-most and right-most being half-open), such that each interval contains the same number (up to rounding errors) of training instances. We then consider each such interval as validation and the remaining data as base sets. We do this for the two most significant PC dimensions and again for bucket size $h \in \{3, 5, 10\}$. Thus, including the three random cross-validations, in total, we generate nine meta-model examples for each original training example.

Table 1. Characteristics of the UCI benchmarks used in the experiments.

Data sets	Airfoil	B. Housing	Combine cycle	Concrete	Diabetes	Yacht
# Observations	1503	506	9568	103	442	308
# Attributes	5	13	4	10	10	6

Lastly, we experiment with two supervised classification techniques to train a meta-model: a support vector machine (SVM) with RBF kernel (one vs. rest decision function for multi-classification) [21], and a random forest classifier [12] with 100 trees and no depth limit. We choose these out-of-the-box models as proof of concept, i.e., we are interested in testing our meta-learning procedure (e.g., meta-features and data-splitting) rather than the performance of the meta-learner. Nonetheless, sophisticated models tune for the specific data sets can lead to better performance of the empirical confidence procedure.

4   Numerical Results

We train primary regression models for six well-known UCI benchmarks [8]. Table 1 shows the characteristics of the selected benchmarks, including the number of observations and attributes. We use three techniques to train a primary regression model: A random forest trainer (RF), a linear regressor (Linear), and a support vector regressor (SVR). These perform differently in each benchmark data set but this is not our primary concern. Rather, we want to find out if our approach creates meta-models that effectively assess when a primary modeled performs similarly to what the training performance led us to expect, and when the primary model is best not to be trusted blindly.

We conduct four different tenfold cross-validations to evaluate the performance of our assessor pipelines. First, we randomly partition the data into equal size buckets and make each bucket the test set, i.e., standard cross-validation (*Random CV*). Secondly, we 20-cluster the data, and then create 10 buckets by joining the smallest and largest cluster, the second smallest and second largest cluster, and so on. Then, we consider each resulting bucket as the test set (called *Cluster*). That is, the test sets are generated by extracting entire clusters of labeled instances from the benchmark and making these the test instances. Note that in this setup the test sets are not guaranteed to be of equal size. For the last sets of evaluations, we project all examples on their first (and third) PC dimension and partition the data in equal size intervals (referred to as *PCA-1* and *PCA-3*, respectively). Again, we consider in turn each interval as the test set. Thus, for each benchmark, we obtain 40 train/test splits. For each of these, we train three primary models (RF, Linear, and SVR), and two respective assessor pipelines (a random forest and an SVM meta-model). In total, we consider 144 10-fold cross-validations to assess the efficacy of our approach.

Figure 2 depicts the competence assessments on a 1-dimensional regression problem. The figure shows only the test set, the training set consisted only of

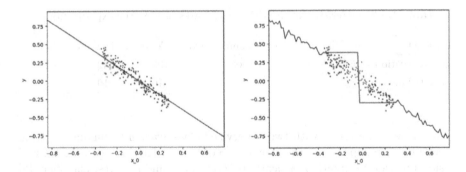

Fig. 2. Scatter plots illustrating competence assessments on 1-D regression with linear regression (left) and random forest (right). (Color figure online)

examples on the top left and lower right of the instances shown in each scatter plot. Our pipeline assesses the competence of a linear regressor and a random forest regressor on this interpolation task. We use green dots for instances for which the primary model is trusted, yellow for caution, and red for not trusted. We can see that the meta-model trusts the linear regression model to perform well on the test set, while the assessor built for the random forest model does not once the test inputs are far from the training data.

The above shows that empirical competence assessment works conceptually, but of course, we are interested in real-world multi-dimensional regression problems. Table 2 shows our results on the UCI benchmarks [8] using as meta-model an off-the-shelf SVM with RBF kernel (no hyper-parameter tuning). For each of the four cross-validation scenarios (i.e., Random CV, PCA-1, PCA-3, and Cluster), we report the average percentage of test examples that fall into each category: trusted (T), cautioned (C), or not trusted (NT), and the geometric mean of the mean squared errors (MSE) over the ten test sets. We can see that the meta-models are effective in identifying when the primary models perform poorly. With few exceptions, when the meta-models label a significant number of test instances as cautioned or not trusted, the mean squared errors of the primary regressors increase. This is an indication that the meta-models we trained can effectively assess when a primary regressor is trustworthy and when not.

Very encouragingly, this works for random cross-validations and the non-i.i.d. splits based on removing entire clusters and PCA-projected intervals. Consider the Airfoil benchmark as an example. On the Random CV, the Linear primary model achieves an average (geometric mean) MSE of 0.46. The meta-model trusts Linear in 61% of the cases with an average MSE of 0.27. For 30% of the test instances, the meta-model cautions that the primary model may be degrading and indeed the average MSE climbs to 0.63. On the remaining 9%, the meta-model no longer trusts the primary model and the MSE is 0.99, more than double the primary's average on all test instances. We observed similar results for the Cluster splits: The average MSE over all ten test sets is 0.46 for the Linear primary model. The meta-model trusts the model in 70% of all test cases with

Table 2. Competence assessor effectiveness on UCI benchmarks.

Benchmark	Class		Random CV				PCA-1				PCA-3				Cluster		
		T	C	NT	all	T	C	NT	all	T	C	NT	all	T	C	NT	all
Airfoil Linear	%	61.11	29.62	9.27	100	64.84	26.60	8.56	100	68.63	23.84	7.53	100	69.77	21.64	8.59	100
	MSE	0.27	0.63	0.99	0.46	0.21	0.95	1.43	0.51	0.30	0.74	1.39	0.48	0.18	0.84	1.90	0.46
SVR	%	42.67	44.46	12.87	100	52.95	33.64	13.41	100	56.83	36.24	6.93	100	57.80	33.82	8.38	100
	MSE	0.18	0.36	0.36	0.31	0.13	0.32	0.45	0.26	0.14	0.31	0.58	0.23	0.08	0.29	0.28	0.19
RF	%	28.11	26.24	45.64	100	45.44	29.75	24.80	100	29.34	25.67	44.98	100	56.12	26.92	16.96	100
	MSE	0.14	0.14	0.33	0.21	0.06	0.11	0.23	0.12	0.07	0.10	0.19	0.14	0.10	0.09	0.20	0.12
Boston Linear	%	51.09	38.61	10.30	100	53.84	37.22	8.94	100	64.69	27.63	7.68	100	57.95	32.11	9.94	100
	MSE	0.14	0.30	0.72	0.26	0.13	0.40	0.70	0.28	0.10	0.31	1.53	0.28	0.12	0.35	0.84	0.27
SVR	%	46.93	41.01	12.06	100	51.16	36.32	12.51	100	51.30	36.82	11.88	100	48.37	39.84	11.79	100
	MSE	0.10	0.12	0.39	0.18	0.12	0.15	0.36	0.18	0.05	0.13	0.81	0.19	0.08	0.17	0.37	0.16
RF	%	41.46	27.49	31.05	100	43.66	29.97	26.37	100	41.52	28.61	29.87	100	39.75	30.87	29.37	100
	MSE	0.07	0.13	0.23	0.17	0.07	0.12	0.33	0.17	0.05	0.08	0.24	0.13	0.08	0.12	0.16	0.15
CCycle Linear	%	62.77	33.22	4.00	100	57.09	33.55	9.36	100	57.55	37.39	5.06	100	57.00	36.99	6.01	100
	MSE	0.04	0.10	0.22	0.07	0.05	0.10	0.15	0.07	0.06	0.09	0.13	0.07	0.06	0.09	0.14	0.07
SVR	%	70.69	27.47	1.84	100	64.06	32.29	3.65	100	60.95	35.00	4.05	100	56.26	37.17	6.57	100
	MSE	0.04	0.08	0.21	0.05	0.05	0.08	0.28	0.06	0.05	0.08	0.11	0.06	0.05	0.07	0.11	0.06
RF	%	55.63	30.78	13.59	100	31.36	38.32	30.31	100	29.20	42.38	28.42	100	26.44	37.97	35.60	100
	MSE	0.02	0.04	0.07	0.04	0.03	0.05	0.08	0.05	0.04	0.05	0.08	0.06	0.03	0.05	0.07	0.05
Concrete Linear	%	64.08	26.21	9.71	100	60.78	25.53	13.69	100	64.27	24.37	11.36	100	57.13	26.80	16.07	100
	MSE	0.24	0.74	1.06	0.42	0.22	0.55	1.00	0.40	0.22	0.65	0.85	0.42	0.19	0.64	1.01	0.44
SVR	%	63.01	23.59	13.40	100	59.03	25.05	15.92	100	53.59	30.10	16.31	100	39.70	42.16	18.14	100
	MSE	0.16	0.35	0.56	0.25	0.13	0.29	0.34	0.21	0.11	0.25	0.41	0.21	0.14	0.23	0.35	0.23
RF	%	32.33	30.87	36.80	100	27.96	31.65	40.39	100	35.83	30.68	33.50	100	14.61	26.00	59.40	100
	MSE	0.07	0.08	0.29	0.15	0.07	0.10	0.30	0.20	0.07	0.08	0.24	0.15	0.05	0.12	0.29	0.21
Diabetes Linear	%	67.63	22.88	9.49	100	61.66	24.75	13.59	100	62.42	26.92	10.66	100	63.77	20.00	16.22	100
	MSE	0.46	0.53	0.36	0.49	0.49	0.52	0.55	0.51	0.49	0.52	0.47	0.50	0.39	0.59	0.46	0.50
SVR	%	59.49	27.17	13.33	100	58.02	27.45	14.54	100	56.72	26.48	16.80	100	51.73	31.20	17.07	100
	MSE	0.46	0.61	0.28	0.52	0.59	0.75	0.52	0.63	0.51	0.62	0.41	0.56	0.44	0.57	0.37	0.53
RF	%	32.34	31.01	36.65	100	28.90	33.85	37.25	100	36.41	27.82	35.77	100	29.50	29.79	40.71	100
	MSE	0.39	0.49	0.73	0.57	0.42	0.54	0.75	0.59	0.41	0.55	0.65	0.57	0.37	0.51	0.77	0.57
Yacht Linear	%	67.86	25.97	6.17	100	69.85	24.63	5.52	100	68.16	26.32	5.52	100	71.90	22.26	5.83	100
	MSE	0.13	0.48	2.04	0.35	0.14	0.52	2.31	0.34	0.13	0.51	2.09	0.35	0.14	0.52	2.19	0.34
SVR	%	51.59	34.75	13.66	100	40.76	45.29	13.95	100	42.70	43.00	14.30	100	41.90	44.40	13.69	100
	MSE	0.02	0.09	1.17	0.22	0.02	0.11	1.17	0.22	0.02	0.12	0.84	0.22	0.03	0.09	1.21	0.22
RF	%	51.95	31.84	16.22	100	48.76	33.12	18.12	100	49.32	33.11	17.57	100	49.52	34.76	15.71	100
	MSE ($\times$ 1e-4)	1.00	15.79	121.72	29.46	0.79	14.41	125.85	31.92	0.62	9.94	144.50	30.87	0.66	13.17	118.62	26.14

an average MSE of 0.18. In 22% of all cases, the meta-model cautions us and the MSE climbs to 0.84. Finally, on 8% of the test instances, the meta-model distrusts Linear and the MSE rises to 1.9, over four times the average error the primary model makes on all test instances.

We observe that forecasting the "cautioned" appears to be more difficult than for the other two cases. Intuitively, assessing whether a test instance may lie in the narrow band of the 80% and 95% error margins is harder than assessing whether an instance belongs to one of the extreme cases.

Figure 3 shows scatter plots on two test sets from the third PC dimension splits, the first using random forest primary model for a middle interval on the Concrete benchmark, the other using Linear as a primary model on the left-most interval for the Airfoil benchmark. The figure shows the forecasted value by the primary model over the true value. The colors indicate the meta-model's assessment: green for trusted, yellow for cautioned, and red for not trusted. Note that this figure shows predicted value over the actual value, so we cannot tell here whether the problematic inputs come from the same feature input region or not. Overall, we observe that the meta-model is effective at identifying test instances on which the primary model has problems, while the vast majority of trusted inputs find themselves within the desired performance bounds.

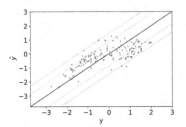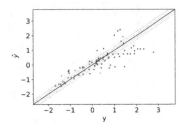

Fig. 3. Competence assessment on test set: linear regression on Airfoil middle PCA-3 interval (left) and Random forest regressor on Concrete leftmost PCA-3 interval (right). (Color figure online)

Table 3 compares our approach, which learns about the *technique* that builds the primal model, with an approach that assesses the primary *model* directly (motivated by Matsumoto et al. [14] approach for ANN primary regressors and Neural Networks Committees as meta-models). Here, we train *one* primary linear regression model on the training set and then, based on the residuals, label training instance as either trusted, cautioned, or warned. As in [14] we use the original benchmark features and the primary model forecast as meta-model features. Thus, this baseline omits our meta-features and the multiple data-splits to construct the meta-model input. We again use an SVM to train the meta-model.

Table 3. Empirical competence assessor (ECM) vs. One-model assessor (Baseline).

Benchmark	Class	Split	Random CV T	W	PCA-1 T	W	PCA-3 T	W	Cluster T	W
Airfoil ECM	%		60.51	39.49	64.24	35.76	68.57	31.43	69.92	30.08
	MSE		0.26	0.86	0.20	1.08	0.31	0.92	0.18	1.19
Baseline	%		50.12	49.88	51.35	48.65	57.29	42.71	51.33	48.67
	MSE		0.31	0.22	0.29	0.79	0.34	0.81	0.23	0.73
Boston ECM	%		51.88	48.12	55.03	44.97	64.49	35.51	57.88	42.12
	MSE		0.15	0.40	0.13	0.49	0.10	0.57	0.11	0.49
Baseline	%		50.93	49.07	59.21	40.79	57.76	42.24	53.80	46.20
	MSE		0.16	0.36	0.14	0.40	0.12	0.49	0.13	0.39
CCycle ECM	%		62.77	37.23	57.55	42.45	57.83	42.17	56.90	43.10
	MSE		0.04	0.12	0.05	0.11	0.06	0.10	0.06	0.09
Baseline	%		48.80	51.20	41.42	58.58	42.83	57.17	42.72	57.28
	MSE		0.05	0.09	0.05	0.09	0.06	0.09	0.06	0.08

Benchmark	Class	Split	Random CV T	W	PCA-1 T	W	PCA-3 T	W	Cluster T	W
Concrete ECM	%		65.15	34.85	60.68	39.32	63.98	36.02	57.91	42.09
	MSE		0.26	0.79	0.23	0.66	0.24	0.75	0.19	0.82
Baseline	%		55.44	44.56	56.89	43.11	55.63	44.37	53.13	46.87
	MSE		0.23	0.67	0.24	0.57	0.23	0.66	0.23	0.68
Diabetes ECM	%		67.40	32.60	61.87	38.13	63.55	36.45	63.06	36.94
	MSE		0.46	0.55	0.51	0.54	0.45	0.57	0.35	0.56
Baseline	%		53.37	46.63	53.01	46.99	52.48	47.52	57.91	42.09
	MSE		0.38	0.59	0.48	0.58	0.40	0.59	0.42	0.65
Yacht ECM	%		68.18	31.82	69.83	30.17	69.76	30.24	72.74	27.26
	MSE		0.13	0.81	0.14	0.84	0.14	0.83	0.14	0.91
Baseline	%		60.39	39.61	60.42	39.58	59.42	40.58	61.79	38.21
	MSE		0.16	0.63	0.15	0.63	0.16	0.62	0.16	0.63

The table shows the percentage of test instance labeled as trusted (T) or warned (W) (i.e., we expect the error to exceed 80% of the average training error), and the respective average MSE (geometric mean over the ten test sets in each cross-validation). Overall, assessing the learning technique works better than assessing the final model. For example, consider the Yacht benchmark. On all cross-validations, our empirical confidence model (ECM) has lower MSE on the trusted instances *while labeling more instances as trusted*. Also, the spread between the MSE on trusted and warned instances is usually higher for Baseline.

We also experimented with a standard random forest as a meta-model, which showed no significant difference to the SVM meta-model. Moreover, we analyzed the importance of our meta-features to characterize how a test input related to the primary model training set. Our analysis showed that the new features are crucial to get good performance. We also found that different features had different importance for different cross-validations and primary models, and, at times, even within the same cross-validation and primary learning technique (results omitted due to space restrictions). Therefore, the six meta-features are important to assess regressor competence in different circumstances.

5   Conclusion

We proposed a new methodology for building competence assessors for regression models. Rather than assessing the model itself, we experiment with the learning technique that generates the model over labeled training data. We showed the effectiveness of the technique, even on non-i.i.d. train/test splits, and when using off-the-shelf classifiers. Assessing primary model competence can thus be learned effectively and automatically. We, therefore, recommend that models that operate in critical environments should be secured using a competence assessor.

We also showed that learning how the primary regression technique behaves on the training set is more effective than aiming to learn only from the final primal model. This procedure obviously incurs higher computational costs, but on the other hand is not excessive either: hyper-parameter optimization of the primal model will require retraining in various cross-validations, so the cross-validations used for training the competence assessor will only add a constant factor to the total computation time.

Future work regards the proposal of other reliability metrics as meta-features. While the six features proposed lead to satisfactory results, we believe that the search for features that characterize the relationship between the run-time feature vector, the training set, and model performance has just begun.

Another aspect that deserves further investigation is the use of active learning. In this setting, we could add targeted training data for the competence assessor where it appears most beneficial. In fact, it is also conceivable that we learn lazily: When a new run-time instance is given that exhibits a very unique relationship to the training data which the competence assessor has not been trained for, we could, at run-time, aim to build new training data to retrain the competence model before it assesses the new input.

References

1. Amodei, D., Olah, C., Steinhardt, J., Christiano, P., Schulman, J., Mané, D.: Concrete problems in ai safety. arXiv preprint arXiv:1606.06565 (2016)
2. Bhattacharyya, S.: Confidence in predictions from random tree ensembles. Knowl. Inf. Syst. **35**(2), 391–410 (2013)
3. Bosnić, Z., Kononenko, I.: Estimation of individual prediction reliability using the local sensitivity analysis. Appl. Intell. **29**(3), 187–203 (2008)
4. Bosnić, Z., Kononenko, I.: Automatic selection of reliability estimates for individual regression predictions. Knowl. Eng. Rev. **25**(1), 27–47 (2010)
5. Bosnić, Z., Kononenko, I.: Correction of regression predictions using the secondary learner on the sensitivity analysis outputs. Comput. Inform. **29**(6), 929–946 (2010)
6. Christiano, P.F., Leike, J., Brown, T., Martic, M., Legg, S., Amodei, D.: Deep reinforcement learning from human preferences. In: Advances in Neural Information Processing Systems, pp. 4299–4307 (2017)
7. Dietvorst, B.J., Simmons, J.P., Massey, C.: Algorithm aversion: people erroneously avoid algorithms after seeing them err. J. Exp. Psychol. Gen. **144**(1), 114–126 (2015)
8. Asuncion, A., Newman, D.: UCI Machine Learning Repository. University of California, School of Information and Computer Science, Irvine, CA (2007)
9. Groce, A., et al.: You are the only possible oracle: effective test selection for end users of interactive machine learning systems. IEEE Trans. Software Eng. **40**(3), 307–323 (2014)
10. Heskes, T.: Practical confidence and prediction intervals. In: Advances in Neural Information Processing Systems, pp. 176–182 (1997)
11. Jiang, H., Kim, B., Guan, M.Y., Gupta, M.: To trust or not to trust a classifier. In: 32nd International Conference on Neural Information Processing Systems, pp. 5546–5557 (2018)
12. Liaw, A., Wiener, M., et al.: Classification and regression by randomforest. R News **2**(3), 18–22 (2002)
13. Lipton, Z.C., Wang, Y.X., Smola, A.: Detecting and correcting for label shift with black box predictors. arXiv preprint arXiv:1802.03916 (2018)
14. Matsumoto, E.Y., Del-Moral-Hernandez, E.: Improving regression predictions using individual point reliability estimates based on critical error scenarios. Inf. Sci. **374**, 65–84 (2016)
15. Nushi, B., Kamar, E., Horvitz, E.: Towards accountable AI: hybrid human-machine analyses for characterizing system failure. In: Sixth AAAI Conference on Human Computation and Crowdsourcing (2018)
16. Papadopoulos, H., Vovk, V., Gammerman, A.: Regression conformal prediction with nearest neighbours. J. Artif. Intell. Res. **40**, 815–840 (2011)
17. Papernot, N., McDaniel, P., Jha, S., Fredrikson, M., Celik, Z.B., Swami, A.: The limitations of deep learning in adversarial settings. In: 2016 IEEE European Symposium on Security and Privacy (EuroS&P), pp. 372–387. IEEE (2016)
18. Pevec, D., Kononenko, I.: Prediction intervals in supervised learning for model evaluation and discrimination. Appl. Intell. **42**(4), 790–804 (2014). https://doi.org/10.1007/s10489-014-0632-z
19. Prahl, A., Van Swol, L.: Understanding algorithm aversion: when is advice from automation discounted? J. Forecast. **36**(6), 691–702 (2017)
20. Shafer, G., Vovk, V.: A tutorial on conformal prediction. J. Mach. Learn. Res. **9**, 371–421 (2008)

21. Suykens, J.A., Vandewalle, J.: Least squares support vector machine classifiers. Neural Process. Lett. **9**(3), 293–300 (1999)
22. Yeomans, M., Shah, A., Mullainathan, S., Kleinberg, J.: Making sense of recommendations. J. Behav. Decis. Making **32**(4), 403–414 (2019)

Selection Driven Query Focused Abstractive Document Summarization

Chudamani Aryal and Yllias Chali[✉]

University of Lethbridge, Lethbridge, AB T1K 3M4, Canada
{chudamani.aryal,yllias.chali}@uleth.ca

Abstract. The current encode-attend-decode paradigm suffers from noisy encoder problem. We implemented a seq2seq model consisting of a novel selective mechanism for query focused abstractive document summarization using neural networks to solve this problem. Selective mechanism was used for the better representation of input (passage) sequence. We conducted experiments on Debatepedia dataset and have demonstrated that our model outperforms the state-of-the-art model in all ROUGE scores.

1 Introduction

Abstractive document summarization is the process of summarizing document by shortening a document to a condensed summary in an abstractive manner where the summary retains the key information about the document. Query-focused abstractive document summarization (QFADS) aims the same while keeping the query in context. However, despite the rise of neural models based on encode-attend-decode paradigm to generic summarization, not enough attention has been given to query focused abstractive document summarization. Although the purpose of attention mechanism is to show the alignment relationship between the input sequence and the output sequence, there is no clear alignment relationship between the two sequences [7]. Our work is based on the hypothesis that the encoder is too confused and noisy with just simple bi-directional RNN unit. We propose a seq2seq model consisting of novel selective mechanism for QFADS task. This selective mechanism can reduce the unnecessary information and enhance the important information to represent a long text in a better way. As the queries are relatively short, the selective encoding is only done on the passage of the document. Our contributions can be summarized as follows: (i) we propose a solution of novel selection mechanism for countering the problem of noisy encoder outputs, (ii) we also demonstrate that our proposed model clearly outperforms current state of art model in all evaluation scores.

C. Goutte and X. Zhu (Eds.): Canadian AI 2020, LNAI 12109, pp. 118–124, 2020.
https://doi.org/10.1007/978-3-030-47358-7_11

2   Proposed Model

The idea of selectively encoding the text was introduced by Lin et al. [2]. Our model is inspired by their model; however, our proposed model is different than their model. First of all, their model was used for the task of generic abstractive document summarization; however, our model is used for the task of QFADS. Along with the passage (document) encoder and the passage (document) attention, our model also includes the query encoder and the query attention. Second, their model uses Inception module to learn the local and global features whereas our model uses Inception Network. Third, their model uses attention mechanism as a post-step to the decoding layer whereas our model uses attention mechanism as a pre-step to the decoding layer. Fourth, the inputs to the passage (document) attention mechanism are different as the query contexts are included in the passage (document) attention along with the addition of separate query attention to the model's attention mechanism. Figure 1 shows the pictorial representation of our model.

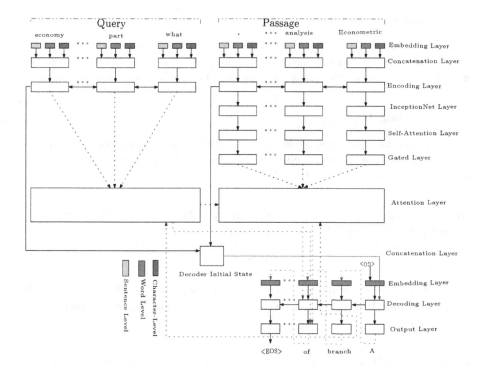

Fig. 1. Query focused abstractive document summarization model.

2.1   Task Description

Given a query $q = q_1, q_2, \ldots, q_k$ containing k words, a passage $p = p_1, p_2, \ldots, p_n$ containing n words, the task is to generate a query focused summary $y = y_1, y_2, \ldots, y_m$ containing m words. This task can be achieved as follows:

$$y^* = argmax_y \prod_{t=1}^{m} p(y_t|y_1, \ldots, y_{t-1}, q, p)$$

This equation can be modeled using neural based seq2seq network.

2.2   Embedding Layer

Our embedding layer consists of three sub-layers: character level embedding layer, word level embedding layer, and sentence level embedding layer. This layer is applied to both passage and query separately.

2.3   Input Concatenation Layer

The input concatenation layer concatenates its multiple input embeddings to one single embedding using linear and highway layers. This results in a single embeddings for the query and a separate single embeddings for the passage.

2.4   Encoding Layer

The encoding layer is responsible for encoding the input sequence (both query and passage separately) via deep bi-directional LSTM. The encoding layer consists of two encoders: Query Encoder and Passage Encoder.

Query Encoder. The query encoder computes the hidden encoder query state in sequential way from left to right for each time-step as:

$$s_i^q = DeepBiLSTM(s_{i-1}^q, e(q_i))$$

where, s_i^q contains the hidden encoder query outputs h_i^q and the hidden cell memory c_i^q and $e(q_i)$ represents the concatenated embedding of the query word at position i.

Passage Encoder. The passage encoder computes the hidden encoder passage state in the similar fashion for each time-step as:

$$s_i^p = DeepBiLSTM(s_{i-1}^p, e(p_i))$$

2.5   InceptionNet Layer

The InceptionNet layer is responsible for learning the local and global features of the document as well as the corpus. This layer is applied only to the passage as the length of passage text is quite long. The main idea of Inception module was introduced by Szegedy et al. [5]. Lin et al. [2] modified the Inception module to match their computational requirements. We are using the same modified Inception module for our proposed model. The InceptionNet layer chains together five Inception modules to form a single network which is used by the model to learn the local and global features of the document and the corpus. The width (the size of the Inception module) of each Inception module is varied to create a filtering effect to the features. Five residual connections are added (one for each Inception module) to the InceptionNet layer. Residual connections help the network to go even deeper. They also help ease the training of deeper neural network. Five outputs, one from each Inception module, are concatenated and followed by linear layer. As a part of regularization process, the concatenated output is followed by batch normalization. The process of InceptionNet layer can be seen in Fig. 2 which can be formulated as:

$$h^{ip} = InceptionNetLayer(h^p)$$

where, $h^p = \{h^p_1, \ldots, h^p_n\}$ is the sequential hidden encoder passage outputs of the first n words from the input passage sequence.

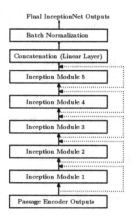

Fig. 2. InceptionNet Layer

2.6   Self-attention Layer

Introduced by [6], self-attention is used to strengthen the learning of the features. This layer can be formulated as:

$$h^{sp} = softmax(\frac{h^{ip}K^T}{\sqrt{d_k}})h^{ip}$$

where, $K = W_{att}h^{ip}$, W_{att} is a learnable matrix, d_k represents the dimension of the InceptionNet outputs which is the same as the RNN size.

2.7 Gated Layer

Gated layer is responsible for filtering the outputs of the passage encoder in order to remove unnecessary and unimportant information and only select the information relevant to the local and global features of the document and the corpus. This gated layer can be formulated as:

$$h^{gp} = h^p \odot \sigma(h^{sp})$$

where, σ and $\odot$ represent sigmoid activation function and Hadamard product, respectively.

2.8 Attention Layer

Attention layer is responsible for aligning the input sequence for the purpose of decoding the output sequence. Our attention layer consists of two sub-layers: Query Attention and Passage Attention.

Query Attention. The query attention mechanism can be formulated as follows:

$$cv_t^q = \sum_{i=1}^{N} Softmax(\frac{Score(h_{t-1}^d, h_i^q)}{\sqrt{d_k}})h_i^q$$

Here, $Score$ represents Luong's multipicative score function [3] to calculate the energy score of the query attention. cv_t^q represents query context vector at time t.

Passage Attention. The passage attention mechanism can be formulated as follows:

$$cv_t^p = \sum_{i=1}^{N} Softmax(\frac{Score(h_{t-1}^d, [h_i^{gp}, cv_t^q])}{\sqrt{d_k}})h_i^{gp}$$

Here, [] represents the concatenation of two vectors followed by a linear layer.

2.9 Final State Concatenation Layer

The final hidden state of both encoders are combined in this final state concatenation layer followed by a linear layer.

2.10 Output Embedding Layer

Output embedding layer is responsible for converting the summary words into vector space using just word level embedding layer.

2.11 Decoding Layer

This layer is responsible for decoding the output sequence via uni-directional LSTM. The decoding layer takes the concatenated final hidden state of both encoders as an initial state to its first recurrent layer. Each decoding step can be formulated as follows:

$$s_t^d = LSTM(s_{t-1}^d, [e(x_t^d), cv_t^p])$$

2.12 Output Layer

The output layer is responsible for predicting the summary words. The output layer can be formulated as below:

$$OV = Linear(Concat(h_t^d, cv_t^q, cv_t^p))$$
$$OD = softmax(Projection(OV))$$
$$y_t = Id2Word(argmax(OD))$$

Here, *Projection* represents converting the dense representation into the sparse representation matching the output size to the vocabulary size. *argmax* function takes the probability distribution and returns the index having highest probability. This index is converted into word by using *Id2Word* function.

3 Experiments and Results

We took the dataset created by Nema et al. [4] from Debatepedia[1]. We used Adam optimizer with initial learning of 0.0001. We used sequence loss function. Gradient clipping technique was used to alleviate the problem of vanishing/exploding gradient. We trained our model on a single Nvidia TITAN X GPU card with 12G RAM and the model took about 24 h to train. We used ROUGE(1, 2, L) [1], METEOR, Embedding Average Cosine Similarity (EACS), Greedy Maching Score (GMS)[2], and Copy Rate(CR) to evaluate the performance of our model.

We compare our model with the current state-of-art model on QFADS model proposed by Nema et al. [4] (Diversity) and vanilla Encode-Attend-Decode (EAD) framework. Both scores are taken from the Diversity paper.

Table 1 summarizes the result of our experiment. It shows that our proposed model outperforms Diversity model for QFADS task in all ROUGE scores.

Table 1. Model Comparison

Models	R-1	R-2	R-L	METEOR	EACS	GMS	CR
EAD	13.73	2.06	12.84	-	-	-	-
Diversity	41.26	18.75	40.43	-	-	-	-
Our	**43.22**	**27.40**	**42.73**	**25.72**	**85.49**	**72.59**	**40.47**

---

[1] https://github.com/PrekshaNema25/DiverstiyBasedAttentionMechanism.
[2] https://github.com/Maluuba/nlg-eval.

4   Conclusion

We implemented a model consisting of a novel selective mechanism for the query based abstractive document summarization. The novel selective mechanism was introduced to reduce the noise present in long text of the input (passage) sequence. Experiments were conducted on Debatepedia dataset which showed that our model outperforms the state-of-the art model in all ROUGE scores.

Acknowledgements. We would like to thank the anonymous reviewers for their useful comments. The research reported in this paper was conducted at the University of Lethbridge and supported by the Natural Sciences and Engineering Research Council (NSERC) of Canada discovery grant and the University of Lethbridge.

References

1. Ganesan, K.: Rouge 2.0: updated and improved measures for evaluation of summarization tasks (2015)
2. Lin, J., Sun, X., Ma, S., Su, Q.: Global encoding for abstractive summarization. In: Proceedings of the 56th Annual Meeting of the Association for Computational Linguistics (Volume 2: Short Papers), pp. 163–169. Association for Computational Linguistics (2018). http://aclweb.org/anthology/P18-2027
3. Luong, T., Pham, H., Manning, C.D.: Effective approaches to attention-based neural machine translation. In: Proceedings of the 2015 Conference on Empirical Methods in Natural Language Processing, pp. 1412–1421. Association for Computational Linguistics (2015). https://doi.org/10.18653/v1/D15-1166, http://aclweb. org/anthology/D15-1166
4. Nema, P., Khapra, M.M., Laha, A., Ravindran, B.: Diversity driven attention model for query-based abstractive summarization. In: Proceedings of the 55th Annual Meeting of the Association for Computational Linguistics (Volume 1: Long Papers), pp. 1063–1072. Association for Computational Linguistics (2017). https://doi.org/ 10.18653/v1/P17-1098, http://aclweb.org/anthology/P17-1098
5. Szegedy, C., et al.: Going deeper with convolutions. In: 2015 IEEE Conference on Computer Vision and Pattern Recognition (CVPR). pp. 1–9, June 2015. https:// doi.org/10.1109/CVPR.2015.7298594
6. Vaswani, A., et al.: Attention is all you need. In: NIPS (2017)
7. Zhou, Q., Yang, N., Wei, F., Zhou, M.: Selective encoding for abstractive sentence summarization. In: Proceedings of the 55th Annual Meeting of the Association for Computational Linguistics (Volume 1: Long Papers), pp. 1095–1104. Association for Computational Linguistics (2017). https://doi.org/10.18653/v1/P17-1101, http:// aclweb.org/anthology/P17-1101

VecHGrad for Solving Accurately Tensor Decomposition

Jeremy Charlier[1]([⊠]) and Vladimir Makarenkov[2]

[1] National Bank of Canada, Montreal, Canada
`jeremy.charlier@bnc.ca`
[2] Université du Québec à Montréal (UQAM), Montreal, Canada
`makarenkov.vladimir@uqam.ca`

Abstract. Tensor decomposition is a collection of factorization techniques for multidimensional arrays. Today's data sets, because of their size, require tensor decomposition involving factorization with multiple matrices and diagonal tensors such as DEDICOM or PARATUCK2. Traditional tensor resolution algorithms such as Stochastic Gradient Descent (SGD) or Non-linear Conjugate Gradient descent (NCG), cannot be easily applied to these types of tensor decomposition or often lead to poor accuracy at convergence. We propose a new resolution algorithm, VecHGrad, for accurate and efficient stochastic resolution over all existing tensor decomposition. VecHGrad relies on the gradient, an Hessian-vector product, and an adaptive line search, to ensure the convergence during optimization. Our experiments on five popular data sets with the state-of-the-art deep learning gradient optimizers show that VecHGrad is capable of converging considerably faster because of its superior convergence rate per step. VecHGrad targets as well deep learning optimizer algorithms. The experiments are performed for various tensor decomposition, including CP, DEDICOM, and PARATUCK2. Although it involves an Hessian-vector update rule, VecHGrad's runtime is similar in practice to that of gradient methods such as SGD, Adam, or RMSProp.

Keywords: Line search · Gradient descent · Loss function

1 Motivation

Tensors are multidimensional, or N-order, arrays. Tensors are able to scale down a large amount of data to an interpretable size using different decomposition types, also called factorizations. The data sets used in machine learning [1] and data mining [2] are multi-dimensional and, therefore, tensors and their decompositions are highly appropriate [3]. Because of the size of modern data sets, Tensor Decompositions (TDs), such as the CP/PARAFAC [4,5] decomposition, later denoted CP, are now challenged by other TDs such as DEDICOM [6] and PARATUCK2 [7,8]. CP decomposes the original tensor as a sum of rank-one tensors, as illustrated in Fig. 1, whereas DEDICOM and PARATUCK2 decompose

© Springer Nature Switzerland AG 2020
C. Goutte and X. Zhu (Eds.): Canadian AI 2020, LNAI 12109, pp. 125–137, 2020.
https://doi.org/10.1007/978-3-030-47358-7_12

the original tensor as a product of matrices and diagonal tensors, as shown in Figs. 2 and 3, respectively. Depending on the TD type, different latent variables are highlighted with their respective asymmetric relationships. Fast and accurate tensor resolutions have however required specific numerical optimization methods known as preconditioning methods.

Well-known in the Machine Learning (ML) and Deep Learning (DL) community, the standard Stochastic Gradient (SGD) optimization method [9,10] is widely used in different ML and DL approaches. It is however losing its momentum over recent preconditioning gradient methods. The latter considers a matrix, a preconditioner, to update the gradient before it is used. Standard preconditioning methods include Newton's method, which employs the exact Hessian matrix, and the quasi-Newton methods, which do not require the exact Hessian matrix [11]. The computational cost of the exact Hessian matrix is one of the major limitations of Newton's method. Introduced to answer some of the challenges in ML and DL, AdaGrad [12] uses as a preconditioner the co-variance matrix of the accumulated gradients. Due to the dimensions of the ML problems, specialized variants were proposed to replace the full preconditioning methods by diagonal approximation methods, such as Adam [13], or by other schemes, such as Nesterov Accelerated Gradient (NAG) [14] or SAGA [15]. The diagonal approximation methods are often preferred because of the super-linear memory consumption of other methods [16].

In this work, we aim at efficient and accurate resolution of high-order TDs for which most of the ML and DL state-of-the-art optimizers fail. We describe how to exploit the quasi-Newton convergence with a diagonal approximation of the Hessian matrix and an adaptive line search. Our algorithm, VecHGrad for Vector Hessian Gradient, returns the tensor structure of the gradient using a separate preconditioner vector. Our analysis relies on the extensions of vector analysis to the tensor world. We show the superior capabilities of VecHGrad for various resolution algorithms, such as Alternating Least Squares (ALS) or Non-linear Conjugate Gradient (NCG) [17], and some popular DL optimizers, such as Adam or RMSProp. Our main contributions are summarized below:

- We propose a new resolution algorithm, VecHGrad, that uses the gradient, the Hessian-vector product and an adaptive line search to perform accurate and fast optimization of the objective function of TD.
- We demonstrate the superior accuracy of VecHGrad at convergence. We compare it, on five popular data sets, to other resolution algorithms, including deep learning optimizers, for the three most common TDs known as CP, DEDICOM, and PARATUCK2.

The paper is structured as follows. We discuss the related work in Sect. 2. We review briefly tensor operations and tensor definitions in Sect. 3. We then describe in Sect. 4 how VecHGrad performs in a numerical optimization scheme applied to tensor structures without the requirement of knowing the Hessian matrix. We highlight the experimental results in Sect. 5. Finally, we conclude and address promising directions for future work.

2   Related Work

VecHGrad uses a diagonal approximation of the Hessian matrix and, therefore, is related to other diagonal approximations such as AdaGrad [12], which is very popular and frequently applied [16]. However, it only uses gradient information, as opposed to VechGrad which uses both gradient and Hessian information. Other approaches extremely popular in ML and DL include Adam [13], NAG [14], SAGA [15], and RMSProp [18]. The non-exhaustive list of ML optimizers is considered in our study case since it offers a strong baseline comparison for VecHGrad.

Since our study case is related to TDs, the methods specifically designed for TDs have to be mentioned. The most popular optimization scheme among the resolution of TDs is the Alternating Least Squares (ALS). Under the ALS scheme [3], one element of the decomposition is fixed. The fixed element is updated using the other elements. Therefore, all the elements are successively updated at each step of the iteration process until a convergence criteria is reached, e.g. a fixed number of iteration. For every TDs, there exists at least one ALS resolution scheme. The ALS resolution scheme was introduced in [4,5] for the CP decomposition, in [19] for the DEDICOM decomposition and in [7] for the PARATUCK2 decomposition. An updated ALS scheme was presented in [20] to solve PARATUCK2. In [6], Bader et al. proposed ASALSAN to solve with non-negativity constraints the DEDICOM decomposition with the ALS scheme. While some matrix updates are not guaranteed to decrease the loss function, the scheme leads to overall convergence. In [21], Charlier et al. have recently proposed a non-negative scheme for the PARATUCK2 decomposition.

Other approaches are specifically designed for one TD using gradient information. In [17,22], an optimized version of NCG for CP is presented, i.e. CP-OPT. In [23], an extension of the Stochastic Gradient Descent (SGD) is described to obtain, as mentioned by the authors, an expected CP TD. The performance on other TDs was, however, not assessed. The comparison to other numerical optimizers in the experiments was rather limited, especially when considering existing popular machine learning and deep learning optimizers. In contrast, VecHGrad is detached from any particular model structure, including the choice of TDs. It only relies on the gradient, the Hessian diagonal approximation and an adaptive line search, crucial for fast convergence of complex numerical optimization. Consequently, VecHGrad is easy to implement and use in practice as it does not require to be optimized for a particular model structure.

3   Background

In this section, we briefly introduce the preliminaries of TDs. Scalars are denoted by lower case letters a, vectors by boldface lowercase letters $\mathbf{a}$, matrices by boldface capital letters $\mathbf{A}$, and N-order tensors by the Euler script notation $\mathfrak{X}$.

3.1 Tensor Operations

Vectorization. The vectorization operator flattens a tensor of n entries to a column vector $\mathbb{R}^n$. The ordering of the tensor elements is not important as long as it is consistent [3]. For a third order tensor $\mathcal{X} \in \mathbb{R}^{I \times J \times K}$, the vectorization of $\mathcal{X}$ is equal to $\text{vec}(\mathcal{X}) = \begin{bmatrix} x_{111}\ x_{112}\ \cdots\ x_{IJK} \end{bmatrix}^T$.

Tensor Norm. The square root of the sum of all squared tensor entries of the tensor $\mathcal{X}$ defines its norm: $\parallel \mathcal{X} \parallel = \sqrt{\sum_{j=1}^{I_1} \sum_{j=2}^{I_2} \cdots \sum_{j=n}^{I_n} x_{j_1, j_2, \dots, j_n}^2}$.

Tensor Rank. The rank-R of a tensor $\mathcal{X} \in \mathbb{R}^{I_1 \times I_2 \times \dots \times I_N}$ is the number of linear components that could fit $\mathcal{X}$ exactly. $\mathcal{X} = \sum_{r=1}^{R} \mathbf{a}_r^{(1)} \circ \mathbf{a}_r^{(2)} \circ \dots \circ \mathbf{a}_r^{(N)}$.

3.2 Tensor Decomposition

The CP decomposition, shown in Fig. 1, has been introduced in [4,5]. The tensor $\mathcal{X} \in \mathbb{R}^{I \times I \times K}$ is defined as a sum of rank-one tensors. The number of rank-one tensors is determined by the rank, denoted by R, of the tensor $\mathcal{X}$. The CP decomposition is expressed as $\mathcal{X} = \sum_{r=1}^{R} \mathbf{a}_r^{(1)} \circ \mathbf{a}_r^{(2)} \circ \mathbf{a}_r^{(3)} \circ \dots \circ \mathbf{a}_r^{(N)}$ where $\mathbf{a}_r^{(1)}, \mathbf{a}_r^{(2)}, \mathbf{a}_r^{(3)}, \dots, \mathbf{a}_r^{(N)}$ are factor vectors of size $\mathbb{R}^{I_1}, \mathbb{R}^{I_2}, \mathbb{R}^{I_3}, \dots, \mathbb{R}^{I_N}$. Each factor vector $\mathbf{a}_r^{(n)}$ with $n \in \{1, 2, \dots, N\}$ and $r \in \{1, \dots, R\}$ refers to one order and one rank of the tensor $\mathcal{X}$.

The DEDICOM decomposition [19], illustrated in Fig. 2, describes the asymmetric relationships between I objects of the tensor $\mathcal{X} \in \mathbb{R}^{I \times I \times K}$. The decomposition groups the I objects into R latent components (or groups) and describes their pattern of interactions by computing $\mathbf{A} \in \mathbb{R}^{I \times R}$, $\mathbf{H} \in \mathbb{R}^{R \times R}$ and $\mathcal{D} \in \mathbb{R}^{R \times R \times K}$ such that $\mathbf{X}_k = \mathbf{A} \mathbf{D}_k \mathbf{H} \mathbf{D}_k \mathbf{A}^T, \forall\, k = 1, \dots, K$. The matrix $\mathbf{A}$ indicates the participation of object $i = 1, \dots, I$ in the group $r = 1, \dots, R$, the matrix $\mathbf{H}$ the interactions between the different components r and the tensor $\mathcal{D}$ represents the participation of the R latent component according to K.

The PARATUCK2 decomposition [7], presented in Fig. 3, expresses the original tensor $\mathcal{X} \in \mathbb{R}^{I \times J \times K}$ as a product of matrices and tensors $\mathbf{X}_k = \mathbf{A} \mathbf{D}_k^A \mathbf{H} \mathbf{D}_k^B \mathbf{B}^T$ with $k = \{1, \dots, K\}$ where $\mathbf{A}$, $\mathbf{H}$ and $\mathbf{B}$ are matrices of size $\mathbb{R}^{I \times P}$, $\mathbb{R}^{P \times Q}$, and $\mathbb{R}^{J \times Q}$. The matrices $\mathbf{D}_k^A \in \mathbb{R}^{P \times P}$ and $\mathbf{D}_k^B \in \mathbb{R}^{Q \times Q}$ $\forall k \in \{1, \dots, K\}$ are the slices of the tensors $\mathcal{D}^A \in \mathbb{R}^{P \times P \times K}$ and $\mathcal{D}^B \in \mathbb{R}^{Q \times Q \times K}$. The columns of the matrices $\mathbf{A}$ and $\mathbf{B}$ represent the latent factors P and Q, and therefore the rank of each object set. The matrix $\mathbf{H}$ underlines the asymmetry between the P latent components and the Q latent components. The tensors $\mathcal{D}^A$ and $\mathcal{D}^B$ measure the evolution of the latent components according to K.

4 VecHGrad for Tensor Decomposition

In this section, we first introduce VecHGrad for the first order tensors, also commonly called vectors. We then present the core of our contribution, VecHGrad for N-order tensor decomposition.

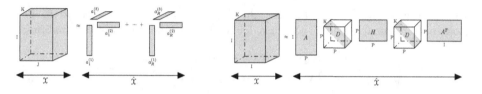

Fig. 1. Third order CP TD   **Fig. 2.** Third order DEDICOM TD

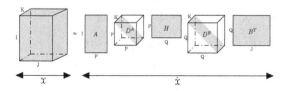

Fig. 3. Third order PARATUCK2 TD

4.1 Introduction to VecHGrad for Vectors

Under Newton's method, the current iterate $\tilde{\mathbf{x}}^t \in \mathcal{C}$ is used to generate the next iterate $\tilde{\mathbf{x}}^{t+1}$ by performing a constrained minimization of the second order Taylor expansion such that:

$$\tilde{\mathbf{x}}^{t+1} = \arg\min_{\mathbf{x}\in\mathcal{C}}\{\frac{1}{2}\left\langle \mathbf{x} - \tilde{\mathbf{x}}^t, \nabla^2 f(\tilde{\mathbf{x}}^t)(\mathbf{x} - \tilde{\mathbf{x}}^t)\right\rangle + \left\langle \nabla f(\tilde{\mathbf{x}}^t), \mathbf{x} - \tilde{\mathbf{x}}^t\right\rangle\}. \tag{1}$$

We recall that ∇f and $\nabla^2 f$ denote the gradient and the Hessian matrix, respectively, of the objective function $f : \mathbb{R}^d \to \mathbb{R}$.

$$\nabla f = \underset{\mathbf{x}\in\mathcal{C}}{grad f}$$
$$\nabla f = [\frac{\partial f}{\partial x_1}, \frac{\partial f}{\partial x_2}, ..., \frac{\partial f}{\partial x_d}] \tag{2}$$

$$\nabla^2 f = \mathbf{Hes}$$

$$\nabla^2 f = \begin{pmatrix} \frac{\partial^2 f}{\partial x_1^2} & \frac{\partial^2 f}{\partial x_1\partial x_2} & \cdots & \frac{\partial^2 f}{\partial x_1\partial x_d} \\ \vdots & \vdots & \ddots & \vdots \\ \frac{\partial^2 f}{\partial x_d\partial x_1} & \frac{\partial^2 f}{\partial x_d\partial x_2} & \cdots & \frac{\partial^2 f}{\partial x_d^2} \end{pmatrix} \tag{3}$$

When $\mathcal{C} \in \mathbb{R}^d$, which is the unconstrained form, the new iterate $\tilde{\mathbf{x}}^{t+1}$ is generated such that:

$$\tilde{\mathbf{x}}^{t+1} = \tilde{\mathbf{x}}^t - [\nabla^2 f(\tilde{\mathbf{x}}^t)]^{-1}\nabla f(\tilde{\mathbf{x}}^t) . \tag{4}$$

We use the strong Wolfe's line search which allows Newton's method to be globally convergent. The line search is defined by the following three inequalities:

$$\text{i) } f(\tilde{\mathbf{x}}^t + \alpha^t \mathbf{p}^t) \leq f(\tilde{\mathbf{x}}^t) + c_1 \alpha^t (\mathbf{p}^t)^T \nabla f(\tilde{\mathbf{x}}^t) \,,$$
$$\text{ii) } -(\mathbf{p}^t)^T \nabla f(\tilde{\mathbf{x}}^t + \alpha^t \mathbf{p}^t) \leq -c_2 (\mathbf{p}^t)^T \nabla f(\tilde{\mathbf{x}}^t) \,, \tag{5}$$
$$\text{iii) } |(\mathbf{p}^t)^T \nabla f(\tilde{\mathbf{x}}^t + \alpha^t \mathbf{p}^t)| \leq c_2 |(\mathbf{p}^t)^T \nabla f(\tilde{\mathbf{x}}^t)| \,,$$

where $0 \leq c_1 \leq c_2 \leq 1$, $\alpha^t > 0$ is the step length and $\mathbf{p}^t = -[\nabla^2 f(\tilde{\mathbf{x}}^t)]^{-1} \nabla f(\tilde{\mathbf{x}}^t)$. Therefore, the iterate $\tilde{\mathbf{x}}^{t+1}$ becomes the following:

$$\begin{cases} \tilde{\mathbf{x}}^{t+1} = & \tilde{\mathbf{x}}^t - \alpha^t [\nabla^2 f(\tilde{\mathbf{x}}^t)]^{-1} \nabla f(\tilde{\mathbf{x}}^t) \,, \\ \tilde{\mathbf{x}}^{t+1} = & \tilde{\mathbf{x}}^t + \alpha^t \mathbf{p}^t \,. \end{cases} \tag{6}$$

Computing the inverse of the exact Hessian matrix, $[\nabla^2 f(\tilde{\mathbf{x}}^t)]^{-1}$, may be difficult. The inverse is therefore computed with a Conjugate Gradient (CG) loop. It has two main advantages: the calculations are considerably less expensive [11] and the Hessian can be expressed by a diagonal approximation. The convergence of the CG loop is defined when a maximum number of iterations is reached or when the residual $\mathbf{r}^t = \nabla^2 f(\tilde{\mathbf{x}}^t) \mathbf{p}^t + \nabla f(\tilde{\mathbf{x}}^t)$ satisfies $\| \mathbf{r}^t \| \leq \sigma \| \nabla f(\tilde{\mathbf{x}}^t) \|$ with $\sigma \in \mathbb{R}^+$. In the CG loop, the exact Hessian matrix is approximated by a diagonal approximation. The Hessian matrix is multiplied by a descent direction vector resulting in a vector which satisfies the requirement of the main optimization loop. Therefore, only the results of the Hessian vector product are needed. The equation is expressed below:

$$\nabla^2 f(\tilde{\mathbf{x}}^t) \, \mathbf{p}^t = \frac{\nabla f(\tilde{\mathbf{x}}^t + \eta \, \mathbf{p}^t) - \nabla f(\tilde{\mathbf{x}}^t)}{\eta} \,. \tag{7}$$

The term η is the perturbation and the term $\mathbf{p}^t$ is the descent direction vector, set to the negative of the gradient at initialization. Consequently, the extensive computation of the inverse of the exact full Hessian matrix is bypassed using only gradient diagonal approximation.

4.2   VecHGrad for Accurate Resolution of Tensor Decomposition

The loss function, or the objective function, is denoted by f. It is equal to:

$$f(\tilde{\mathbf{x}}) = \min_{\hat{\mathcal{X}}} ||\mathcal{X} - \hat{\mathcal{X}}||. \tag{8}$$

The tensor $\mathcal{X}$ is the original tensor and the tensor $\hat{\mathcal{X}}$ is the approximated tensor built from the decomposition. If we consider, for instance, that the CP TD applied on a third order tensor, the tensor $\hat{\mathcal{X}}$ is the tensor built with the factor vectors $\mathbf{a}_r^{(1)}, \mathbf{a}_r^{(2)}, \mathbf{a}_r^{(3)}$ for $r = 1, ..., R$ initially randomized such as:

$$\hat{\mathcal{X}} = \sum_{r=1}^{R} \mathbf{a}_r^{(1)} \circ \mathbf{a}_r^{(2)} \circ \mathbf{a}_r^{(3)}. \tag{9}$$

The vector $\tilde{\mathbf{x}}$ is a flattened vector containing all the entries of the decomposed tensor $\hat{\mathcal{X}}$. If we consider a third order tensor $\hat{\mathcal{X}}$ of rank R factorized with the CP TD, we obtain the following vector $\tilde{\mathbf{x}} \in \mathbb{R}^{d=R(I+J+K)}$ such that:

$$\tilde{\mathbf{x}} = \text{vec}(\hat{\mathcal{X}}) = [\mathbf{a}_1^{(1)}, \mathbf{a}_2^{(1)}, ..., \mathbf{a}_I^{(R)}, \mathbf{a}_1^{(2)}, \mathbf{a}_2^{(2)}, ..., \mathbf{a}_J^{(R)}, \mathbf{a}_1^{(3)}, \mathbf{a}_2^{(3)}, ..., \mathbf{a}_K^{(R)}]^T. \tag{10}$$

Since the objective is to propose a universal approach for any TDs, we rely on finite difference method to compute the gradient of the loss function of any TDs. Thus, the method can be transposed to any decomposition just by changing the decomposition equation. The approximate gradient is based on a fourth order formula (11) to ensure reliable approximation [24]:

$$\frac{\partial}{\partial x_i} f(\tilde{\mathbf{x}}^t) \approx \frac{1}{4!\eta} \big(2[f(\tilde{\mathbf{x}}^t - 2\eta\mathbf{e}_i) - f(\tilde{\mathbf{x}}^t + 2\eta\mathbf{e}_i)] \\ + 16[f(\tilde{\mathbf{x}}^t + \eta\mathbf{e}_i) - f(\tilde{\mathbf{x}}^t - \eta\mathbf{e}_i)]\big) \tag{11}$$

In Eq. (11), the index i is the index of the variables for which the derivative is to be evaluated. The variable $\mathbf{e}_i$ is the ith unit vector. The term η, the perturbation, is set to a small value to achieve the convergence of the process.

The Hessian diagonal approximation is evaluated as described in the previous section, using Eq. (7). Our approach is therefore free of the extensive computation of the inverse of the exact Hessian matrix. We finally reached the core objective of describing VecHGrad for tensors, summarized in Algorithm 1.

Algorithm 1: VecHGrad, tensor case

1   Select tensor decomposition equation: $g : \begin{cases} \mathbb{R}^d \to \mathbb{R}^{I_1 \times I_2 \times ... \times I_n} \\ \tilde{\mathbf{x}}^t \mapsto \tilde{\mathcal{X}} \end{cases}$

2   Select randomly $\tilde{\mathbf{x}}^0 = \text{vec}(\hat{\mathcal{X}})$

3   **repeat**

4   Select loss function: $f : \begin{cases} \mathbb{R}^d \to \mathbb{R} \\ \tilde{\mathbf{x}}^t \mapsto \| \mathcal{X} - g(\tilde{\mathbf{x}}^t) \| \end{cases}$

5   Compute gradient $\nabla f(\tilde{\mathbf{x}}^t) \in \mathbb{R}^d$

6   Fix $\mathbf{p}_0^t = -\nabla f(\tilde{\mathbf{x}}^t)$

7   **repeat**

8   Update $\mathbf{p}_k^t$ with CG loop: $\mathbf{r}_k = \nabla^2 f(\tilde{\mathbf{x}}^t)\mathbf{p}_k^t + \nabla f(\tilde{\mathbf{x}}^t)$

9   **until** $k = cg_{maxiter}$ OR $\| \mathbf{r}_k \| \leq \sigma \| \nabla f(\tilde{\mathbf{x}})^t) \|$

10     $\alpha^t \leftarrow$ Wolfe's line search

11   Update parameters: $\tilde{\mathbf{x}}^{t+1} = \tilde{\mathbf{x}}^t + \alpha^t \mathbf{p}_{opt}^t$

12 **until** $t = maxiter$ OR $f(\tilde{\mathbf{x}}^t) \leq \epsilon_1$ OR $\| \nabla f(\tilde{\mathbf{x}}^t) \| \leq \epsilon_2$

---

5   Experiments

We investigate the convergence behavior of VecHGrad and compare it to other popular optimizers inherited from both the tensor and machine learning communities. We compare VecHGrad with ten different algorithms applied to the three main TDs, CP, DEDICOM, and PARATUCK2, with increasing linear algebra complexity:

- ALS, Alternating Least Squares [20,21];
- SGD, Stochastic Gradient Descent [11];
- NAG, Nesterov Accelerated Gradient [14];
- CP-OPT and the Non-linear Conjugate Gradient (NCG) [17,22];
- Adam [13];
- RMSProp [18];
- SAGA [15];
- AdaGrad [12];
- L-BFGS [25].

Data Availability and Code Availability. We highlight the performance of VecHGrad using 5 popular data sets CIFAR10, CIFAR100, MNIST, LFW and COCO, all available online. Each data set has different intrinsic characteristics such as size or sparsity. A quick overview of their characteristics is presented in Table 1. We chose to use different data sets as the performance of different optimizers might vary depending on the data. The overall conclusion for the experiments is therefore independent of one particular data set. The implementation and the data for the experiments are available on GitHub[1].

Table 1. Description of the data sets used (K: thousands).

Data Set	Labels	Size	Batch Size
CIFAR-10	Image × pixels × pixels	50K × 32 × 32	64
CIFAR-100	Image × pixels × pixels	50K × 32 × 32	64
MNIST	Image × pixels × pixels	60K × 28 × 28	64
COCO	Image × pixels × pixels	123K × 64 × 64	32
LFW	Image × pixels × pixels	13K × 64 × 64	32

Experimental Setup. In our experiments, we use the standard parameters for the popular ML and DL gradient optimization methods. We use $\eta = 10^{-4}$ for SGD, $\gamma = 0.9$ and $\eta = 10^{-4}$ for NAG, $\beta_1 = 0.9, \beta_2 = 0.999, \epsilon = 10^{-8}$ and $\eta = 0.001$ for Adam, $\gamma = 0.9, \eta = 0.001$ and $\epsilon = 10^{-8}$ for RMSProp, $\eta = 10^{-4}$ for SAGA, $\eta = 0.01$ and $\epsilon = 10^{-8}$ for AdaGrad. We use the Hestenes-Stiefel update for the NCG resolution. Furthermore, the convergence criteria is reached when $f^{i+1} - f^i \leq 0.001$ or when a maximum number of iterations is reached. We use 100,000 iterations for gradient-free methods, 10,000 iterations for gradient methods and 1,000 iterations for Hessian-based methods. Additionally, we fixed the number of iterations to 20 for VecHGrad's inner CG loop, used to determine the descent direction. We invite the reader to see our code available on GitHub[1] for further insights about the parameters used in our study. The simulations

---

[1] The code is available at https://github.com/dagrate/vechgrad.

were conducted on a server with 50 Intel Xeon E5-4650 CPU cores and 50GB of RAM. All the resolution schemes have been implemented in Julia and are compatible with the ArrayFire GPU accelerator library.

Results and Discussion. First, we performed an experiment to identify visually the strengths of each of the optimization algorithms. The Fig. 4 depicts the resulting error of the loss function for each of the methods at convergence for PARATUCK2 TD. We fixed latent components for which the differences between the optimizers are easily noticeable. The error of the loss function depends on how blurry the picture is, measured by the error at convergence. Some of the numerical methods, ALS, RMSProp and VecHGrad, offered the best observable image quality at convergence, given our choice of parameters. Other popular schemes, including NAG and SAGA, however failed to converge to a solution resulting in a noisy image, far from being close to the original one.

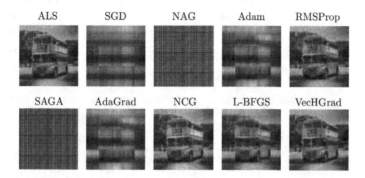

Fig. 4. Visual simulation of the accuracy at convergence of the different optimizers for the PARATUCK2 decomposition. The accuracy is highlighted by how blurry the image is (error at convergence ALS: 1018, SGD: 4082, NAG: 6469, Adam: 2996, RMSProp: 1184, SAGA: 6378, AdaGrad: 2961, NCG: 1569, L-BFGS: 1771, VecHGrad: 599). The popular gradient optimizers AdaGrad, NAG and SAGA failed to converge to a solution close to the original image, contrarily to VecHGrad or RMSProp.

In a second experiment, we compared in Table 2 and Table 3 the loss function errors and the calculation times of the optimizers for the five ML data sets described in Table 1 for the three TDs aforementioned. Both the loss function errors and the calculation times were computed based on the mean of the loss function errors and the mean of the calculation times over all batches. The numerical schemes of the NAG, SAGA and AdaGrad algorithms failed to minimize the error of the loss function accurately. The ALS scheme furthermore offers a good compromise between the resulting errors and the required calculation times, explaining its major success across TD applications. The gradient descent optimizers, Adam and RMSProp, and the Hessian based optimizers, VecHGrad and L-BFGS, were capable to minimize the loss function the most accurately. The NCG method achieved satisfying errors for the CP and the DEDICOM decomposition but its performance decreases significantly when trying to solve the PARATUCK2 decomposition. Surprisingly, the calculation times of the Adam and RMSProp gradient descents were greater than the calculation times of VecHGrad. VecHGrad was capable to outperform the gradient descent schemes for both accuracy and speed thanks to the use of the strong Wolfe's line search and the vector Hessian approximation, inherited from gradient information. This result is reported in Table 4 presenting the experimental mean convergence rate, defined such that $q \approx \log \left| \frac{f^{t+1}-f^t}{f^t-f^{t-1}} \right| \left[\log \left| \frac{f^t-f^{t-1}}{f^{t-1}-f^{t-2}} \right| \right]^{-1}$

Table 2. Mean of the loss function errors at convergence over all batches. The lower, the better. The strong Wolfe's line search is crucial for the VecHGrad's performance.

Decomposition	Optimizer	CIFAR-10	CIFAR-100	MNIST	COCO	LFW
CP	ALS	318.667	428.402	897.766	485.138	4792.605
CP	SGD	2112.904	2825.710	2995.528	3407.415	7599.458
CP	NAG	4338.492	5511.272	4916.003	8187.315	18316.589
CP	Adam	1578.225	2451.217	1631.367	2223.211	6644.167
CP	RMSProp	127.961	128.137	200.002	86.792	4205.520
CP	SAGA	4332.879	5501.528	4342.708	6327.580	13242.181
CP	AdaGrad	3142.583	4072.551	2944.768	4921.861	10652.488
CP	NCG	41.990	37.086	23.320	76.478	4130.942
CP	L-BFGS	195.298	525.279	184.906	596.160	4893.815
CP	VecHGrad	**≤1.000**	**≤1.000**	**≤1.000**	**≤1.000**	**≤1.000**
DEDICOM	ALS	1350.991	1763.718	1830.830	1894.742	3193.685
DEDICOM	SGD	435.780	456.051	567.503	406.760	511.093
DEDICOM	NAG	4349.151	5722.073	4415.687	6325.638	9860.454
DEDICOM	Adam	579.723	673.316	575.341	743.977	541.515
DEDICOM	RMSProp	63.795	236.974	96.240	177.419	33.224
DEDICOM	SAGA	4285.512	5577.981	4214.771	5797.562	8128.724
DEDICOM	AdaGrad	1962.966	2544.436	1452.278	2851.649	3033.965
DEDICOM	NCG	550.554	321.332	171.181	583.430	711.549
DEDICOM	L-BFGS	423.802	561.689	339.284	435.188	511.620
DEDICOM	VecHGrad	**≤1.000**	**≤1.000**	**≤1.000**	**≤1.000**	**≤1.000**
PARATUCK2	ALS	408.724	480.312	1028.250	714.623	658.284
PARATUCK2	SGD	639.556	631.870	1306.869	648.962	495.188
PARATUCK2	NAG	4699.058	6046.024	5168.824	8205.223	14546.438
PARATUCK2	Adam	512.725	680.653	591.156	594.687	615.731
PARATUCK2	RMSProp	133.416	145.766	164.709	134.047	174.769
PARATUCK2	SAGA	4665.435	5923.178	4934.328	6350.172	8847.886
PARATUCK2	AdaGrad	1775.433	2310.402	1715.316	2752.348	2986.919
PARATUCK2	NCG	772.634	1013.032	270.288	335.532	15181.961
PARATUCK2	L-BFGS	409.666	522.158	464.259	467.139	416.761
PARATUCK2	VecHGrad	**≤textbf1.000**	**≤1.000**	**≤1.000**	**≤1.000**	**≤1.000**

Table 3. Mean calculation times (sec.) to reach convergence over all batches.

Decomposition	Optimizer	CIFAR-10	CIFAR-100	MNIST	COCO	LFW
CP	ALS	5.289	4.584	2.710	5.850	4.085
CP	SGD	1060.455	1019.432	0.193	2335.060	6657.985
CP	NAG	280.432	256.196	0.400	1860.660	1.317
CP	Adam	2587.467	2771.068	2062.562	6667.673	6397.708
CP	RMSProp	2013.424	2620.088	2082.481	5588.660	4975.279
CP	SAGA	1141.374	1160.775	0.191	3504.593	3692.471
CP	AdaGrad	1768.562	2324.147	959.408	3729.306	6269.536
CP	NCG	315.132	165.983	4.910	778.279	716.355
CP	L-BFGS	2389.839	2762.555	2326.405	5936.053	5494.634
CP	VecHGrad	200.417	583.117	644.445	1128.358	1866.799
DEDICOM	ALS	21.280	70.820	14.469	55.783	158.946
DEDICOM	SGD	1826.214	1751.355	1758.625	1775.100	1145.594
DEDICOM	NAG	30.847	25.820	240.587	43.003	49.518
DEDICOM	Adam	2105.825	2128.626	1791.295	2056.036	1992.987
DEDICOM	RMSProp	1233.237	1129.172	993.429	1140.844	1027.007
DEDICOM	SAGA	27.859	30.970	64.440	28.319	32.154
DEDICOM	AdaGrad	196.208	266.057	1856.267	2020.417	2027.370
DEDICOM	NCG	2524.762	644.067	236.868	1665.704	4219.446
DEDICOM	L-BFGS	1568.677	1519.808	1209.971	1857.267	1364.027
DEDICOM	VecHGrad	592.688	918.439	412.623	607.254	854.839
PARATUCK2	ALS	225.952	209.978	230.392	589.437	625.668
PARATUCK2	SGD	1953.609	2625.722	2067.727	3002.172	2745.380
PARATUCK2	NAG	48.468	48.724	285.679	76.811	72.068
PARATUCK2	Adam	2628.211	2657.387	2081.996	2719.519	2709.638
PARATUCK2	RMSProp	1407.752	1156.370	1092.156	1352.057	1042.899
PARATUCK2	SAGA	74.248	70.952	120.861	71.398	86.682
PARATUCK2	AdaGrad	2595.478	2626.939	2073.777	292.564	2795.260
PARATUCK2	NCG	150.196	1390.013	928.071	1586.523	82.701
PARATUCK2	L-BFGS	2780.658	2656.062	2188.253	3522.249	2822.661
PARATUCK2	VecHGrad	885.246	1149.594	1241.425	1075.570	1222.827

Table 4. PARATUCK2 experimental mean convergence rate per step.

ALS	SGD	NAG	Adam	RMSProp	SAGA	AdaGrad	NCG	L-BFGS	VecHGrad
0.958	1.004	1.010	1.009	0.992	0.994	0.983	1.376	1.452	**1.551**

for PARATUCK2 TD. The latter underlines the best differences between the optimizers. The highest convergence rate was obtained by VecHGrad, followed by L-BFGS and NCG. Similar values were obtained for the other decomposition. Thus, we can conclude that VecHGrad is capable to solve accurately and efficiently TDs outperforming popular machine learning gradient descent algorithms.

6   Conclusion

We introduced VecHGrad, a Vector Hessian Gradient optimization method, to solve accurately linear algebra error minimization problems. VecHGrad uses a strong Wolfe's line search, crucial for fast convergence and accurate resolution, with partial information of the second derivative. We conducted experiments on five real data sets, CIFAR10, CIFAR100, MNIST, COCO and LFW, very popular in machine learning and deep learning. We highlighted that VecHGrad is capable to outperform widely-used gradient-based resolution methods, such as Adam, RMSProp or Adagrad, on three different tensor decompositions, CP, DEDICOM and PARATUCK2, offering different levels of linear algebra complexity. We emphasized our experiments with machine learning optimizers since VecHGrad can be easily applied to solve machine learning error minimization problems. Surprisingly, the runtimes of the gradient-based and the Hessian-based optimization methods were very similar as the runtime per step of the gradient-based methods was slightly faster, but their convergence per step was lower. Therefore, gradient-based optimization methods require more iterations to converge. Furthermore, the accuracy of some of the popular schemes, such as NAG, was fairly poor while requiring a similar runtime than that of the other methods. Future work will concentrate on the influence of the adaptive line search to other machine learning optimizers. We observed that the performance of the algorithms is strongly correlated to the performance of the adaptive line search optimization. Simultaneously, we will look to reduce the memory cost of the adaptive line search as it has a crucial impact on a GPU implementation. We will finally provide a Python and PyTorch public implementations of our method to answer the need of the machine learning and deep learning communities.

References

1. Globerson, A., Chechik, G., Pereira, F., Tishby, N.: Euclidean embedding of co-occurrence data. J. Mach. Learn. Res. (2007)
2. Zhou, S., Erfani, S.M., Bailey, J.: Online CP decomposition for sparse tensors (2018)
3. Kolda, T.G., Bader, B.W.: Tensor decompositions and applications. SIAM Rev. **51**, 455–500 (2009)
4. Carroll, J.D., Chang, J.J.: Analysis of individual differences in multidimensional scaling via an N-way generalization of "Eckart-Young" decomposition. Psychometrika **35**, 283–319 (1970)

5. Harshman, R.A.: Foundations of the parafac procedure: models and conditions for an explanatory multimodal factor analysis (1970)
6. Bader, B.W., Harshman, R.A., Kolda, T.G.: Temporal analysis of semantic graphs using ASALSAN. In: ICDM. IEEE (2007)
7. Harshman, R.A., Lundy, M.E.: Uniqueness proof for a family of models sharing features of tucker's three-mode factor analysis and PARAFAC/CANDECOMP. Psychometrika **61**, 133–154 (1996)
8. Charlier, J.H.J., Falk, E., Hilger, J., et al.: User-device authentication in mobile banking using APHEN for PARATUCK2 tensor decomposition. In: ICDMW. IEEE (2018)
9. Robbins, H., Monro, S.: A stochastic approximation method. Ann. Math. Stat. **22**, 400–407 (1951)
10. Kiefer, J., Wolfowitz, J., et al.: Stochastic estimation of the maximum of a regression function. Ann. Math. Stat. **23**, 462–466 (1952)
11. Wright, S., Nocedal, J.: Numerical optimization. Springer, New York (1999). https://doi.org/10.1007/978-0-387-40065-5
12. Duchi, J., Hazan, E., Singer, Y.: Adaptive subgradient methods for online learning and stochastic optimization. J. Mach. Learn. Res. (2011)
13. Kingma, D.P., Ba, J.: Adam: a method for stochastic optimization. arXiv preprint arXiv:1412.6980 (2014)
14. Nesterov, Y., et al.: Gradient methods for minimizing composite objective function (2007)
15. Defazio, A., Bach, F., Lacoste-Julien, S.: SAGA: a fast incremental gradient method with support for non-strongly convex composite objectives. In: Advances in Neural Information Processing Systems (2014)
16. Gupta, V., Koren, T., Singer, Y.: Shampoo: preconditioned stochastic tensor optimization. arXiv preprint arXiv:1802.09568 (2018)
17. Acar, E., Dunlavy, D.M., Kolda, T.G.: A scalable optimization approach for fitting canonical tensor decompositions. J. Chemometr. (2011)
18. Hinton, G., Srivastava, N., Swersky, K.: RMSprop: divide the gradient by a running average of its recent magnitude. In: Neural Networks for Machine Learning, Coursera Lecture 6e (2012)
19. Harshman, R.A.: Models for analysis of asymmetrical relationships among n objects or stimuli. In: First Joint Meeting of the Psychometric Society and the Society of Mathematical Psychology, Hamilton, Ontario (1978)
20. Bro, R.: Multi-way analysis in the food industry: models, algorithms, and applications. Ph.D. thesis, Københavns UniversitetKøbenhavns Universitet (1998)
21. Charlier, J., Hilger, J., et al.: Non-negative paratuck2 tensor decomposition combined to LSTM network for smart contracts profiling. In: BigComp. IEEE (2018)
22. Acar, E., Kolda, T.G., Dunlavy, D.M.: All-at-once optimization for coupled matrix and tensor factorizations. arXiv preprint arXiv:1105.3422 (2011)
23. Maehara, T., Hayashi, K., Kawarabayashi, K.I.: Expected tensor decomposition with stochastic gradient descent. In: AAAI, pp. 1919–1925 (2016)
24. Schittkowski, K.: NLPQLP: a new Fortran implementation of a sequential quadratic programming algorithm for parallel computing. Report, Department of Mathematics, University of Bayreuth (2001)
25. Liu, D.C., Nocedal, J.: On the limited memory BFGS method for large scale optimization. Math. Programm. **45**, 503–528 (1989)

Sensitivity to Risk Profiles of Users When Developing AI Systems

Robin Cohen, Rishav Raj Agarwal$^{(\boxtimes)}$, Dhruv Kumar, Alexandre Parmentier, and Tsz Him Leung

School of Computer Science, University of Waterloo, Waterloo, Canada
{rcohen,rragarwal,d35kumar,aparment,th4heung}@uwaterloo.ca

Abstract. The AI community today has renewed concern about the social implications of the models they design, imagining the impact of deployed systems. One thrust has been to reflect on issues of fairness and explainability before the design process begins. There is increasing awareness as well of the need to engender trust from users, examining the origins of mistrust as well as the value of multiagent trust modelling solutions. In this paper, we argue that social AI efforts to date often imagine a homogenous user base and those models which do support differing solutions for users with different profiles have not yet examined one important consideration upon which trusted AI may depend: the risk profile of the user. We suggest how user risk attitudes can be integrated into approaches that try to reason about such dilemmas as sacrificing optimality for the sake of explainability. In the end, we reveal that it is challenging to be satisfying the myriad needs of users in their desire to be more comfortable accepting AI solutions and conclude that trade-offs need to be examined and balanced. We advocate reasoning about these tradeoffs concerning user models and risk profiles, as we design the decision making algorithms of our systems.

Keywords: Position paper · Trusted AI · Risk profiles · Explainability

1 Introduction and Background

This position paper argues that deciding how to design AI systems which will be accepted by human users is a process where the specific profile of the user needs to be considered. We frame this discussion in terms of engendering trust in AI and advocate for reasoning about user risk profiles, when deciding whether to sacrifice accuracy for explainability. We note that several researchers have already suggested that differing user profiles may come into play, when studying mistrust in AI. For example, Salem et al. [20] assessed trust in robots as the participants' willingness to cooperate with the robot when it makes several unusual and usual requests. The nature of the task and the participant's personality were both considered to be deciding factors. Rossi et al. in [18,19] delved further into which aspects of personality may be necessary: agreeableness, conscientiousness, stable

© Springer Nature Switzerland AG 2020
C. Goutte and X. Zhu (Eds.): Canadian AI 2020, LNAI 12109, pp. 138–150, 2020.
https://doi.org/10.1007/978-3-030-47358-7_13

emotionality all suggested higher disposition for assuming benevolence and thus, a tendency to adopt a trusting stance. Extroverts were also more likely to trust a robot. How trusted agents differ from trusted humans is another theme. De Melo et al. [15,16] indicate that humans experience less guilt when dealing with agent partners than humans, and also reveal that humans are concerned when agents adopt different values during bargaining. That humans are less tolerant of mistakes made by AI agents than by humans has also been observed [25]. These papers draw out some of the differing emotional reactions to AI from users. A final paper that helps to emphasize the importance of considering personalized solutions, when developing AI approaches that consider human acceptance, is that of Anjomshoae et al. [1]. This work draws out the critical value of providing explanations to non-experts but also reveals that few proposals have addressed the significant concern of personalization and context-awareness. Indeed, without sufficient explanation, users may ascribe unfounded mental states to robots that are operating on their behalf [10], so that it is essential to address explainability well. Increased confidence can also arise with sufficient explanation [17], which allows the receiver to perceive the sender's state of mind.

The fact that risk profiles may hold the key to determining how to design our systems is a theme that emerges from the work of Mayer et al. [14] exploring organizational trust. The authors suggest that perceived trustworthiness is comprised of the components of benevolence (belief in wanting good), integrity (adhering to desired principles) and ability (having necessary skills). They raise the point that risk-taking actions may influence future perceptions of trustworthiness.

2   Desiderata for Trusted AI

As outlined in [4], there are a variety of reasons why individuals and organizations may be concerned about employing AI solutions for decision making. These include issues of fairness, explainability, ethics and safety. One question to consider is whether multiple concerns can be addressed simultaneously, with any of the efforts today to adjust our AI solutions in order to engender trust. For instance, are current approaches to address issues of fairness for AI systems in use also helping to increase transparency? We return to discuss the challenge of taking multiple concerns into consideration at once, towards the end of the paper.

The proposal of [4] is to make better use of existing frameworks for modelling agent trustworthiness, precisely in order to compare competing proposals for enabling trusted AI solutions for users. Trust modelling algorithms can identify less reliable sources either by learning from direct experience, together with a prediction of future behaviour or by interpreting reports received from other agents in the environment, judged according to their inherent reliability. The authors sketch how the use of trust modelling, injected into an environment for judging competing explanations from differing sources, can assist in determining which overall consensus is most dependable. This may be effective for settings such as gauging the value of supervised learning. A specific proposal outlining how to integrate trust modelling into a system aimed at promoting trusted AI is not offered in this short paper. The paper at least argues that not all efforts to address anyone of the desiderata (fairness, accountability, transparency) are equally valuable, and some comparisons are thus necessary.

Our view, however, is that for any attempt to improve outcomes regarding even one of these concerns, there will be a tradeoff. Furthermore, ultimately, user preferences should determine how those tradeoffs are modelled, and thus which competing goal should take precedence. One researcher with the desire to identify user needs for the sake of improved acceptance from users is Kambhampati [12]. References such as [22] suggest that planning and explainability should go hand in hand. Which kinds of explanations provide the best outcomes depend at times on the perceived model differences of the users [26]. There is also an effort to describe problems in a space of plan interpretability [2] where the computational capability of the observer is an issue. Agents can opt to make intentions clear to users or to obfuscate. Explicable planning can be viewed as an effort to minimize the distance from the plan expected by the human as well [2] so that mappings of plan distances need to be estimated and reasoned with. True human-aware planning [22] makes adjustments during the plan generation process itself, effectively beginning to reason about trading off the cost and computation time of plans in a way that serves the human observer best. In essence, the main tradeoff is between sacrificing optimality against the cost of an explanation. What would be especially valuable to consider are different ways in which these sacrifices can be decided based on the particular user at hand and their specific preferences. One relevant dimension that we choose to explore here is that of the user's risk attitude.

3   Reasoning with Differing User Preferences for Trusted AI

In order to support our position that it is valuable for efforts towards trusted AI to consider specific risk preferences of users, we offer three primary arguments. The first two arguments are framed within the context of research which suggests that agents should adopt less optimal but safer plans in scenarios where users opt to observe the plans or the execution of plans of intelligent agents [12]. The first argument opens up the process of requesting observability, clarifying that despite a user's initial risk profile, they may progressively update their preferences about the Agent's plan and its need for explainability. The second argument distinguishes the concept of user trust in an agent's plan and the notion of user's inherent risk profile. We outline how cautious users may still prefer less safe plans under certain circumstances. After discussing how risk profiles can influence trusted AI solutions for explainability, we then briefly explain how fairness decisions may also be influenced by risk profiles of users, as our third argument. We conclude with suggested steps forward to continue to map out user-specific approaches for engendering user trust.

3.1 Game-Theoretic Reasoning

We are considering one especially relevant starting point for framing an avenue to integrate reasoning about user preferences as the detailed proposal of [21], which reflects on the kind of costs that users may need to bear within the setting of a game-theoretic model of trust. Interestingly, the risks associated with robots or humans, making certain decisions with respect to their partnership for executing the real-world plan are discussed within this paper as well. The framework at least suggests that each user may have differing preferences, so that decisions about actions such as observing the Robot or not will incur costs that users may willingly opt to endure, again with certain consequences of doing so with respect to optimality. Borrowing a similar game-theoretic framework, we begin to study how user risk profiles may end up suggesting differing outcomes for the trusted AI effort. Our game-theoretic framework for studying AI behaviour under assumptions of risk is as follows.

We first note that in the model of [21], the Robot has a model of the Human's assessment of the Robot. The Human has an option to monitor the Robot's behaviour and stop execution if needed. Since meta-models of human behaviour may be challenging to learn and interpret, we recast the goals and constraints of each actor based on the risk profiles of the humans and also use the word Agent instead of Robot to represent the AI entity. Finally, we re-imagine the cost of inconvenience for the Human as the risk that the Human takes for allowing the Agent to create and execute the plan. We believe that the success of a plan will depend on the explainability, i.e. how well the Human understands the solution given by the Agent. The Agent should have a cost of explainability. Our framework, outlined below, assumes that the Human arrives with a known general risk profile and that the Agent primarily aims to avoid the cost of not achieving the goal at hand. This formulation thus moves beyond the more vague concept of Human mental model, which [21] assumes to be the basis for the Agent's reasoning. For now, we imagine that explainability incurs an additional computational cost, beyond that of making or executing a plan. Our framework proceeds as follows.

1. Agent is the artificial intelligence agent who has the following properties:
   (a) Agent is uncertain of Human's risk assessment of the Agent but knows that their risk profile of the Human is the space of all possible risk profiles. Thus, it has a perceived risk profile that we denote as $\mathcal{R}^H$.
   (b) Agent has plan π_p. Agent incurs some cost for not achieving the goal.
2. Human is the human actor who requests some information from the Agent.
   (a) Human has a risk profile R^H.
   (b) If Human believes a plan is risky, it can observe and stop at any time. It incurs the cost of observing the partially executed plan and a cost of the inconvenience of not achieving goal.
   (c) Human has non zero cost of observation.
   (d) Human is rational, i.e. they only stop execution if the plan is too risky.

3. **Plan**
    (a) A plan π_p is a set of sequential decisions made by the agent. π_s is the safest plan that satisfies all of the models of R^H and π_r satisfies none.

The Human has following strategies: Observe, stop execution at that time $(S_{O,\sim E})$; observe at some time but not stop the execution $(S_{O,E})$; not observe and not stop $(S_{\sim O,E})$; and not observe and stop at any time$(S_{\sim O,\sim E})$.

The payoffs for the Human and Agent for a plan are given in Table 1. Note there is a negative sign as the actor incurs a penalty equal to the cost. These calculations are based on the following utilities.

1. **Agent**
    (a) Cost of making plan is $C_P^A(\pi_p)$.
    (b) Cost of explaining is $C_E^A(\pi_p)$.
    (c) We can say that $C_E^A(\pi_p) > C_P^A(\pi_p)$ i.e. cost of explaining a plan is greater than cost of making the plan.
    (d) Cost of explaining until a partial plan $(\hat{\pi}_p)$ will be less than cost of explaining entire plan (π_p) i.e., $C_E^A(\pi_p) > C_E^A(\hat{\pi}_p)$
    (e) Cost of not achieving goal (G) is $C_{\hat{G}}^A$. We can assume that the safest plan doesn't have a cost of failure i.e., $C_{\hat{G}}^A = 0$ when $\pi_p = \pi_s$.
2. **Human**
    (a) Cost of observing the plan until some plan $(\hat{\pi}_p)$ has been executed is $C_E^H(\hat{\pi}_p)$.
    (b) Cost of observing at the end is $C_E^H(\pi_p)$.
    (c) Cost of not achieving goal is $C_{\hat{G}}^H$.
        $C_{\hat{G}}^H = 0$ when $\pi_p = \pi_s$.
    (d) Risk of executing a plan (π_p) is $R^H(\pi_p)$.
        We can assume that there is no risk in the safest plan i.e., $R^H(\pi_s) = 0$.
    (e) Risk of executing a full plan (π_p) is more than that of a partial plan $(\hat{\pi}_p)$ i.e. $R^H(\pi_p) > R^H(\hat{\pi}_p)$.

Table 1. Normal form game matrix for our formulation. The top line is the human's payoff and the bottom line is the agent's payoff.

	$S_{O,\sim E}$	$S_{\sim O,\sim E}$	$S_{\sim O,E}$	$S_{O,E}$
π_p	$-C_E^H(\hat{\pi}_p) - C_{\hat{G}}^H$	$-C_{\hat{G}}^H$	$-R^H(\pi_p)$	$-C_E^H(\hat{\pi}_p) - R_E^H(\pi_p - \hat{\pi}_p)$
	$-C_P^A(\pi_p) - C_{\hat{G}}^A - C_E^A(\hat{\pi}_p)$	$-C_P^A(\pi_p) - C_{\hat{G}}^A$	$-C_P^A(\pi_p) - C_E^A(\pi_p)$	$-C_P^A(\pi_p) - C_E^A(\pi_p)$
π_s	$-C_E^H(\pi_s)$	NA	0	$-C_E^H(\pi_s)$
	$-C_P^A(\pi_s) - C_E^A(\pi_s)$		$-C_P^A(\pi_s) - C_E^A(\pi_s)$	$-C_P^A(\pi_s) - C_E^A(\pi_s)$

Note that, if the Human completely trusts the system $(\pi_p = \pi_s)$, then the Nash Equilibrium is when the Human selects $S_{\sim O,E}$, i.e. the plan is executed without observing. If the Human completely distrusts the system $(\pi_p = \pi_r)$, then the Nash Equilibrium is when the Human selects $S_{\sim O,\sim E}$, i.e. they do not execute the plan without observing. In the discussion that follows, we explore how a Human's risk profile can influence the Agent's reasoning (moves in the game), equating the risk profile with certain assumptions about whether to require explainability.

1. **Risk Averse:** In this case, we can assume that $R^H(\pi_p) > C_E^H(\pi_p) + C_{\hat{G}}^H$, i.e. it is riskier to let any plan run its course without observing. The human will prefer either $S_{O,\sim E}$ or $S_{O,E}$. In order for the goal to be achieved, the Agent's costs $C_{\hat{G}}^A > C_E^A(\pi_p) - C_P^A(\hat{\pi}_p)$, i.e. the cost of achieving the goal must at least be greater than the cost of explaining the rest of the task.
2. **Risk Taking:** It is most important to get to the goal $R^H(\pi_p) < C_{\hat{G}}^H$. In this case, the Agent does not need to change its strategy. However, the Agent can make a more explainable plan to start with such that the Human does not need to observe at all.

We begin to sketch one way in which the Agent's decision procedure could be represented. The Agent needs to reason about whether to execute its plan at hand or to adjust its plans based on the user's perceived risk profile. The Agent will focus on explainability at the expense of accuracy (i.e. optimality), with a risk-averse human. It may forgo attempts at explainability to promote accuracy if the user is more risk-seeking. One interesting question is whether the Agent believes it can achieve the kind of explainability desired by the Human. The reasoning could proceed in the following fashion. If the Agent believes that the plan is more difficult to explain than it is worth trying to do so, it could integrate a kind of mixed-initiative dialogue with the user, asking for more direction. The Agent may view this as a scenario where the user's risk threshold may be exceeded, if that Agent proceeds with the plan at hand. The Human could be presented with options to either dismiss the Agent's execution of actions on its behalf (take manual control) or could instruct the Agent to begin reformulating a new plan from the current state (making more of an effort with the explainability). This procedure is intended to give more agency to the Human, and to prevent unintended outcomes that the Human's risk profile suggests should be avoided.

One interesting option is to have the Agent reflect on whether it should continue to execute its plan, each time a new step is taken. Such a decision procedure could be run as follows. At each decision element $p_i \in \pi_p$, the Agent reflects on whether the cost of achieving its goal is more than the cost of explanation. If so, it prompts the user. The user's risk profile can then be progressively learned during the dialogue as well. If the user opts to observe an execution, this can cause the Agent to increase its belief that the user is risk-averse. If the user proceeds for some time without requesting any observations, the Agent may opt to reduce its view of the user as risk-averse.

We view the primary decision-making of the Agent in a cycle of execution as one of assessing $\mathcal{R}^H(\pi)$ as $\mathcal{R}^H(\pi) > C_E^H(\pi) + G_{\hat{G}}^H$ and increasing the cost by $cost = cost + C_E^A(\pi)$. When steps of a plan have begun executing without a request for observability from the user, the Agent can reduce $\mathcal{R}^H$ (its view of the Human's risk profile); likewise, as Humans engage in observing, the Agent may increase its belief that the Human is risk-averse.

3.2   Distinguishing Trust and Risk Averseness

In Sengupta et al.'s [21] model, the human H has three strategies, to only observe the planning process (OP), to only observe the execution process (OE), or to not observe the robot at all (NO-OB). They argue that it is the Human's lack of trust in the Agent, which causes the Agent to opt for a safer plan (abandoning one which may be more accurate but also riskier). We note that while risk averseness of a user can translate into a distinct consideration for an Agent to reason about explaining its actions, it is also valuable to consider distinguishing between the Human's trust of the Agent and their inherent risk profile. In Sengupta et al.'s, original model, the trust boundary of the Human is derived to ensure that the Agent will never execute the risky plan. However, in some situations, the risky plan might be desired by the Human, as it incurs a lower cost. Thus, the Human chooses a mixed strategy $q = [q_P, q_E, (1 - q_P - q_E)]^T$ over the actions OP, OE and NO-OB respectively. For the risky plan to be worth executing, the expected utility given trust boundary q has to be higher for π_p than for π_s by α. $\alpha \in (0, 1]$ and we must consider the cost of execution of the plan $C_E^A(\pi_p)$

$$\mathbb{E}_q[U(\pi_s)] < \mathbb{E}_q[U(\pi_p)] \times \alpha \quad \text{i.e.}$$

$$
\begin{aligned}
(-C_P^A(\pi_s) - C_E^A(\pi_s)) < (&((-C_P^A(\pi_p) - C_E^A(\pi_p) - C_{\hat{G}}^A) \times q_P \\
&+ (-C_P^A(\pi_p) - C_E^A(\hat{\pi}_p) - C_{\hat{G}}^A) \times q_E \\
&+ (-C_P^A(\pi_p) - C_E^A(\pi_p)) \times (1 - q_P - q_E)) \times \alpha
\end{aligned}
\tag{1}
$$

If the Agent knows the user's risk profile and trust boundary, using the above equation would enable reasoning about whether the risky plan is worth executing instead of the safe plan. It is important to note, however, that while the user's risk profile (obtained through some initial questionnaire, for instance) is relatively stable, the user's trust boundary (determined by noticing how often they observe the planning process or execution) is constantly changing. If a user observes the Agent planning or executing a risky plan, for instance, then their trust in the Agent may be lowered. If a user observes the Agent executing a safe plan, their trust in the Agent may increase. As such, a progressive update of the model of the Human is necessary for the Agent's decision procedure. But merely relying on a modelling of the Human's trust may cause the Agent to be overly cautious in its planning.

4   Fairness and Explainability

In the previous section, we studied how designing for trusted AI may require reasoning about differing preferences from users with respect to accuracy and explainability. In this section, we delve further into the consideration of fairness, as another pillar of trusted AI for which distinct user risk profiles may suggest alternative designs for reasoning about costs and tradeoffs. We begin by exploring further the kinds of concerns that exist today regarding the fairness of AI systems used for decision making. We reflect on whether solutions for fairness can also satisfy a desire for explainability. We then discuss how to reason about individual preferences when trying to balance these two considerations. One view is that risk-averse individuals may be willing to accept systems that have been demonstrated to be fair, even if the methods for achieving this fairness cannot be fully explained.

As motivation for this position about tradeoffs for trusted AI, consider the case where an organization is running AI algorithms in order to make decisions on whether to hire a new employee. One might imagine setting the risk profile of the organization initially to be extremely risk-averse with respect to investment in fairness. This means that a solution that has inadequately considered fairness is problematic, as it could result in the organization being charged with discrimination. Consider as well the context where the hiring decisions are derived from modelling various features of successful employees in the past so that the solution is data-driven machine learning. In this case, the data used for training, as well as the reasoning about which features constitute an ideal employee, must both be under the microscope with respect to fairness. Now suppose that cases of clear discrimination do not arise, but that explanations for failing to hire a particular individual are desired, and these are difficult to articulate clearly, as they are tied to some sophisticated deep learning methodology. One might imagine being disappointed in the failure of explainability, but willing to put up with this consequence, if the required fairness has at least been attained. We begin with the first observation that the concern at hand may be with respect to the fairness of the data or with respect to the decision-making algorithm, which is making use of the data that is provided.

There are three major approaches towards achieving fairness in decision-making algorithms. We can modify the input data distribution to reduce bias at the source and thus train our model on the cleaner data. This method is called pre-processing. An alternate approach is to instead regulate the loss function of the classifier by adding fairness measures as regularization terms. This helps to control the tradeoff between fairness and the overall accuracy of the system. [9] showed that machine learning models amplified representation disparity over time and proposed to alter the loss function to minimize the worst-case risk for the minority groups. Notice that in the above approaches, we need to have access to the underlying data, which is not possible in many cases. Thus, in an alternative approach, one may first use the original data to train a classifier and then generate another classifier. This new classifier is independent of the original data and is created using just the original classifier and the protected attribute. However, we then need to ensure that the new classifier is fair by some definition of fairness. This approach is referred to as post-processing. One way to formalize this idea was proposed by [8], where they learn a probability distribution which controls whether to change the value of the predicted output from the original classifier or to keep it the same. The probability distribution is learnt by solving an optimization problem, which ensures that the desired fairness constraints are met while keeping the accuracy close to that of the original classifier.

Even if we have a user who is risk-averse with respect to explainability, there may be different options for addressing their needs. [6] uses crowd-sourcing in the form of Amazon Mechanical Turks to provide insights on how different styles of explanation impact people's fairness judgment of ML systems. They show people certain explanations of a model and ask them to differentiate between global explanations (describing the whole model) and local explanations (justifying a particular decision). They argue that it depends on the kinds of fairness issues and user profiles and that there is no one-size-fits-all solution for an effective explanation. Finally, they show that individuals' prior positions on algorithmic fairness also influence how they react to explanations. They argue for providing personalized forms of explanations to users. The authors do not, however, provide specific insights into how this personalization can be achieved.

Tradeoffs and Alternative Definitions for Fairness

In order to map out a decision procedure to respect user preferences with respect to fairness, the same concerns of Kambhampati and his coauthors [12, 21, 22], namely tradeoffs between accuracy and explainability, seem to arise. The point is that efforts to examine the success of the classification algorithm, with respect to the data on which it has been trained, may fail to consider bias, so that obtaining what appears to be well-respected performance accuracy, may still reflect critical failure with respect to fairness. A small example here helps to draw out why this might be the case.

Imagine we have a task to determine if a student will be successful in graduate school using some test scores. For simplicity, let us assume that students belong to two demographic groups. Students of one demographic group might be in the majority. If for certain reasons, students belonging to this majority demographics have higher scores (they are rich and give exams multiple times) than the students in the minority group. Therefore a classifier trained to get the best accuracy might more often reject the students belonging to the minority group. Thus, a more accurate solution does not mean a fair solution. Optimizing for average errors fits the majority error. We can obtain Pareto improvements by using group memberships. Current models tend to be inherently designed to be more accurate and not fair. They pick up the bias present in the data and, in some cases, amplify it [27]. [5] showed that if we have to satisfy even a single fairness criterion, we will sacrifice on the utility (accuracy).

We note as well that in some cases, users may not be risk-averse regarding fairness (e.g. they believe that this is not an issue for their particular application), while they may instead be much less forgiving of a failure to explain (e.g. unwilling to run software in their firm which cannot be justified to shareholders). For these individuals, the cost profile has changed and, all of this still needs to be considered against the primary aim of producing a system that provides important overall accuracy.

It is also important to note that there are alternate definitions of fairness, so that any effort to achieve this aim for trusted AI, may need to be attuned to a more precise measurement. Three widely used ways of modelling fairness are disparate impact, individual fairness or equalized odds. It may, therefore, be important to determine which of these considerations is paramount for the user, for their given context, before proceeding to reason about how best to design and where the tradeoffs between costs are best determined. Our discussion below also points out some challenges in achieving each of these differing perspectives, to date. Disparate Impact is measured by $Pr(Y = 1|A = 0)/Pr(Y = 1/A = 1) <= t$, where t is a threshold value, Y represents an outcome (1 is a positive event) and A represents the protected attribute (0 is the minority/protected class). Thus, it means that the probability of the classification event (getting admitted) should be independent of the protected attribute (race). For our scenario of admission to graduate school, getting admitted should, for instance, be independent of race. This measure may not ensure fairness, however, as the notion permits that we accept qualified applicants in the demographic $A = 1$ but unqualified individuals in $A = 0$, as long as the percentages of acceptances match. In addition, demographic parity may cripple utility in cases where the target variable Y is correlated with A. Individual Fairness [7] defines the notion that similar individuals should be treated similarly. How best to frame this metric is currently unresolved. Equalized Odds measure was defined by [8]. The constraint requires that the classifier(represented by Y) has equal true (y=1) and false (y=0) positive rates across the two demographics A=0 and A=1. $Pr(Y = 1|A = 0, C = y) = Pr(Y = 1|A = 1, C = y), y \in 0, 1$ C represents the true label present in the dataset. However, equalized odds enforces that the accuracy is equally high in all demographics, punishing models that perform well only on the majority.

5   Acquiring and Updating User Risk Profiles

A question to resolve is how best to represent a user's risk profile. A broad classification of the user as risk-averse or risk-seeking would enable an initial step forward with the reasoning proposed in this paper, namely to vary the outcomes of algorithms aiming to achieve trusted AI. We argue that it is desirable to make these adjustments. Any effort to model a user requires decisions along several fronts per the longstanding field of user modeling [13]: representing the user (what to model and how to represent), deciding when to update the user model, deciding how to reason with the user model to adjust an intelligent agent's decision making and how to acquire the user model. It is only the first of these elements that we have focused on so far.

Some research has explored the methods best able to elicit user risk profiles [3]. These authors mention a suggestion that individuals attempt to maximize some specific utility functions, as well. They ultimately caution against using conclusions within one limited context in order to predict what users will prefer in other scenarios. The fact that users may deviate from expected utility theory is also mentioned in the work of [11]. These authors then suggest that solutions for eliciting preferences should be able to function well, for cases where users are instead of making decisions based on cumulative prospect theory [24]. One element sketched in this paper is an expected minimax regret (EMR) heuristic, leading to the selection of queries for the user based on maintaining the lowest expected pairwise maximum regret, between pairs of possible decisions. It is therefore quite important to acknowledge the challenges of acquiring an effective representation of a user's risk profile, in order to then reason with that information when making higher-level decisions about how best to respect the various elements of trusted AI. One critical point is the fact that user risk profiles are not static, but may dynamically change as the user can experience various outcomes. Moreover, user risk profiles may also be quite varied, depending on the specific context. This point has been acknowledged well already in the multiagent trust modelling community, where trying to engender trust in a user may need to differ considerably, depending on which features are most important to that user (e.g. quality or cost, for e-commerce transactions) [23]. While the work above reinforces the point that it is important to develop complex strategies for properly determining the risk profiles of users, it is still the case that if some risk attitudes were known, AI system choices could be adjusted to increase acceptability from users.

6   Conclusion and Future Work

This paper proposes new directions for addressing trustworthiness of AI, presenting a particular viewpoint for designing future AI systems. We reveal that there are tradeoffs when aiming for trusted AI, that user risk profiles matter and can be integrated into decisions about how to design our systems, and that we should be considering solutions where different costs are more central, depending on the user. We have also reflected on basic considerations of accuracy, explainability and fairness, revealing that distinct needs and definitions may be at play; we have also sketched the process for reasoning about costs and risk profiles. While we comment on how broad, straightforward classification of user risk profiles enable an initial solution, we also acknowledge that this consideration is considerably more complex, requiring a collection of more complex reasoning processes.

The most obvious first step for future work is to expand upon representing user risk aversion, reflecting further on how these profiles are best known or acquired, and allowing finer-grained distinctions (for example, enabling users to be risk-averse in certain specified contexts, and more forgiving for other tasks being executed by the AI system). We have sketched only one small proposal for designing AI systems based on risk profiles and costs; ours is embedded in the game-theoretic analysis. Future work should proceed to calibrate gains for trusted AI using our approach. We should also consider many other desiderata of AI systems and continue to determine how decisions made by AI systems can be modulated by these considerations in a way that is faithful to individual user needs. We also acknowledge that reasoning more broadly about user preferences rather than risk profiles per se might open up a deeper set of approaches. But our position is that focusing on risk aversion as the critical element which may require sacrificing accuracy for explainability (or other trusted AI concerns) is a very powerful and effective stand-in,[1] one that shows promise as we continue our dialogue with those invested in securing better acceptance of AI from non-experts.

References

1. Anjomshoae, S., Främling, K., Najjar, A.: Explanations of black-box model predictions by contextual importance and utility. In: Calvaresi, D., Najjar, A., Schumacher, M., Främling, K. (eds.) EXTRAAMAS 2019. LNCS (LNAI), vol. 11763, pp. 95–109. Springer, Cham (2019). https://doi.org/10.1007/978-3-030-30391-4_6
2. Chakraborti, T., Kulkarni, A., Sreedharan, S., Smith, D.E., Kambhampati, S.: Explicability? legibility? predictability? transparency? privacy? security? the emerging landscape of interpretable agent behavior. In: Proceedings of the International Conference on Automated Planning and Scheduling, vol. 29, pp. 86–96 (2019)
3. Charness, G., Gneezy, U., Imas, A.: Experimental methods: eliciting risk preferences. J. Econ. Behav. Organ. **87**, 43–51 (2013)
4. Cohen, R., Schaekermann, M., Liu, S., Cormier, M.: Trusted AI and the contribution of trust modeling in multiagent systems. In: Proceedings of AAMAS, pp. 1644–1648 (2019)
5. Corbett-Davies, S., Pierson, E., Feller, A., Goel, S., Huq, A.: Algorithmic decision making and the cost of fairness. In: Proceedings of the 23rd ACM SIGKDD International Conference on Knowledge Discovery and Data Mining, pp. 797–806. ACM (2017)
6. Dodge, J., Liao, Q.V., Zhang, Y., Bellamy, R.K., Dugan, C.: Explaining models: an empirical study of how explanations impact fairness judgment. In: Proceedings of the 24th International Conference on Intelligent User Interfaces, pp. 275–285. ACM (2019)
7. Dwork, C., Hardt, M., Pitassi, T., Reingold, O., Zemel, R.: Fairness through awareness. In: Proceedings of the 3rd Innovations in Theoretical Computer Science Conference, pp. 214–226. ACM (2012)
8. Hardt, M., Price, E., Srebro, N., et al.: Equality of opportunity in supervised learning. In: Advances in Neural Information Processing Systems, pp. 3315–3323 (2016)
9. Hashimoto, T.B., Srivastava, M., Namkoong, H., Liang, P.: Fairness without demographics in repeated loss minimization. arXiv preprint arXiv:1806.08010 (2018)
10. Hellström, T., Bensch, S.: Understandable robots-what, why, and how. Paladyn J. Behav. Robot. **9**(1), 110–123 (2018)
11. Hines, G., Larson, K.: Preference elicitation for risky prospects. In: Proceedings of AAMAS, pp. 889–896 (2010)
12. Kambhampati, S.: Synthesizing explainable behavior for human-ai collaboration. In: Proceedings of AAMAS. Richland, SC, pp. 1–2 (2019)
13. Kass, R., Finin, T.: Modeling the user in natural language systems. Comput. Linguist. **14**(3), 5–22 (1988)
14. Mayer, R.C., Davis, J.H., Schoorman, F.D.: An integrative model of organizational trust. Acad. Manag. Rev. **20**(3), 709–734 (1995)

---

[1] In contrast with the more general term of human mental models proposed in [12].

15. Melo, C.D., Marsella, S., Gratch, J.: People do not feel guilty about exploiting machines. ACM Trans. Comput. Hum. Interac. (TOCHI) **23**(2), 8 (2016)
16. de Melo, C.M., Marsella, S., Gratch, J.: Do as I say, not as I do: challenges in delegating decisions to automated agents. In: Proceedings of AAMAS, pp. 949–956 (2016)
17. Nomura, T., Kawakami, K.: Relationships between robot's self-disclosures and human's anxiety toward robots. In: Proceedings of the 2011 IEEE/WIC/ACM International Conferences on Web Intelligence and Intelligent Agent Technology-vol. 03, pp. 66–69. IEEE Computer Society (2011)
18. Rossi, A., Dautenhahn, K., Koay, K.L., Walters, M.L.: The impact of peoples' personal dispositions and personalities on their trust of robots in an emergency scenario. Paladyn J. Behav. Robot. **9**(1), 137–154 (2018)
19. Rossi, A., Holthaus, P., Dautenhahn, K., Koay, K.L., Walters, M.L.: Getting to know pepper: effects of people's awareness of a robot's capabilities on their trust in the robot. In: Proceedings of the 6th International Conference on Human-Agent Interaction, pp. 246–252. ACM (2018)
20. Salem, M., Lakatos, G., Amirabdollahian, F., Dautenhahn, K.: Would you trust a (faulty) robot?: effects of error, task type and personality on human-robot cooperation and trust. In: Proceedings of the Tenth Annual ACM/IEEE International Conference on Human-Robot Interaction, pp. 141–148. ACM (2015)
21. Sengupta, S., Zahedi, Z., Kambhampati, S.: To monitor or to trust: observing robot's behavior based on a game-theoretic model of trust. In: Proceedings of the Trust Workshop at AAMAS (2019)
22. Sreedharan, S., Kambhampati, S., et al.: Balancing explicability and explanation in human-aware planning. In: 2017 AAAI Fall Symposium Series (2017)
23. Tran, T.T., Cohen, R., Langlois, E., Kates, P.: Establishing trust in multiagent environments: realizing the comprehensive trust management dream. TRUST@ AAMAS **1740**, 35–43 (2014)
24. Tversky, A., Kahneman, D.: Advances in prospect theory: cumulative representation of uncertainty. J. Risk Uncertainty **5**(4), 297–323 (1992)
25. Yuksel, B.F., Collisson, P., Czerwinski, M.: Brains or beauty: how to engender trust in user-agent interactions. ACM Trans. Internet Technol. (TOIT) **17**(1), 2 (2017)
26. Zahedi, Z., Olmo, A., Chakraborti, T., Sreedharan, S., Kambhampati, S.: Towards understanding user preferences for explanation types in model reconciliation. In: 2019 14th ACM/IEEE International Conference on Human-Robot Interaction (HRI), pp. 648–649. IEEE (2019)
27. Zhao, J., et al.: Men also like shopping: reducing gender bias amplification using corpus-level constraints. In: Proceedings of the 2017 Conference on Empirical Methods in Natural Language Processing (2017)

Forecasting Seat Counts in the 2019 Canadian Federal Election Using Twitter

Shainen M. Davidson[(✉)] and Kenton White

Advanced Symbolics, 109 York Street, Ottawa, ON, Canada
{shainen.davidson,kenton.white}@advancedsymbolics.com

Abstract. Previous attempts to predict elections using social media data have attempted to emulate traditional polling by predicting the share of votes received by major parties. However, in parliamentary elections, such as those held in Canada, the party who wins the most seats in parliament forms government (which may not be the party with the most votes nationally). In this paper, a method is presented which predicts seat counts using supervised learning with Twitter, polling, and historical election data. The model was trained on the 2015 Canadian federal election and was able to accurately predict the outcome of the 2019 Canadian federal election (a Liberal minority government, despite the Conservative Party winning the plurality of votes nationally). The model was designed before the 2019 election, and predictions were made public before election day. It is demonstrated that Twitter data about local candidates is more predictive than incumbency.

1 Introduction

Forecasting a Canadian federal election is a uniquely complex challenge. To form government in the first-past-the-post parliamentary system, a political party must win the most seats. A seat is won by gaining the plurality of votes in a geographic region (known colloquially in Canada as a "riding"). At the federal level, there are currently 338 ridings in Canada, ranging in population from 26,728 (Labrador, NL) to 132,000 (Brantford—Brant, ON)[1]. Thus accurately forecasting the winner of the 2019 Canadian federal election essentially requires forecasting 338 local elections, each with local candidates and unique local issues. Both traditional methods (i.e. polling), and newer social media-based prediction algorithms [3] generally tackle the more manageable task of forecasting national or provincial-wide vote ratios. While the party with the largest national vote total frequently does win the most seats, this is not always the case: in the 2019 Canadian federal election, the Conservative Party won the plurality of the national vote, while the Liberal Party won the most seats and thus was able to form government. This paper presents a model which overcomes this weakness by using supervised learning to forecast vote shares in each riding, and

---

[1] https://www.elections.ca/content.aspx?section=res&dir=cir/list&document=index338&lang=e.

C. Goutte and X. Zhu (Eds.): Canadian AI 2020, LNAI 12109, pp. 151–162, 2020.
https://doi.org/10.1007/978-3-030-47358-7_14

thus determine the seat totals of the major parties. The forecast of the model for the 2019 Canadian federal election accurately predicted a Liberal minority government.

In Canada, major polling firms will publicly release national, and sometimes regional, polling numbers in the lead-up to an election; however, they do not publish polling in individual ridings (due to the cost and difficulty of running hundreds of individual polls)[2]. Since riding level results are needed to accurately predict electoral outcomes, into this information vacuum step popular poll aggregation websites, such as the CBC Poll Tracker [9] and 338Canada.com [5]. These websites predict likely seat counts based on current regional polling aggregations, historical riding voting results, and other local factors that the model constructor adds at various weights using their political experience. This is essentially a data-driven version of how a seasoned political observer would determine the odds of a party winning a particular riding: take in the regional popularity of the party from the polls, and weigh it against the historical voting behaviour of the riding and the details of the local candidates.

In this paper the seasoned political observer is replaced with supervised learning. Similar to other seat models, it uses as input regional level polling; here the regional polling is forecasted with Twitter data using a method previously presented by White [18]. The model then combines the regional polling forecast with historical electoral data as well as locally relevant Twitter data to predict riding level results. The model deviates from previous seat models by weighting the effect of the inputs to the forecast by training on previous elections, as opposed to adding local effects on an ad-hoc basis.

The question of which information from social media is relevant for electoral outcomes is inconclusive, and some question its relevance at all. In the 2009 German election, the volume of twitter messages mentioning each party was closely correlated to the election results [16]; however, in the 2017 French presidential election, sentiment was correlated to the result, rather than volume [17]. In other cases, more complex relationships between the Twitter data and elections have been found [15], including the use of non-linear supervised models [1]. While many try to replace polling entirely with social media data, polling and Twitter data can be used in conjunction to interpolate polling results to regions and cases where polling information is lacking [2]. Note that all of these methods are predicting total vote-share, not seat counts, even in the case of parliamentary systems. When seat counts are given as a prediction, it is using an ad-hoc method to convert the vote-share prediction to seat-level election results [4].

Metaxas et al. [14] express skepticism of predictive power of social media data for elections, noting that most election studies choose the best method and features after the election has happened. The present work addresses this concern, as the model was chosen before the 2019 Federal Election, and the

---

[2] In the UK, the pollster YouGov has used multi-level regression and post-stratification with census and historical data to construct district level estimates from polling data with some success in the 2015 UK election [12].

forecast of the model was made public days before the election on the internet and on national television[3].

2   Twitter Data

All Twitter data is gathered from a sample of 273,830 Canadians on Twitter that was gathered using the network crawling algorithm Conditional Independence Coupling (CIC) [13]. White [18] demonstrated that a CIC sample generated for Toronto was broadly representative by gender, age, and race, with notable under-representation for ages 5–14 and 65+, and Filipinos.

The posting histories of the users in the sample are gathered using the Twitter API, and stored in an ElasticSearch database to allow for efficient querying based on post contents. As opposed to pulling Twitter data directly from the Twitter stream API, pulling from a sample of users has several advantages.

First, the Twitter stream is down-sampled and at peak times is rate limited, so unknown biases may creep into this data. In the case of the CIC sample, the results for any query are the full set of relevant posts of a representative sample of Twitter users.

Second, the full background of Twitter volume of the sample is known, and thus any tweets volume of interest may be normalized against this background. This especially relevant to take into account seasonal and daily patterns in Twitter use. E.g., Twitter use goes down substantially over the weekend, and so an actual increase in the interest in a political party on Saturday may still manifest as a lower absolute volume on that day; however, measuring the volume against the background rate of the sample will capture an increase in the ratio.

Third, the model here attempts to capture features at the riding level, and thus needs to identify the riding in which users reside. For locating users within cities, it is often possible to use the self-reported location field included in Twitter user-level data. However, there are typically many ridings within a city, and thus precise lat/long coordinates are required to place a user within a riding. A Twitter user may choose to enable a feature that publishes the geocoordinates of Tweets they make when possible; however, even then, only some tweets by the user are typically geotagged in this manner. By having the full history of users in the sample, it is possible to determine the likely riding of residence of a user if any of the tweets in their history are geotagged. Out of the 273,830 users in the full sample, 29,590 are pinpointed to ridings in this way, about 11%. For users who have tweeted in multiple ridings, a probability of them being located in a particular riding is used proportional to the ratio of their tweets in that riding.

An additional methodology to identify users' locations uses the text contents of users' histories; however, the attempted location resolution is generally at the city level [10]. To use text data to isolate a users ridings, a geographical construct relevant only to a particular election, the localising text would generally have to also be related to the same election. The obvious choice would be the candidates

---

[3] https://twitter.com/CTV_PowerPlay/status/1185310000501145601.

for election in the particular riding. In this study candidate mentions are used as a proxy for popularity of the local candidates in the model, so this method is not used for localising users in the present work.

3   Riding Prediction Model

There are two supervised learning models we use to predict the voter ratio within each riding, henceforth called the local vote ratio. First, Vector Autoregression with Exogeneous Variables (VARX) is used to forecast the voter ratio at the regional level (henceforth called the global vote ratio). Then the forecasted global vote ratio is distributed down to the riding level with linear regression, using historical and Twitter data to account for local variation.

3.1   Global Vote Ratio Forecasting

To start, a model is needed to forecast the global vote ratio for the new election:

$$V_p^e,\tag{1}$$

where p is the political party and e is the election.

The model follows the methodology presented in [18]. In brief, a Vector Autoregression with Exogeneous Variables (VARX) model is trained with time-series aggregated polls as the endogenous variable and time-series Twitter data as the exogeneous variables.

The Twitter data is the ratio of the number of users tweeting about any of the major parties in a given day, divided by the total number of users in the sample tweeting on that day. The aggregated poll numbers are from the CBC Poll Tracker [9]. The VARX model is implemented in the R package DSE [8].

3.2   Riding Distribution Model

Once a prediction is made for the global vote ratio, a prediction is needed for how these votes will be distributed among the ridings, where the ratio translates to seats.

To get the vote ratio in each riding, the local riding "lean" is forecasted, ie the difference between the global vote ratio per party and what the party receives in a particular riding [11]:

$$\text{vote ratio in riding } r \text{ for party } p = V_p^e + L_{p,r}^e,\tag{2}$$

where $L_{p,r}^e$ is the lean for party p in riding r in election e. This is the variable that contains all the unique local character of the riding, that which makes it different from the "average" riding: local candidate effects, local demographics, the history of political party activities in the area, local issues, etc. This is a good variable to look at, as riding lean is relatively stable between elections even when there is a change in government (Fig. 1). Any sophisticated model can be compared to the simplest assumption (call it $m1$) to forecast the local voter share:

$$L^c_{p,r} = L^o_{p,r}, \qquad (m1)$$

i.e. the riding lean in the current election c will be the same as it was in an old election o. More advanced models can be constructed as a perturbation on this simple model.

By building the distribution model in this way, it is assumed that there is a linear relationship between the global vote ratio and the local vote ratio, i.e. if the global vote total for party x goes up by 3% between elections, then the vote share of party x in a particular riding will also go up by 3%, at least as a first approximation. This assumption is driven by empirical rather than theoretical considerations. Perhaps a simpler distribution model is that the riding's share of the total votes received by the party is constant. I.e. if a riding gets 0.3% of the votes for party x in a previous election, then in another election where party x doubles their vote count, the riding still gets 0.3% of that doubled total vote, and thus we expect the vote ratio for that party to roughly double. This distribution model has the benefit that given a value for total votes, the distributed votes will add up to the total votes (not the case for riding lean). However, empirically this metric is not as consistent between elections as riding lean (Fig. 1). This implies a model of the electorate where the difference between ridings is largely due to groups of voters who vote consistently for the same party, and these voters give the riding its consistent lean.

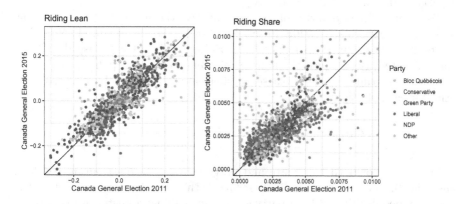

Fig. 1. On the left the riding lean for each of 338 federal ridings is plotted as it was in the 2011 compared to the 2016 federal elections. The diagonal line indicates where a riding with exactly the same lean in both elections would lie. On the right, the riding share of party votes is plotted for the two elections. The riding share deviates further between elections than riding lean.

Having defined the ad-hoc baseline model $m1$, the question is, can a simple supervised model do better? In the following, three models of increasing complexity are presented. Due to a paucity of data, all models are linear regression models to minimise over-fitting by keeping complexity to a minimum. As social media data is used, the models can only be fitted to data in the recent past where use of Twitter was common. Also, to keep the modelling simple, the models are only fitted to elections of the same type: predicting a future election based on data from a previous election of the same type is already an out-of-sample prediction, and thus risks over-fitting.

The first model adds features derived from the global vote ratio forecasted for the current election, as well as the old global vote ratio. Interactions between global vote share and voter lean are accounted for with quadratic terms. This model ($m2$) has equation

$$L_{p,r}^c \propto L_{p,r}^o + V_p^o + V_p^c + (V_p^o)(L_{p,r}^o) + (V_p^c)(L_{p,r}^o). \qquad (m2)$$

In addition to electoral ratios, incumbency, i.e. having a local candidate running for reelection, is known to elevate vote ratio [7], so that is added in model $m3$:

$$L_{p,r}^c \propto m2 + I_{p,r}, \qquad (m3)$$

where the incumbency $I_{p,r}$ is 1 if the candidate running is the current representative for the riding, and 0 otherwise.

Finally, twitter may be used to capture some local riding effects. To begin with, a proxy for the local standing of each party may be gleaned by counting the number of people in the sample in each riding who have discussed each party or party leader in the last 30 days, $N_{p,r}^P$.

To account for popularity of local candidates, the number of people in the sample mentioning each candidate in the past 30 days is measured, $N_{p,r}^C$. The results are not confined to a particular riding, to be able to use data from non-geolocated users. Since the number of users engaged in politics online has increased over time (Table 1), these numbers are normalized by converting to ratios for each election:

$$CR_{p,r} = \frac{N_{p,r}^C}{\sum_{p,r} N_{p,r}^C}, \qquad PR_{p,r} = \frac{N_{p,r}^P}{\sum_{p,r} N_{p,r}^P}. \qquad (3)$$

In the case of both candidate and party counts, the distribution has long tails (Fig. 2). In the case of candidate mentions this is particularly evident, as the number of people talking about party leaders and cabinet ministers is many times greater than those talking about even popular candidates with only local name recognition. To take this into account, the log of the distribution is calculated:

$$CL_{p,r} = \log(CR_{p,r} + 10^{-10}) + 10, \qquad (4)$$
$$PL_{p,r} = \log(PR_{p,r} + 10^{-10}) + 10. \qquad (5)$$

Both versions of the data are added to the final model ($m4$):

$$L_{p,r}^c \propto m3 + CR_{p,r} + CL_{p,r} + PR_{p,r} + PL_{p,r}. \qquad (m4)$$

Table 1. Percent of users in sample active in 30 days before election who mention a leader of a political party in that period. Comparing the 2015 and 2019 elections, the percent of users mentioning a party leader increases 50%, illustrating the increase in political speech on Twitter.

Party	2015 Election	2019 Election
Any Party	11.3%	16.9%
Liberal	8.2%	16.8%
Conservative	7.5%	13.3%
NDP	3.9%	8.8%
PPC	0.0%	4.7%
Green Party	1.4%	3.7%
Bloc Québécois	0.9%	1.1%

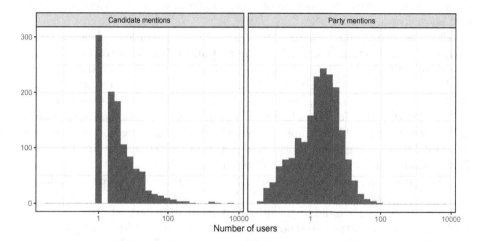

Fig. 2. Histogram of the number of users in the sample mentioning either a local riding candidate (left) and the number of users identified in a particular riding mentioning a particular party. Note that there may be fractional values for geolocated users, as users who tweet in more than one riding are split proportional to the number of tweets. The distributions have long tails (note the log scale).

By using both the pure ratio and the log rescaled values, the model may take into account two regimes where these values may have an effect: one expects a large change in voter behaviour when a truly famous candidate is up for election, and this effect can be represented by a linear response to $CR_{p,r}$. However, there will also be an effect due to locally known candidates which can be modelled as a linear response to $CL_{p,r}$.

4   Training and Testing the Model

The goal is to predict the 2019 Canadian federal election. Thus the model is fit using as labelled data the results of the next most recent election of the same type, the 2015 Canadian federal election.

The same fitted model parameters are used for all political parties[4], so the training data labels are the riding lean results in each of the 338 federal ridings for six different parties: Liberal, Conservative, NDP, Bloc Québécois[5], Green Party, and Other. Liberal, Conservative, and NDP had candidates in all 338 ridings, the Green Party in 336, the Bloc Québécois in all 78 ridings in Québec, and Other was a factor in 214 ridings. As the same fit is applied to all parties, there are 1,642 labelled data points.

To predict the 2015 Canadian federal election, the model uses historical data from the previous elections, in this case the 2012 Canadian federal election. In the case where a party was not present in a riding in a previous election, a lean of 0 is used as a default.

Electoral history data for the 2012 and 2015 Canadian federal elections are gathered from Elections Canada[6], at the poll and riding level (each riding has many polling regions within). Riding boundaries changed between the 2012 and 2015 elections, so the poll level data is used to interpolate the results from the 2012 elections into the riding boundaries used in the 2015 and 2019 elections.

The Twitter data for candidate mentions and party mentions within ridings are retrieved from a CIC sample generated for Canada with a sample size of 273,830. The data is gathered for messages within the 30 days leading up to the 2015 Canadian federal elections, on October 19, 2015.

The models $m2$, $m3$, and $m4$ are fit using this data with ordinary-least-squares linear regression. In Table 2, the P-values for the models are presented. The log-normalized candidate mentions in particular have strong predictive value, which is in contrast to studies of Gayo-Avello [6] and Metaxas [14] which report that tweet volume was not predictive of the 2010 US Congressional election. This may be due to the trend of increased political speech on Twitter since that election, as indicated by Table 1.

The model is used to forecast the 2019 Canadian Federal Election, which took place on October 21, 2019. The results presented here are using Twitter data gathered on October 20, 2019, the day before the election, from the same CIC sample of 273,830 users as the training data. In total there were 24,226 mentions by unique users of local candidates, and 6,320 mentions of parties by unique users located to ridings (out of 34,488 total mentions of parties).

---

[4] It would also be possible to fit a model for each party separately, but that is not done in this case to avoid over-fitting.

[5] This party only has candidates in the province of Québec.

[6] https://elections.ca/.

Table 2. P-values for the various models, fitted with historical data o from the 2012 Canadian federal election to predict current election c the 2015 Canadian federal election. While the most predictive feature is the old riding leans $L^o_{p,r}$ as expected, the next most predictive feature is from Twitter, the log-normalised local candidate mentions $CL_{p,r}$.

Term	$m2$	$m3$	$m4$
$L^o_{p,r}$	4×10^{-96}	6×10^{-92}	2×10^{-74}
V^o_p	4×10^{-1}	1×10^{-1}	4×10^{-2}
V^c_p	4×10^{-1}	9×10^{-1}	1×10^{-2}
$(V^o_p)(L^o_{p,r})$	4×10^{-2}	8×10^{-3}	2×10^{-2}
$(V^c_p)(L^o_{p,r})$	4×10^{-4}	3×10^{-4}	2×10^{-6}
$I_{p,r}$		1×10^{-7}	2×10^{-5}
$CR_{p,r}$			2×10^{-2}
$CL_{p,r}$			5×10^{-16}
$PR_{p,r}$			4×10^{-1}
$PL_{p,r}$			7×10^{-2}

The most likely seat counts per party predicted by the various models compared to the actual result are presented in Table 3. Note that this is not simply taking the prediction for the vote share in each riding and selecting the party with the highest vote share as having won the seat; this would be the prediction for the most likely specific vote share result in each riding. Since there is uncertainty involved in the prediction of each party vote share in each riding (estimated with the standard error), this must be taken into account. E.g., if the model predicts the two parties with greatest vote share in a particular riding are well within the prediction error, then that seat could really go to either party. To get the most likely seat count, the full distribution of possible vote share ratios with probabilities determined by the model must be integrated over. Here this is approximated, with probabilities for voter ratio for each party in each riding modeled as a Gaussian with mean the point estimate from the model and using the standard error from the model for the standard deviation, and is numerically integrated via Monte Carlo integration.

The results for the models are presented in Table 3, compared to the actual results, as well as the predictions of two popular polling aggregator websites. In Table 4, the sum of the absolute error of the seat counts for the models is compared.

Comparing the models, the greatest improvement is found in simply going from an ad-hoc model ($m1$) to the simplest supervised learning model ($m2$) (already we achieve better results than both poll aggregators, which while more sophisticated than $m1$, use ad-hoc parameters instead of fitting parameters through training). As the model gets more complicated, the accuracy improves. Of note, while the addition of the local twitter variables only increases the accuracy by about three seats, it is much more effective than including incumbency (a feature already well-known in the literature to be highly predictive [7]). This result, along with measure of model fit in Table 2, bolsters the case for local Twitter data to be an effective tool to fine-tune predictive election models. The accuracy of the final model is quite striking, accurately predicting the seat count of the winning party to within a single seat.

Table 3. The actual seat count for the 2019 Canadian Federal Election compared to the average seat count per party for the various models along with error at 95% CI. For comparison, we include the predictions of popular poll aggregating websites CBC Poll Tracker [9] and 338canada.com [5].

Party	Actual	$m1$	$m2$	$m3$	$m4$	CBC	338Canada
Liberal	157	146 ± 9	153 ± 10	159 ± 10	156 ± 9	137	142
Conservative	121	119 ± 8	120 ± 8	117 ± 8	117 ± 8	124	125
NDP	24	39 ± 6	32 ± 6	31 ± 6	31 ± 6	35	35
Bloc Québécois	32	30 ± 5	30 ± 5	28 ± 6	30 ± 5	39	33
Green Party	3	2 ± 2	2 ± 1	2 ± 1	2 ± 1	1	2
Other	1	1 ± 0	1 ± 0	1 ± 0	1 ± 0	1	0
PPC	0	1 ± 0	0 ± 0	0 ± 0	1 ± 0	1	1

Table 4. Total of absolute error on average seat count for each model. There is a large decrease in error when moving from the ad-hoc model $m1$ to the simplest trained model $m2$. There is again a smaller decrease in error when local Twitter data is added to the model ($m4$). Curiously, incumbency does not appear to have much effect on accuracy in this case ($m3$).

Model	Sum of absolute error
$m1$	31.25
$m2$	17.45
$m3$	17.46
$m4$	14.68

5   Conclusion and Future Work

A supervised learning model was presented that uses a VARX prediction of global vote ratios, historical election data, and Twitter data to predict the number of seats won by a party in an election. Thus the model forecasts the party that will form government, as opposed to the winner of the popular vote (as is done by traditional polling). The model was used to forecast the result of the 2019 Canadian Federal Election in advance, and was successful in predicting the overall result (Liberal minority), as well as predicting the seat total for the winning party within one seat.

This model generalises beyond the Canadian context to other parliamentary systems (e.g. Great Britain). Beyond that, this work can be expanded upon to forecast the results in elections which have a local structure but are not parliamentary systems. For example, the US Presidential election involves winning the majority of "electoral colleges", which translates to winning the popular vote in each of the 50 states. This model will serve as a template for a similar model to predict the winner of the 2020 US Presidential election.

References

1. Attarwala, A, Dimitrov, S., Obeidi, A.: How efficient is Twitter: predicting 2012 U.S. presidential elections using support vector machine via Twitter and comparing against IOWA electronic markets. In: 2017 Intelligent Systems Conference (IntelliSys). IEEE (2017)
2. Beauchamp, N.: Predicting and interpolating state-level polls using Twitter textual data. Am. J. Political Sci. **61**(2), 490–503 (2017)
3. Bilal, M, et al.: Predicting elections: social media data and techniques. In: 2019 International Conference on Engineering and Emerging Technologies (ICEET). IEEE (2019)
4. Burnap, P., et al.: 140 characters to victory?: Using Twitter to predict the UK 2015 general election. Electoral Stud. **41**, 230–233 (2016)
5. Fournier, P.J.: Electoral projection for 43rd Canadian general election. 338Canada.com. https://338canada.com/#2019. Accessed 15 Jan 2020
6. Gayo-Avello, D., Metaxas, P.T., Mustafaraj, E.: Limits of electoral predictions using Twitter. In: Proceedings of the Fifth International AAAI Conference on Weblogs and Social Media. Association for the Advancement of Artificial Intelligence (2011)
7. Gelman, A., King, G.: Estimating incumbency advantage without bias. Am. J. Political Sci. **34**(4), 1142–1164 (1990)
8. Gilbert, P.D.: Brief User's Guide: Dynamic Systems Estimation (2006). http://cran.r-project.org/web/packages/dse/vignettes/Guide.pdf
9. Grenier, É.: CBC Canada Poll Tracker. https://newsinteractives.cbc.ca/elections/poll-tracker/canada/. Accessed 15 Jan 2020
10. Han, B., Cook, P., Baldwin, T.: Text-based Twitter user geolocation prediction. J. Artif. Intell. Res. **49**, 451–500 (2014)
11. Jackman, S.: The predictive power of uniform swing. PS Political Sci. Polit. **47**(2), 317–321 (2014). https://doi.org/10.1017/S1049096514000109
12. Lauderdale, B.E., Bailey, D., Blumenau, J., Rivers, D.: Model-based pre-election polling for national and sub-national outcomes in the US and UK. Int. J. Forecast. **36**(2), 399–413 (2019)
13. Li, G.: Sampling graphical networks via conditional independence coupling of Markov chains. In: Khoury, R., Drummond, C. (eds.) AI 2016. LNCS (LNAI), vol. 9673, pp. 298–303. Springer, Cham (2016). https://doi.org/10.1007/978-3-319-34111-8_36
14. Metaxas, P.T., Mustafaraj, E., Gayo-Avello, D.: How (not) to predict elections. In: Privacy, Security, Risk and Trust (PASSAT) and 2011 IEEE Third Inernational Conference on Social Computing (SocialCom), pp. 165–171. IEEE (2011)

15. Singh, P., Sawhney, R.S., Kahlon, K.S.: Predicting the outcome of Spanish general elections 2016 using Twitter as a tool. In: Singh, D., Raman, B., Luhach, A.K., Lingras, P. (eds.) Advanced Informatics for Computing Research. CCIS, vol. 712, pp. 73–83. Springer, Singapore (2017). https://doi.org/10.1007/978-981-10-5780-9_7

16. Tumasjan, A., Sprenger, T.O., Sandner, P.G., Welpe, I.M.: Predicting elections with Twitter: what 140 characters reveal about political sentiment. In: ICWSM, vol. 10, pp. 178–185 (2010)

17. Wang, L., Gan, J. Q.: Prediction of the 2017 French election based on Twitter data analysis. In: 2017 9th Computer Science and Electronic Engineering (CEEC). IEEE (2017)

18. White, K.: Forecasting Canadian elections using Twitter. In: Khoury, R., Drummond, C. (eds.) AI 2016. LNCS (LNAI), vol. 9673, pp. 186–191. Springer, Cham (2016). https://doi.org/10.1007/978-3-319-34111-8_24

Adapting Ensemble Neural Networks to Clinical Prediction in High-Dimensional Settings

Simon de Montigny[1,2(✉)] and Philippe Broët[1,2,3]

[1] CHU Sainte-Justine Research Center, Montreal, QC, Canada
[2] School of Public Health, University of Montreal, Montreal, QC, Canada
simon.de.montigny@umontreal.ca
[3] University Paris-Saclay, University Paris-Sud, CESP, INSERM, Paris, France

Abstract. Neural networks have been investigated as models for survival data using a training criterion similar to that of the Cox proportional hazards model, a criterion not designed for clinical prediction. In this paper, we develop a new survival learning algorithm where a neural network ensemble minimizes the integrated Brier score. We compare the results obtained with this method to a standard implementation of random survival forests in R and to an ensemble of linear units.

Keywords: Neural networks · Survival analysis · Predictive models

1 Introduction

Neural networks (NNs) have been discussed for clinical use and survival analysis starting in the mid 90s, but early works had serious shortcomings [1]. Many survival deep learning models have now been proposed [2–8], with a clear focus on regularization and validation. Predictive accuracy of these NN models are usually assessed with the C-index [9] or the Brier score [10]. Limitations remain for clinical applications: these NNs have loss functions that don't measure predictive accuracy, and they are not well suited for high-dimensional data. In this work, we propose a new survival learning algorithm which combines predictions from an ensemble of NN models minimizing the integrated Brier score, optionally with L_1 penalization. We compare this procedure to the state-of-the-art ensemble approach which is the Random Survival Forest [11], and to a baseline ensemble of linear units that maximize partial likelihood under L_1 penalization. To evaluate performance in the high-dimensional setting, we created different survival data sets by adding non-informative covariates to the well-known Primary Biliary Cirrhosis (PBC) dataset [12].

2 Probabilistic Survival Model

The health status of a patient is measured until a certain event occurs or until he is lost to follow-up. Let the random variables T and C be the time-to-event

© Springer Nature Switzerland AG 2020
C. Goutte and X. Zhu (Eds.): Canadian AI 2020, LNAI 12109, pp. 163–169, 2020.
https://doi.org/10.1007/978-3-030-47358-7_15

and the censoring time, respectively. We define $X = \min(T, C)$ as the observed follow-up time and $\delta = 1_{(X=T)}$ as the event indicator. We assume noninformative and independent censoring for T and C [13]. The survival function of T is defined by $S(t) = P[T > t]$ $(t \geq 0)$, the hazard function by $\lambda(t) = -\left(\frac{d}{dt}S(t)\right)/S(t)$, and the cumulative hazard function by $\Lambda(t) = \int_0^t \lambda(s)ds$; we have $S(t) = \exp(-\Lambda(t))$.

To take into account that some patients are not susceptible to the event of interest, we use an improper survival function $S(t)$ such as $\lim_{t\to\infty} S(t) = \epsilon$ where ϵ $(0 < \epsilon < 1)$ is the tail defect; we then have $\Lambda(t) \leq -\ln \epsilon$. Broadly speaking, the random variable T takes the value ∞_+ for non-susceptible patients. In this context, we consider an improper semi-parametric model given by $S(t \mid Z) = \exp\left\{-\theta\exp[\phi(Z)]\left[1 - A(t)^{\exp[\psi(Z)]}\right]\right\}$ where $Z = (Z_1; \ldots; Z_p)$ is a p-dimensional vector of covariates, where $A(t)$ can be any function decreasing with time from one to zero, and where θ is a positive parameter. This type of model is a useful alternative to the standard Cox model which allows to investigate survival effects evolving in time. Here, $\phi(Z)$ and $\psi(Z)$ are two risk functions that correspond to the long-term effect (linked to the tail defect) and the short-term effect (linked to the time-to-event survival distribution for susceptible patients), respectively. The tail defect is given by $\epsilon = \exp[-\theta\exp(\phi(Z))]$. We define θ and $A(t)$ based on the Nelson-Aalen estimator of the cumulative hazard rate, noted $H(t)$, as follows. We set $\theta = \max\{H(t)\}$ and, given $H^-(t) = \max\{H(t)1_{(H(t)<\theta)}\}$ and $H^*(t) = H(t)1_{(H(t)<\theta)} + H^-(t)1_{(H(t)=\theta)}$, we set $A(t) = 1 - \theta^{-1}H^*(t)$. Moreover, for small values of $\psi(Z)$, $S(t|Z)$ can be re-expressed as a time-dependent proportional hazard model [14].

2.1   Neural Network Architecture Proposal

We propose to model the risk functions $\phi(Z)$ and $\psi(Z)$ with a NN having a p-dimensional input and a two-dimensional output $(o_{3,1}; o_{3,2})$. The network, shown in Fig. 1A, is described by $o_{a,b} = h_a\left(w_{a,b,0} + \sum_{j=1}^{10} w_{a,b,j}o_{a-1,j}\right)$ for layers $a = 2, 3$, and by $o_{1,b} = h_{1,b}\left(w_{1,b,0} + \sum_{j=1}^{p} w_{1,b,j}z_j\right)$ for layer 1. We use $h_1(x) = h_2(x) = \text{selu}(x)$, a scaled exponential linear unit [15], and $h_3(x) = 5\tanh(x)$, a scaled hyperbolic tangent. The resulting survival function is noted $\hat{S}(t|Z)$. A variant of the network, where input variables are subjected to L_1 penalization, is described in Fig. 1B. In this case, the equation for the first layer is given by $o_{1,b} = \phi_1\left(w_{1,b,0} + \sum_{j=1}^{p} w_{1,b,j}o_{0,j}\right)$ with $o_{0,j} = w_{0,j}z_j$, where $w_{0,j}$ is the weight of the jth unit of the penalization layer (note that these units have no bias term).

We base the loss function of the network on the integrated Brier score [16], defined by $\text{IBS} = \frac{1}{\tau}\int_0^\tau \text{BS}(t)\,dt$ where $\tau = \max(X_i\delta_i)$ is the time of the last uncensored event, and where $\text{BS}(t)$ is the Brier score at time t, a pointwise mean square error between $\hat{S}(t|Z)$ and what is observed. The observation variable takes value 1 if the event did not occur up to time t, value 0 if the event did occur, and it does not exist in case of censoring. To account for this third case, the

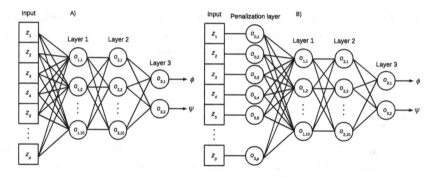

Fig. 1. A) Three layered NN. B) Modified NN with penalization layer.

error is weighted by the inverse probability of censoring. Thus, we have $\mathrm{BS}(t) = \frac{1}{n}\sum_{i=1}^{n}\left\{\left[\hat{S}\left(t|Z_i\right)\right]^2\hat{G}^{-1}(X_i)1_{(X_i\leq t,\delta_i=1)} + \left[1-\hat{S}\left(t|Z_i\right)\right]^2\hat{G}^{-1}(t)1_{(X_i>t)}\right\}$. The function $\hat{G}(t)$ is the nonparametric Kaplan-Meier estimate of the censoring distribution. The square root $\sqrt{\mathrm{BS}(t)}$ represents the deviation between the predicted outcome and the true event status. In the modified network, a penalization term $\lambda_1\sum_{j=1}^{p}|w_{0,j}|$ is added to the IBS, where λ_1 is the penalization parameter.

2.2 Classical Approaches

The baseline model (ensemble of linear units) that we use in our experiments is derived from the hazard $\lambda(t|Z) = \nu(t)e^{\phi(Z)}$, with $\nu(t)$ a baseline hazard, and from the partial likelihood function $L = \prod_{i=1}^{n} e^{\phi(Z_i)}\delta_i/\left(\sum_{j=1}^{n} e^{\phi(Z_j)}1_{(X_j\geq X_i)}\right)$. Model parameters in $\phi(Z)$ are adjusted to maximize L. Equivalently, we can minimize $\ell = -\sum_{i=1}^{n}\left(\phi(Z_i)\delta_i - \sum_{j=1}^{n}\phi(Z_j)1_{(X_j\geq X_i)}\right)$, that is the negative partial log-likelihood. We use ℓ as the loss for each unit of the ensemble. Applications of NNs to survival analysis have also focused on minimizing ℓ or its variants.

Random Survival Forest (RSF) is one of the most effective machine learning approaches for survival prediction. Broadly speaking, the RSF builds a series of binary decision trees from which a final prediction is obtained by combining the predictions from each individual tree. These latter tree-based learners are nonparametric approaches that partition recursively the predictor space into disjoint sub-regions that are homogeneous according to the outcome of interest. These partitions are obtained from a splitting criterion, usually the logrank statistic, that can be expressed as a score test from the partial likelihood function.

3 Experiment

3.1 Simulated Dataset

The PBC dataset has $n = 312$ observations and $p = 17$ covariates. To test the capacity of the models to select relevant covariates, we generated two modified

versions of the PBC dataset. For the second version, we added 500 uninformative variables (each of them, for every patient, generated randomly following an
uniform distribution on the interval 0–1), resulting in a dataset with $p = 517$
covariates. For the third version, we added 5000 uninformative variables in the
same manner instead of 500, resulting in a dataset with $p = 5017$ covariates.

3.2 Models

We tested four models on the dataset: a survival NN ensemble (SNNE), a SNNE
with L_1 penalization (SNNE-L_1), a RSF, and an ensemble of linear units (baseline). The survival random forest model is generated with the rfsrc function
(with default values) from the R package randomForestSRC. We implemented
the three other models in Python with Keras and TensorFlow. The ensemble
method comprises bagging with 1000 bootstrap samples for all four models.

The prediction of NN ensembles for a patient is the average of the survival
curves $\hat{S}(t|Z)$ from every network where the patient was out-of-bag. Note that
$H(t)$, θ, $A(t)$, $\hat{G}(t)$ and τ are computed in-bag. The process is similar for the
baseline model: the survival estimate for each bootstrap sample is given by
$\hat{S}(t|Z) = [K(t)]^{\exp\left[h\left(w_{1,0}+\sum_{j=1}^{p} w_{1,j}w_{0,j}z_j\right)\right]}$, where $w_{1,j}$ for $j = 0,\ldots,p$ are the
weights of the linear unit, where $w_{0,j}$ are the penalization weights, and where
$K(t) = \exp[-H(t)]$ is the Fleming-Harrington estimator.

For the SNNE model, we normalized the inputs (in-bag) and we used the
Glorot uniform initializer. We then trained each NN for 200 epochs with mini-
batches (size 32) with the default Adam optimizer, and we selected the best
weights with 15% in-bag validation. In addition, for the SNNE-L_1 model, we used
$\lambda_1 = 0.01$ and we initialized the penalization layer with a uniform distribution
(0.95–1.05 interval). For the baseline model, we used the same training setup
(with $\lambda_1 = 0.01$ for penalization), expect that we used the batch mode of training
(no validation set), because ℓ is not a sum of individual error terms (mini-batches
with validation have not been studied in the literature for partial likelihood).

Table 1. Out-of-bag prediction error, computed with $\tau = 4191$ (time of the last uncensored event). SNNE-L_1 shows best performance (values highlighted in bold). These
values do not include the penalization term for the SNNE-L_1 and baseline models.

Model	IBS ($p = 17$)	IBS ($p = 517$)	IBS ($p = 5017$)
SNNE	0.1217	0.1545	0.1898
SNNE-L_1	**0.1151**	**0.1310**	**0.1316**
RSF	0.1252	0.1550	0.1855
Baseline	0.2270	0.1956	0.2147

The out-of-bag IBS for all models and for the three datasets is given in
Table 1. The SNNE yields a slightly lower IBS value that the RSF, but this

advantage is lost in the presence of uninformative variables. The SNNE-L_1 has
the overall best performance. The baseline model performs notably worse that
the other models due to batch training without validation.

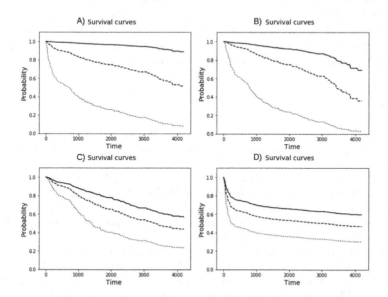

Fig. 2. Survival stratification for A) **SNNE** model, B) **SNNE-L_1** model, C) **RSF**
model, D) **baseline** model (solid curve for low-risk group, dashed curve for mid-risk
group, dotted curve for high-risk group)

To highlight the differences between models, we stratified the out-of-bag survival estimates (for the second version of the PBC dataset) into three groups
based on the survival probability value at the time of the last uncensored event:
patients in the upper quartile form the low-risk group, patients in the interquar-
tile range form the mid-risk group, and patients in the lower quartile form the
high-risk group. The groupwise survival curves obtained with each model are
shown in Fig. 2. Despite having similar performance, the SNNE and RSF models have very noticeably different survival curves, with the RSF model having
more pessimistic survival for the low-risk group and more optimistic survival for
the high-risk group. The SNNE-L_1 model makes a compromise between SNNE
and RSF for the survival of the low-risk group, whereas it predicts low survival for the high-risk group, like SNNE. The baseline model generates survival
curves that clearly display the proportional hazards assumption, and its predictions show a trend similar to those of RSF: survival is pessimistic in the low-risk
group and optimistic in the high-risk group.

Our results indicate that there is potential in using NNs for survival prediction based on the integrated Brier score. In particular, they allow penalization
strategies via modifications of the loss function. We showed that this strategy is
well suited to situations where few relevant predictors are expected.

4   Conclusion

In this paper, We have shown that an ensemble of NNs provides a valuable tool for survival prediction in high dimensional setting. The proposed strategy shows better predictive performance than survival random forests on the PBC dataset. The originality of the proposed model lies in its choice of loss function to train an NN ensemble with regularization. Future work will evaluate the interest of such approach in ultra-high dimensional genomics datasets.

References

1. Schwarzer, G., Vach, W., Schumacher, M.: On the misuses of artificial neural networks for prognostic and diagnostic classification in oncology. Stat. Med. **19**(4), 541–561 (2000)
2. Luck, M., Sylvain, T., Cardinal, H., Lodi, A., Bengio, Y.: Deep learning for patient-specific kidney graft survival analysis. CoRR abs/1705.10245 (2017)
3. Chapfuwa, P., et al.: Adversarial time-to-event modeling. In: ICML (2018)
4. Fotso, S.: Deep neural networks for survival analysis based on a multi-task framework. CoRR abs/1801.05512 (2018)
5. Giunchiglia, E., Nemchenko, A., van der Schaar, M.: RNN-SURV: a deep recurrent model for survival analysis. In: Kůrková, V., Manolopoulos, Y., Hammer, B., Iliadis, L., Maglogiannis, I. (eds.) ICANN 2018. LNCS, vol. 11141, pp. 23–32. Springer, Cham (2018). https://doi.org/10.1007/978-3-030-01424-7_3
6. Katzman, J., Shaham, U., Cloninger, A., Bates, J., Jiang, T., Kluger, Y.: DeepSurv: personalized treatment recommender system using a cox proportional hazards deep neural network. BMC Med. Res. Methodol. **18**, 24 (2018)
7. Manyam, R.B., Zhang, Y., Keeling, W.B., Binongo, J., Kayatta, M., Carter, S.: Deep learning approach for predicting 30 day readmissions after coronary artery bypass graft surgery. CoRR abs/1812.00596 (2018)
8. Nezhad, M.Z., Sadati, N., Yang, K., Zhu, D.: A deep active survival analysis approach for precision treatment recommendations: application of prostate cancer. Expert Syst. Appl. **115**, 16–26 (2019)
9. Harrell, F.E., Califf, R.M., Pryor, D.B., Lee, K.L., Rosati, R.A.: Evaluating the yield of medical tests. JAMA **247**(18), 2543–2546 (1982)
10. Brier, G.W.: Verification of forecasts expressed in terms of probability. Mon. Weather Rev. **78**(1), 1–3 (1950)
11. Ishwaran, H., Kogalur, U., Blackstone, E., Lauer, M.: Random survival forests. Ann. Appl. Stat. **2**(3), 841–860 (2008)
12. Therneau, T., Grambsch, P.: Modeling Survival Data: Extending the Cox Model. Statistics for Biology and Health. Springer, New York (2000). https://doi.org/10.1007/978-1-4757-3294-8
13. Fleming, T.R., Harrington, D.P.: Counting Processes and Survival Analysis. Wiley, Hoboken (2005)

14. Broët, P., De Rycke, Y., Tubert-Bitter, P., Lellouch, J., Asselain, B., Moreau, T.: A semiparametric approach for the two-sample comparison of survival times with long-term survivors. Biometrics **57**(3), 844–852 (2001)
15. Klambauer, G., Unterthiner, T., Mayr, A., Hochreiter, S.: Self-normalizing neural networks. CoRR abs/1706.02515 (2017)
16. Graf, E., Schmoor, C., Sauerbrei, W., Schumacher, M.: Assessment and comparison of prognostic classification schemes for survival data. Stat. Med. **18**(17–18), 2529–2545 (1999)

A Cost Skew Aware Predictive System for Chest Drain Management

Nicholas J. Denis[1], Danny French[2], Sebastien Gilbert[3], and Nathalie Japkowicz[4(✉)]

[1] Department of Mathematics, University of Ottawa, Ottawa, ON, Canada
[2] Department of Surgery, Dalhousie University, Halifax, NS, Canada
[3] Division of Thoracic Surgery, The Ottawa Hospital, Ottawa, ON, Canada
[4] Department of Computer Science, American University, Washington D.C., USA
japkowic@american.edu

Abstract. Many real world classification problems are skewed in terms of the cost of potential misclassifications. Clinical health care for individuals following pulmonary resection involves chest drainage tube management where the decision to remove or maintain a patient's chest drain post-surgery is one such skewed classification problem. This is because the associated cost of premature removal is significantly higher than delayed removal in terms of health risks, discomfort and economic factors. While recognition of a cost differential in a problem is straightforward, its implementation in a predictive system is not, because there is no simple way to quantify cost. We addressed this issue through the design of an evolutionary based optimization approach for cost matrices. In order to test our approach, we compared three different settings: one with no cost matrix, one where the cost matrix used is provided by the thoracic surgeons, and one where the cost matrix is optimized through our evolutionary algorithm. The results show that our optimization method for cost matrices yields a large improvement over the other two settings on most performance measures, including an almost 20% increase in overall accuracy. This is a surprising result since it suggests that cost matrices provided by experts may not be as useful as those derived by a computational optimization approach.

Keywords: Cost sensitive learning · Evolutionary algorithms · Health care

1 Introduction

After lung resection a chest drain is required to drain air and liquid from the pleural space. Patients typically stay in hospital until the chest drain is removed, however some patients requiring prolonged drainage can be sent home with a portable drainage device (pneumostat). Experienced health care providers manage chest drain considering many factors about the patient and the features of the drain. (Gilbert et al. 2015). Using retrospective patient data indicating what action was taken and whether or not that action was a good one, machine learning has the potential to create a clinical decision support system that can help health care providers optimize management of chest drains ultimately

© Springer Nature Switzerland AG 2020
C. Goutte and X. Zhu (Eds.): Canadian AI 2020, LNAI 12109, pp. 170–176, 2020.
https://doi.org/10.1007/978-3-030-47358-7_16

resulting in earlier discharge of patients while preserving healthcare safety (Shah et al. 2019). Currently, no chest drain management algorithm has been universally accepted.

In this study various machine learning approaches were explored with the aim of providing clinical decision support for health care staff to improve accuracy and precision of chest drain management decisions. This classification problem is complicated by unequal costs of adverse events. For example, delay in removal of the chest drain results in the patient remaining in hospital longer than necessary often causing patient dissatisfaction and poor use of hospital resources. On the other hand, premature removal of the chest drain requires re-insertion causing the patient pain and ultimately delaying discharge. Premature removal of the chest drain is, thus, considered to be costlier than delayed removal. However, setting a cost matrix to achieve the most desirable outcome is not a simple task. The objective of this study is to propose a method for improving the performance of machine learning algorithms by optimizing the cost matrix provided to them.

The particular paradigm that we chose for this optimization problem is that of Evolutionary Algorithm because of its simplicity and general measure of success. In particular, we used a genetic algorithm to optimize the cost matrix over multiple objectives. The approach was tested on clinical data collected prospectively in compliance with the research ethics board protocol used at the Ottawa Hospital.

The remainder of the paper is divided into five sections. Section 2 presents previous work used in our study. Section 3 describes our algorithm. Section 4 describes the domain in detail along with the pre-processing of the data. Section 5 presents our results. Section 6 concludes the paper and suggests some avenues for future work.

2 Previous Work

Our research falls into the area of Cost-Sensitive Learning. Cost-sensitive learning addresses the problem of skewed misclassification costs (Elkan 2001). Several approaches have been proposed to convert arbitrary cost insensitive classifiers into cost sensitive ones able to use a cost matrix. (Domingos 1999; Fan et al. 1999; Margineantu 2002; Zadrozny et al. 2003). In this work, instead of using a given cost matrix, we use an Evolutionary Algorithm (EA) to search for an optimal cost-matrix. EAs mimic the evolutionary process, where a specific trait is selected for through iterative breeding of populations. The iterative process in an EA is a stochastic approach to optimization. An EA approach was chosen because it requires less *a priori* knowledge of the range that the cost matrix entries are likely to belong to.

3 Cost Matrix Genetic Algorithm Optimization

Cost matrices are the tools used to account for the skewed nature of misclassification costs. In our clinical setting, for example, a cost matrix could reflect the fact that it is much more 'costly' for a health care provider to prematurely remove the final chest tube of a patient versus maintaining a chest tube for an excessive amount of time. Cost matrix construction may, thus, rely on domain specific knowledge. However, this study chose to use an agnostic approach allowing for a stochastic optimization protocol. Optimizing the entries of the cost matrix followed an EA approach. The EA approach focused on optimizing the actions: *remove* and *discharge with a pleural drain* (DPD) over multiple objectives. Here, the objectives we optimized were the following performance measures: false positive rate (FPR), precision, positive likelihood and negative likelihood.

We performed cost matrix EA optimization as follows. An initial population of 100 cost matrices was constructed at random. The fitness of the population was determined following 5-fold Cross-Validation of the cost-sensitive implementation of the classifier implementation. The associated cost matrix implementations were ranked based on the above mentioned performance measures, and the top five cost matrices for each measure were selected for survival and propagation into the next generation. The 40 surviving matrices were then randomly coupled together with each couple surviving, unchanged, into the following generation, while also producing three offspring matrices. For each surviving couple, progeny cost matrices were generated as follows: first, one offspring matrix was produced by taking the mean of its parents' entries, which can be interpreted as a cross-over event between parental genotypes. The other two children were clones of each of the individual parent matrices that underwent a mutation implemented by adding Gaussian noise with mean zero and variance $1/10^{th}$ of the cost matrix entries' current value. This process was repeated for a set number of generations. Finally, the matrix with the highest precision with respect to class label Remove, here denoted, $precision_{remove}$, was selected for further analysis.

Our multi-performance measure optimized approach was interpreted as optimizing over multiple performance measures, in some blended sense. This was important, as initial experiments optimizing over a single performance measure were often shown to succeed at the detriment of other as important metrics. Hence, this approach was interpreted as selecting a better balanced cost-matrix over multiple performance measures, while not necessarily optimizing for a single performance measure.

4   Domain Description and Pre-processing Methods

The data set used in this study consisted of preoperative, intraoperative and postoperative features collected from 67 patients. Data was obtained under the approval of the Ottawa Hospital Research Ethics Board. Preoperative features include variables such as age, BMI, gender, results of lung function testing and other clinically relevant patient-related factors. Intraoperative features are events that occurred during the surgery such as specifying which lobe of the lung was removed and how many staples were used. Postoperative features include data collected in the post-operative period such as chest x-ray findings, and a discrete time series representing the mean airflow from the chest drainage tube recorded every 10 min. At each recorded point in time, three actions can be taken relative to the chest drain: *Maintain, Remove, Discharge with Pleural Drain (DPD)*. Maintain and Remove are self-explanatory. DPD applies to patients requiring prolonged drainage but who can be sent home with a portable drainage device named pneumostat. We constructed a data set appropriate for machine learning methods as follows.

Instance Generation. Data instances were generated as follows. The raw data representing a patient comprises preoperative and intraoperative features as well as a time series of T airflow measurements, each occurring at discrete ten minute intervals. Hence, from a single patient's time series, every discrete ten-minute event post-surgery represents an instance where a classification can be made. Using the entire time series for each patient, we combined the pre- and intraoperative features together with the steps of the time series up until the present time, t. For example, a patient staying in the hospital for 72-hours post-surgery comprises of $72 * 6 = 432$ discrete events, each with a class label attributed to the action to be taken.

Instance Labeling. The data instances were assigned class labels corresponding to one of the three actions described earlier or a fourth label corresponding to the case where the label is uncertain (*Maybe*). Retrospective data include the time of chest drain removal, the time of discharge with a portable pleural drain, and the presence of adverse events. For example, if a removal was successful, the instance corresponding to that time of removal would be labeled as 'Remove'. On the other hand, if this removal led to adverse events, then this instance and the following ones would be labeled as 'Maintain'. 'Maybe' labels correspond to instance for which the clinical drain removal criteria was reached, t_{DR}, but the actual removal was not done. We thus don't know whether the removal would have been successful or not at these times.

Sampling. Since, given the nature of the data, the vast majority of class instances are of the classes Maintain and Maybe. Not all of these instances are necessary. Therefore, only a sub-sample were utilized. We used an exponential decay sampling technique, where the 1st most recent time point (with respect to the upper time boundary) is used, as well as the 2nd, 4th, 8th, 16th, etc. In this way, overfitting to individuals with longer hospital stays was reduced.

Feature Construction. As in current clinical practice, new postoperative features representing descriptive statistics were constructed from the data to summarize the time series. These include the minimum, maximum, mean, median, and total airflow volume over the prior 1, 2, 4, 6, 8, 10 and 12 h. Thus, each instance was represented by 56 features (44 pre- and intra-operative, and the 12 post-operative features just mentioned). From this dataset, a second dataset denoted "Mass", was constructed, where the postoperative features were normalized by the patient's body mass.

Feature Selection. To determine which of the provided or constructed features hold discriminative information, we implemented a strong majority rule voting ensemble of feature selection methods. Our approach, denoted FS1, pools multiple base feature selection approaches together and maintains only the features selected in at least 80% of the feature selection algorithms considered. The base feature selection algorithms used were correlation-based feature selection, information gain, Chi squared attribute evaluation, and classifier subset evaluation. Another feature selection approach (FS2/GilParam) selected pre- and intra- operative features determined by surgical professionals. We also ran experiments with no feature selection for comparison.

Class Imbalances. A patient may be required to have a chest drain in place for up to several weeks if a prolonged air leak from the lung tissues is present. Given that every 10-minute interval acts as an instance, almost all the instances in the training set are of class

Maintain. In order to deal with the class imbalance, two methods were implemented in addition to keeping the imbalance: under sampling from the majority classes and SMOTE (Chawla et al. 2002), to augment the minority classes.

Classifier Selection. The classifiers chosen for this study were J48, JRIP and PART. These classifiers were chosen for their model transparency which is of great importance in the medical domain. Additionally, bagging and boosting ensemble methods were implemented for each base classifier mentioned so as to verify that the human readable constraint is not overly limiting with respect to predictive ability.

5   Experimental Framework

Prior to evaluating our cost matrix optimization scheme we conducted a large study aimed at testing the various settings discussed in Sect. 4 and identifying the optimal combinations when no confusion matrix is used (no-CM). The results of this investigation revealed that the most effective and stable combinations of factors for this problem were: {SMOTE, PART}, {FS1, SMOTE, PART}, {Mass, SMOTE, PART} and {Mass, FS2, SMOTE, PART} (results not shown due to space limitations). We then sought to compare the performance of these four implementations, to two additional situations: when a cost matrix derived from domain experts is provided (Heuristic/H-CM) and when using a cost matrix optimized by our EA algorithm (CM-Opt).

The table below displays our results. The best result obtained for each metric and for each implementation is highlighted in bold. As can be seen from the table, the EA optimization approach outperforms the same implementations without the use of the EA optimization in virtually all cases, showing a large improvement over the No-CM and Heuristic CM approaches across most performance measures, including an almost 20% increase in overall accuracy. Another interesting observation is that in most cases performance measures were improving from No-CM to Heuristic CM to the EA approach, suggesting that the performance of classifiers using Heuristic CM were generally equal to or greater than No-CM implementations. This provides further evidence to suggest that using cost matrices was well suited for this learning problem.

For each given implementation, in order to determine if there was any significant difference in performance between the No-CM, Heuristic CM or EA optimization approaches the Wilcoxon's signed-rank test was performed. At $\alpha = 0.0005$, the EA optimization approach was significantly differently (better) to both No-CM and Heuristic CM for each of the four implementations considered. This is shown with a (*).

	{SMOTE, PART}			{FS1, SMOTE, PART}			{MASS, SMOTE, PART}		
	No-CM	H-CM	CM-Opt[*]	No-CM	H-CM	CM-Opt[*]	No-CM	H-CM	CM-Opt[*]
fpr_{re}	0.0649	0.0613	**0**	0.0720	0.0720	**0**	0.0649	0.0649	**0**
Pr_{re}	0.4303	0.4482	**1**	0.3987	0.3987	**1**	0.3767	0.3767	**1**
Sp_{re}	0.9351	0.9387	**1**	0.9280	0.9280	**1**	0.9351	0.9351	**1**
tpr_{DPD}	0.9000	0.9000	**1**	0.9000	0.9000	**1**	0.8000	0.8000	**1**
fpr_{DPD}	**0.0050**	**0.0050**	**0.0017**	**0.0149**	**0.0149**	0.0270	0.0116	0.0116	**0.0032**
Sp_{DPD}	0.9950	0.9950	**0.9983**	0.9851	0.9851	0.9730	0.9884	0.9884	**0.9968**
tpr_{Main}	0.8697	0.8720	**0.9823**	0.8228	0.8228	**0.9971**	0.8193	0.8193	**0.9867**
fpr_{Main}	0.1745	0.1745	**0.0174**	0.1712	0.1712	**0.0254**	0.1610	0.1610	**0.0591**
Pr_{main}	0.8664	0.8668	**0.9836**	0.8649	0.8649	**0.9791**	0.8665	0.8665	**0.9537**
Re_{mai}	0.8697	0.8720	**0.9823**	0.8228	0.8228	**0.9971**	0.8193	0.8193	**0.9867**
Sp_{Man}	0.8255	0.8255	**0.9827**	0.8288	0.8228	**0.9746**	0.8390	0.8390	**0.9409**
Pr_{May}	0.8611	0.8646	**0.9241**	0.7895	0.7861	**0.9952**	0.7735	0.7735	**0.9804**
Sp_{May}	0.9274	0.9263	**0.9534**	0.8891	0.8866	**0.9973**	0.8739	0.8739	**0.9856**
Acc	0.8126	0.8158	**0.9603**	0.7711	0.7695	**0.9602**	0.7774	0.7774	**0.9622**

6 Conclusion

This study is based on clinical data from patients admitted to hospital for pulmonary resection, after which clinically implemented chest drain management guidelines have been followed. The particular focus of this study was to test an evolutionary algorithm approach to optimize the cost matrix of the domain. Our algorithm is designed to learn from the errors of the health care providers and in so doing to improve upon the clinical decision to maintain or remove the chest drain. While the main novelty of the approach resides in the search for an optimal cost-matrix, significant data issues needed to be handled prior to considering the cost-matrix. The results show that our optimization method yields a large improvement over settings with no confusion matrix or with a heuristically constructed one.

There are several avenues to consider before testing our approach with clinicians. First, this study was conducted on only $N = 67$ patients. New patients will be added. Second, since this was a proof-of-concept study, we limited the computation time used during the evolutionary algorithm. A population size of 100 cost matrices was used, for a total of 100 generations. In future work, we plan to use more powerful machines to speed-up the process and explore a deeper search space. We also intend to experiment with different mating and mutation strategies. A third consideration regards the choice of classifiers. While we chose to stay away from 'black box' classifier, our future work will include stronger classifiers. Finally, there are a number of practical obstacles to be overcome including ways to ensure that the approach we suggest speeds up actual patient release time, taking into consideration hospital delays as well as patient preparation time prior to chest drain removal. Our future work will examine such issues as well.

References

Chawla, N.V., Bowyer, K.W., Hall, L.O., Kegelmeyer, W.P.: SMOTE: synthetic minority over-sampling technique. JAIR **16**, 321–357 (2002)

Domingos, P.: MetaCost: a general method for making classifiers cost-sensitive. In: KDD 1999, pp. 155–164 (1999)

Elkan, C.: The foundations of cost-sensitive learning. In: IJCAI 2001, pp. 973–978 (2001)

Fan, W., Stolfo, S., Zhang, J., Chan, P.: AdaCost: misclassification cost-sensitive boosting. In: ICML 1999, pp. 97–105 (1999)

Gilbert, S., et al.: Randomized trial of digital versus analog pleural drainage in patients with or without pulmonary air leak after lung resection. J. Thorac. Cardiovasc. Surg. **150**(5), 1243–1251 (2015)

Margineantu, Dragos D.: Class probability estimation and cost-sensitive classification decisions. In: Elomaa, T., Mannila, H., Toivonen, H. (eds.) ECML 2002. LNCS (LNAI), vol. 2430, pp. 270–281. Springer, Heidelberg (2002). https://doi.org/10.1007/3-540-36755-1_23

Shah, P., et al.: Artificial intelligence and machine learning in clinical development: a translational perspective. NPJ Digit. Med. **2**, 69 (2019)

Zadrozny, B., Langford, J., Abe, N.: Cost-sensitive learning by cost - proportionate example weighting. In: ICDM 2003, pp. 435–442 (2003)

Topological Data Analysis for Arrhythmia Detection Through Modular Neural Networks

Meryll Dindin[1,2](✉), Yuhei Umeda[2], and Frederic Chazal[3]

[1] Ecole Centrale Paris, Paris, France
meryll_dindin@berkeley.edu
[2] Fujitsu Laboratories Ltd., Kawasaki, Japan
Umeda.yuhei@fujitsu.com
[3] INRIA, Paris, France
frederic.chazal@inria.fr

Abstract. This paper presents an innovative and generic deep learning approach to monitor heart conditions from ECG signals. We focus our attention on both the detection and classification of abnormal heartbeats, known as arrhythmia. We strongly insist on generalization throughout the construction of a shallow deep-learning model that turns out to be effective for new unseen patient. The novelty of our approach relies on the use of topological data analysis to deal with individual differences. We show that our structure reaches the performances of the state-of-the-art methods for both arrhythmia detection and classification.

Keywords: Topological data analysis · Deep learning · Arrhythmia · Auto-encoder · Multi-channel · Convolution networks · Classification

1 Introduction

Healthcare is among the most thriving domains since the democratized usage of artificial intelligence. Tasks such as monitoring, diagnostic and aided clinical decisions are relying on sharpened machine learning algorithms. As heart attack and strokes are among the five first causes of death in the US, it comes to no surprise that heart monitoring is of particular importance. Developing wearable medical devices would help to deal with a larger proportion of the population, and reduce the time used by cardiologists to make their diagnosis. This paper focuses on both the detection and classification of arrhythmia, an umbrella term for group of conditions describing irregular heartbeats. Detection deals with spotting abnormal heartbeats, while classification consists in giving the right label to the detected abnormal heartbeats.

Among the several existing studies, some developed descriptive temporal features to feed SVM [12] or neural networks [22], sometimes mixed with optimization methods [12,23]. The general approach of those papers enables arrhythmia classification through machine learning. However, most papers [11,13,14]

© Springer Nature Switzerland AG 2020
C. Goutte and X. Zhu (Eds.): Canadian AI 2020, LNAI 12109, pp. 177–188, 2020.
https://doi.org/10.1007/978-3-030-47358-7_17

either reduce the classification task to groups of conditions or to a few ones only. On the other hand, [10] sought to improve multi-class classification. But the more the classes, the faster the performances did vanish. To overcome this issue, [5,10,13] have introduced deep learning methods based on convolutional networks. Other teams focused on unsupervised learning, such as auto-encoders [16], with promising results. Nonetheless, the methods presented so far have low performance for unknown patient, due to obvious individual differences. Generalization, which means robustness regarding individual differences, is a serious issue for any application in the healthcare sector.

The approach we propose consists in the analysis of ECG through a modular multi-channel neural network whose originality is to include a new channel relying on topological data analysis to capture new robust patterns in the ECG signals. That information describes at best the geometry of each heartbeat, independently of the values of the signal or the individual heart rhythms. By combining topological data analysis, handcrafted features and deep-learning, we reached better generalization compared to existing literature.

Our paper is organized as follows: After presenting Topological Data Analysis, we condensed our approach in the presentation of the datasets, our preprocessing methodology and the general deep-learning architecture. We then expand our testing procedure, which is key to quantify generalization. Follows the introduction of our benchmark for arrhythmia classification, underlining the strengths of topological data analysis and auto-encoders to tackle the issue of individual differences. The last two sections provide comparisons with existing state-of-the-art performances. Finally, remarks and thoughts are provided as conclusion at the end of the paper.

2   Topological Data Analysis

Among the main challenges to reach generalization, we find individual differences, and specifically bradycardia and tachycardia. We dealt with it by bridging Topological Data Analysis with a modular deep-learning architecture. Topological Data Analysis (TDA) is a recent and fast growing field that provides mathematically well-founded methods [3] to efficiently exhibit topological patterns in data and to encode them into quantitative and qualitative features. In our setting, TDA, and more precisely *persistent homology theory* [6], powerfully characterizes the shape of the ECG signals in a compact way, avoiding complex geometric feature engineering. Thanks to fundamental stability properties of persistent homology [4], the TDA features appear to be very robust to the deformations of the patterns of interest in the ECG signal, especially expansion and contraction in the time axis direction. This makes them particularly useful to overcome individual differences and rhythms diversity due by bradycardia and tachycardia.

Persistence Homology. To characterize the heartbeats, we consider the persistent homology of the so called sub-level (resp. upper-level) sets filtration of the considered time series. Seeing the signal as a function f defined on an interval I and given a threshold value α, we consider the connected components of $F_\alpha = \{t \in I : f(t) \leq \alpha\}$ (resp. $F^\alpha = \{t \in I : f(t) \geq \alpha\}$). As α increases (resp. decreases) some components appear and some others get merged together. Persistent homology keeps track of the evolution of these components and encodes it in a *persistence barcode*, i.e. a set of intervals - see Fig. 1 for an example of barcode computation on a simple example. The starting point of each interval corresponds to a value α where a new component is created while the end point corresponds to the value α' where the created component gets merged into another one. In our practical setting, the function f is the piecewise linear interpolation of the ECG time series and persistence barcodes can be efficiently computed in $O(n \log n)$ time, using, e.g., the GUDHI library [18], where n is the number of nodes of the time series.

To clarify the construction of a persistence barcode, one may observe Fig. 1 with the following notations: $y = f(t)$: for $\alpha < \alpha_1$, F_α is empty. A first component appears in F_α as α reaches α_1, resulting in the beginning of an interval. Similarly when α reaches α_2 and then α_3, new components appear in F_α giving birth to the starting point of new intervals. When α reaches α_4, the two components born at α_1 and α_3 get merged, resulting in the "death" of the most recently born component (persistence rule), i.e. the one that appeared at α_3 and creation of the interval $[\alpha_3, \alpha_4]$ in the persistence barcode. Similarly when α reaches α_5 the interval $[\alpha_2, \alpha_5]$ is added to the barcode. The component appeared at α_1 remains until the end of the sweeping-up process, resulting in the interval $[\alpha_1, \alpha_6]$.

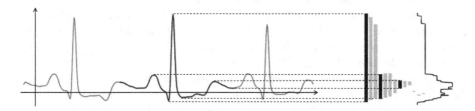

Fig. 1. Heartbeat transformation in a persistence barcode and Betti curve

Betti Curves. As an unstructured set of intervals, the persistence barcodes are not suitable for direct integration in machine-learning models. To tackle this issue, we use a specific representation of the barcode diagrams, the so-called Betti curves [24]: for each α, the Betti curve value at α is defined as the number of intervals containing the value α. See Fig. 1 for an a specific Betti curve construction based on an ECG signal. The Betti curves are computed and discretized on the interval delimited by the minimum and maximum of the birth and death values of each persistent diagram, both for the time-series and its opposite (in order to study the sub-levels and upper-levels of the signal). One may observe that a fundamental property of Betti curves of 1D signal that follows from the definition of barcodes is their stability with respect to time re-parametrization and signal value rescaling, as stated in the following theorem. This allows us to build an uniform input for classical 1D convolutional deep-learning models, thus tacking the main issue of individual differences.

Theorem: Time Independence of Betti Curves

Given a function $f : I \to \mathbb{R}$ and a real number $a > 0$ the Betti curves of $t \to f(t)$ and $t \to f(at)$ are the same.

Moreover, if $g(t) = bf(t)$ for some $b > 0$, then the Betti curves of f and g are related by $BC_g(\alpha) = BC_f(\frac{\alpha}{b})$.

This theorem is a particular case of a more general statement resulting from classical properties of general persistence theory [4]. Intuitively, the invariance to time rescaling follows from the observation that persistence intervals measure the relative height of the peaks of the signal and not their width. The value-rescaling of the signal by a factor b results in a stretching of the persistence intervals by the same factor resulting in the above relation between the Betti curves of the signal and its rescaled version.

3   Deep-Learning Approach

3.1   Datasets

To facilitate comparison to other existing methods, our experiments are based on a family of open-source data sets that have been extensively used in the literature. Those are provided by the Physionet platform, and named after the diseases they describe: *MIT-BIH Normal Sinus Rhythm Database* [7], *MIT-BIH Arrhythmia Database* [7,19], *MIT-BIH Supraventricular Arrhythmia Database* [7,9], *MIT-BIH Malignant Ventricular Arrhythmia Database* [7,8] and *MIT-BIH Long Term Database* [7]. Those databases present single-channel ECGs, each sampled at 360 Hz with 11-bit resolution over a 10 mV range. Two or more cardiologists independently annotated each record, whose disagreements were resolved to obtain the reference annotations for each beat in the databases. Each heartbeat is annotated independently, making peak detection thus unnecessary.

3.2   Preprocessing

Regarding preprocessing, we focused on the standardization of all the available ECG. To reduce both noise and bias, we re-sampled the signals at 200 Hz, removed the baselines and applied filters, based on both a FIR filter and a Kalman filter. The signal is then rescaled between 0 and 1 before being translated to get a mean of the signal close to 0, for distribution concerns.

Baseline Wander. Removing baselines is motivated by the fact it generally corresponds to muscular and respiratory artifacts. To do so, we used the established baseline drift technique [2], based on the Daubechies wavelets. Using consecutive convolution processes of decomposition and reconstruction of the signal, it removes the outlying components and enable us to identify and suppress the baseline component in the signal.

Filtering. The first applied filter to each ECG is a FIR (Finite Impulse Response) filter. It performs particularly well on ECG, and wavelets-based signals in general. It behaves basically as a band-filter. We chose 0.05 Hz and 50 Hz as cut frequencies to minimize the resulting distortion [25].

Heartbeats Slicing. Once standardized, each ECG is segmented into partially overlapping elementary sequences made of a fix number of consecutive heartbeats. Each sequence is extracted according to the previous and next heartbeat. This extraction being patient-dependent, it reduces the influence of diverging heartbeat rhythms, *e.g* bradycardia and tachycardia. The number of consecutive heartbeats in that window is a controlled parameter. The labels are attributed by the central peak (whose index is the integer value of half the number of peaks). Once the windows are defined, we use interpolation to standardize the vectors, making them suitable for deep-learning purposes.

Feature Engineering. Once those heartbeats are extracted, we began building relative features. Literature [2, 17, 20] screening brought us to the discrete Fourier transform of each window, the linear relationships between each temporal components (P, Q, R, S, T), and the statistical values given by the extrema, mean, standard deviation, kurtosis, skewness and entropy measures, crossing-overs and PCA window reduction to 10 components.

3.3   Auto-Encoder

One of our major motivation being generalization, we quickly faced the issue of uneven distribution of labels, as well as extreme minority classes. This was true for both the binary and multi-class classification tasks.

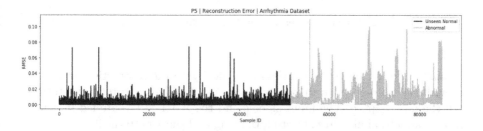

Fig. 2. Reconstruction error on heartbeat signals

We decided to take advantage of the large amount of normal samples over abnormal samples, thus exploiting unsupervised learning with auto-encoders [1]. Our structure has six fully-connected hidden layers, developed in a symmetric fashion, with an input dimension of size 400 (equals to the signal shape) and a latent space of size 20 (experimentally leading to the best performances). We trained the model on all the normal beats available, minimizing the reconstruction error defined by the *mean squared error*. Once its weights frozen, this model is integrated into our larger architecture. The reconstruction error, represented in Fig. 2, is used as a new feature for anomaly detection, unfortunately not descriptive enough for classification.

Regarding binary classification, such model may be used in two different ways: by either using the embedding as a set of features, or by using the reconstruction error with a subtraction layer linking input to reconstruction. Those solutions are respectively referred to by *encoder* and *auto-encoder* in our architectures. Another way of using this structure is to directly integrate it into the deep-learning model. The concurrent optimization of two models is thus necessary, building a relational encoding space relative to the task. This is the strategy that has been applied for multi-class classification.

3.4 Architecture

Once the signals standardized, we undertook the construction of our deep-learning model to exploit the multi-modality of inputs. Our first objective was to determine whether the heartbeats are normal or abnormal, before classifying them. The motivation was to avoid the issue of great imbalance between normal and abnormal samples, while initially focusing on an easier task. A representation of our architecture is given in Fig. 3.

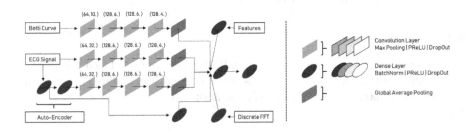

Fig. 3. General overview of deep-learning architecture

Channels. We use a convolution channel to deal with the subtracting layer dealing with the auto-encoder output, while we use a fully-connected layer to deal with the latent space. The input signals and the Betti curves are fed in convolution channels, aiming at extracting the long and short patterns [5,10,15]. The other inputs (both features and discrete Fast Fourier Transform coefficients) are injected into fully connected networks.

Annealed DropOut. As we launched a first battery of tests, we were confronted to the unexpected strong influence of the DropOut parameter over the performances. Since DropOut is of great help for generalization, we sought a way to reach greater robustness. A solution came from the annealing dropout technique [21], which stabilized our results.

4   Experimental Results

From the problem presentation, we highlighted two issues: imbalanced datasets and individual differences. The fewer the patients and the bigger the imbalance, the greater the influence of individual differences. To deal with the issue of imbalance, we introduced our auto-encoder architecture, while dealing with the individual differences by introducing Topological Data Analysis. Once we established our solution to both imbalance and individual differences, we aimed at developing our own approach and validation. As we mentioned earlier, two ways have been explored, both for performance enhancement and reduction of the influence of imbalance. The first one has been to detect whether a heart-beat is normal or abnormal, in order to get a first classification. The second one has been multi-class classification (13 classes) on the arrhythmic heartbeats only. Our objective is to introduce a new benchmark to attest that TDA (and auto-encoders) do improve generalization for arrhythmia classification.

Training Parameters. Different methods have been used for the model training and optimization. Firstly, all the channels described previously are concatenated into one fully-connected network, dealing with all the obtained feature maps concurrently. Secondly, all the activation layers used are *PReLU*, initialized with *he_normal*. Thirdly, the dropout has been configured according to the strategy of the *annealing dropout*, from a rate of 0.5 to a rate of 0.0 after 100 epochs. Concerning the training losses, we used *categorical_crossentropy* or *binary_crossentropy* for the classification model, and *mean_squared_error* for the auto-encoder structure. *Adadelta* was used for optimization with an initial learning rate of *1.0*.

Testing Methodology. Dealing with a medical diagnosis, the testing methodology has to be rigorously defined to accurately analyze the performances. A great importance was given to generalization abilities of the developed models. For that purpose, our strategy aimed at performing **patient-based cross-validation**, which means that for each model, train and validation sets were build on a fraction of the available patients, while the remaining patients constituted the test set. This way, validation score indicates the ability of the model to dissociate arrhythmias on known patients, while the test score demonstrates the ability of the model to detect arrhythmias on new patients. By using permutations of all the available patients, we were able to train, validate and test each model on all patients. The results presented in the following parts stem from a cross-validation keeping 5 unique patients for testing at each cross-validation permutation.

4.1   Channel Comparison

We first quantified the importance of each channel, by turning them off and on. That strategy allowed us to specifically quantify the importance of the introduced TDA channel in terms of generalization improvement. Once again, we tested our architecture through a patient-based cross-validation. This time, we made 10 experiments with the underlying objective of generalization: **for each experiment**, 225 patients are used for both training (70%) and validation (30%), while 15 patients are kept for testing. Each subset of 15 patients is not overlapping between experiments.

Table 1. Weighted test accuracy for channel comparisons.

ID	Arrhythmia detection		Arrhythmia classification	
	With TDA	Without TDA	With TDA	Without TDA
0	**0.99**	0.98	**0.73**	0.68
1	**0.96**	0.90	**0.75**	0.69
2	0.85	**0.86**	**0.68**	0.65
3	0.94	**0.95**	0.95	**0.96**
4	**0.85**	0.80	**0.97**	0.97
5	**0.87**	0.77	**0.96**	0.93
6	0.78	**0.80**	**0.94**	0.93
7	**0.81**	0.63	**0.90**	0.80
8	**0.79**	0.65	**0.85**	0.78
9	0.84	**0.86**	**0.68**	0.47

A closer look at Table 1 supports the importance of TDA regarding generalization. Its role is emphasized for multi-class classification, with an important improvement of performances. With this right combination of channels, we started testing our model through patient-based cross-validation for both binary and multi-class classification. **For the purpose of the demonstration, the scores are weighted in order to compensate for the general imbalance.** Moreover, multi-class classification is not biased by normal samples since they have been put aside beforehand. This finally supports the generalization role of TDA, that is expected to bring improvements combined with other deep-learning architectures as well.

4.2   Arrhythmia Detection

Our first benchmark dealt with arrhythmia detection (binary classification). It consisted in using our architecture, enhanced with the (auto-) encoder trained in an unsupervised manner on normal beats. The model determined by channel comparison has thus been used for cross-validation. Each instance of cross-validation has been made by randomly undersampling the majority class to obtain balanced datasets. It takes approximately 10 h to train on a GPU (GeForce GTX). We used the data structure previously presented to test over the 240 patients we have in our datasets. Moreover, to tackle the issue of anomaly detection, and accelerate the process of validation, each cross-validation round is respectively built out of a set of 5 unknown patients. The mean accuracy score is **98%** for validation and **90%** for test. This approach shows great generalization abilities. With a closer look on the results, the low performances appear on patients for which it was hard to recognize their normal beats. It comforts us as more patients will improve the generalization abilities of the model. However, its performances on the validation results prove its abilities to learn about specific patients, thus making it suitable for personalized monitoring devices. Unfortunately, no other paper do use those test settings for comparison.

4.3   Arrhythmia Classification

The same strategy has been applied for multi-class classification. The greatest channel influence are the ones linked to the TDA and the encoder. As a consequence, we reduced the original model to the one composed of four channels in the same fashion than we did for anomaly detection. Moreover, the influence of those channels is greater than observed for binary classification. Since the previous approach was not enough, we went further with 13-class classification. The models proved their ability to learn about heartbeat condition through cross-validation, with a mean validation score of **97.3%**, while being able to generalize this acquired knowledge on patients it never saw, with a mean testing accuracy of **80.5%**. Once again, literature does not provide comparable settings. The use of cross-validation focused on the generalization ability of the model. By also removing the normal beats, we focus on differences between the different arrhythmias, and remove the influence of imbalance that is generally found in the scores presented in the literature.

5   Benchmarks Comparison

5.1   Premature Ventricular Heartbeats Detection

Once our model established, we ran through evaluation of its performances against existing benchmarks. Our claim is enhanced generalization thanks to both the usage of TDA and the auto-encoder. To support it, our first comparison has been made with [13], which focuses on the detection of premature ventricular contractions (PVC). The classification of a given condition is a particular case of anomaly detection (one-vs-all), for which our architecture is built for. The results we obtained are presented in Tables 2 and 3, and support the generalization ability of our model. Those are obtained using the same methodology and the same scoring system than the quoted paper (*PPV* stands for

Table 2. Detection of PVC on the MIT-BIH database

Paper	Acc	PPV	Sen
Proposed	99.2%	95.1%	95.5%
Li et al.	99.4%	93.9%	98.2%

Table 3. Detection of PVC on the augmented databases.

Paper	Acc	PPV	Sen
Proposed	96.1%	97.0%	96.0%
Li et al.	95.6%	94.1%	92.7%

Positive Predictive Value, Acc for *Accuracy* and *Sen* for *Sensitivity*). The first experiment (Table 2) only concerns the MIT-BIH Arrhythmia database. Out of the 48 initial patients in the MIT-BIH Arrhythmia Database, 4 patients are discarded. The remaining patients are split in two groups: 22 are used for training and validation, 22 are used for testing. Such approach focuses on the premature ventricular contractions, which are the most reported labels among the available arrhythmias.

The second experiment (Table 3) aggregates the five databases. This time they split the group of 240 patients in two, both 120 for training and validation, and 120 for testing. With more patients, the experiment configuration imply a greater need for generalization, which emphasizes our model ability to learn the concept of PVC and apply it to unknown patients. This performance difference is emphasized in our results, and highlight our greater generalization performance.

5.2   8-Classes Classification

Table 4. 8-classes classification - MIT-BIH database

Paper	Classes	Acc	PPV	Sen
Proposed	8	99.0%	99.0%	98.5%
Jun et al. (2017)	8	99.0%	98.5%	97.8%
Kiranyaz et al. (2017)	5	96.4%	79.2%	68.8%
Güler et al. (2018)	4	96.9%	–	96.3%
Melgani et al. (2008)	6	91.7%	–	93.8%

The previous experiment is a specific use-case for our architecture. In this second comparison, we focus on 8-classes classification [14]. Once again, our claim is better generalization, which is done thanks to patient-based cross-validation. Nonetheless, their settings imply a limitation to the MIT-BIH Arrhythmia Database, from which they select 8 classes, comprising normal beats. Unexpectedly, this selection does not correspond to the majority classes. Our performances are compared in Table 4, extending the results they present in their paper. We pinpoint, thanks to those results, the generalization ability of our model, which has better positive predictive value (here precision) and sensitivity. It finally underlines a more efficient classification.

6   Conclusion

We developed a new approach to deal with the issue of generalization in arrhythmia detection and classification. Our innovative architecture uses common source of information, Topological Data Analysis and auto-encoders. We supported our claim of improved generalization with scores reaching the performances of state of the art methods, and above. Our experiments pinpoint the strengths of TDA and auto-encoders to improve generalization results. Moreover, the modularity of such model allows us to build and add new channels, such as a possible channel based on the Wavelet transform, which also gives a good description of the ECG time-series. Finally, we give a new benchmark on five open-source datasets, and get excited by performance improvement through the release of larger datasets such as presented in [5].

References

1. Baldi, P.: Autoencoders, unsupervised learning, and deep architectures. In: Proceedings of the ICML Workshop on Unsupervised and Transfer Learning, PMLR, vol. 27, pp. 37–49 (2012)
2. Blanco-Velasco, M., Weng, B., Barner, K.E.: ECG signal denoising and baseline wander correction based on the empirical mode decomposition. Comput. Biol. Med. **38**, 1–13 (2008). https://doi.org/10.1016/j.compbiomed.2007.06.003
3. Chazal, F., Michel, B.: An introduction to topological data analysis: fundamental and practical aspects for data scientists arXiv:1710.04019 (2017)
4. Chazal, F., de Silva, V., Glisse, M., Oudot, S.: The Structure and Stability of Persistence Modules. SM. Springer, Cham (2016). https://doi.org/10.1007/978-3-319-42545-0
5. Clifford, G.D., et al.: AF classification from a short single lead ECG recording: the PhysioNet computing in cardiology challenge 2017. In: Conference: 2017 Computing in Cardiology Conference (2017). https://doi.org/10.22489/CinC.2017.065-469
6. Edelsbrunner, H., Harer, J.: Computational Topology: An Introduction. AMS Press, Naga (2009)
7. Goldberger, A.L., et al.: PhysioBank, PhysioToolkit, and PhysioNet: components of a new research resource for complex physiologic signals. Circulation **101**(23), E215–E220 (2000). https://doi.org/10.1161/01.cir.101.23.e215
8. Greenwald, S.D.: Development and analysis of a ventricular fibrillation detector. Thesis (M.S.) - Massachusetts Institute of Technology, Department of Electrical Engineering and Computer Science (1986)
9. Greenwald, S.D.: Improved detection and classification of arrhythmias in noise-corrupted electrocardiograms using contextual information. Harvard-MIT Division of Health Sciences and Technology (1990). https://doi.org/10.1109/CIC.1990.144257
10. Hannun, Y., et al.: Cardiologist-level arrhythmia detection and classification in ambulatory electrocardiograms using a deep neural network. Nat. Med. **25**, 65–69 (2019). https://doi.org/10.1038/s41591-018-0268-3
11. Hassanien, A.E., Kilany, M., Houssein, E.H.: Combining support vector machine and elephant herding optimization for cardiac arrhythmias. CoRR, abs/1806.08242 (2018)

12. Houssein, E., Kilany, M., Hassanien, A.E.: ECG signals classification: a review. Int. J. Med. Eng. Inform. 5(4), 376–396 (2017). https://doi.org/10.1504/IJIEI. 2017.10008807
13. Li, J.: Detection of premature ventricular contractions using densely connected deep convolutional neural network with spatial pyramid pooling layer (2018)
14. Jun, T.J., Nguyen, H.M., Kang, D., Kim, D., Kim, Y.-H., Kim, D.: ECG arrhythmia classification using a 2-D convolutional neural network (2017)
15. Kachuee, M., Fazeli, S., Sarrafzadeh, M.: ECG heartbeat classification: a deep transferable representation. In: IEEE International Conference on Healthcare Informatics (2018). https://doi.org/10.1109/ICHI.2018.00092
16. Weijia, L., Shuai, J., Shuyan, G., Xue, J.: Method to annotate arrhythmias by deep network. In: IEEE International Conference on Internet of Things (iThings) (2018). https://doi.org/10.1109/Cybermatics_2018.2018.00307
17. Luz, E.J.D.S., Schwartz, W.R., Cámara-Chávez, G., Menotti, D.: ECG-based heartbeat classification for arrhythmia detection: a survey. Comput. Methods Programs Biomed. 127, 144–164 (2016). https://doi.org/10.1016/j.cmpb.2015.12.008
18. Maria, C., Boissonnat, J.-D., Glisse, M., Yvinec, M.: The Gudhi library: simplicial complexes and persistent homology. In: Hong, H., Yap, C. (eds.) ICMS 2014. LNCS, vol. 8592, pp. 167–174. Springer, Heidelberg (2014). https://doi.org/10.1007/978-3-662-44199-2_28
19. Moody, G.B., Mark, R.G.: The impact of the MIT-BIH arrhythmia database. IEEE Eng. Med. Biol. Mag. 20 (2001). https://doi.org/10.1109/51.932724
20. Pyakillya, B., Kazachenko, N., Mikhailovsky, N.: Deep learning for ECG Classification. J. Phys. Conf. Ser. 913 (2017). https://doi.org/10.1088/1742-6596/913/1/012004
21. Rennie, S.J., Goel, V., Thomas, S.: Annealed dropout training of deep networks. In: IEEE Spoken Language Technology Workshop (2014). https://doi.org/10.1109/SLT.2014.7078567
22. Shirin, S., Behbood, M.: A new personalized ECG signal classification algorithm using block-based neural network and particle swarm optimization. Biomed. Signal Process. Control 25, 12–23 (2016). https://doi.org/10.1016/j.bspc.2015.10.008
23. Abd-Elazim, S.M., Ali, E.S.: A hybrid particle swarm optimization and bacterial foraging for optimal power system stabilizers design. Int. J. Electr. Power Energy Syst. 46, 334–341 (2013). https://doi.org/10.1016/j.ijepes.2012.10.047
24. Umeda, Y.: Time series classification via topological data analysis. Trans. Jpn. Soc. Artif. Intell. 32(3), 1–12 (2016). https://doi.org/10.1527/tjsai.D-G72
25. Upganlawar, I.V., Chowhan, H.: Pre-processing of ECG signals using filters. Int. J. Comput. Trends Technol. 11(4) (2014). https://doi.org/10.14445/22312803/IJCTT-V11P135

Big Players: Emotion in Twitter Communities Tweeting About Global Warming

Dennis J. Drown[1(✉)], Roger Villemaire[1], and Serge Robert[2]

[1] Dép. d'informatique, Université du Québec à Montréal, Montreal, Canada
drown.dennis@courrier.uqam.ca
[2] Dép. de philosophie, Université du Québec à Montréal, Montreal, Canada

Abstract. This work considers exceptionally active users on Twitter, the "big players," and analyzes the correlation between the level of emotion these users express in communications tagged with the hashtag *#globalwarming* and the levels expressed by the Twitter community as a whole. Using an emotion lexicon incorporating four of the eight base human emotions according to Plutchik: anger, fear, sadness, and joy, we identify to what extent a small group of these big players may predict the emotion expressed by their online community in their tweets.

1 Introduction

The significant world-wide rise in temperatures is a major challenge facing humanity today. As assessed by the Intergovernmental Panel on Climate Change (IPCC) [26], human-induced changes in atmospheric composition are leading to a rise in global temperatures that has had and will continue to have a severe impact on the earth's climate. Rising temperatures not only challenge human health and food supplies, they are also a threat to biodiversity. Global warming therefore represents a serious social and political issue, and much has been done in recent years to better understand related public perception and behaviour.

Research indicates that there is a complex relation between environmental concern, informal education, experience, and behaviour [20]. Furthermore, environmental risk perception and policy support is in fact strongly influenced by sentiment (positive/negative reactions), emotional responses, imagery, and values [15]. Emotions are a driving force in modifying behaviour in order to avoid risk in dangerous situations [29], and they may be key to understanding how people reason and respond to information about global warming [17].

This work considers emotion expressed on climate-related issues in the context of the social media site Twitter,[1] evaluating affect using sentiment analysis. This relatively new subfield of natural language processing (NLP) is becoming increasingly popular largely due to the enormous amount of opinion that the

---

[1] https://twitter.com/.

C. Goutte and X. Zhu (Eds.): Canadian AI 2020, LNAI 12109, pp. 189–200, 2020.
https://doi.org/10.1007/978-3-030-47358-7_18

public expresses online in today's world [16]. Sentiment analysis is often concerned with polarity, determining if a message is positive or negative. Here we focus on affect using a type of analysis commonly called emotion mining.

Of particular interest is the behaviour of Twitter users showing exceptionally high participation. One might, for instance, wonder whether a small number of the "top" high-activity users could be particularly representative of the overall community. Hence, we present affective models to predict emotion expressed in the general community[2] from the emotion expressed by these "big players" in their tweets. We also compare these models with predictive models from the emotion expressed by a random set of users of the same size in order to ascertain whether the top high-activity users have a distinct predictive capability.

We show that, while the big players are not necessarily more representative of the whole community than other groups of the same size, they do indeed show a distinct predictive capability. Moreover, this effect sharply depends on the emotion being considered.

2 Related Work

When seeking to identify individuals filling specific roles in their communities, a common approach is to analyze centrality in graph-based representations of the relationships between users. Bigonha et al. [2] create a model to find influencers based on three elements: sentiment polarity in users' tweets; two types of graph representations ("who follows whom" and "who reacts to whose tweets"); and grade-level readability of the messages. Aleahmad et al. [1] propose an algorithm called OLFinder to identify major topics for a given domain in a set of tweets and determine which users are "opinion leaders" for those topics based on the users' calculated competency in that domain and a popularity score determined by graphs of follower relations. A study by Eliacik and Erdogan [6] seeks to boost graph-based methods using a calculated measure of trust that others extend to a user based on her relations, expertise, and activity in a topic-centred community. In research pertaining to climate change, Cody et al. [5] use sentiment analysis to examine changes in polarity in tweets with respect to climate-related events.

The studies above involve measuring the polarity in tweets, but there are also a number of projects which focus on emotion as we do in the present work. Mitchell et al. [19] look at finding the happiest and saddest states and cities in the United States using a large corpus of tweets tagged with geolocation information. Preoţiuc-Pietro et al. [25] work with regression models for predicting the income of Twitter users based partly on polarity and emotion content in their tweets. Finally, Halse et al. [12] use affect models during a crisis to determine if tweets are trustworthy and contain information that may be useful to first responders.

---

[2] We use the word *community* to indicate the users sending tweets or being referred to in tweets with a given hashtag.

Contrary to research involving graph-based representations of a given role between certain users and their followers, we focus our attention on users set apart by their unusually high level of activity in order to determine to what extent their behaviour can predict that of the general *#globalwarming* community.

3   Methodology

To determine which top high-activity users will be the big players, we look at the correlation between the emotion expressed in the online activity of a tentative set of big players and the emotion in tweets published by "regular players," the rest of the Twitter community talking about global warming.

Tweet Dataset: We made use of the Twitter developer platform API[3] to collect 414,035 tweets from 239,590 users, published between January 1, 2018 and August 31, 2019 and incorporating the hashtag *#globalwarming*. We call the analysis of the tweets from this dataset over a fixed period of time a tracking run. We are interested in communications on climate change generally, but this work specifically tracks the hashtag *#globalwarming*. According to a study by Williams et al. [30], this hashtag is more regularly used by Twitter users from both pro-science and skeptic communities, compared to *#climatechange*, which more often appears in tweets from climate activists than from skeptics and deniers. The dataset contains only tweets in English as flagged by accompanying metadata from the Twitter API.

Emotion Lexicon: We analyze tweets using four base emotions from Robert Plutchik's model [23]: *anger, fear, sadness,* and *joy*. We treat the tweet text as a bag of words to be checked against the Affect Intensity Lexicon from the National Research Council of Canada (NRC). We chose this lexicon as we see potential in the method known as Best-Worst Scaling, which the NRC used to create it [21].[4] The lexicon is available via the Affective Tweets plugin[5] for the machine learning platform Weka [10]. Our algorithm uses the plugin to standardize usernames and web URLs and to calculate floating-point values in a tweet's emotion vector, summing intensity levels for each of the four emotions across all words in the text that are contained in the lexicon. When analyzing a tracking run, the first step is to determine this emotion vector for each tweet tagged with *#globalwarming*.

NLP Tools: The Affective Tweets library [21] provides an NLP tokenizer specifically designed for tweets [11] to delimit words, user names, and web links in a user's text. The library also provides other NLP preprocessing tools, allowing the modeller to specify a stopword list to remove common words devoid of analytical value and a stemming algorithm to reduce words to their grammatical roots.

---

[3] https://developer.twitter.com/.

[4] Plutchik's model also includes *anticipation, trust, surprise,* and *disgust*. The NRC is in the process of expanding the lexicon to include all eight emotions.

[5] https://github.com/felipebravom/AffectiveTweets.

When creating our models, we experimented with the English stopword list from Apache's information retrieval package Lucene[6] as well as the Snowball Porter stemmer [24] for the English language.[7] However, our results did not improve substantially, and therefore in this paper we present results obtained without a stop list and without stemming.

The lack of improvement should not be too surprising considering that the utility of traditional NLP tools may suffer when analyzing human language on social media. Users are non-professional writers, tending to express ideas with little thought towards clear content. Tweets contain frequent abbreviations, slang, and (often intentional) errors in grammar and spelling. Twitter is particularly problematic due to the small size of the texts [8].

Communication Categories: There are various ways to demonstrate high activity. We rank users in terms of tweet count for the following categories:

1. **Original Tweeters** (*oter*): Users publishing personally-authored messages.
2. **Retweeters** (*rter*): Users retweeting (resending tweets written by another user) using Twitter's popular "*RT @author ...*" syntax.[8]
3. **Retweeted Authors** (*rted*): Users whose tweets are retweeted by others.
4. **Mentioned Authors** (*tmed*):[9] Users specifically mentioned in the tweets of others via Twitter's "*@user*" notation. This syntax usually serves to address specific users or attract their attention [14].

A single tweet may be accounted for multiple times. For example, if user U_1 publishes tweet T_1, which he has personally authored and which mentions user U_2, then T_1 contributes to U_1's participation in the *oter* category and also to U_2's participation in *tmed*. Similarly, if user U_2 sends a retweet R_2 originally published by user U_1 that mentions user U_3, then R_2 contributes to U_2's participation in *rter*, to U_1's participation in *rted*, and to U_3's participation in *tmed*.

Each category represents a distinct type of participation. For instance, in the last example the original tweet from U_1 would not be considered at all if it was sent before the tracking run. Nevertheless, if others continue retweeting it often enough, then U_1 may become a *rted* big player. Likewise, when considering the category of mentioned authors, one should keep in mind that these users may not have actively participated in the tracking run. For example, the top-mentioned account in our collected *#globalwarming* tweets is *realDonaldTrump*; however, this famous user authored no tweets with this hashtag during the tracking run.

We may also think of *oter* and *rter* as categories of active participation, similarly considering *rted* and *tmed* as representing a passive form of participation. Yet, while the concept of passive big players may be useful, it is not valid in every sense. Twitter users may actively work to be retweeted [3], and users often mention each other in tweets when establishing communication threads [14].

---

[6] https://lucene.apache.org/.

[7] http://snowball.tartarus.org/.

[8] The syntax for retweeting is not standardized, and alternatives do exist. The Twitter API sends metadata that identifies retweets and indicates the retweeted author.

[9] Note that the *tmed* code is backwards: "*mentioned in tweet*.".

"Top N" Big Players: For each of the four communication categories described above, we identify a set number of users who rank highest with respect to the type of activity that the category represents: the Top N. Note that we are essentially considering four types of big players, and it is possible for a specific user to belong to more than one big player group.

For this study we evaluate a series of big player groups with sizes ranging from 5 to 25. Our reasoning is that N should be small enough that the total number of big players will not be overwhelming to a researcher who must, for example, examine users' account profiles. Ideally, we can identify a size N that demonstrates a significant correlation between emotion levels in tweets from the big players and emotion levels in the general community.

We define the regular players as users who are not in any of the four big player categories and who have published at least one original tweet with the hashtag *#globalwarming* during the tracking run. We are endeavouring to predict the emotion levels expressed by the regular players in these original tweets. It stands to reason that tweets authored and published by the general *#globalwarming* community are likely a meaningful reflection of what that community is feeling.

Machine Learning Models: We conducted experiments with the following models: linear regression, Gaussian processes, decision lists using separate-and-conquer (M5Rules), random forests, and support-vector machines with first and second degree polynomial kernels (SMOreg using PolyKernel with exponents set to 1 and 2). For each of these we used the implementations in Weka [31] with the default settings. Linear regression set itself apart by consistently showing better results. It also has the advantage of being a "white-box" modelling technique, enabling the modeller to understand how it arrives at its predicted values. Henceforth, we will only report on the linear regression models.

Emotion Models: We create regression models for four target values, which are the variation from one week to the next in levels for the emotions anger, fear, sadness, and joy as measured in tweets from regular players in the *#globalwarming* Twitter community. To predict these values, we use 16 independent attributes that describe the variation, week by week, in the average levels of each of the four emotions for big player tweets across the four communication categories. We name these attributes and the targets of the models using a three-part syntax to indicate the community group, the communication category, and the emotion:

$$\begin{pmatrix} big \\ reg \end{pmatrix} - \begin{pmatrix} oter \\ rter \\ rted \\ tmed \end{pmatrix} - \begin{pmatrix} anger \\ fear \\ sadness \\ joy \end{pmatrix}$$

Independent attributes begin with *big* as these are values representing big players. For example, *big_rter_fear* gives the variation in the level of fear in retweets from users in the big retweeters category, and *big_tmed_joy* is the variation in joy measured in tweets that include frequently-mentioned authors. Target attributes start with the community code *reg* as these are values we are predicting for the regular players.[10] As an example, the following represents a typical regression

---

[10] For targets we consider only the original tweets (*oter*) for the four base emotions.

model to predict the variation in the level of anger measured in regular players' tweets for a given week with respect to the previous week:

$$
\begin{aligned}
reg\_oter\_anger = \ &0.218 \times big\_oter\_anger \\
&- 0.105 \times big\_oter\_fear \\
&- 0.056 \times big\_tmed\_fear \\
&- 0.001
\end{aligned}
\tag{1}
$$

Data Preparation: To prepare tracking run data for a regression model, our system runs all the tweets tagged with *#globalwarming*, collected over the tracking period, through the Affective Tweets Weka filter to determine their associated emotion vectors. We keep running averages for all users with respect to each of the four emotions for all four communication categories. Once we have the emotion vector for a tweet, we apply it to the user either as an original tweet (*oter*) or a retweet (*rter*). For a retweet, we also apply the emotion vector to the tweet's original author (*rted*). Finally, we search the tweet text for all mentioned users and apply the emotion vector to their accounts as well (*tmed*). Applying the vector to a user means that the system incorporates the emotion levels for the tweet into an emotion vector representing the running averages of the emotions on the day the tweet was published for the communication type being processed. The system increments the user's counter for the communication activity, used to track the user's position in the ranking for that category.

After the tweets have been processed, the system finds the big players. For N ranging from 5 to 25, we create player activity rankings for the four communication categories and select the Top N players in each case. Note that as players are ranked, there is an occasional tie in activity levels. For example, when determining the Top 10 original tweeters, the players ranked #10 and #11 may have both sent the same number of tweets. In this case we accept both players, and the Top 10 big players for one category will incorporate 11 users. As our models target the variation from one week to the next for the levels of emotion in the regular players' original tweets, the group of regular players is defined as all accounts which have authored and sent one or more tweets during the tracking run but are not included in the four big player groups.

After identifying the big players in the four communication categories and the remaining accounts that make up the regular players, our system then creates five groups (one for each player type) and calculates an average of the emotion vectors across all users in a group for each day of the tracking run. It then sorts the day groupings and bundles them into super-groups representing a week (seven consecutive day groups). The system reduces all the emotion vectors in each week-long grouping into a single vector corresponding to the average emotion levels measured in the tweets for all players in a community (*big* or *reg*) and communication category (*oter*, *rter*, *rted*, or *tmed*) over that week.

At this point we create a preliminary set of data instances, which each contain levels of emotion intensity for one week of tweets. As a final step, we subtract the attribute values for each instance from those of the previous one to obtain the variation in emotion intensity from week to week.[11] Thus, variations in the levels of the four emotions in the four big player groups become the 16 independent attributes (*big_oter_anger*, etc.). These values remain the same across all affect models for a given N. The target attribute for each model represents the weekly variation in the average affect intensity expressed by all regular players for one emotion. Each element of an emotion vector in the grouping for regular players (e.g., *reg_oter_fear*) represents the dependent attribute for one affect model. Its value for a given data instance is computed from a pair of consecutive weeks from the regular player group.

4   Results and Analysis

This section presents results from the models predicting the variation in level from one week to the next for anger, fear, sadness, and joy in the regular players' original tweets for the *#globalwarming* Twitter community. For each emotion we ran a series of models for the Top N big players with N ranging from 5 to 25. Our dataset contains 20 full months of tweets tagged with *#globalwarming*, beginning on January 1, 2018 and ending on August 31, 2019. For each set of experiments we ran 9 tracking runs with each run using 12 months of data, starting on midnight of the first day of month M_i, for i ranging from 1 to 9, and ending at 23:59:59.999 on the last day of month M_{i+12}. As each tracking run shifts the starting month by one, we are essentially sweeping a 12-month window across the 20 months of Twitter data.

In addition to being a natural choice with respect to the calendar, the one-year window provides sufficient data to train our models, given that each data instance represents a whole week of Twitter activity. For each 12-month period we used the first nine months (75%) for training and tested the models on the last three months (25%). We report averages of the Pearson Correlation Coefficient (PCC) across the 9 tracking runs in Table 1 for each of the four emotions. The PCC is a value between -1 and 1, with 1 indicating a total positive correlation between the model's predictions and the measured values in the test data.

To determine whether a group of big players of a given size is significant with respect to predicting emotion in the general community, we compare the results for the big players with results for reference groups of the same size. In order to build a meaningful model, however, we must ensure that users in the reference groups have a minimal level of participation. Therefore, we generate reference groups in the following way. We pick random users to form four groups (*oter*, *rter*, *rted*, and *tmed*).[12] These groups are the same size as the big player groups,

---

[11] For a given attribute, A_Δ, we compute $A_\Delta = A_i - A_{i-1}$, where A_i is the emotion level for the current week, and A_{i-1} is the level for the previous week. One might consider using instead the relative change, dividing our A_Δ by A_{i-1}, to get a percentage. This was not possible here because if a player group does not express some given emotion in a week's worth of tweets, the relative change will be undefined.

[12] We use the Erlang *rand* library's implementation (*exrop*) of the Xoroshiro116+ pseudorandom number generator [28] with 58 bits of precision and a period of 2^{116-1}.

and they must have at least 40 tweets in the communication category for their group. To avoid biased results wherein a few big players represent the dominant contribution, we require that users in the reference groups not be in any of the four big player groups. After creating the groups, we process the tweets and create regression models as explained above, replacing the big players with these reference groups. For each tracking run, and for each group size of N reference users (analogous to the Top N players), we repeat this procedure 20 times.

The "ref" columns in Tables 1 and 2 report the averages over these 180 models (20 sets of reference groups × 9 tracking runs). In these tables, a field containing asterisks (*****) indicates that for all 180 models, there was at least one week for which a reference group had no tweets for one or more communication categories. This generally occurs for small values of N and for larger training periods (12 months as opposed to 9). By definition, the reference users are less active than the big players, and if there are too few of them in a given group, together their activity may not be consistent enough to cover every week in the tracking run.[13]

Table 1. Correlation (PCC) for models predicting emotion in the last 3 months.

N	Anger BIG	Anger ref.	Fear BIG.	Fear ref.	Sadness BIG	Sadness ref.	Joy BIG	Joy ref.
5	0.1376	*****	0.0129	*****	-0.0992	*****	0.1130	*****
6	0.1660	*****	-0.0818	*****	-0.1369	*****	0.1473	*****
7	0.1465	0.3301	0.0509	0.2432	-0.1539	-0.2223	0.1680	-0.0852
8	0.3415	0.0751	0.0429	-0.1945	-0.0758	-0.3385	0.0170	0.1602
9	0.2727	0.0090	-0.0337	-0.1281	-0.1429	0.2542	0.1183	-0.0504
10	0.2294	-0.0753	0.0204	0.1587	-0.1407	-0.0119	0.1291	0.1278
11	0.2534	0.0951	0.0031	0.0112	-0.2119	0.0155	0.1066	0.0938
12	0.3189	0.0595	-0.0009	-0.0340	-0.2380	-0.0701	0.1058	0.0816
13	0.3035	0.0144	0.0197	0.0425	-0.1542	0.0436	0.0935	0.0766
14	0.3024	-0.0786	0.1568	0.0128	-0.1876	0.0682	0.0854	-0.0219
15	0.2793	-0.0599	0.2358	0.0334	-0.1989	-0.0854	0.0115	-0.0606
16	0.2770	-0.0229	**0.2972**	-0.0332	-0.1221	0.0127	0.0632	-0.0007
17	**0.3869**	0.0139	0.2566	0.0099	-0.1812	0.0468	0.0945	0.0639
18	0.3787	0.0580	0.1282	0.0530	-0.2420	-0.0236	-0.0110	0.0829
19	0.3402	-0.0351	-0.0134	0.0053	-0.1857	-0.0609	0.2517	-0.0044
20	0.2764	0.0046	0.0389	0.0204	-0.1725	-0.0063	0.0016	0.0721
21	0.2917	0.0233	0.0914	0.0677	-0.0572	0.0015	0.0531	-0.0348
22	0.2042	0.0524	0.0934	-0.0289	-0.1444	0.0032	0.0942	-0.0314
23	0.1969	-0.0066	0.1732	0.0245	-0.1179	0.0222	0.1064	-0.0126
24	0.1816	-0.0012	0.1331	0.0177	-0.0533	0.0185	0.1061	0.0595
25	0.1201	0.0552	0.0594	0.0241	-0.0897	-0.0442	0.0522	0.0115

Table 2. Correlation (PCC) for models evaluated using 10-fold cross-validation.

N	Anger BIG	Anger ref.	Fear BIG.	Fear ref.	Sadness BIG	Sadness ref.	Joy BIG	Joy ref.
5	0.3521	*****	0.0721	*****	0.3203	*****	0.1580	*****
6	0.3620	*****	0.2865	*****	0.3309	*****	0.1559	*****
7	0.4067	*****	0.3893	*****	0.3138	*****	0.2326	*****
8	0.4127	-0.0685	0.4050	0.1910	0.2369	0.1227	0.2317	0.5470
9	0.4576	0.3461	0.3075	0.2587	0.2970	-0.0342	0.2418	0.3372
10	0.4545	0.1041	0.2906	0.2317	0.3216	0.1892	0.2215	0.2865
11	0.4322	0.2651	0.3815	0.2895	0.3102	0.2942	0.2428	0.3487
12	0.4327	0.1883	0.3369	0.2678	0.3000	0.1924	0.1974	0.3406
13	0.3800	0.1766	0.3189	0.1528	0.2297	0.1536	0.2233	0.3262
14	0.4299	0.2770	0.3341	0.3137	0.2917	0.4133	0.2367	0.2675
15	0.4364	0.2358	0.3238	0.3180	0.2610	0.3765	0.2432	0.2508
16	0.4098	0.2712	0.3625	0.2686	0.2416	0.3788	0.1705	0.3115
17	0.3740	0.2361	0.2448	0.2638	0.2651	0.2903	0.1393	0.2719
18	0.3823	0.3284	0.2670	0.2845	0.2697	0.2582	0.1425	0.2410
19	0.3550	0.2701	0.3248	0.2804	0.2871	0.2609	0.1194	0.2724
20	0.3955	0.2926	0.2482	0.3146	0.2509	0.2686	0.0858	0.2840
21	0.4236	0.3049	0.3583	0.2927	0.3052	0.3601	0.1160	0.3011
22	0.3716	0.2828	0.3044	0.2811	0.2900	0.3102	0.1401	0.2419
23	0.4306	0.3191	0.3106	0.2623	0.3086	0.3263	0.1364	0.3048
24	0.4248	0.3189	0.3235	0.3178	0.3080	0.2934	0.1673	0.3004
25	0.4201	0.3382	0.2678	0.2700	0.2740	0.3054	0.1909	0.2849

Comparing columns 2 and 3 from Table 1, we see the PCC for anger (variation) predictions from big players closer to 1 and higher than the PCC for the reference groups. Interestingly, as we follow N, looking at larger groups of big players, the PCC increases to 0.3869 before decreasing again for even larger groups. The best results are obtained with a group of around 17 big players.

From columns 4 and 5 of the same table we observe that while the fear models for big players do not always outperform the reference models, they do show a better linear correlation for N between 14 and 18, where the PCC

---

[13] Setting a higher minimum tweet limit would help to correct this problem; however, raising the minimum past 40 means there may not be a large enough pool of candidate users to fill the reference groups for larger values of N.

for the big players reaches values up to 0.2972. The big players perform better than the reference groups near the middle of the table, but not when we create models using smaller groups or larger ones. For anger and fear, one may conclude that the big players are indeed significant as a group. Furthermore, the PCC is maximal for N in the range 16–18, showing that interested researchers can focus their attention on a reasonable number of the top players.

For sadness and joy, the results are much less clear as Table 1 shows. Models for sadness do not show a significant level of correlation for the big players, nor for the reference groups. As for joy, one notable PCC of 0.2517 for $N = 19$ does not seem significant.

In order to determine if this lack of correlation for sadness and joy might be a consequence of the limitations of linear regression, we experimented with other learners as mentioned in Sect. 3. These include Gaussian processes, decision lists, random forests, and support-vector machines with first and second degree polynomial kernels. None of these algorithms outperformed the linear model.

We also used the affect models to predict variation in levels of emotion expressed in the Twitter community *during* the twelve-month period itself, rather than predicting over the last three months. To test this scenario, we evaluated the models using 10-fold cross-validation for each of the nine 12-month periods. From Table 2 one can see that the PCC does not vary significantly across values of N, and therefore it is difficult to identify a value of N of particular interest. Even more significant is the fact that the PCC obtained by the big players and those for the reference groups are similar. This indicates that the big players are comparable to other groups of the same size with respect to their predictive value within the twelve-month periods. This finding is in stark contrast to their predictive value for the last 3 months of these periods.

As an additional test, we repeated the experiment, but rather than using cross-validation, we created an independent test dataset by randomly selecting three-month's worth of data instances throughout the 12-month period, removing those instances from the training dataset. This method is of interest since it more closely parallels the methodology we followed when using the final three months of the period for model evaluation. The results with this independent test dataset were similar to those we obtained using cross-validation.

5   Discussion

When examining ways in which our results may help to further research on climate change, three hypotheses give likely interpretations of the correlation the models show for anger and fear:

1. Emotion expressed by a big player is representative of the larger community.
2. A big player's tweets are influencing the emotional state of the community. (High Twitter activity may indicate a user is seeking to gain an online presence or communicate a specific message to a perceived audience [3,18].)
3. A big player and the community are each influencing each other's emotional state. (Users interact online mainly with like-minded individuals [9,30].)

In each case, the big player may potentially be of high interest. The fact that the model identifies a relatively small number of big players greatly reduces the work effort involved in looking up user profiles and following specific chains of tweets. Furthermore, since models take into account different types of big player activity as well as a set of base emotions, they may be useful for organizations aiming to evaluate various types of high-level participation in order to improve communication methods which use emotion-based message framing.

Furthermore, researchers are exploring the relation between emotion at a social level and people's response to the dangers of climate change. Anger and fear are of particular interest, and our results show that big players can be a group of significant interest in the context of #globalwarming on Twitter. For instance, studies are looking into how fear affects people's reactions to information about climate change [13] and the role fear can play when framing messages intended to promote climate change advocacy [22].

Our models for sadness and joy, however, do not show any significant correlation. This does not seem to be an artifact of linear regression as the other algorithms do not produce better results. Hence, techniques to model sadness and joy may differ sharply from those for anger and fear. Further research on sadness in the context of high-activity online users is certainly warranted because this emotion is an important aspect in studies of human reactions to climate change. One example is the study by Farbotko and McGregor [7] exploring the influence which sadness can have on shaping international policy on climate change.

Our initial interpretations indicate that models analyzing sadness and joy may need to handle additional complications. We would not generally expect joy in particular to be a clear, unblurred emotion in messages about global warming. For instance, Sulis et al. [27] demonstrate that high levels of sadness may be found in tweets expressing irony, while joy occurs frequently in tweets expressing sarcasm. We would expect these emotions to be particularly difficult to model for communications on climate change, and it is intriguing to speculate on the extent to which irony and sarcastic remarks are influencing our models.

Limitations: With statistical models we must remember that finding a correlation does not mean we understand the causes behind the phenomenon we are studying. We present a method for predicting emotion levels in tweets about global warming, but we cannot say that the elements that we are considering as big player activity is causing the expression of emotion. We must also exercise a measure of restraint as we interpret our results. Tweets are a noisy, extremely informal, and non-standard use of language that traditional NLP techniques often find problematic [8]. Users may repeatedly send the same tweet (or retweet) numerous times; they may alter the original author's text when retweeting [3]; and they may use the "@" sign for purposes other than addressing another user. Additionally, emotions are only a part of the complex system that is human cognition. When using affective models to study how best to talk to people to inform them about climate change and work with them to mitigate its effects, we must continually be conscious of the underlying complexities and, as much as possible, avoid oversimplifying human understanding and behaviour [4].

6 Conclusion

This work shows that top high-activity users in the *#globalwarming* community on Twitter do not demonstrate a general predictive capacity compared to other groups of the same size. However, high-activity users do show a distinct predictive capability when predicting for the three months following the training period for anger and fear, two particularly relevant emotions with respect to climate change. Furthermore, this correlation occurs for the Top N players in groups small enough to allow researchers to follow up on them if needed. In contrast, this is not the case for sadness and joy, indicating that modelling these emotions is not a completely straightforward process in the context of online communications about global warming.

References

1. Aleahmad, A., Karisani, P., Rahgoza, M., Oroumchian, F.: OLFinder: finding opinion leaders in online social networks. J. Inf. Sci. **42**(5), 659–674 (2016)
2. Bigonha, C., Cardoso, T.N.C., Moro, M.M., Gonçalves, M.A., Almeida, V.A.F.: Sentiment-based influence detection on Twitter. J. Braz. Comput. Soc. **18**(3), 169–183 (2011). https://doi.org/10.1007/s13173-011-0051-5
3. Boyd, D., Golder, S., Lotan, G.: Tweet, tweet, retweet: conversational aspects of retweeting on Twitter. In: Proceedings of the 43rd Hawaii International Conference on System Sciences, pp. 1–10 (2010)
4. Chapman, D., Lickel, B., Markowitz, E.: Reassessing emotion in climate change communication. Nat. Clim. Change **5**(12), 850–852 (2017)
5. Cody, E.M., Reagan, A.J., Mitchell, L., Dodds, P.S., Danforth, C.M.: Climate change sentiment on Twitter: an unsolicited public opinion poll. PLoS ONE **10**(8), e0136092 (2015)
6. Eliacik, A.B., Erdogan, N.: Influential user weighted sentiment analysis on topic based microblogging community. Expert Syst. Appl. **92**, 403–418 (2018)
7. Farbotko, C., Mcgregor, H.V.: Copenhagen, climate science and the emotional geographies of climate change. Aust. Geogr. **41**(2), 159–166 (2010)
8. Farzindar, A., Inkpen, D.: Natural Language Processing for Social Media. Morgan & Claypool, San Rafael (2015)
9. Fersini, E.: Sentiment analysis in social networks: a machine learning perspective. In: Pozzi, F.A., Fersini, E., Messina, E., Liu, B. (eds.) Sentiment Analysis in Social Networks, pp. 91–111. Morgan Kaufmann, Burlington (2017)
10. Frank, E., Hall, M.A., Witten, I.H.: The WEKA Workbench. Online Appendix for "Data Mining: Practical Machine Learning Tools and Techniques", 4th edn. Morgan Kaufmann, Burlington (2016)
11. Gimpel, K., et al.: Part-of-speech tagging for Twitter: annotation, features, and experiments. In: Proceedings of the 49th Annual Meeting of the Association for Computational Linguistics: Human Language Technologies: Short Papers-Volume 2, pp. 42–47 (2011)
12. Halse, S.E., Tapia, A., Squicciarini, A., Caragea, C.: An emotional step toward automated trust detection in crisis social media. Inf. Commun. Soc. **21**(2), 288–305 (2018)

13. Haltinner, K., Sarathchandra, D.: Climate change skepticism as a psychological coping strategy. Soc. Compass **12**(6), 84–85 (2018)
14. Honeycutt, C., Herring, S.C.: Beyond microblogging: conversation and collaboration via Twitter. In: Proceedings of the 42nd Hawaii International Conference on System Sciences (2009)
15. Leiserowitz, A.: Climate change risk perception and policy preferences: the role of affect, imagery, and values. Clim. Change **77**(1), 45–72 (2006). https://doi.org/10.1007/s10584-006-9059-9
16. Liu, B.: Sentiment Analysis: Mining Opinions, Sentiments, and Emotions. Cambridge University Press, Cambridge (2015)
17. Lu, H., Schuldt, J.P.: Exploring the role of incidental emotions in support for climate change policy. Clim. Change **131**(4), 719–726 (2015). https://doi.org/10.1007/s10584-015-1443-x
18. Marwick, A.E., Boyd, D.: I tweet honestly, I tweet passionately: Twitter users, context collapse, and the imagined audience. New Media Soc. **13**(1), 114–133 (2011)
19. Mitchell, L., Frank, M.R., Harris, K.D., Dodds, P.S., Danforth, C.M.: The geography of happiness: connecting Twitter sentiment and expression, demographics, and objective characteristics of place. PLoS ONE **8**(5), e64417 (2013)
20. Mobley, C., Vagias, W.M., DeWard, S.L.: Exploring additional determinants of environmentally responsible behavior: the influence of environmental literature and environmental attitudes. Environ. Behav. **42**(4), 420–447 (2010)
21. Mohammad, S.M., Bravo-Marquez, F.: Emotion intensities in tweets. In: Proceedings of the Joint Conference on Lexical and Computational Linguistics, pp. 65–77 (2017)
22. Nabi, R.L., Gustafson, A., Jensen, R.: Framing climate change: exploring the role of emotion in generating advocacy behavior. Sci. Commun. **40**(4), 442–468 (2018)
23. Plutchik, R.: The nature of emotions: human emotions have deep evolutionary roots, a fact that may explain their complexity and provide tools for clinical practice. Am. Sci. **89**(4), 344–350 (2001)
24. Porter, M.F.: An algorithm for suffix stripping. Program **40**(3), 211–218 (1980/2006)
25. Preoţiuc-Pietro, D., Volkova, S., Lampos, V., Bachrach, Y., Aletras, N.: Studying user income through language, behaviour and affect in social media. PLoS ONE **10**(9), e0138717 (2015)
26. Stocker, T., et al. (eds.): Climate Change 2013: The Physical Science Basis. Contribution of Working Group I to the Fifth Assessment Report of the Intergovernmental Panel on Climate Change. Cambridge University Press, Cambridge, United Kingdom and New York (2013)
27. Sulis, E., Irazú Hernández Farías, D., Rosso, P., Patti, V., Ruffo, G.: Figurative messages and affect in Twitter: differences between #irony, #sarcasm and #not. Knowl.-Based Syst. **108**, 132–143 (2016)
28. Vigna, S.: Further scramblings of Marsaglia's xorshift generators. J. Comput. Appl. Math. **315**, 175–181 (2016)
29. Weber, E.U.: Experience-based and description-based perceptions of long-term risk: why global warming does not scare us (yet). Clim. Change **77**(1), 103–120 (2006). https://doi.org/10.1007/s10584-006-9060-3
30. Williams, H.T., Mcmurray, J.R., Kurz, T., Lambert, F.H.: Network analysis reveals open forums and echo chambers in social media discussions of climate change. Glob. Environ. Change **32**, 126–138 (2015)
31. Witten, I.H., Frank, E.: Data Mining: Practical Machine Learning Tools and Techniques, 2nd edn. Morgan Kaufmann, Burlington (2005)

Using Topic Modelling to Improve Prediction of Financial Report Commentary Classes

Karim El Mokhtari, Mucahit Cevik[(✉)], and Ayşe Başar

Data Science Lab, Ryerson University, Toronto, Canada
{elmkarim,mcevik,ayse.bener}@ryerson.ca
https://www.ryerson.ca/dsl

Abstract. We consider the task of predicting the class of commentaries associated with financial discrepancies between actual and estimated sales data. Such analysis of the financial data is helpful in meeting targets and assessing the overall performance of the company. While generating a commentary and its associated class is the task of an analyst, these manual operations might be erroneous and as a result, might lead to a diminished performance for the employed prediction model due to wrong class labels. Accordingly, we propose using topic modelling, namely Latent Dirichlet Allocation (LDA), for automated extraction of the classes of the commentaries. In addition, we use feature selection strategies to improve the accuracy of the prediction models. Our analysis with various time series classification methods points to improved performance due to LDA and feature selection.

Keywords: Time series classification · LDA · NLP

1 Introduction

On an annual basis, companies make budgets to set targets for sales, revenues and expenses. Financial reports that are generated through various transactions are then used to understand the discrepancies between actual performance and financial forecast. These reports are generated from different sources including the daily sales transactions, inventories, cash flows, and supplier transactions. Financial analysts within the company examine the variances between projected outcomes and the actuals in certain time intervals (e.g. weekly, monthly, or quarterly), and provide management with insights. Accordingly, it is determined whether some adjustments are needed to be made to the targets or some other actions are needed to be taken in various departments to meet these targets.

The analyst report usually involves a summary information that relates each variance with a short commentary that serves as a baseline for top management to take immediate actions. While this process relies on the analyst's knowledge of the business areas, the data coming from different silos of the company might

C. Goutte and X. Zhu (Eds.): Canadian AI 2020, LNAI 12109, pp. 201–207, 2020.
https://doi.org/10.1007/978-3-030-47358-7_19

lead to erroneous assessments. As such, there is a need to learn from existing commentaries and build a prediction model that helps the automatic generation of the commentaries and their associated classes.

In our previous work [5], we tried to predict the commentaries using an encoder/decoder structures that predict jointly from the variance dataset and the expert commentaries. However, the reduced set of commentaries and a high variety among commentaries in terms of writing and wording, made it hard to predict meaningful commentaries. Later, we simplified the problem and aimed to predict the category of the commentary rather than trying to generate the original text [6]. Our analysis with various machine learning models including Support Vector Machines (SVMs), Random Forests (RFs) and Long Short Term Memory Neural Networks (LSTMs) indicated a positive predictive performance.

We note that, while generating a commentary and its associated class is the task of an analyst, such manual operations might be erroneous and as a result might lead to a diminished performance for the employed prediction model due to wrong class labels. Accordingly, we propose applying topic modelling, namely Latent Dirichlet Allocation (LDA) on the commentaries dataset in order to cluster commentaries based on their similarity, and then predict the topic of the commentary rather than the full text. The model learns from the variance dataset and predicts the topic, which helps automating the overall process.

Topic modelling aims to identify abstract or hidden structures of the text bodies which are called topics. There are various topic modelling techniques in the literature such as Latent Semantic Analysis (LSA) [3], probabilistic latent semantic analysis (pLSA) [7] and LDA [2]. While LSA is a simple dimensionality reduction technique with SVD, pLSA and LDA consider a statistical approach that builds a probabilistic language model from documents considered as a mixture of topics. Unlike pLSA that ignores how the topic mixture is generated for a document [9], LDA uses the Dirichlet distribution to determine this topic distribution. Liu et al. [10] used sentiment as topics and applied pLSA to model weblogs entries and predict product sales performance. LDA was successfully applied for Pseudo Relevance Feedback as well, where Miao et al. [11] introduced a topic space to evaluate the reliability of each candidate feedback document, then they used the reliability scores to adjust the weights of terms. LDA was also applied in identifying medical prescription patterns [12] and modelling the correlations of news items with stock price movements [8].

2   Methodology

2.1   Topic Modelling

Data is provided by our partner company in two datasets that include 34 months, from January 2016 to October 2018. The first dataset, COM (see Table 2), contains the commentaries written by the analyst for every customer and brand. The second dataset, VAR (see Table 1), includes the difference between forecasts and actuals in millions of dollars for every customer and brand.

Table 1. The VAR dataset records the discrepancy between forecast and actuals by customer and brand in Millions of CAD. Sample for one customer.

Customer 1				
	Jan. 2016	Feb. 2016	...	Sep. 2018
B1	+0.10	−0.01	...	+0.05
B2	−0.08	−0.05	...	+0.12
...	...		...	...

Table 2. Commentaries are written monthly for brands with discrepancies higher than 0.2 Millions CAD. Sample commentaries for January 2016

January 2016		
Brand	Var.	Commentary
B3	+0.50	Customer10: caused by over delivery
B15	+0.63	Customer25: declining faster than seen in the market
...	...	...

The analyst labelled each commentary with one of the following five labels: 1. "Promo" when a promotion did not perform as expected, 2. "SP&D" related to special offers including multiple products sold as a package, 3. "POS" related to sales that showed unexpected highs or lows at retail stores, 4. "Other" for rare topics, and 5. "NoComm" when no commentary was provided. The process of labelling commentaries is challenging as it often happens that a commentary has a main topic and one or two secondary topics. Further, it leads to an imbalanced dataset having NoComm as a majority class and Other as a minority class. We apply undersampling for NoComm and oversampling for the remaining classes to balance the labels' occurrences.

We process the VAR dataset in a way that for every commentary emitted for a customer and a brand in any given month, we associate a time series of variances (i.e. *forecasts − actuals*) recorded for the customer and the brand for all previous 12 months. Therefore, we build a new dataset VAR-COM where each row is associated with a brand b, a customer c and a month m. The input is in the form of a vector of 13 elements that represents the variance of the brand/customer pair (b, c) during the month m and the 12 preceding months: $\{v^{(b,c)}_{m-12}, \ldots, v^{(b,c)}_{m-1}, v^{(b,c)}_m\}$. The output is a one-hot-encoded representation of the label of the commentary generated by the expert for the tuple (b, c, m).

In addition, we pre-process all commentaries by converting them to lowercase then removing numbers, punctuation signs and common English stop-words. We also build a context-related list of stop-words such as customer name, product description and months. The comment after pre-processing is supposed to contain only useful keywords. We later apply lemmatizing and stemming to transform all word variants to standardized roots. Then, we apply *TF-IDF* to the resulting commentaries according to the formula $TF\text{-}IDF(t, c) = TF(t, c) \cdot \log\left(\frac{N}{DF(t)}\right)$ where t denotes a term in a commentary c, $TF(t, c)$ is the frequency of the term t in the commentary c, N is the number of non-empty commentaries in the dataset, and $DF(t)$ the document frequency defined as the number of documents containing the term t. Therefore, *TF-IDF* associates a higher score to the relevant words. Every commentary is converted into a set of tokens associated with their *TF-IDF* scores. We choose to use TF-IDF scoring instead of embedding vectors as the vocabulary size and the number of commentaries are relatively low (around 521 commentaries and 441 unique words after pre-processing).

Finally, we apply LDA to all commentaries by using three topics. The one with the highest probability is considered as the main topic of the commentary. It may happen that all words in a commentary are removed in the pre-processing phase, in this case, we associate a label "General" to the commentary. Empty commentaries are labelled with NoComm. Consequently, LDA results in five labels, which is equal in number to the labels provided by analysts, but different in nature as they are produced by clustering (i.e. LDA). Each commentary is assigned one of the following labels: "Topic1", "Topic2", "Topic3", "General" or "NoComm". We experiment with a different number of topics as well.

2.2 Learning Model

We train five models, each one is based on a different classification approach: RF, Gradient Boosting (XGB), SVM, LSTM and one-nearest neighbour dynamic time warping (1-NN DTW) [1]. Previously, we observed that not all months in the dataset are important in the time series [4], thus we consider feature selection based on RF feature importance by keeping 4 months out of 13 months, namely v_{m-12}, v_{m-11}, v_{m-1} and v_m where m is the month the commentary is issued.

3   Results

The commentaries dataset contains 2607 data points where only 521 are associated with commentaries. The average number of words in a commentary is nine. After pre-processing this number is reduced to four. Around 6% of the commentaries ended up being empty after pre-processing. We carry out three different experiments: (1) with analyst and LDA labels using 3 topics and no feature selection, (2) with analyst and LDA labels using 3 topics and with feature selection, (3) with analyst and LDA labels using 2 topics and with feature selection. After splitting the dataset into a train and test sets with an 80/20 ratio, we run all models 25 times with random oversampling and downsampling. We report the F1-score macro resulting from each run.

Figure 1 indicates that there is no significant improvement when learning with the original labels and the ones generated by LDA without feature selection. We note that, after feature selection, results are improved for XGB and SVM. With 2 topics, LDA-based models perform better than with the original labels as shown in Fig. 2. This can be explained by the fact that the number of labels with LDA is smaller, however, from a practical perspective, such results are still interesting for the consumer goods company as the model is able to predict the commentary topic and provide insightful keywords for every topic. Figure 3 illustrates the

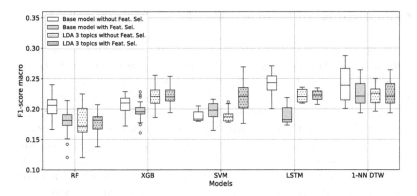

Fig. 1. F1-scores for different models before and after applying LDA using 3 topics (i.e. 5 labels), with and without feature selection.

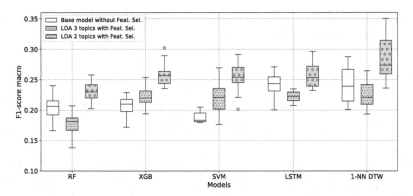

Fig. 2. F1-scores for considering different number of topics selected by LDA.

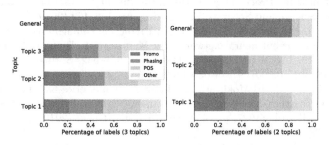

Fig. 3. Percentage of each expert label in LDA topics.

distribution of each analyst label by LDA topic. If we discard the "General" topic, the figure does not reveal a clear association between topics and labels which may suggest that the labels may not be related to specific keywords as was assumed in this analysis. Alternatively, this might be considered as evidence for data quality issues, if the expectation is to have some correlation between the labels and the associated commentaries.

4   Conclusions and Future Work

In this paper, we incorporate LDA into the classification of financial time series data into categories of commentaries. We examine the impact of feature selection on prediction quality as well. Our analysis with various time series classification methods shows that LDA in combination with feature selection leads to improved results. Moreover, automated identification of class labels from commentaries enables experimenting with different numbers of class labels, which further improves the predictive performance. 1-NN DTW performed the best in predicting labels after applying LDA. However, no significant correlation is observed between analysts' labels and LDA topics. Future work includes acquiring more data to test out our approaches as well as employing an active learning mechanism to guide the expert in manually labelling the instances so that number of manually labelled instances and the associated errors will be reduced.

Acknowledgement. This work is supported by Smart Computing For Innovation (SOSCIP) consortium, Toronto, Canada.

References

1. Bagnall, A., Lines, J., Bostrom, A., Large, J., Keogh, E.: The great time series classification bake off: a review and experimental evaluation of recent algorithmic advances. Data Min. Knowl. Disc. **31**(3), 606–660 (2016). https://doi.org/10.1007/s10618-016-0483-9
2. Blei, D., Ng, A., Jordan, M.: Latent dirichlet allocation. JMLR **3**, 993–1022 (2003)
3. Deerwester, S., Dumais, S.T., Furnas, G.W., Landauer, T.K., Harshman, R.: Indexing by latent semantic analysis. J. Am. Soc. Inf. Sci. **41**(6), 391–407 (1990)
4. El Mokhtari, K., Higdon, B., Başar, A.: Interpreting financial time series with SHAP values. In: Proceedings of the 29th Annual International Conference on Computer Science and Software Engineering, pp. 166–172 (2019)
5. El Mokhtari, K., Maidens, J., Bener, A.: Predicting commentaries on a financial report with recurrent neural networks. In: Meurs, M.-J., Rudzicz, F. (eds.) Canadian AI 2019. LNCS (LNAI), vol. 11489, pp. 531–542. Springer, Cham (2019). https://doi.org/10.1007/978-3-030-18305-9_56
6. Peachey Higdon, B., El Mokhtari, K., Başar, A.: Time-series-based classification of financial forecasting discrepancies. In: Bramer, M., Petridis, M. (eds.) SGAI 2019. LNCS (LNAI), vol. 11927, pp. 474–479. Springer, Cham (2019). https://doi.org/10.1007/978-3-030-34885-4_39
7. Hofmann, T.: Probabilistic latent semantic indexing. In: Proceedings of the 22nd Annual International ACM SIGIR Conference on Research and Development in Information Retrieval, pp. 50–57 (1999)
8. Asadi Kakhki, S.S., Kavaklioglu, C., Bener, A.: Topic detection and document similarity on financial news. In: Bagheri, E., Cheung, J.C.K. (eds.) Canadian AI 2018. LNCS (LNAI), vol. 10832, pp. 322–328. Springer, Cham (2018). https://doi.org/10.1007/978-3-319-89656-4_34
9. Lee, S., Baker, J., Song, J., Wetherbe, J.C.: An empirical comparison of four text mining methods. In: 2010 43rd Hawaii International Conference on System Sciences, pp. 1–10, January 2010. https://doi.org/10.1109/HICSS.2010.48

10. Liu, Y., Huang, X., An, A., Yu, X.: ARSA: a sentiment-aware model for predicting sales performance using blogs. In: Proceedings of the 30th Annual International ACM SIGIR Conference on Research and Development in Information Retrieval, pp. 607–614 (2007)
11. Miao, J., Huang, J.X., Zhao, J.: TopPRF: a probabilistic framework for integrating topic space into pseudo relevance feedback. ACM TOIS **34**(4), 1–36 (2016)
12. Park, S., Choi, D., Kim, M., Cha, W., Kim, C., Moon, I.C.: Identifying prescription patterns with a topic model of diseases and medications. J. Biomed. Inform. **75**, 35–47 (2017)

Wise Sliding Window Segmentation: A Classification-Aided Approach for Trajectory Segmentation

Mohammad Etemad[1]([✉]), Zahra Etemad[3], Amílcar Soares[1], Vania Bogorny[4], Stan Matwin[1,2], and Luis Torgo[1]

[1] Institute for Big Data Analytics, Dalhousie University, Halifax, Canada
etemad@dal.ca
[2] Institute for Computer Science, Polish Academy of Sciences, Warsaw, Poland
[3] Bu-Ali Sina University, Hamedan, Iran
[4] PPGCC - Universidade Federal de Santa Catarina (UFSC), Florianópolis, Brazil

Abstract. Large amounts of mobility data are being generated from many different sources, and several data mining methods have been proposed for this data. One of the most critical steps for trajectory data mining is segmentation. This task can be seen as a pre-processing step in which a trajectory is divided into several meaningful consecutive subsequences. This process is necessary because trajectory patterns may not hold in the entire trajectory but on trajectory parts. In this work we propose a supervised trajectory segmentation algorithm, called Wise Sliding Window Segmentation (WS-II). It processes the trajectory coordinates to find behavioral changes in space and time, generating an error signal that is further used to train a binary classifier for segmenting trajectory data. This algorithm is flexible and can be used in different domains. We evaluate our method over three real datasets from different domains (meteorology, fishing, and individuals movements), and compare it with four other trajectory segmentation algorithms: OWS, GRASP-UTS, CB-SMoT, and SPD. We observed that the proposed algorithm achieves the highest performance for all datasets with statistically significant differences in terms of the harmonic mean of purity and coverage.

Keywords: Trajectory segmentation · Spatio-temporal segmentation · Trajectory partition · Supervised trajectory segmentation

1 Introduction

An essential task for mobility data mining is trajectory segmentation. Different to classical data mining, in trajectory data mining, the attributes/features are extracted from subtrajectory parts. The partitioning is necessary because a mobility pattern, in general, does not hold for the entire trajectory, but for subtrajectory parts. Therefore, the segmentation process becomes one of the most critical pre-processing steps for trajectory data mining.

C. Goutte and X. Zhu (Eds.): Canadian AI 2020, LNAI 12109, pp. 208–219, 2020.
https://doi.org/10.1007/978-3-030-47358-7_20

Trajectory segmentation is the process of splitting a given trajectory into several homogeneous segments regarding some criteria. This task plays a pivotal role in trajectory mining since it affects the features of each segment, as the features may depend on the size of the trajectory segment, independently of the application domain, such as fishing detection [15], animal behavior [7,8], tourism [6], traffic dynamics [5,7,13,16], vessel movement patterns [2] etc.

A trajectory is a sequence of points located in space and time, and different criteria can be used to split trajectories. There are several approaches that can be used for trajectory segmentation such as CB-SMoT [12], SPD [17], WK-Means [10], GRASP-UTS [15], TRACLUS [9], OWS [4], etc. Different to previous approaches where no training step is performed, we propose in this paper a supervised strategy to segment trajectory data. To the best of our knowledge this is the first approach that actually learns partitioning positions (i.e., the last trajectory point of a segment) from trajectory data characteristics for a given application domain. The main advantage of this supervised strategy is that the transitioning characteristics of a behavior change can be learned from the training data. The model built to forecast partitioning positions is further used to segment trajectories. After that, a majority vote strategy decides the proper location to place a partitioning position.

In summary, the main contributions of this work include: (i) a method for producing training data from partitioning positions on a labeled trajectory; (ii) a method to decide when a partitioning position occurs in a trajectory; and (iii) an empirical study comparing WS-II and several baselines for segmentation.

This paper is organized as follows. Section 2 shows the definitions necessary to describe our trajectory segmentation method. In Sect. 3, the related works are described. In Sect. 4, we propose WS-II with details. In Sect. 5, we applied the proposed method and other trajectory segmentation algorithms on three datasets and reported their performance results. Finally, we conclude our work in Sect. 6.

2   Definitions

In this section we present the basic concepts related to trajectories and used throughout this paper.

The trajectory of a moving object o can be described by a time ordered sequence of locations the object has visited. We call these locations, *trajectory points*.

Trajectory Point. A *trajectory point*, l_i^o, is the location of object o at time i, and is defined as,

$$l_i^o = \langle x_i^o, y_i^o \rangle \tag{1}$$

where x_i^o is the longitude of the location which varies from $0°$ to $\pm 180°$, while y_i^o is the latitude which varies from $0°$ to $\pm 90°$.

Raw Trajectory. A *raw trajectory*, or simply *trajectory*, is a time-ordered sequence of trajectory points of some moving object o,

$$\tau^o = \langle l_0^o, l_1^o, .., l_n^o \rangle \tag{2}$$

Segment or Subtrajectory is a set of consecutive trajectory points belonging to a raw trajectory $\tau^o = \langle l_0^o, l_1^o, .., l_n^o \rangle$,

$$s^o = \langle l_j^o, \cdots , l_k^o \rangle, \quad j \geq 0, \; k \leq n \text{ and } s^o \subset \tau^o \tag{3}$$

The process of generating segments from a trajectory is called *Trajectory Segmentation (TS)*. The most common way of defining TS involves splitting a raw trajectory into a set of non-overlapping segments. More formally:

Trajectory Segmentation. Given a raw trajectory $\tau^o = \langle l_0^o, l_1^o, .., l_n^o \rangle$, we define a sequence of segments $S = \langle s_0^o, \cdots , s_k^o \rangle$, such that

$$\forall_{s_i^o, s_{i+1}^o \in S} \; s_i^o = \langle l_p^o, \cdots , l_{p+t}^o \rangle, \; s_{i+1}^o = \langle l_{p+t+1}^o, \cdots , l_{p+t+u}^o \rangle \tag{4}$$

and

$$s_0^o = \langle l_0^o, \cdots , l_i^o \rangle, \; s_k^o = \langle l_j^o, \cdots , l_n^o \rangle \tag{5}$$

Equation 6 shows the input and output of the trajectory segmentation process, where τ is a raw trajectory which contains n trajectory points, and S is the set of all segments generated from τ using TS.

$$TS : \tau \longrightarrow S, \; |\tau| = n+1, \; |S| = k+1 \tag{6}$$

In this notation, $n+1$ is the number of trajectory points and $k+1$ is the number of segments resulting from applying TS to the trajectory.

We call a trajectory point at the end of each segment as *partitioning position*. This means that the result of applying TS to a trajectory, S contains k partitioning positions.

Problem Definition. Given a raw trajectory τ^o, we would like to generate a sequence of segments $S = \langle s_0^o, \cdots , s_k^o \rangle$ so that each s_i^o satisfies a certain homogeneity criteria for a given application domain. To evaluate the performance of the generated S, we rely on the knowledge of an expert user to provide a set of semantic tuples $sl_i = (sid, label)$ where sid identifies a segment s_i of a trajectory, generated by the expert user, and $label$ is a semantic label attached by the expert to this segment, such as for instance, a transportation mode or status of fishing or non-fishing.

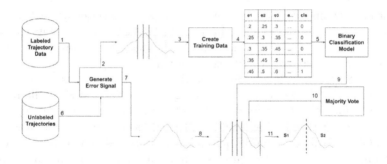

Fig. 1. A high level view of wise sliding window segmentation algorithm.

3   Related Works

In this section, we give an overview of several methods for trajectory segmentation. Warped K-Means (WK-Means), which is a general-purpose segmentation algorithm based on K-Means [11], is introduced in [10]. It modifies the K-Means algorithm by minimizing a quadratic error (cost function) while imposing a sequential constraint in the segmentation step. Since WK-Means imposes a hard sequential constraint, segments can be updated while new samples arrive without affecting too much the previous clustering configuration [10]. This algorithm receives the number of segments to be found on the data (k). Having such input parameter is the main limitation of using it in domains where the number of segments is not pre-defined or is dynamic.

The Stay Point Detection (SPD) [17] is a simple algorithm that follows the idea that between each two-movements, there is a stop. SPD applies a distance threshold (θ_d) and a time threshold θ_t so that a moving object which spends more than θ_t time in the neighborhood of θ_d belongs to a stay point. Hence, each stay point identifies a segment, and the trajectory points between two stay points are generated in another segment.

An extension of DB-SCAN [3], CB-SMoT detects stops and moves segments in a trajectory [12]. The original definitions of a $\epsilon-$neighborhood and minimum points in DB-SCAN are altered so that CB-SMoT utilizes spatial and temporal aspects of trajectories. CB-SMoT works based on the trajectory speed, and the stop points are consecutive trajectory points where the moving object has a lower speed.

TRACLUS [9] detects dense regions with the same line segment characteristics. This clustering algorithm has two steps: (i) partitioning of the trajectory to line segments; and (ii) clustering these lines. A cost function based on the Minimum Description Length (MDL) principle is applied in the first step to split a trajectory into its line segments. It considers three trajectory segment's attributes: (i) parallel distance, (ii) perpendicular distance, and (iii) angular distance. Clustering line segments using DB-SCAN is run in the next step [9].

GRASP-UTS is an unsupervised trajectory segmentation algorithm that benefits from the Minimum Description Length (MDL) principle to build the most homogeneous segments. First, GRASP-UTS generates random landmarks.

Then, it builds homogeneous segments by swapping the trajectory points across temporally-ordered segments and adjusting the landmarks based on its cost function's value [15]. GRASP-UTS can apply additional features on top of the raw trajectories to perform trajectory segmentation.

The OWS (Octal Window Segmentation) algorithm is based on computing the error signal generated by measuring the deviation of a middle point of an octal window [4]. The intuition behind OWS is that when a moving object changes its behavior, this shift may be detected using only its geolocation over time [4]. OWS uses interpolation methods to find the estimated position of the moving object, i.e., where it is supposed to be if its behavior does not change. Then, OWS compares the real position of the moving object with the estimated one, creating an error signal. With such a procedure, it is possible to determine where the moving object changed its behavior and to use this information to create segments.

In this work, we extend the idea of OWS by using a configurable sliding window for interpolating points and a supervised strategy for deciding where partitioning positions should be placed. Unlike all previous segmentation algorithms, WS-II is supervised. This means that WS-II is able to learn the variations in the error signal generated by interpolation techniques which characterize partitioning positions over consecutive segments, avoiding in this way the decision of choosing an error threshold value (i.e., an epsilon value in the OWS) that relies on the characteristics of the domain where trajectories were collected.

4   The Proposed Method

Figure 1 shows an overview of the Wise Sliding Window Segmentation (WS-II) method, which has four core procedures: Generate Error Signal, Create Training Data, Binary Classification Model, and Majority Vote. First, the WS-II creates the error signal from the labeled dataset, which is detailed in Sect. 4.1. The second step is to generate the training data using the error signal, by sliding a window over its values and adding the presence or absence of a partitioning position. This part is detailed in Sect. 4.2. The third step is to train a binary classifier to recognize the partitioning positions over the sequence of error signals. This part is detailed in Sect. 4.3. Finally, unlabeled trajectories can then be segmented based on the model learned in the previous step and using the majority vote, as detailed in Sect. 4.4.

4.1   Generating the Error Signal

The first step of our proposal is similar to the OWS algorithm [4], which creates a sliding window over a trajectory to compute a signal error between trajectory points. For each sliding window, the error is generated by calculating the deviation of the interpolated midpoint of the window from the actual midpoint. This process is repeated by sliding the window by one point forward, so receiving a new trajectory point, it adds the newer point to the window set and removes the oldest point from the set. An example of this process is shown in Fig. 2a.

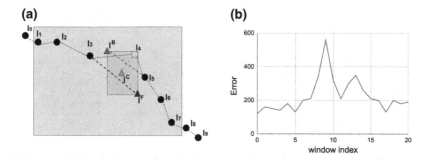

Fig. 2. (a) Example of error calculation where seven trajectory points (e.g., l_1 to l_7) are selected as the *current sliding window* (e.g., green box). (b) Example of error signal generated in meters for trajectory data. (Color figure online)

In Fig. 2a, the green rectangle is a sliding window of size 7, the l^B (red triangle), and l^F (blue triangle) are the interpolated positions. l^F is generated using extrapolation on the first three points (l_1, l_2, l_3) and l^B is generated using the last three points inside the window (l_5, l_6, l_7). The green dot (l_4) is assumed to be the missing point in the sliding window, while the l^C (orange triangle) is generated as a middle point between l^B and l^F. The distance between the midpoint (l^C) and the missing point (l_4) is called the *error value* of this window. In the example of Fig. 2a, the haversine distance from the estimated position l^C to the real position l_i is visible. This may indicate that the moving object's behavior has changed at position l_4.

An example of the error signal from a trajectory is shown in Fig. 2b. A raw trajectory with 26 points that forms 20 sliding window of size 7 (generating error for point index 4 to 23) is displayed in this example. The first three and last three error values are dropped. Window index is the index of the middle trajectory point in each window. Figure 2b illustrates a situation in which they are several trajectory points (e.g., around trajectory points 8 and 12) along the raw trajectory where the estimated positions were far from the real trajectory positions. These boundaries are considered as potential partitioning positions for creating trajectory segments.

4.2   Creating Training Data

The second core procedure of WS-II is to create a training dataset using the sequential error values extracted in the previous step. First, we create an array of size q of error signals that will belong to the first training sample, and we use the ground truth information (i.e., if in this particular region there was a change in the behavior) to annotate the label of this sample. If this window includes a partitioning position, it is labeled as 1 and 0 otherwise. By receiving every new trajectory point, we remove one point from the start of our window and add the new point to the end of the window. Then we create our next sample by

	e_1	e_2	e_3	e_4	e_5	e_6	e_7	label
...	...	...	...	...	...	...	...	...
w_1	120	160	150	140	180	130	200	0
w_2	160	150	140	180	130	200	210	0
w_3	150	140	180	130	200	210	340	0
w_4	140	180	130	200	210	340	**560**	1
w_5	180	130	200	210	340	**560**	320	1
w_6	130	200	210	340	**560**	320	210	1
$w...$	...	...	...	...	...	...	...	...
w_{10}	**560**	320	210	120	320	273	200	1
w_{11}	320	210	120	320	273	200	130	0

(a) Example of a training set generated by WS-II.

	e_1	e_2	e_3	e_4	e_5	e_6	e_7	b_{cls}	m_{cls}
$w...$	...	...	...	...	...	...	...	...	...
w_m	100	150	230	160	170	320	400	0	...
w_{m+1}	150	230	160	170	320	400	390	1	...
w_{m+2}	230	160	170	320	400	390	320	0	...
w_{m+3}	160	170	320	400	390	320	380	0	0
w_{m+4}	170	320	400	390	320	380	330	0	1
w_{m+5}	320	400	390	320	380	330	490	1	0
w_{m+6}	400	390	320	380	330	490	450	1	0
w_{m+7}	390	320	380	330	490	450	230	1	...
w_{m+8}	320	380	330	490	450	230	160	0	...
w_{m+9}	380	330	490	450	230	160	190	0	...
$w...$	...	...	...	...	...	...	...	...	...

(b) Example of the majority vote mechanism.

Fig. 3. Examples of data tables generated by our algorithm.

applying the same step of labeling 1 when a partitioning position is present in the sliding window, and 0 if it is not. This procedure is repeated until all the error signals are evaluated.

To understand how the labeling process works, we show an example in Fig. 3a. In this example, the training data are created for the sliding window built with seven (e_1 to e_7) trajectory points over eleven slides (i.e., w_1 to w_{11}). As can be seen in Fig. 3a, from w_1 to w_3, there was no big change in the error signal (ranging from 120 to 340 m). In w_4, the value of 560 characterizes a high jump in the estimated error and actually reflects a real change in the behavior of the moving object, resulting in a positive example (i.e., there is a partitioning position) in the training data. Examples from w_4 to w_{10} are labeled as positive due to the presence of partitioning position in the sliding window. From w_{11}, the samples are again labeled as negative examples due to the absence of a partitioning position in the data.

4.3   Binary Classification Model

A binary classifier is used by WS-II to categorize each error signal sample into either a partitioning position or not. The labeled trajectory data created in the previous step is used to generate training samples for this binary classifier so that it can classify signal samples into a class where a sliding window has a partitioning position (e.g., value 1) or a class when it does not have a partitioning position (e.g., value 0).

It was observed that the error signal has its minimum fluctuations far from a partitioning position, and it has its maximum fluctuations while transitioning from one segment to a new one. Therefore, detecting the area that includes partitioning positions is an indicator that the behavior has changed. We apply the binary classifier to identify these areas over a trajectory that has the highest likelihood of containing partitioning positions. In this work, we used a Random Forest classifier [1] to benefit from its bagging power while processing long window sizes faster by limiting the number of features. However, we emphasize that any classification model can be used in this step. After forecasting these transitioning areas, we use a majority vote mechanism to decide precisely where to place a partitioning position, explained in the next section.

4.4 Majority Vote

At this step, we use the same sliding window of size q to decide if a partitioning position occurred. Since we are using a window slide point by point, each trajectory point can be part of q sliding windows, and we classify each window using the binary classifier. This means that we have q outputs that the binary classifier generates for a trajectory point belongs to the q windows. Using a majority vote mechanism for these q outputs leads us to the final decision: the trajectory point is a partitioning position if more than 50% of the sampled signals are labeled as a partitioning position.

Leveraging this feature and applying the voting technique, we can have a more robust evaluation to support if a point is a partitioning position or not. The decision to identify a trajectory point as a partitioning position is supported by q results, each of which contributes $1/q$ to the final decision. This means a misclassification of the binary classifier weights $1/q$. Although increasing q can make the algorithm more robust to noise, it will make it fail to identify segments with a length smaller than q. Furthermore, the algorithm is more robust against noisy points, which may happen in trajectory data due to device collection errors.

An example of the advantages of the majority vote mechanism are exemplified in Fig. 3b, where a window with $q = 7$ was used. In Fig. 3b, the column b_{cls} was forecast by the binary classifier for w_m to w_{m+9}. It is possible to see in Fig. 3b that w_{m+3} is decided by evaluating the b_{cls} column values from w_m to w_{m+6} $(0, 1, 0, 0, 0, 1, 1)$. The decision regarding a majority vote for w_{m+3} is equal to 0 since $|\#0| = 4$ and $|\#1| = 3$. For deciding the final value of w_{m+4} the lines from w_{m+1} to w_{m+7} are used. The evaluation of the set $(1, 0, 0, 0, 1, 1, 1)$ through a majority vote $(|\#0| = 3$ and $|\#1| = 4)$ results in the decision of 1 (i.e., a partitioning position occurred). As previously stated, such strategy makes WS-II robust against spatial jumps due to GPS error in the data collection process.

5   Experimental Evaluation

In this section, we evaluate the proposed method and compare it to state-of-the-art approaches. In Sect. 5.1, we describe the datasets. In Sect. 5.2, we detail the experimental setup and we report the results in Sect. 5.3.

5.1   Datasets

We evaluate our method on three datasets. The first is a fishing dataset containing 5190 trajectory points and 153 segments, where fishing activity labels (e.g., fishing or not-fishing) were provided by specialists and used to create trajectory segments. The second is the Atlantic hurricane dataset, which contains

216 M. Etemad et al.

1990 trajectory points and 182 segments. The Saffir-Simpson scale was used to determine the type of hurricane, and the transitions from one hurricane-level to another was used for creating trajectory segments. Finally, a subset of the Geolife dataset containing 12,955 trajectory points and 181 segments was used as a third dataset. For this dataset, we use the transportation mode as the ground truth for creating the segments. The reason that we did not use the full Geolife data set was that some of the segmentation algorithms, such as GRASP-UTS were not able to provide segments in a reasonable time. We create a sample Automatic Identification System (AIS) data to debug our algorithm and test our code and made it available to public[1].

5.2 Experimental Setup

In this work we measure the trajectory segmentation performance using Harmonic mean of Purity and Coverage, introduced in [4]. The use of purity and coverage for trajectory segmentation performance measurement originally is introduced in [15]. We do not use clustering measures such as completeness and homogeneity since the segmentation task is different from clustering. In trajectory segmentation, the order of the segments is essential, and adjacent segments can come from the same cluster. For example, an object moving to a shopping store and going back home characterizes two segments, that would be in the same "walk" cluster.

In each experiment, we divided the dataset into ten folds, one of which is applied to tuning/training the algorithm and the rest to testing its performance. Each fold contains different trajectories of different moving objects; therefore, we individually segment each trajectory and report the average results.

Since we divide data into ten folds, we calculate ten values for the Harmonic means. A boxplot is used to show the visual difference between these ten values for each algorithm, Fig. 4. Although the boxplot can show the difference between the performance of algorithms, we perform a Mann Whitney U test (having only ten numbers, we could not prove the data follows normal distribution, so we did not use T-test) to show that the difference between the median of each set is not generated randomly.

The state of the art methods that we compare to our approach requires some parameterization. The input parameter values estimation for GRASP-UTS was using a grid search with all combinations of values reported in [15]. For the SPD algorithm, we used the suggested parameters on the original paper for the subset of Geolife dataset, and for the rest of datasets we used a grid search to find the best parameters. For CB-SMoT, we applied a grid search to tune parameters using the parameter tuning fold. For OWS, we have tested the four kernels (e.g., random walk, kinematic, linear, and cubic) and used the same strategy reported in [4] to find the best value of *epsilon*. We decided only to report the random walk kernel findings since it obtained good results for all datasets. For a fair comparison between OWS and WS-II, we only report the WS-II results with the

---

[1] https://github.com/metemaad/WS-II.

random walk kernel. Details regarding the input parameter values ranges for all algorithms can be found in the following link[2].

5.3   Results and Discussion

Figure 4.a displays the results of executing different segmentation algorithms on the Fishing dataset. A Mann Whitney U test indicated that WS-II produces statistically significant higher median ($mean = 94.32$, $std = 0.9$) harmonic mean for trajectory segmentation comparing to OWS with random walk kernel ($p_{value} = 9.133e{-}05$, $mean = 89.04$, $std = 1.03$). Therefore, the proposed method achieved better performance in comparison to other trajectory segmentation methods.

A fishing activity is characterized by several ship turns. We believe that WS-II had a better result when compared with the other algorithms because of its capability to analyze not only a single trajectory point, but a larger region (i.e., a larger sliding window size). By analyzing a larger region, WS-II will only place a partitioning position when a partitioning position actually occurred (e.g., learned from the training data). Since a single turn is not enough to characterize a fishing activity, WS-II's strategy of analyzing a larger window is more robust in learning such behavior.

In this experiment, we compare five trajectory segmentation algorithms: CB-SMoT, SPD, GRASP-UTS, OWS with Random Walk kernel, and our proposed trajectory segmentation algorithm (WS-II) on Atlantic hurricane dataset. Figure 4.b shows that WS-II performed better than all other algorithms. A Mann Whitney U test indicated that WS-II produces statistically significant higher median ($mean = 94.68$, $std = 2.23$) harmonic mean for trajectory segmentation comparing to OWS with random walk kernel ($p_{value} = 9.1e{-}05$, $mean = 85.67$, $std = 0.59$).

In this experiment, we applied all the segmentation algorithms on a subset of Geolife containing ten different users. Each user's trajectory creates a fold and we use one fold to tune up our algorithm each time. Figure 4.c depicts our experiment results. Moreover, a Mann Whitney U test supports the claim that WS-II produces statistically significant higher median ($mean = 92.8$, $std = 2.11$) harmonic mean for trajectory segmentation comparing to OWS with random walk kernel ($p_{value} = 0.00065$, $mean = 88.94$, $std = 5.06$).

In the Geolife dataset, there are two major types of movement: (1) fast movements of buses, trains, and cars; and (2) slow movements of walk and bike, which have a random nature. The selection of a random walk seems to be a reasonable decision in this dataset because as long as a moving object moves slowly, the random walk kernel seems to reproduce the random nature of the movement. On the other hand, for the moving object that travels fast, the behavior of random walk is similar to a linear interpolation kernel in terms of direction because the direction variation decreases.

---

[2] https://github.com/metemaad/WS-II.

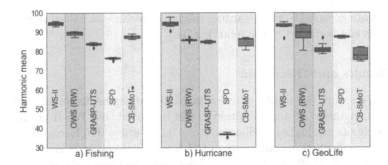

Fig. 4. Our proposed method outperforms CB-SMoT, SPD, GRASP-UTS, and OWS with Random Walk kernel on Fishing, Hurricane and Geolife dataset

6 Conclusions

In this paper we presented a supervised method for trajectory segmentation named Wise Sliding Window Segmentation (WS-II), that uses a trained model for deciding where partitioning positions should be placed. With the majority voting strategy the method becomes more robust to noise points and avoiding unnecessary partitions. The experimental results show that WS-II achieves better performance in terms of a harmonic mean of purity and coverage when compared with state-of-art trajectory segmentation algorithms in three datasets of different domains. One limitation of WS-II, which is a limitation for all learning methods, is that several domains do not have a labeled dataset where the patterns of movement behavior change can be learned. Although there are tools in the literature that encourage and assist the user in the process of labeling trajectory data [14], most trajectory datasets still do not provide any type of ground truth for validating supervised methods. As future work, we would like test how this algorithm performs with different sample sizes for training, i.e., how large the labeled data needs to be to find good results.

Acknowledgments. This work was financed by the Brazilian Agencies CNPq, CAPEs (Project Big Data Analytics [CAPES/PRINT process number 88887.310782/2018-00]), FAPESC (Project Match co-financing of H2020 Projects - Grant 2018TR 1266); the European Union's Horizon 2020 research and innovation programme under Grant Agreement 777695 (MASTER1); and the Natural Sciences and Engineering Research Council of Canada (NSERC).

References

1. Breiman, L.: Random forests. Mach. Learn. **45**(1), 5–32 (2001). https://doi.org/10.1023/A:1010933404324
2. Carlini, E., de Lira, V.M., Soares, A., Etemad, M., Machado, B.B., Matwin, S.: Uncovering vessel movement patterns from AIS data with graph evolution analysis **2020** (2020)

3. Ester, M., Kriegel, H.P., Sander, J., Xu, X., et al.: A density-based algorithm for discovering clusters in large spatial databases with noise. In: KDD, vol. 96, pp. 226–231 (1996)

4. Etemad, M., Hoseyni, A., Rose, J., Matwin, S.: A trajectory segmentation algorithm based on interpolation-based change detection strategies. In: EDBT/ICDT Workshops (2019)

5. Etemad, M., Soares Júnior, A., Matwin, S.: Predicting transportation modes of GPS trajectories using feature engineering and noise removal. In: Bagheri, E., Cheung, J.C.K. (eds.) Canadian AI 2018. LNCS (LNAI), vol. 10832, pp. 259–264. Springer, Cham (2018). https://doi.org/10.1007/978-3-319-89656-4_24

6. Feng, S., Cong, G., An, B., Chee, Y.M.: POI2Vec: geographical latent representation for predicting future visitors. In: Thirty-First AAAI Conference on Artificial Intelligence (2017)

7. Júnior, A.S., Renso, C., Matwin, S.: Analytic: an active learning system for trajectory classification. IEEE Comput. Graphics Appl. **37**(5), 28–39 (2017)

8. Junior, A.S., Times, V.C., Renso, C., Matwin, S., Cabral, L.A.: A semi-supervised approach for the semantic segmentation of trajectories. In: 19th IEEE International Conference on Mobile Data Management (MDM), pp. 145–154. IEEE (2018)

9. Lee, J.G., Han, J., Whang, K.Y.: Trajectory clustering: a partition-and-group framework. In: Proceedings of the 2007 ACM SIGMOD International Conference on Management of Data, pp. 593–604. ACM (2007)

10. Leiva, L.A., Vidal, E.: Warped k-means: an algorithm to cluster sequentially-distributed data. Inf. Sci. **237**, 196–210 (2013)

11. MacQueen, J., et al.: Some methods for classification and analysis of multivariate observations. In: Proceedings of the Fifth Berkeley Symposium on Mathematical Statistics and Probability, Oakland, CA, USA, vol. 1, pp. 281–297 (1967)

12. Palma, A.T., Bogorny, V., Kuijpers, B., Alvares, L.O.: A clustering-based approach for discovering interesting places in trajectories. In: Proceedings of the 2008 ACM Symposium on Applied Computing, pp. 863–868. ACM (2008)

13. Soares, A., et al.: CRISIS: integrating AIS and ocean data streams using semantic web standards for event detection. In: International Conference on Military Communications and Information Systems (2019)

14. Soares, A., Rose, J., Etemad, M., Renso, C., Matwin, S.: VISTA: a visual analytics platform for semantic annotation of trajectories. In: EDBT, pp. 570–573 (2019)

15. Soares Júnior, A., Moreno, B.N., Times, V.C., Matwin, S., Cabral, L.D.A.F.: GRASP-UTS: an algorithm for unsupervised trajectory segmentation. Int. J. Geograph. Inf. Sci. **29**(1), 46–68 (2015)

16. Varlamis, I., Tserpes, K., Etemad, M., Júnior, A.S., Matwin, S.: A network abstraction of multi-vessel trajectory data for detecting anomalies. In: EDBT/ICDT Workshops (2019)

17. Zheng, Y., Zhang, L., Ma, Z., Xie, X., Ma, W.Y.: Recommending friends and locations based on individual location history. ACM Trans. Web (TWEB) **5**(1), 5 (2011)

Using Deep Reinforcement Learning Methods for Autonomous Vessels in 2D Environments

Mohammad Etemad[1]([✉]), Nader Zare[1], Mahtab Sarvmaili[1], Amílcar Soares[1], Bruno Brandoli Machado[1], and Stan Matwin[1,2]

[1] Institute for Big Data Analytics, Dalhousie University, Halifax, Canada
etemad@dal.ca
[2] Institute for Computer Science, Polish Academy of Sciences, Warsaw, Poland

Abstract. Unmanned Surface Vehicles technology (USVs) is an exciting topic that essentially deploys an algorithm to safely and efficiently performs a mission. Although reinforcement learning is a well-known approach to modeling such a task, instability and divergence may occur when combining off-policy and function approximation. In this work, we used deep reinforcement learning combining Q-learning with a neural representation to avoid instability. Our methodology uses deep q-learning and combines it with a rolling wave planning approach on agile methodology. Our method contains two critical parts in order to perform missions in an unknown environment. The first is a path planner that is responsible for generating a potential effective path to a destination without considering the details of the root. The latter is a decision-making module that is responsible for short-term decisions on avoiding obstacles during the near future steps of USV exploitation within the context of the value function. Simulations were performed using two algorithms: a basic vanilla vessel navigator (VVN) as a baseline and an improved one for the vessel navigator with a planner and local view (VNPLV). Experimental results show that the proposed method enhanced the performance of VVN by 55.31% on average for long-distance missions. Our model successfully demonstrated obstacle avoidance by means of deep reinforcement learning using planning adaptive paths in unknown environments.

Keywords: Deep reinforcement learning · Path planning · Obstacle avoidance · Maritime autonomous surface vessels

1 Introduction

Ship collision avoidance and path planning is a fundamental research topic for autonomous navigation. Several methods have been proposed in the literature to this end. However, deep reinforcement learning strategies have empowered models for automatic maneuverability of vessels [4]. Autonomous vehicles have been developed and improved in different areas, ranging from unmanned aerial

© Springer Nature Switzerland AG 2020
C. Goutte and X. Zhu (Eds.): Canadian AI 2020, LNAI 12109, pp. 220–231, 2020.
https://doi.org/10.1007/978-3-030-47358-7_21

(e.g., planes and drones) [8], to underwater (e.g., ships, remotely operated underwater vehicles (ROVs), submarine gliders, and unmanned suface vessels (USVs)) [1].

Path planning has the objective of generating a path between an initial location and the desired destination with an optimal or near-optimal performance under specific constraints. Avoiding obstacles in real or simulated environments is an essential task for safely driving USVs towards a target without human intervention [5]. For example, in marine application scenarios, it is of extreme importance to avoid obstacles such as rocks, floaters, debris, and other ships [4].

In this paper, we develop a new method for path planning and obstacle avoidance in marine environments by using deep reinforcement learning (DRL) and local view strategy, namely Vessel Navigator with Planner and Local View (VNPLV). Unlike previous method as proposed in [16], which has a single environment with a fixed origin and destination points, we developed a methodology that can surpass traditional global approaches in unknown environments without any limitation of various origin and destination points. Basically, we improve the performance of the model by feeding it with CNN features and reducing the number of states using the Ramer–Douglas–Peucker algorithm. Results show that we can further enhance our model by using the idea of rolling wave planning and that our method benefits from combining the path planner for longterm planning and local view for short term decisions [9].

We summarize the contributions of our work as follows. Inspired by agile methodology, we implemented the idea of rolling wave planning by using a deep reinforcement learning model for short-term decision making and a planner for longterm general planning. (i) We created 2D marine environments by extracting information of geographical layers. (ii) We developed a deep reinforcement learning method for agents that simulates USV movements. Agents trained by our method are able to autonomously plan their path and avoid obstacles in a simulated 2D marine environment. (iii) We performed extensive experiments by means of simulations comparing our method and a baseline and propose a metric named Rate of Arrival to Destination (RATD) to evaluate the performance of our method.

The rest of this paper is organized as follows. Section 2 presents some related works in the area of path planning, reinforcement learning, and deep reinforcement learning. In Sect. 3, we provide definitions used across our work and detail our four proposed methods. Section 4 demonestrates the performance of our method in a simulated marine environment. Finally, Sect. 5 concludes the work and also discusses future works.

2   Related Works

In recent years USVs have attracted a great deal of attention from several maritime companies and research groups all over the world. There exist several approaches developed and applied for the USVs, which are mainly divided into path planning, obstacle avoidance, and intelligent optimization methods. We divide this section into three subsections to which our work is related.

Traditional Path Planning Approaches. Several research methods on collision avoidance have been developed for path planning based on A* and Artificial Potential Field (APF). The algorithm A* is a global heuristic search strategy that takes into account both the start position and the destination. However, the algorithm is considered in the literature to be an inefficient search in a large grid map. The works of [3] and [17] proposed hierarchical path planning strategies to improve the efficiency of A*. In contrast to A* algorithm, APF employs repulsive fields to model the environment with higher efficiency [10]. Such a strategy can create a smoother path when it compares with A*.

Reinforcement Learning. Some articles have employed reinforcement learning (RL) for the collision avoidance and path planning task to improve the autonomy of the obstacle avoidance system. Reinforcement learning is a classical machine learning method, first proposed by Sutton in 1984, and widely explored in the '90s. It has been widely used in the artificial intelligence field. Although the main algorithm used for path planning is Q-Learning [13], many methods in the literature are hybrid of RL with other methods [11]. In general, the two main drawbacks of RL approaches are the high cost of the learning process that depends on the environment and the user condition, and the degrees of freedom (DOF). Although the reinforcement learning algorithms have shown successful performance in variety of domains, their applicability has been restricted to fully observable low-dimensional state spaces domains.

Deep Reinforcement Learning. DRL is a novel topic which has been emerged to address the challenges of using RL in complex, high dimensional problems. In addition to the outstanding performance of Deep RL models in other domains, they have attracted a lot of attention in ship collision avoidance topic. The work of [4] proposed a deep reinforcement learning obstacle avoidance approach with the deep Q-network architecture for unmanned marine vessels in unknown environment. The authors presented a learning policy for obstacle avoidance at a safe distance in unknown environments with 3-DOF. They used the replay buffer and self-play trials to learn the control behaviors. Recently, [13] presented a prototype of multiple autonomous vessels controlled by deep q-learning. Their reward function and the training process were designed with respect to the maneuverability of the vessel, including speed, and acceleration. The incorporated navigation rules employed the conversion to navigational limitation by polygons or lines.

It is also important to point out that there exist two kinds of analysis regarding the enviroment exploration: (i) the global view, and (ii) the local view. Environment exploration is linked to the performance of the approaches, but there is no strict definition related to the size of the local view. The most noteworthy approach for local view strategy is based on line-of-sight (LOS), presented by [14]. Moreover, utilizing a local view became a best practice in some strategic game solvers such as [15].

In this paper we present a path planning method using deep q-learning with unknown position of dynamically generated obstacles in the environment. Our DRL-based method is focused on a local view strategy which reduces the number of states. Our policy is obtained using four different iterations of evolving our proposed work evaluated in simulation experiments detailed in Sect. 3.2.

3   A New DRL Method for Unmmaned Surface Vessels

In this section we go through the details of the proposed method. First, we define the main concepts used to model our agent-environment approach (Sect. 3.1). After that, we describe our proposed method for path planning and obstacle avoidance in the maritime domain (Sect. 3.2).

3.1   Definitions

In this work, an **agent** is a vessel voyaging from an origin in the environment with the objective of safely arriving at a desired destination in the environment. The **environment** is a bounding box area that contains a body of water that an agent can voyage through and variety of lands which the agent cannot travel through. The environment has access to a layer of obstacles, such as vessels moving in the environment. Our agent can take some actions from a set of directions to move from its current location to its next location in a direction for $0.001°$, which is about $100\,\mathrm{m}$ of distance traveled in the real world.

An **origin point** is the position of an agent at the first moment of the training or testing phases of our methods. This means that our agent is positioned at the location of the origin point at the start of each experiment or training episode. We can define an origin point, $OP = (x_s, y_s)$, as a tuple where x_s is the latitude of the agent at the start of the experiment or training episode, and y_s is the longitude of the agent.

The **destination point** is the final geographical position that an agent should arrive at. We define $DP = (x_d, y_d)$ as a tuple where x_d is the latitude of the desired location that the agent aims to arrive at, and y_d is the longitude of that location.

We use eight discrete **actions** $A = \{N, S, E, W, NE, NW, SE, SW\}$, representing the directions North, South, East, West, Northeast, Northwest, Southeast, and Southwest, respectively.

In this work, we use five discrete **outcomes** for an action is taken by an agent $O = \{hit\ an\ obstacle,\ hit\ land,\ arrive\ at\ target,\ vanish\ target,\ normal\ movement\}$. To **hit an obstacle** means that the agent hits one of the vessels moving in the environment. To **hit land** means the agent took a direction that moved it to a geographic area that has land, which is not suitable for the vessel. To **arrive at target** means that our agent successfully reached its destination point. To **vanish target** means that the agent is farther from the target than the distance threshold. Finally, a **normal movement** is an output where the agent has not finished its mission, but there is no reason to stop the voyage.

Each action and its outcome for our agent in the environment is considered as one **step** $s_i = (a_i, o_i)$, where $s_i \in S$, $a_i \in A$, $o_i \in O$, and S is the set of all possible steps. Therefore, each step moves our agent from a current state to a future state.

An **episode** is a set of consecutive steps with a fixed origin and destination point. In an episode, the agent voyages from the origin point with the objective of arriving at its destination point.

An episode $e_j = (OP_j, DP_j, < s_1, s_2, ..., s_n >)$ - where OP_j is an origin point, DP_j is its destination point, $< s_1, s_2, ..., s_n >$ - is a sequence of steps, and $e_j \in E$, where E is the set of all possible episodes. The outcome of an episode is the outcome of the last step of that episode, which is s_n. When the outcome of an episode is to *hit an obstacle, go to land, vanish target*, this is considered to be a **failed episode**. If the outcome of an episode is *arrive at target*, this is considered to be a **successful episode**. In this work, we define a **maximum number of steps** for each episode because we want to encourage our agent to arrive at its destination as fast as possible and to avoid repetitive actions. If the number of steps in each episode exceeds this number, we call that episode a failed episode as well.

A **plan**, which is defined as $p_k = (OP_k, DP_k, < e_0, e_1, ..., e_m >)$ where, OP_k is an origin point, DP_k is its destination point, and $< e_1, e_2, ..., e_m >$ is a set of episodes. The destinations in each episode $(e_0, e_1, ..., e_m)$ of a plan p_k are called **intermediary goals**, except for the last episode, which is the final destination reached by a plan. If the outcome of a plan is not reached the destination, we call that plan a **failed plan**. The *arrive at target* outcome means the agent has (i) arrived at target and (ii) the agent is in its destination. If the outcome of a plan is to reach the destination, we call that plan a *successful plan*.

We limit the knowledge our agent has about the environment so that the agent is only able to observe within the boundary around itself. This approach of creating a local view has been used in [15]; however, the local view of our work is not a limitation for an agent. In this way, we force the agent to learn general rules of movement, without memorizing the whole environment and the best paths. Furthermore, some details in the environment, such as dynamic obstacles that are far from an agent, can move to other locations by the time our agent arrives there.

Having access to full information about the environment would encourages our agent to memorize the environment. When an intermediary goal is outside of our agent's local view, a subset of our environment, we make an abstract

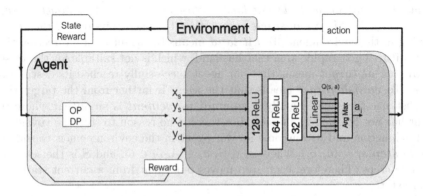

Fig. 1. Vanilla Vessel Navigator (VVN)

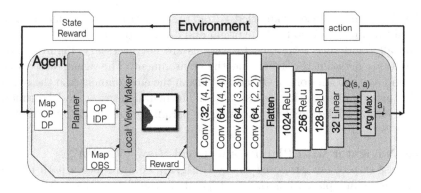

Fig. 2. Vessel Navigator with planner and localview (VNPLV)

line from the agent position to its intermediary goal and find the shadow of the intermediary goal inside the local view with some margin of freedom.

3.2 Baseline and Our Proposed Method

In Figs. 1 and 2, we summarize the baseline method and our proposed one, respectively. In these figures, x_s and y_s are the origin point coordinates, x_d, y_d are destination point coordinates, a_i is the action suggested by the method, $Q(s, a)$ is the action value function that predicts the reward given to an agent if it selects an action a in state s, $ReLU$ is a rectified linear unit, $Linear$ is a linear activation function, $Flatten$ is a flattening process that involves transforming the entire pooled feature map matrix into a single column, and $Conv$ is a convolutional neural network. We detail how these methods work below.

Vanilla Vessel Navigator (VVN)

This method was implemented based on the work of [16], but we addressed the two limitations of their algorithm for a fair comparison with our method. The first limitation is that the model's destination is a static point, and as a result, the model is able to learn only routes to a single destination. The second limitation is that using the QLearning approach with a large and dynamic environment makes it impossible for QLearning to train.

In our implementation of the VVN method, the agent receives its origin and destination point with freedom inside the environment.

Figure 1 shows the architecture of this method where the origin point (x_s and y_s), and destination point (x_d and y_d) are the only inputs that feed our model. The output of this architecture, which is the $argmax$ of the outputs of our neural network, determines the action of our agent. Since the objective of this paper is not to search for the best model for this architecture, we select one model with reasonably good results on training using a trial-and-error approach.

In the original model presented by [16], by changing the destination point, the agent would need to be trained again. Therefore agent trained by the original method is not able to use its knowledge from previous training. Our modification removed this limitation so that the origin and destination points can be selected dynamically, and such retraining limitations are not necessary. Figure 1 shows the reactions between the agent and the environment. The agent receives two origin and destination points, OP and DP, from the environment and the reward that the agent gained during its last action. The neural network model estimates the next action of the agent by updating its parameters using the reward of the previous action.

Vessel Navigator with Planner and Local View (VNPLV)

In the first step of our proposed method (Fig. 2), we introduce a path planner that makes a full plan from the start point to the destination without considering any dynamic obstacle in its way using the Floyd Algorithm [6]. The Floyd algorithm has the objective of finding the shortest paths in a weighted graph. This algorithm is computationally expensive but we only run it once for an environment and store the calculations results. This can be seen as high-level planning without considering the details of a plan. In the second step, we reduce this high-level plan by removing similar intermediary goals applying the Ramer–Douglas–Peucker algorithm [7], in our work $\epsilon = 0.001$ geomtry degree. The Ramer–Douglas–Peucker algorithm has the objective of simplifying a curve composed of a line of segments to a similar curve with fewer points. This reduction gives our agent more flexibility in making local decisions. In the third step, our agent decides on the details of the plan in the near future to arrive at the shadow of its intermediary goal residing in the local view. The shadow is a point in the direction to intermediary goal inside a definde margin of the local view.in This work we use a margin of 3 pixcels. Our agent observes its local view and finds an abstract destination point, which is the shadow of the nearest intermediate destination provided by the path planner. This first destination point is the destination point of the first episode in this plan, which resides in the local view. The agent starts moving towards its abstract destination point, which is residing inside the local view. After achieving the intermediary goals (i.e., the destination point of that episode), the environment updates the destination point of the next episode in its plan. The idea of using CNN is not a novel idea and has been introduced in reinforcement learning by the work of [12]. Moreover, the idea of using a local view also is introduced in [15]. To the best of our knowledge, we are the first to use a combination of path planning and local view for implementing the rolling wave planning approach for this problem. In Fig. 2, our agent benefits from two planning modules. First, the path planner provides a longterm plan without considering any dynamic obstacles on the way to arriving at the destination. Second, a DRL decision-making approach provides details for short-term decisions actions in detail. This part is responsible for avoiding obstacles and moving the agent safely in the near future.

4   Experiments

In this section we describe the dataset and evaluation metrics (Sect. 4.1), the setup for our algorithm (Sect. 4.2) and the experiments performed (Sect. 4.3).

4.1   Dataset Creation and Evaluation Metrics

We selected a region in the area of Halifax, Nova Scotia (Canada) with a bounding box (*longitude, latitude*) starting from $(-63.69, 44.58)$ and ending at $(-63.49, 44.73)$. We created a water and a land layer using public data on earth elevation using the NOAA[1] dataset.

Then we drew a buffer of 50 m around the land and used it as the environment. We also randomly added some moving objects to play the role of obstacles in this simulated 2D environment. In this work, we dynamically generated the obstacles during an episode. However, the design we have in mind for the future is based on the assumption that these obstacles can be dynamic and can move based on traffic patterns, such as can be extracted from Automatic Identification System (AIS) messages.

We also defined a metric called RATD to measure the performance of a method as follows. When we test a method, we randomly generate a set of origin and destination points N. Such information is provided to the method being tested as an episode or a plan.

Then we observe if the method can successfully place the agent at its destination (i.e., a successful episode or a successful plan) or not (i.e., a failed episode or a failed plan). We count how many times a model successfully conducts the agent to its destination and call it P. The RATD is calculated as $RATD = \frac{|P|}{|N|} * 100$.

4.2   Training and Testing Setup

In our experiments, we use the reward function $R(o_i \in O)$ detailed in Eq. 1, where Δ_d is the distance in geometry degree that an agent moved from the last state, ψ is set at 1,000, Δ_{od} is the distance of our agent from its nearest obstacle in geometry degree, and ϕ is set at 20. In Eq. 1 we deducted κ from the reward to encourage the agent to find the nearest path, and it was set at 0.01. The values of ψ, ϕ, and κ were manually tuned using a trial and error approach and the values reported are those that provided the optimal performance for all methods.

$$R(x) = \begin{cases} -5, & \text{Vanish target, Obstacle collision, Land collision} \\ +5, & \text{Arrive target} \\ \psi * \Delta_d - \phi * \Delta_{od} - \kappa, & \text{Other} \end{cases} \quad (1)$$

In this work, we use a *target network* to adjust the action-values (Q) iteratively towards the target values as in [12]. The application of target network application is to reduce correlations with the target. In the work of [12], a reply

---

[1]   https://coast.noaa.gov/dataviewer/.

buffer is proposed to eliminate relationships in the observation chain and to soften fluctuations in the data distribution, and we use the same idea here.

In our training phase, a *training step* includes 1,000 episodes or plans randomly executed to update our neural network parameters. Each of these episodes or plans includes an undefined number of training steps. We store (i) the agent's state, (ii) the agent's next state (iii) the reward, and (iv) the action, in a reply buffer with a size of 100,000 in the same way as introduced in [12].

We retrieve 3,000 items from this reply buffer after every 100 training steps. Every 200 training steps, the parameters of our model are copied to our target network. These values were configured empirically by trial and error tests.

In the training phase of our models we use the idea of exploit and explore which means the action of our agent is based on two types of learning [2]. First, exploring the environment randomly and measure the gained rewards. Second, exploiting the learned knowleadge by using the parameters of the trained neural network. In this work, we assigned the weight of 0.9 for the explore part and 0.1 weight for the exploit part. The weight of explore increases during the training so that in traing step 25,000 of our training, the weight of exploit becomes 0.9 and the weight of explore becomes 0.1. After training step 25,000, these weights are not changed.

4.3   Result Analysis and Discussion

An experiment was conducted to answer the two following research questions. First, what is the rate of arriving at the destination if we randomly select some sets of origin and destination points for episodes? Second, how does this rate change by increasing the distance between the origin and destination points? We increased the distance between the origin and destination points using the following distances in degrees $[0.01, 0.02, 0.04, 0.08, 0.16, 0.32]$, which is roughly equal to $[1364.59, 2729.11, 5457.90, 13004.68, 21824.01, 43627.70]$ in meters.

Figure 3 shows the results of our experiment for each training step.

Therefore, each point in Fig. 3 represents the results of 100 tests while the x-axis shows the progress on the number of training steps in a 10^2 scale. Between each two points, we update the parameters of our models using 1,000 executions of episodes or plans so that they update the parameters of our network. In this experiment, we dynamically randomly add some vessels on the body of water to play the role of obstacles.

The results for VVN (Fig. 3(a)) show that by increasing the distance between the origin point and the destination point, the percentage of successful trips decreases. In the experiment with a distance of $0.32°$ from origin to destination, the mean of RATD decreased from 79.44% (using VNPLV) to 24.13% (using VVN). These results show a weakness of VVN, as it is only good for near distance situations and cannot perform well in long-distance path planning. This is because (i) the search space of the agent increases when the distance is longer, and (ii) the probability of selecting actions from a loop of actions can increase so that the agent can just move back and forth without arriving at its destination.

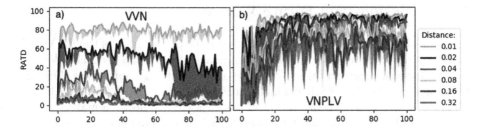

Fig. 3. Results of training of each model for different sets of distances between the start point and destination.

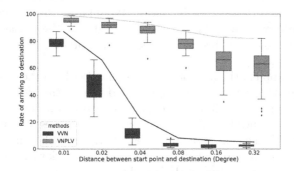

Fig. 4. Comparing the two vessel navigation methods using RATD for six category of distances after training models for 100,000 training steps

In Fig. 3(a) the shortest distance, shown by yellow, achieved the highest RATD. As can be seen, by increasing the distance, the VVN performance declined so that the mean of RATD was 24.13% for a 0.32° distance.

Figure 3(b) shows the result for the VNPLV.

The results shows that the performance of VNPLV is improved considerably, 55.31% on mean of RATD for a 0.32° distance, in comparison to VVN. The proposed method is more stable and learns better to navigate our agent, even with fewer episodes for training. This enhancement is because VNPLV is equipped with two crucial modules to benefit from rolling wave planning. First, the longterm planner provides a potential optimal path without considering the details of the movement. Second, the local view makes the decisions related to navigating the agent in the near future to avoid dynamic obstacles.

Figure 4 presents a high level comparison of all methods developed in this work. Figure 4 shows boxplots of the RATD for the distances of 0.01, 0.02, 0.04, 0.08, 0.16, and 0.32°. As can be seen, VNPLV has the best performance values for all categories of distances. We used a T-test to compare the results of VVN and VNLPV methods which supports that there is a statistically significant difference between VVN and VNPLV p_value = 0.000019. The results show that the mean of the RATD for these two methods are statistically different, confirming that the performance of VNPLP is higher than the VVN for this experimental setup.

5   Conclusions and Future Works

In this work, we proposed a method to improve the performance of the RL Vessel Navigator model using the concept of rolling wave planning in agile methodology.

Our proposed method takes advantage of two planning approaches: (1) long-term planning using a path planner with the assumption of no obstacles to generate a potential efficient path, and (2) a short-term decision-maker that is the output of our reinforcement learning model to avoid dynamically generated obstacles and to navigate the agent in the near future.

Our experiments show that the use of a local view improves the performance of our basic model. However, its performance reduces when the distance between the origin point and destination increases. We address this weakness using a path planner to provide a potentially efficient path for the whole trip without the detail of movements. This is followed by a short-term decision-maker to navigate agents safely.

Although we applied the idea of adaptive planning from the agile methodology for autonomous navigation, it can be applied to any other domain to increase the agent's performance. We intend to expand our work in various ways. We want to connect our dynamically generated obstacles to AIS messages received from vessels and test our agent in an unknown environment with dynamic obstacles. In this way, we would like to change the core neural network with sequence model networks such as a combination of RNN and CNN to have a memory of the past agent actions. We also would like to add similar agents to the environment and define a global task to perform it simultaneously in the environment.

Acknowledgments. The authors would like to thank NSERC (Natural Sciences and Engineering Research Council of Canada) for financial support, and Jennifer Strang, GIS Analyst at the Dalhousie University Libraries, for her help in preparing the geographic data for analysis. Computations were performed on the DeepSense computing platform. DeepSense is funded by ACOA, the Province of Nova Scotia, The Centre for Ocean Ventures & Entrepreneurship (COVE), Ocean Frontier Institute (OFI), IBM and Dalhousie University.

References

1. Wynn, R.B., et al.: Autonomous underwater vehicles (AUVs): their past, present and future contributions to the advancement of marine geoscience. Marine Geol. **352**, 451–468 (2014)
2. Brafman, R.I., Tennenholtz, M.: R-max-a general polynomial time algorithm for near-optimal reinforcement learning. J. Mach. Learn. Res. **3**, 213–231 (2002)
3. Cheng, L., Liu, C., Yan, B.: Improved hierarchical a-star algorithm for optimal parking path planning of the large parking lot. In: 2014 IEEE International Conference on Information and Automation (ICIA), pp. 695–698. July 2014. https://doi.org/10.1109/ICInfA.2014.6932742
4. Cheng, Y., Zhang, W.: Concise deep reinforcement learning obstacle avoidance for underactuated unmanned marine vessels. Neurocomputing **272**, 63–73 (2018)

5. Elkins, L.E.A.: The autonomous maritime navigation (AMN) project: field tests, autonomous and cooperative behaviors, data fusion, sensors, and vehicles. J. Field Robot. **27**(6), 790–818 (2010). https://doi.org/10.1002/rob.20367

6. Floyd, R.W.: Algorithm 97: shortest path. Communi. ACM **5**(6), 345 (1962)

7. Hershberger, J., Snoeyink, J.: An O(nlogn) implementation of the Douglas-Peucker algorithm for line simplification. In: Proceedings of the Tenth Annual Symposium on Computational Geometry, pp. 383–384 (1994)

8. Kanistras, K., Martins, G., Rutherford, M.J., Valavanis, K.P.: Survey of unmanned aerial vehicles (UAVs) for traffic monitoring. Handbook of Unmanned Aerial Vehicles, pp. 2643–2666 (2015)

9. Larman, C.: Agile and Iterative Development: A Manager's Guide. Addison-Wesley Professional, Boston (2004)

10. Lyu, H., Yin, Y.: Fast path planning for autonomous ships in restricted waters. Appl. Sci. **8**(12), 2592 (2018)

11. Magalhães, J., Damas, B., Lobo, V.: Reinforcement learning: the application to autonomous biomimetic underwater vehicles control. In: IOP Conference Series: Earth and Environmental Science, vol. 172, p. 012019 (2018). https://doi.org/10.1088/1755-1315/172/1/012019

12. Mnih, V., et al.: Human-level control through deep reinforcement learning. Nature **518**(7540), 529 (2015)

13. Shen, H., Hashimoto, H., Matsuda, A., Taniguchi, Y., Terada, D., Guo, C.: Automatic collision avoidance of multiple ships based on deep q-learning. Appl. Ocean Res. **86**, 268–288 (2019). https://doi.org/10.1016/j.apor.2019.02.020

14. Tran, N.-H., Choi, H.-S., Baek, S.-H., Shin, H.-Y.: Tracking control of an unmanned surface vehicle. In: Zelinka, I., Duy, V.H., Cha, J. (eds.) AETA 2013: Recent Advances in Electrical Engineering and Related Sciences. LNEE, vol. 282, pp. 575–584. Springer, Heidelberg (2014). https://doi.org/10.1007/978-3-642-41968-3_57

15. Vinyals, O.: Grandmaster level in StarCraft II using multi-agent reinforcement learning. Nature **575**(7782), 350–354 (2019)

16. Wang, C., Zhang, X., Li, R., Dong, P.: Path planning of maritime autonomous surface ships in unknown environment with reinforcement learning. In: Sun, F., Liu, H., Hu, D. (eds.) ICCSIP 2018. CCIS, vol. 1006, pp. 127–137. Springer, Singapore (2019). https://doi.org/10.1007/978-981-13-7986-4_12

17. Wang, H., Zhou, J., Zheng, G., Liang, Y.: Has: hierarchical a-star algorithm for big map navigation in special areas. In: 2014 5th International Conference on Digital Home, pp. 222–225, November 2014. https://doi.org/10.1109/ICDH.2014.49

CB-DBSCAN: A Novel Clustering Algorithm for Adjacent Clusters with Different Densities

Gashin Ghazizadeh[1(✉)], Mirerfan Gheibi[1], and Stan Matwin[1,2]

[1] Faculty of Computer Science, Dalhousie University, Halifax, Canada
g.gashin@gmail.com
[2] Institute of Computer Science, Polish Academy of Sciences, Warsaw, Poland

Abstract. Density-based clustering is well-known for finding clusters that have different shapes and sizes, but they have unsatisfactory results on adjacent clusters with different densities. In this paper, we propose a novel algorithm that combines DBSCAN with centroid-based algorithms to address this issue. Our algorithm uses DBSCAN to form mini-clusters, which will be merged based on their density and center distances. We test the new algorithm on synthetic and real datasets to show the significant improvement in the results.

Keywords: Clustering · DBSCAN · Centroid-based · CB-DBSCAN

1 Introduction

Clustering has been one of the major tasks in the field of Data Mining that tries to separate data points into groups or objects that share the same characteristics. Density-based spatial clustering of applications with noise (DBSCAN) [3] was introduced for spatial clustering problem and quickly became popular due to its excellent performance and features. One of the requirements that DBSCAN cannot satisfy is cluster analysis on adjacent clusters with different densities [11]. In this paper, we introduce a novel algorithm based on both DBSCAN and centroid-based clustering algorithms to overcome the problem of multi-density clustering.

2 Preliminaries

2.1 DBSCAN

DBSCAN is a clustering algorithm that tries to find clusters based on the density of different regions of the data. **MinPts** and **Eps** are two parameters of DBSCAN, and based on them, clusters are formed. Different categories of points in DBSCAN are explained as follows [3]:

C. Goutte and X. Zhu (Eds.): Canadian AI 2020, LNAI 12109, pp. 232–237, 2020.
https://doi.org/10.1007/978-3-030-47358-7_22

- **Core point**: any point in the data that has at least **MinPts** in its **Eps**-neighborhood.
- **Border point**: any point that is not a core point but is directly density reachable from a core point.
- **Noise point**: any point that is neither a core point nor a border point.

DBSCAN forms clusters in a recursive way. First, it chooses a random unvisited point, and if it is a core point, all of its directly density reachable points will be visited. The same steps will be done for all other points.

2.2   Multi-density Clustering

Multi-density clustering is a cluster analysis task on data that contains clusters with different densities, which DBSCAN does not perform well on.

The density of a cluster c is defined in Eq. (1) with N being the total number of points in that cluster and V as the total volume it occupies. If we assume that the density of the cluster is same over all of its volume, the density of it can be calculated with $nEps$ and $vEps$, which are respectively number of points and volume of an **Eps** neighbourhood. Let us assume that b is a boundary point which is directly density reachable from core points of both clusters C_1 and C_2 (C_1 is the denser cluster so $d_1 > d_2$). Since the volume in any Eps neighborhood is same, number of points decides the density. As D_b, the density around b, is between the two other densities, they will be connected and detected as one cluster using DBSCAN.

$$Density(c) = \frac{N}{V} = \frac{nEps}{vEps}$$
$$d_1 = \frac{n_1}{v_1}, d_2 = \frac{n_2}{v_2}, v_1 = v_2 = v_b, \ d_2 < d_1 \tag{1}$$
$$D_b = \frac{n_b}{v_b} = \frac{\frac{n_1}{2} + \frac{n_2}{2}}{v_b} \rightarrow D_b = \frac{d_1 + d_2}{2} \rightarrow d_2 < D_b < d_1$$

3   Related Work

One of the main ways of improving DBSCAN that has been used is using Grid-based partitioning for determining the densities of the grids [12]. GMDBSCAN [12] is one of the papers that have used this concept. PACA-DBSCAN [5] is another proposed algorithm that finds the appropriate parameters by defining a density for each point and portioning the data into K subsets. PACA-DBSCAN has a high time complexity, and finding the proper parameters for it can be complicated [6].

Another algorithm that tries to solve the problem of multi-density clustering with partitioning is kAA-DBSCAN [6]. Their algorithm builds a tree called DLT. The algorithm splits the data into cells of the DLT tree and finds approximate adaptive **Eps**, which can be used for different densities.

HDBSCAN [2] is another extension of DBSCAN. This algorithm adds hierarchy to DBSCAN and makes flat clustering out of it. HDBSCAN needs a lot of data for higher dimensions that might not always be available.

Some other algorithms have attempted to solve the problem by finding the proper parameters using mathematical computations. OPTICS [1] has two additional parameters, "reachability distance" and "core distance". It finds a representation of the data based on ordered points with the additional parameters it has. OPTICS does not need global parameters for the density of the clusters.

4   CB-DBSCAN

Our proposed algorithm centroid-based DBSCAN (CB-DBSCAN) contains several basic steps to do the cluster analysis:

- Cluster analysis with DBSCAN to extract mini-clusters.
- Merging mini-clusters based on their centroid and density distance.
- Assigning noise points to clusters based on the specified thresholds.

4.1   Mini-Clustering

Mini-clusters are small dense regions of the data that will form the final clusters. To extract mini-clusters, we need to find the right parameters for DBSCAN. The mini-clusters should be big enough to form the final clusters in a reasonable amount of time, but also small enough that the algorithm's ability to find clusters of different shapes is preserved.

If we choose **Eps** and **MinPts** in a way that both of the clusters are split into several smaller clusters, depending on the parameters, it might be possible to correct the results by merging them into bigger clusters. The optimal set of parameters is a set that breaks only the sparser cluster so that the merging step can correct the result easily. We should avoid choosing a set of parameters that combines the adjacent clusters because the results cannot be corrected with merging steps.

4.2   Centroid-Based Merging

CB-DBSCAN merges the mini-clusters in two different levels and has two parameters that are called **Cdis** and **Ddis**. **Cdis** is the threshold for centroid distance and **Ddis** is the threshold for density difference.

For the first step of merging the mini-clusters, we compare the centroid distances of the mini-clusters and start with the closest pair of them. If the density difference between this pair of mini-clusters is lower than **Ddis**, they are going to be added to a group. Once we have the first mini-clusters in the group, we search and add all the mini-clusters that are close enough to one of them. By doing so, the mini-clusters will be added to the group and then merged in a chain way. One example of how mini-clusters are merged in the first level of merging is shown in Fig. 1.

Later, second level of merging is done. We added this step to avoid the separation of boundary points from their clusters due to the lower density in the border areas. For doing so, we will merge clusters pairwise instead of merging a group of clusters. The algorithm of CB-DBSCAN is presented in Algorithm 1.

Algorithm 1. CB-DBSCAN

Input: $Data, \mathbf{Eps}, \mathbf{MinPts}, Cdis, Ddis$
Output: $FinalClusters$
 1: $mini - clusters = \text{DBSCAN}(Data, \mathbf{Eps}, \mathbf{MinPts})$
 2: Set all of the mini-clusters to $UNVISITED$
 3: $FinalClusters = \{\}$
 4: **while** Any mini-cluster is $UNVISITED$ **do**
 5:    Create a new $CLUSTER$
 6:    $StartMCs =$ closest pair in $mini - clusters$
 7:    **if** Distance($StartMCs$) $< Cdis$ and Density($StartMCs$) $< Ddis$ **then**
 8:       Add $StartMCs$ to $CLUSTER$ and set them to $VISITED$
 9:    **end if**
10:    **for** $mc \in CLUSTER$ **do**
11:       **for** $m \in mini - clusters$ **do**
12:          **if** Distance(mc, m) $< Cdis$ and Density(mc, m) $< Ddis$ **then**
13:             Add m to $CLUSTER$ and set it to $VISITED$
14:          **end if**
15:       **end for**
16:    **end for**
17:    Add $CLUSTER$ TO $FinalClusters$
18: **end while**
19: **for** I and $J \in FinalClusters$ **do**
20:    **if** Distance(I, J) $< \frac{Cdis}{2}$ and Density(I, J) $< 3Ddis$ **then**
21:       Combine I and J
22:       Update $FinalClusters$
23:    **end if**
24: **end for**
25: **return** $FinalClusters$

---

5 Clustering Evaluation

In order to compare our results to some other algorithms that aimed to solve the multi-density clustering problem of DBSCAN, we have used five different labeled datasets from two different sources [4,9], which are widely used for comparing different clustering algorithm.

The comparison between the performance of state of the art algorithms that have been proposed for multi-density clustering [6] and CB-DBSCAN is shown in Table 1. CB-DBSCAN has outperformed all the other methods in almost all of the cases with a noticeable difference in F-measure.

Furthermore, we used CB-DBSCAN for a clustering task of Twitter data. Two hand-classified topics are clustered that have different densities [9]. In this task, each tweet is represented by calculating the mean of all the word vectors [7] and applying PCA [8] on them. The improvement in all of the metrics versus DBSCAN's results, confirms the satisfactory performance that our algorithm achieved, Table 2.

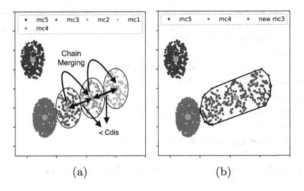

Fig. 1. (a) Merging mini-clusters based on density and their centroid distance to form final clusters. (b) New clusters.

Table 1. Comparison of CB-DBSCAN and state of the art algorithms' results on synthetic and real datasets.

Dataset	F-measure				
	DBSCAN	CB-DBSCAN	PACA-DBSCAN	kAA-DBSCAN	OPTICS
Flame	0.89	**0.99**	0.98	0.98	0.91
Seeds	0.56	**0.9**	0.72	0.85	0.8
Path based	0.66	0.91	0.92	**0.99**	0.69
Breast	0.61	**0.97**	0.75	0.73	0.66
Wine	0.67	**0.94**	0.72	0.72	0.71

Table 2. Results of CB-DBSCAN and DBSCAN on Twitter data.

Algorithm	Accuracy	F-measure	Silhouette coefficient
DBSCAN	0.91	0.47	0.38
CB-DBSCAN	**0.98**	**0.65**	**0.51**

6   Time Complexity Analysis

The time complexity of the algorithm depends on two main parts of it, mini-clustering and merging. The worst-case time complexity of mini-clustering step (DBSCAN) is $\mathcal{O}(n^2)$ [10].

The merging part has two different levels. The first merging level goes over the mini-clusters once to combine them, which means that the time complexity of this step is linear ($\mathcal{O}(n)$). The second level of merging compares the mini-clusters pairwise, which leads to worst-case time complexity of $\mathcal{O}(n^2)$.

The time complexity of all the steps is lower than or equal to $\mathcal{O}(n^2)$. Hence, the overall worst-case of CB-DBSCAN has a time complexity of $\mathcal{O}(n^2)$.

7   Conclusion and Future Work

In this paper, a novel algorithm was proposed for solving the issue of multi-density clustering. CB-DBSCAN uses DBSCAN to produce mini-clusters that are going to be merged in a chain way to form bigger clusters with different densities. One of the future improvements that can be done for CB-DBSCAN is automating the process of choosing the proper parameters for the DBSCAN to produce the mini-clusters.

Acknowledgment. We acknowledge the support of the Natural Sciences and Engineering Research Council of Canada for this research.

References

1. Ankerst, M., Breunig, M.M., Kriegel, H.P., Sander, J.: Optics: ordering points to identify the clustering structure. ACM SIGMOD Rec. **28**, 49–60 (1999)
2. Campello, R.J.G.B., Moulavi, D., Sander, J.: Density-based clustering based on hierarchical density estimates. In: Pei, J., Tseng, V.S., Cao, L., Motoda, H., Xu, G. (eds.) PAKDD 2013. LNCS (LNAI), vol. 7819, pp. 160–172. Springer, Heidelberg (2013). https://doi.org/10.1007/978-3-642-37456-2_14
3. Duan, L., Xu, L., Guo, F., Lee, J., Yan, B.: A local-density based spatial clustering algorithm with noise. Inf. Syst. **32**(7), 978–986 (2007)
4. Fränti, P., Sieranoja, S.: K-means properties on six clustering benchmark datasets (2018). http://cs.uef.fi/sipu/datasets/
5. Jiang, H., Li, J., Yi, S., Wang, X., Hu, X.: A new hybrid method based on partitioning-based DBSCAN and ant clustering. Expert Syst. Appl. **38**(8), 9373–9381 (2011)
6. Kim, J.H., Choi, J.H., Yoo, K.H., Nasridinov, A.: AA-DBSCAN: an approximate adaptive DBSCAN for finding clusters with varying densities. J. Supercomput. **75**(1), 142–169 (2019)
7. Mikolov, T., Sutskever, I., Chen, K., Corrado, G.S., Dean, J.: Distributed representations of words and phrases and their compositionality. In: Advances in Neural Information Processing Systems, pp. 3111–3119 (2013)
8. Pearson, K.: LIII. On lines and planes of closest fit to systems of points in space. Lond. Edinb. Dubl. Phil. Mag. J. Sci. **2**(11), 559–572 (1901)
9. Sanders, A.: Sanders sentiment analysis dataset (2013). https://github.com/zfz/twitter_corpus. Accessed 30 Sept 2019
10. Viswanath, P., Pinkesh, R.: l-DBSCAN: a fast hybrid density based clustering method. In: 18th International Conference on Pattern Recognition (ICPR 2006), vol. 1, pp. 912–915. IEEE (2006)
11. Wang, X., Hamilton, H.J.: DBRS: a density-based spatial clustering method with random sampling. In: Whang, K.-Y., Jeon, J., Shim, K., Srivastava, J. (eds.) PAKDD 2003. LNCS (LNAI), vol. 2637, pp. 563–575. Springer, Heidelberg (2003). https://doi.org/10.1007/3-540-36175-8_56
12. Xiaoyun, C., Yufang, M., Yan, Z., Ping, W.: GMDBSCAN: multi-density DBSCAN cluster based on grid. In: 2008 IEEE International Conference on e-Business Engineering, pp. 780–783. IEEE (2008)

Anomaly Detection and Prototype Selection Using Polyhedron Curvature

Benyamin Ghojogh$^{(\boxtimes)}$, Fakhri Karray , and Mark Crowley

Electrical and Computer Engineering, University of Waterloo, Waterloo, ON, Canada
{bghojogh,karray,mcrowley}@uwaterloo.ca

Abstract. We propose a novel approach to anomaly detection called Curvature Anomaly Detection (CAD) and Kernel CAD based on the idea of polyhedron curvature. Using the nearest neighbors for a point, we consider every data point as the vertex of a polyhedron where the more anomalous point has more curvature. We also propose inverse CAD (iCAD) and Kernel iCAD for instance ranking and prototype selection by looking at CAD from an opposite perspective. We define the concept of anomaly landscape and anomaly path and we demonstrate an application for it which is image denoising. The proposed methods are straightforward and easy to implement. Our experiments on different benchmarks show that the proposed methods are effective for anomaly detection and prototype selection.

Keywords: Anomaly detection · Prototype selection · Polyhedron curvature · Curvature Anomaly Detection (CAD)

1 Introduction

Anomaly detection, instance ranking, and prototype selection are important tasks in data mining. Anomaly detection refers to finding outliers or anomalies which differ significantly from the normal data points [1]. There exist many applications for anomaly detection such as fraud detection, intrusion detection, medical diagnosis, and damage detection [2].

Ranking data points (instances) according to their importance can be useful for better representation of data, omitting the dummy or noisy points, better discrimination of classes in classification tasks, etc. [3]. Prototype selection is referred to finding the best data points in terms of representation of data, discrimination of classes, information of points, etc. [4]. It can also be useful for better storage and processing time efficiency. Prototype selection can be done either using ranking the points and then discarding the less important points or by merely retaining a portion of data and discarding the others.

In this paper, we propose Curvature Anomaly Detection (CAD) and inverse CAD (iCAD) for anomaly detection and prototype selection, respectively. We also propose their kernel versions which are Kernel CAD (K-CAD) and Kernel

© Springer Nature Switzerland AG 2020
C. Goutte and X. Zhu (Eds.): Canadian AI 2020, LNAI 12109, pp. 238–250, 2020.
https://doi.org/10.1007/978-3-030-47358-7_23

iCAD (K-iCAD). The idea of proposed algorithms is based on polyhedron curvature where every point is imagined to be the vertex of a hypothetical polyhedron defined by its neighbors. We also define anomaly landscape and anomaly path which can have different applications such as image denoising. In the following, we mention the related work for anomaly detection and prototype selection. Then, we explain the background for polyhedron curvature. Afterwards, the proposed CAD, K-CAD, iCAD, and K-iCAD are explained. Finally, the experiments are reported.

Anomaly Detection: Local Outlier Factor (LOF) [5] is one of the important anomaly detection algorithms. It defines a measure for local density of every data point according to its neighbors. It compares the local density of every point with its neighbors and find the anomalies. One-class SVM [6] is another method which estimates a function which is positive on the regions of data with high density and negative elsewhere. Therefore, the points with negative values of that function are considered as anomalies. If the data are assumed to have Gaussian distribution as the most common distribution, Elliptic Envelope (EE) can be fitted to data [7] and the points having low probability in the fitted envelope are considered to be anomaly. Isolation forest [8] is an isolation-based anomaly detection method [9] which isolates the anomalies using an ensemble approach. The ensemble includes isolation trees where the more depth of tree for isolating a point is a measure of its normality.

Prototype Selection: Prototype selection [4] is also referred to as instance ranking and numerosity reduction. Edited Nearest Neighbor (ENN) [10] is one of the oldest prototype selection method which removes the points having most of its neighbors from another class. Decremental Reduction Optimization Procedure 3 (DROP3) [11] has the opposite perspective and removes a point if its removal improves the k-Nearest Neighbor (k-NN) classification accuracy. Stratified Ordered Selection (SOS) [12] starts with boundary points and then recursively finds the median points noticing that boundary and median points are informative. Shell Extraction (SE) [13] introduces a reduction sphere and removes the points falling in this hyper-sphere in order to approximate the support vectors. Principal Sample Analysis (PSA) [3], which is extended for regression and clustering tasks in [14], considers the scatter of data as well as the regression of prototypes for better representation. Instance Ranking by Matrix Decomposition (IRMD) [15] decomposes the matrix of data and makes use of the bases of decomposition. The more similar points to the bases are considered to be more important.

2   Background on Polyhedron Curvature

A *polytope* is a geometrical object in $\mathbb{R}^d$ whose faces are planar. The special cases of polytope in $\mathbb{R}^2$ and $\mathbb{R}^3$ are called *polygon* and *polyhedron*, respectively. Some examples for polyhedron are cube, tetrahedron, octahedron, icosahedron, and dodecahedron with four, eight, and twenty triangular faces, and twelve flat faces, respectively [16]. Consider a polygon where τ_j and μ_j are the interior and

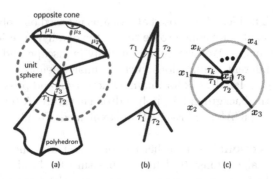

Fig. 1. (a) Polyhedron vertex, unit sphere, and the opposite cone, (b) large and small curvature, (c) a point and its neighbors normalized on a unit hyper-sphere around it.

exterior angles at the j-th vertex; we have $\tau_j + \mu_j = \pi$. A similar analysis holds in $\mathbb{R}^3$ for Fig. 1-a. In this figure, a vertex of a polyhedron and its opposite cone are shown where the opposite cone is defined to have perpendicular faces to the faces of the polyhedron at the vertex. The intersection of a unit sphere centered at the vertex and the opposite cone is shown in the figure. This intersection is a geodesic on the unit sphere. According to Thomas Harriot's theorem proposed in 1603 [17], if this geodesic on the unit sphere is a triangle, its area is $\mu_1 + \mu_2 + \mu_3 - \pi = 2\pi - (\tau_1 + \tau_2 + \tau_3)$. The generalization of this theorem from a geodesic triangular polygon (3-gon) to an k-gon is $\mu_1 + \cdots + \mu_k - k\pi + 2\pi = 2\pi - \sum_{a=1}^{k} \tau_a$ [17], where the polyhedron has k faces meeting at the vertex.

The Descartes's *angular defect* at a vertex x of a polyhedron is [18]: $\mathcal{D}(x) := 2\pi - \sum_{a=1}^{k} \tau_a$. The total defect of a polyhedron is defined as the summation of the defects over the vertices. It can be shown that the total defect of a polyhedron with v vertices, e edges, and f faces is: $\mathcal{D} := \sum_{i=1}^{v} \mathcal{D}(x_i) = 2\pi(v - e + f)$. The term $v - e + f$ is Euler-Poincaré characteristic of the polyhedron [19,20]; therefore, the total defect of a polyhedron is equal to its Euler-Poincaré characteristic. According to Fig. 1-b, the smaller τ angles result in sharper corner of the polyhedron. Therefore, we can consider the angular defect as the *curvature* of the vertex.

3   Anomaly Detection

3.1   Curvature Anomaly Detection

The main idea of the *Curvature Anomaly Detection (CAD)* method is as follows. Every data point is considered to be the vertex of a hypothetical polyhedron (see Fig. 1-a). For every point, we find its k-Nearest Neighbors (k-NN). The k neighbors of the point (vertex) form the k faces of a polyhedron meeting at that vertex. Then, the more curvature that point (vertex) has, the more anomalous it is because it is far away (different) from its neighbors. Therefore, *anomaly score* s_A is proportional to the curvature.

Since, according to the equation of angular effect, the curvature is proportional to minus the summation of angles, we can consider the anomaly score to be inversely proportional to the summation of angles. Without loss of generality, we assume the angles to be in range $[0, \pi]$ (otherwise, we take the smaller angle). The less the angles between two edges of the polyhedron, the more their cosine. As the anomaly score is inversely proportional to the angles, we can use cosine for the anomaly score: $s_A(\boldsymbol{x}_i) \propto 1/\tau_a \propto \cos(\tau_a)$. We define the anomaly score to be the summation of cosine of the angles of the polyhedron faces meeting at that point: $s_A(\boldsymbol{x}_i) := \sum_{a=1}^{k} \cos(\tau_a) = \sum_{a=1}^{k} (\breve{\boldsymbol{x}}_a^\top \breve{\boldsymbol{x}}_{a+1})/(||\breve{\boldsymbol{x}}_a||_2 ||\breve{\boldsymbol{x}}_{a+1}||_2)$ where $\breve{\boldsymbol{x}}_a := \boldsymbol{x}_a - \boldsymbol{x}_i$ is the a-th edge of the polyhedron passing through the vertex $\boldsymbol{x}_i$, $\boldsymbol{x}_a$ is the a-th neighbor of $\boldsymbol{x}_i$, and $\breve{\boldsymbol{x}}_{a+1}$ denotes the next edge sharing the same polyhedron face with $\breve{\boldsymbol{x}}_a$ where $\breve{\boldsymbol{x}}_{k+1} = \breve{\boldsymbol{x}}_1$.

Note that finding the pairs of edges which belong to the same face is difficult and time-consuming so we relax this calculation to the summation of the cosine of angles between all pairs of edges meeting at the vertex $\boldsymbol{x}_i$:

$$s_A(\boldsymbol{x}_i) := \sum_{a=1}^{k-1} \sum_{b=a+1}^{k} \frac{\breve{\boldsymbol{x}}_a^\top \breve{\boldsymbol{x}}_b}{||\breve{\boldsymbol{x}}_a||_2 ||\breve{\boldsymbol{x}}_b||_2}, \tag{1}$$

where $\breve{\boldsymbol{x}}_a := \boldsymbol{x}_a - \boldsymbol{x}_i$, $\breve{\boldsymbol{x}}_b := \boldsymbol{x}_b - \boldsymbol{x}_i$, and $\boldsymbol{x}_a$ and $\boldsymbol{x}_b$ denote the a-th and b-th neighbor of $\boldsymbol{x}_i$. In Eq. (1), we have omitted the redundant angles because of symmetry of inner product. Note that the Eq. (1) implies that we normalize the k neighbors of $\boldsymbol{x}_i$ to fall on the unit hyper-sphere centered at $\boldsymbol{x}_i$ and then compute their cosine similarities (see Fig. 1-c).

The mentioned relaxation is valid for the following reason. Take two edges meeting at the vertex $\boldsymbol{x}_i$. If the two edges belong to the same polyhedron face, the relaxation is exact. Consider the case where the two edges do not belong to the same face. These two edges are connected with a set of polyhedron faces. If we tweak one of the two edges to increase/decrease the angle between them, the angle of that edge with its neighbor edge on the same face also increases/decreases. Therefore, the changes in the additional angles of relaxation are consistent with the changes of the angles between the edges sharing the same faces.

After scoring the data points, we can sort the points and find a suitable threshold visually using a scree plot of the scores. However, in order to find anomalies automatically, we apply K-means clustering, with two clusters, to the scores. The cluster with the larger mean is the cluster of anomalies because the higher the score, the more anomalous the point.

For finding anomalies for out-of-sample data, we find k-NN for the out-of-sample point where the neighbors are from the training points. Then, we calculate the anomaly score using Eq. (1). The K-means cluster whose mean is closer to the calculated score determines whether the point is normal or anomaly. It is noteworthy that one can see anomaly detection for out-of-sample data as novelty detection [21].

3.2 Kernel Curvature Anomaly Detection

The pattern of normal and anomalous data might not be linear. Therefore, we propose *Kernel CAD (K-CAD)* to work on data in the feature space. In K-CAD, the two stages of finding k-NN and calculating the anomaly score are performed in the feature space. Let $\phi : \mathcal{X} \rightarrow \mathcal{H}$ be the pulling function mapping the data $x \in \mathcal{X}$ to the feature space $\mathcal{H}$. In other words, $x \mapsto \phi(x)$. Let t denote the dimensionality of the feature space, i.e., $\phi(x) \in \mathbb{R}^t$ while $x \in \mathbb{R}^d$. Note that we usually have $t \gg d$. The kernel over two vectors x_1 and x_2 is the inner product of their pulled data [22]: $\mathbb{R} \ni k(x_1, x_2) := \phi(x_1)^\top \phi(x_2)$. The Euclidean distance in the feature space is [23]: $\|\phi(x_i) - \phi(x_j)\|_2 = \sqrt{k(x_i, x_i) - 2k(x_i, x_j) + k(x_j, x_j)}$. Using this distance, we find the k-NN of the dataset in the feature space.

After finding k-NN in the feature space, we calculate the score in the feature space. We pull the vectors $\breve{x}_a$ and $\breve{x}_b$ to the feature space so $\breve{x}_a^\top \breve{x}_b$ is changed to $k(\breve{x}_a, \breve{x}_b) = \phi(\breve{x}_a)^\top \phi(\breve{x}_b)$. Let $K_i \in \mathbb{R}^{k \times k}$ denote the kernel of neighbors of x_i whose (a,b)-th element is $k(\breve{x}_a, \breve{x}_b)$. The vectors in Eq. (1) are normalized. In the feature space, this is equivalent to normalizing the kernel $k(\breve{x}_a, \breve{x}_b) := k(\breve{x}_a, \breve{x}_b)/\sqrt{k(\breve{x}_a, \breve{x}_a) k(\breve{x}_b, \breve{x}_b)}$ [24]. If K_i' denotes the normalized kernel K_i, the anomaly score in the feature space is:

$$s_A(x_i) := \sum_{a=1}^{k-1} \sum_{b=a+1}^{k} K_i'(a, b), \tag{2}$$

where $K_i'(a, b)$ is the (a, b)-th element of the kernel. The K-means clustering and out-of-sample anomaly detection are similarly performed as in CAD.

Our observations in experiments showed that the anomaly score in K-CAD is ranked inversely for some kernels such as Radial Basis Function (RBF), Laplacian, and polynomial (different degrees) in various datasets. In other words, for example, in K-CAD with linear (i.e., CAD), cosine, and sigmoid kernels, the more anomalous points have greater score but in K-CAD with RBF, Laplacian, and polynomial kernels, the smaller score is assigned to the more anomalous points. We conjecture that the reason lies in the characteristics of the kernels. We defer more investigations for the reason as a future work. In conclusion, for the mentioned kernels, we should either multiply the scores by -1 or take the K-means cluster with smaller mean as the anomaly cluster.

3.3 Anomaly Landscape and Anomaly Paths

We define *anomaly landscape* to be the landscape in the input space whose value at every point x_i in the space is the anomaly score computed by Eq. (1) or (2). The point x_i in the space can be either the training or out-of-sample point but the k-NN is obtained from the training data. We can have two types of anomaly landscape where all the training data points or merely the non-anomaly training points are used for k-NN. In the latter type, the training phase of CAD or K-CAD are performed before calculating the anomaly landscape for the whole input space.

We also define the *anomaly path* as the path that an anomalous point has traversed from its not-known-yet normal version to become anomalous. Conversely, it is the path that an anomalous point should traverse to become normal. In other words, *an anomaly path can be used to make a normal sample anomalous or vice-versa.* At every point on the path, we calculate the k-NN again because the neighbors may change slightly during the path. For anomaly path, we use the second type of anomaly landscape where the path is like going up/down the mountains in this landscape. For finding the anomaly path for every anomaly point, we use gradient descent where the gradient of the Eq. (1) is used:

$$\frac{\partial s_A(\boldsymbol{x}_i)}{\partial \boldsymbol{x}_i} = \sum_{a=1}^{k-1} \sum_{b=a+1}^{k} \left[\frac{1}{||\breve{\boldsymbol{x}}_a||_2 ||\breve{\boldsymbol{x}}_b||_2} \left[-(\breve{\boldsymbol{x}}_a + \breve{\boldsymbol{x}}_b) + \breve{\boldsymbol{x}}_a^\top \breve{\boldsymbol{x}}_b \left(\frac{\breve{\boldsymbol{x}}_a}{||\breve{\boldsymbol{x}}_a||_2^2} + \frac{\breve{\boldsymbol{x}}_b}{||\breve{\boldsymbol{x}}_b||_2^2} \right) \right] \right],$$

(3)

whose derivation is eliminated for brevity (will be submitted to arXiv). The anomaly path can be computed in CAD and not K-CAD because the gradient in K-CAD cannot be computed analytically. The anomaly path can have many applications one of which is image denoising as explained in our experiments.

4   Prototype Selection

4.1   Inverse Curvature Anomaly Detection

If the anomaly detection uses scores, we can see instance ranking and numerosity reduction in the opposite perspective of anomaly detection. Therefore, the ranking scores can be considered as the anomaly scores multiplied by -1: $s_R(\boldsymbol{x}_i) := -1 \times s_A(\boldsymbol{x}_i) = -\sum_{a=1}^{k-1} \sum_{b=a+1}^{k} (\breve{\boldsymbol{x}}_a^\top \breve{\boldsymbol{x}}_b)/(||\breve{\boldsymbol{x}}_a||_2 ||\breve{\boldsymbol{x}}_b||_2)$. We sort the ranking scores in descending order. The data point with larger ranking score is more important. As the order of ranking scores is inverse of the order of anomaly scores, we name this method as *inverse CAD (iCAD)*.

Prototype selection can be performed in two approaches: (I) the data points are sorted and a portion of the points having the best ranks is retained, or (II) a portion of data points is retained as prototypes and the rest of points are discarded. Some examples of the fist approach is IRMD, PSA, SOS, and SE. DROP3 and ENN are examples for the second approach. The iCAD can be used for both approaches. The first approach is ranking the points with the ranking score. For the second approach, we apply K-means clustering, with two clusters, to the ranking scores and take the points of the cluster with larger mean.

4.2   Kernel Inverse Curvature Anomaly Detection

We can perform iCAD in the feature space to have *Kernel iCAD (K-iCAD)*. The ranking score is again the anomaly score multiplied by -1 to reverse the ranks of scores: $s_R(\boldsymbol{x}_i) := -1 \times s_A(\boldsymbol{x}_i) = -\sum_{a=1}^{k-1} \sum_{b=a+1}^{k} \boldsymbol{K}'_i(a,b)$. Again, we have two approaches where the points are ranked or K-means is applied on the scores. Note that for what was mentioned before, we do not multiply by -1 for some kernels including RBF, Laplacian, and polynomial. Note that iCAD and

K-iCAD are task agnostic and can be used for data reduction in classification, regression, and clustering. For classification, we apply the method for every class while in regression and clustering, the method is applied on the entire data.

5 Experiments

5.1 Experiments for Anomaly Detection

Synthetic Datasets: We examined CAD and iCAD on three two-dimensional synthetic datasets, i.e., two moons and two homogeneous and heterogeneous clusters. Figure 2 shows the results for CAD and K-CAD with RBF and polynomial (degree three) kernels. As expected, the abnormal and core points are correctly detected as anomalous and normal points, respectively. The boundary points are detected as anomaly in CAD while they are correctly recognized as normal points in K-CAD. In heterogeneous clusters data, the larger cluster is correctly detected as normal in CAD but not in K-CAD; however, if the threshold is manually changed (rather than by K-means) in K-CAD, the larger cluster will also be correctly recognized. As seen in this figure, the scores are reverse in EBF and polynomial kernels which is consistent to our previous note in the paper. We also show the anomaly landscape and anomaly paths for CAD in Fig. 2. The K-CAD does not have anomaly paths as mentioned before. The landscapes in this figure are of the second type and the paths are shown by red traces which simulates climbing down the mountains in the landscape.

Real Datasets: We did experiments on several real datasets of anomaly detection. The datasets, which are taken from [25], are speech, opt. digits, arrhythmia, wine, and musk with 1.65%, 3%, 15%, 7.7%, and 3.2% portions of anomalies, respectively. The sample size of these datasets are 3686, 5216, 452, 129, and 3062 and their dimensionality are 400, 64, 274, 13, and 166, respectively. We compared CAD and K-CAD with RBF and polynomial (degree 3) kernels to Isolation forest, LOF, one-class SVM (RBF kernel), and EE. We used $k = 10$ in LOF, CAD, and K-CAD. The average area under the ROC curve (AUC) and the average time for both training and test phases over 10-fold Cross Validation (CV) are reported in Table 1. For wine data, because of small sample size, we used 2-fold CV. The system running the methods was Intel Core i7, 3.60 GHz, with 32 GB RAM. In most cases, K-CAD has better performance than CAD; although CAD is useful and effective as we will see for anomaly path and also instance ranking. For speech and optdigits datasets, RBF kernel has better performance than polynomial and for other datasets, polynomial kernel is better. Mostly, K-CAD is faster in both training and test phases because K-CAD uses kernel matrix and normalizing the matrix rather than element-wise cosines in CAD. In speech and optdigits datasets, we outperform all the baseline methods in both training and test AUC rates. In arrhythmia data, K-CAD with polynomial kernel has better results than isolation forest. For wine dataset, K-CAD with polynomial kernel is better than isolation forest, SVM, and EE. In musk data, K-CAD with both RBF and polynomial kernels is better than isolation forest and SVM.

For experimenting the effect of k in CAD and K-CAD, we report the results of $k \in \{3, 10, 20\}$ for arrhythmia dataset in Table 2. For CAD, where the cosine is done element-wise, time increases by k as expected. Overall, the accuracy, especially the training AUC, increases in CAD by k. K-CAD is more robust to change of k in terms of accuracy and time.

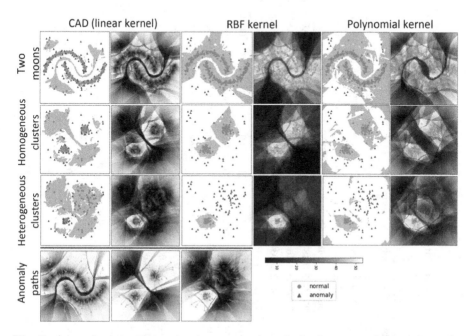

Fig. 2. Anomaly detection, anomaly scores, anomaly landscape, and anomaly paths for synthetic datasets. In the gray and white plots, the gray and white colors show the regions determined as normal and anomaly, respectively. The gray-scale plots are the anomaly scores.

An Application; Image Denoising: One of the applications for anomaly path is image denoising where several similar reference images exist; for example, in video where neighbor frames exist. For experiment, we used the first 100 frames of Frey face dataset. We selected one of the frames and applied different types of noises, i.e., Gaussian noise, Gaussian blurring, salt & pepper impulse noise, and JPEG blocking to it all with the same mean squared errors (MSE = 625). For a more difficult experiment, we removed the non-distorted frame from dataset. Figure 3 shows the iterations of denoising for different noise types where $k = 3$ is used.

Table 1. Comparison of anomaly detection methods. Rates are AUC percentage and times are in seconds.

			CAD	K-CAD (rbf)	K-CAD (poly)	Iso forest	LOF	SVM	EE
Speech	Train	Time	14.84 ± 0.21	13.54 ± 0.21	13.87 ± 0.33	2.82 ± 0.06	6.53 ± 0.02	6.12 ± 0.02	23.35 ± 0.25
		AUC	34.78 ± 0.15	76.15 ± 2.08	63.69 ± 1.48	48.37 ± 1.38	53.99 ± 1.75	46.63 ± 1.59	49.16 ± 1.84
	Test	Time	7.16 ± 0.03	1.38 ± 0.03	1.42 ± 0.03	0.06 ± 00.01	7.31 ± 0.10	0.24 ± 0.01	0.01 ± 0.00
		AUC	42.07 ± 10.48	71.23 ± 13.18	56.15 ± 10.48	45.23 ± 12.17	53.53 ± 12.29	44.55 ± 12.09	47.25 ± 12.54
Opt digits	Train	Time	13.27 ± 0.11	26.96 ± 0.31	25.86 ± 0.48	0.81 ± 0.01	1.84 ± 0.02	3.10 ± 0.01	0.96 ± 0.04
		AUC	32.67 ± 1.53	87.52 ± 1.76	77.79 ± 1.45	68.38 ± 4.64	60.84 ± 1.67	50.52 ± 3.81	39.04 ± 2.44
	Test	Time	11.24 ± 0.12	2.67 ± 0.01	2.59 ± 0.03	0.03 ± 0.01	2.13 ± 0.08	0.15 ± 00.00	00.01 ± 00.00
		AUC	26.28 ± 7.10	88.22 ± 5.62	79.72 ± 4.81	68.36 ± 8.11	61.12 ± 11.65	37.49 ± 7.41	38.84 ± 4.29
Arrhythmia	Train	Time	4.76 ± 0.02	2.75 ± 0.06	2.53 ± 0.03	0.20 ± 0.01	0.07 ± 00.00	0.13 ± 00.00	0.85 ± 0.02
		AUC	52.89 ± 0.96	48.87 ± 0.51	73.92 ± 1.12	62.43 ± 2.05	91.04 ± 0.66	88.56 ± 00.87	80.59 ± 0.65
	Test	Time	1.59 ± 0.01	0.30 ± 0.01	0.28 ± 00.00	0.02 ± 0.01	0.08 ± 0.00	0.01 ± 00.00	00.01 ± 00.00
		AUC	48.02 ± 9.06	48.56 ± 5.38	71.88 ± 9.23	63.07 ± 11.55	90.57 ± 5.47	90.03 ± 5.63	80.32 ± 4.73
Wine	Train	Time	0.28 ± 0.00	0.03 ± 0.03	0.03 ± 0.01	0.09 ± 0.00	0.01 ± 0.00	0.01 ± 0.00	0.05 ± 0.02
		AUC	25.59 ± 4.28	27.04 ± 10.66	92.11 ± 7.06	79.56 ± 10.59	98.70 ± 1.29	68.59 ± 4.25	59.56 ± 37.15
	Test	Time	0.18 ± 00.00	0.02 ± 00.00	0.01 ± 0.00	0.02 ± 0.00	0.01 ± 0.00	0.01 ± 0.00	0.01 ± 0.00
		AUC	23.65 ± 14.45	40.17 ± 13.12	86.97 ± 2.96	76.09 ± 10.11	92.58 ± 5.69	91.13 ± 3.83	57.70 ± 38.84
Musk	Train	Time	11.15 ± 0.20	9.67 ± 0.47	9.42 ± 0.03	0.98 ± 0.01	1.16 ± 0.03	3.16 ± 0.00	10.16 ± 0.32
		AUC	40.69 ± 2.97	69.68 ± 2.97	93.45 ± 1.46	99.91 ± 00.00	41.93 ± 3.34	57.99 ± 7.34	99.99 ± 00.00
	Test	Time	4.89 ± 0.02	0.98 ± 0.02	0.98 ± 0.01	0.03 ± 0.00	1.24 ± 0.01	0.17 ± 0.00	0.01 ± 0.00
		AUC	30.30 ± 10.37	50.00 ± 0.00	93.80 ± 3.77	99.95 ± 0.00	39.00 ± 10.55	5.71 ± 3.63	100 ± 0.00

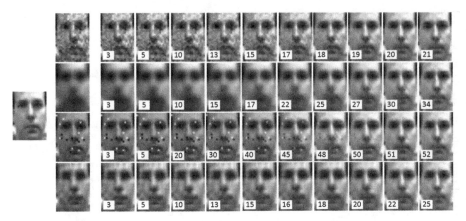

Fig. 3. Image denoising using anomaly paths: the most left image is the original image and the first to fourth rows are for Gaussian noise, Gaussian blurring, salt & pepper impulse noise, and JPEG blocking. The numbers are the iteration indices.

5.2   Experiments for Prototype Selection

Synthetic Datasets: The performances of iCAD and K-iCAD with RBF and polynomial (degree 3) kernels are illustrated in Fig. 4 for the three synthetic datasets where the larger markers show the more important points. We can see that the points are ranked as expected.

Real Datasets: We performed experiments on several real datasets, i.e., pima, image segment, Facebook metrics, and iris datasets, from the UCI machine learn-

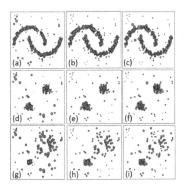

Fig. 4. Instance ranking for synthetic datasets where larger markers are for more important data points. The first to third columns correspond to iCAD, K-iCAD with RBF kernel, and K-iCAD with polynomial kernel, respectively.

Table 2. Comparison of CAD and K-CAD performance on arrhythmia dataset for different k values.

			CAD	K-CAD (rbf)	K-CAD (poly)
$k = 3$	Train	Time	0.55 ± 0.00	3.25 ± 0.15	3.06 ± 0.16
		AUC	37.01 ± 1.00	49.83 ± 0.63	63.61 ± 1.36
	Test	Time	1.12 ± 0.02	0.37 ± 0.01	0.35 ± 0.01
		AUC	45.51 ± 6.21	49.32 ± 1.69	61.67 ± 10.18
$k = 10$	Train	Time	4.76 ± 0.02	2.75 ± 0.06	2.53 ± 0.03
		AUC	52.89 ± 0.96	48.87 ± 0.51	73.92 ± 1.12
	Test	Time	1.59 ± 0.01	0.30 ± 0.01	0.28 ± 00.00
		AUC	48.02 ± 9.06	48.56 ± 5.38	71.88 ± 9.23
$k = 20$	Train	Time	16.56 ± 0.07	3.41 ± 0.13	3.24 ± 0.24
		AUC	56.87 ± 1.02	49.33 ± 0.91	76.83 ± 1.07
	Test	Time	2.88 ± 0.01	0.39 ± 0.01	0.36 ± 0.01
		AUC	47.90 ± 9.18	49.27 ± 7.06	74.88 ± 8.47

Table 3. Comparison of instance selection methods in classification. Classification, regression, and clustering rates are accuracy, mean absolute error, and adjusted rand index (best is 1), respectively, and times are in seconds. The left and right columns are for rank-based and retaining-based methods. Numbers in parentheses show the percentage of retained data.

Pima dataset

		iCAD	K-iCAD (rbf)	K-iCAD (poly)	IRMD	PSA	SOS	SE	SDM	iCAD	K-iCAD (rbf)	K-iCAD (poly)	ENN	DROP3	NR
1NN	Time	1.98E+0	4.41E-1	4.14E-1	9.52E-2	5.87E+0	2.13E-2	1.59E-2	4.51E-3	2.09E+0	4.73E-1	4.73E-1	1.87E-2	5.00E+0	×
	20% data	70.69%	67.70%	61.97%	67.28%	66.53%	67.06%	44.40%	62.87%	(40.79%)	(4.35%)	(12.76%)	(69.40%)	(13.44%)	68.48%
	50% data	70.83%	66.01%	64.96%	66.64%	64.44%	67.04%	50.64%	66.39%	65.23%	66.93%	65.88%	71.74%	64.06%	
	70% data	69.26%	67.56%	65.74%	67.95%	65.75%	66.91%	65.62%	67.81%						
LDA	20% data	75.64%	67.82%	72.25%	70.81%	76.17%	73.56%	54.55%	65.09%	65.62%	71.74%	66.14%	77.07%	75.12%	77.73%
	50% data	76.42%	76.03%	76.43%	73.81%	76.81%	76.43%	71.35%	73.68%	74.98%					
	70% data	76.94%	76.82%	76.81%	75.38%	76.82%	76.55%	76.55%	74.98%						
SVM	20% data	63.28%	57.65%	62.50%	59.62%	57.68%	63.52%	42.04%	63.78%	64.84%	63.92%	63.93%	67.18%	53.12%	64.57%
	50% data	65.22%	62.25%	57.94%	61.57%	61.57%	62.25%	48.15%	60.01%						
	70% data	62.76%	61.28%	56.40%	55.97%	60.42%	62.63%	55.23%	64.98%						

Image segment dataset

		iCAD	K-iCAD (rbf)	K-iCAD (poly)	IRMD	PSA	SOS	SE	SDM	iCAD	K-iCAD (rbf)	K-iCAD (poly)	ENN	DROP3	NR
1NN	Time	6.25E+0	1.20E+0	1.04E+0	3.64E-2	3.28E+2	5.86E-2	4.85E-2	1.38E-2	6.13E+0	1.58E+0	1.44E+0	1.30E-1	3.13E+2	×
	20% data	90.34%	81.42%	83.72%	78.00%	90.25%	89.26%	83.41%	80.99%	(9.54%)	(3.97%)	(3.14%)	(95.25%)	(6.78%)	96.45%
	50% data	93.41%	89.43%	92.85%	86.83%	94.76%	94.19%	84.71%	87.05%	47.22%	49.13%	49.43%	94.45%	60.34%	
	70% data	94.84%	92.20%	95.75%	91.03%	96.06%	95.41%	90.69%	90.00%						
LDA	20% data	89.26%	85.97%	90.86%	82.90%	90.47%	90.51%	85.23%	86.36%	60.60%	58.70%	59.91%	67.57%	79.35%	91.55%
	50% data	90.64%	90.04%	91.94%	86.32%	91.08%	91.16%	86.27%	87.83%						
	70% data	90.90%	90.99%	91.64%	88.26%	91.42%	91.34%	90.30%	89.48%						
SVM	20% data	73.41%	77.74%	77.40%	71.42%	78.44%	78.35%	71.73%	71.86%	39.43%	40.38%	38.70%	78.52%	55.10%	85.23%
	50% data	80.90%	81.08%	80.30%	80.60%	77.18%	82.85%	70.60%	81.34%						
	70% data	84.71%	84.71%	83.67%	78.87%	82.20%	85.67%	86.79%	84.32%						

Facebook Dataset

		iCAD	K-iCAD (rbf)	K-iCAD (poly)	IRMD	PSA	SOS	SE	SDM	iCAD	K-iCAD (rbf)	K-iCAD (poly)	ENN	DROP3	NR
LR	Time	1.30E+0	6.33E-1	1.01E+0	7.01E-3	1.08E-1	1.40E-2	8.40E-3	4.81E-3	1.26E-1	3.18E-1	3.18E-1	×	×	×
	20% data	7.57E+3	6.00E+3	7.45E+3	5.28E+3	9.98E+3	6.07E+3	1.89E+4	7.50E+3	(49.96%)	(6.95%)	(34.61%)			5.81E+3
	50% data	6.11E+3	5.51E+3	5.86E+3	4.85E+3	6.30E+3	6.25E+3	6.59E+3	5.69E+3	6.10E+3	1.56E+4	6.41E+3			
	70% data	5.70E+3	6.00E+3	5.72E+3	4.95E+3	5.53E+3	5.90E+3	5.83E+3	5.30E+3						
RF	20% data	8.54E+3	7.87E+3	6.85E+3	5.53E+3	7.12E+3	6.56E+3	1.09E+4	6.58E+3	6.36E+3	9.82E+3	6.46E+3			6.17E+3
	50% data	6.41E+3	7.08E+3	6.28E+3	5.02E+3	6.20E+3	6.76E+3	6.10E+3	7.19E+3						
	70% data	5.92E+3	6.95E+3	6.19E+3	5.03E+3	5.86E+3	6.26E+3	6.32E+3	6.03E+3						
MLP	20% data	1.348E+4	1.61E+4	1.02E+4	7.11E+3	2.12E+4	7.46E+3	4.70E+4	2.31E+4	6.11E+3	5.06E+4	8.31E+3			5.72E+3
	50% data	6.10E+3	5.44E+3	5.57E+3	4.75E+3	6.93E+3	6.02E+3	5.62E+3	6.64E+3						
	70% data	6.31E+3	6.00E+3	5.66E+3	5.14E+3	5.75E+3	5.95E+3	5.86E+3	6.00E+3						

Iris dataset

		iCAD	K-iCAD (rbf)	K-iCAD (poly)	IRMD	PSA	SOS	SE	SDM	iCAD	K-iCAD (rbf)	K-iCAD (poly)	ENN	DROP3	NR
K-means	Time	4.96E-1	5.30E-2	4.46E-2	1.60E-3	2.83E-1	3.70E-3	3.00E-3	1.90E-3	4.68E-1	6.07E-2	5.95E-2	×	×	×
	20% data	6.87E-1	1.56E-1	2.34E-1	1.32E-1	6.18E-1	6.92E-1	5.02E-1	4.89E-1	(60.59%)	(84.07%)	(84.07%)			7.03E-1
	50% data	6.98E-1	7.13E-1	8.33E-1	5.93E-1	7.01E-1	7.38E-1	4.85E-1	5.77E-1	7.34E-1	8.06E-1	8.20E-1			
	70% data	7.18E-1	7.83E-1	8.20E-1	6.53E-1	7.18E-1	7.35E-1	6.27E-1	7.12E-1						
Birch	20% data	6.09E-1	0.00E+0	1.26E-1	1.07E-1	6.97E-1	6.41E-1	5.10E-1	2.90E-1	7.71E-1	6.60E-1	6.85E-1			5.93E-1
	50% data	6.48E-1	6.58E-1	7.15E-1	6.35E-1	6.35E-1	5.34E-1	5.11E-1	5.92E-1						
	70% data	6.67E-1	7.17E-1	7.37E-1	6.61E-1	6.48E-1	5.79E-1	6.23E-1	6.04E-1						

ing repository. The first two datasets are used for classification, the third for regression, and the last one for clustering. The sample size of datasets are 768, 2310, 500, and 150, and their dimensionality are 8, 19, 19, and 4, respectively. The number of classes/clusters in pima, image segment, and iris are 2, 7, and 3, respectively. We used 1-Nearest Neighbor (1-NN), Linear Discriminant Analysis (LDA), SVM, Linear Regression (LR), Random Forest (RF), Multi-Layer Perceptron (MLP) with two hidden layers, K-means and Birch clustering methods in experiments. Table 3 reports the average accuracy and time over 10-fold CV and comparison to IRMD (with QR decomposition), PSA, SOS, Sorted by Distance from Mean (SDM), ENN, DROP3, and No Reduction (NR).

The iCAD and K-iCAD are reported in both rank-based and retaining-based versions of prototype selection. For pima and image segment datasets, iCAD and K-iCAD are both performing equally well but in other datasets, K-iCAD is mostly better. In terms of time, we outperform PSA and DROP3. In pima, we outperform all other baselines. In image segment, we are better than IRMD, SE, and SDM. In facebook data, we are mostly better than SOS, SE, and SDM, and in some cases better than PSA. In iris data, we outperform all the baselines. In some cases, we even outperform using the entire data. In retaining-based iCAD and K-iCAD, mostly, K-iCAD with RBF kernel retains the least, then K-iCAD with polynomial kernel, and then CAD.

6   Conclusion and Future Direction

This paper proposed a new method for anomaly detection, named CAD, and its kernel version. The main idea was to consider every point as a vertex of a polyhedron with the help of its neighbors and measure its curvature. Moreover, with an opposite view to CA, iCAD and K-iCAD were proposed for prototype selection. Different experiments as well as an application in image denoising were also reported. As a possible future work, we will try the idea of curvature for manifold embedding to propose a curvature preserving embedding method.

References

1. Emmott, A.F., Das, S., Dietterich, T., Fern, A., Wong, W.K.: Systematic construction of anomaly detection benchmarks from real data. In: Proceedings of the ACM SIGKDD Workshop on Outlier Detection and Description, pp. 16–21. ACM (2013)
2. Chandola, V., Banerjee, A., Kumar, V.: Anomaly detection: a survey. ACM Comput. Surv. (CSUR) **41**(3), 15 (2009)
3. Ghojogh, B., Crowley, M.: Principal sample analysis for data reduction. In: 2018 IEEE International Conference on Big Knowledge (ICBK), pp. 350–357. IEEE (2018)
4. Garcia, S., Derrac, J., Cano, J.R., Herrera, F.: Prototype selection for nearest neighbor classification: taxonomy and empirical study. IEEE Trans. Pattern Anal. Mach. Intell. **34**(3), 417–435 (2012)
5. Breunig, M.M., Kriegel, H.P., Ng, R.T., Sander, J.: LOF: identifying density-based local outliers. In: ACM SIGMOD Record, vol. 29, pp. 93–104 (2000)

6. Schölkopf, B., Williamson, R.C., Smola, A.J., Shawe-Taylor, J., Platt, J.C.: Support vector method for novelty detection. In: Advances in Neural Information Processing Systems, pp. 582–588 (2000)
7. Rousseeuw, P.J., Driessen, K.V.: A fast algorithm for the minimum covariance determinant estimator. Technometrics **41**(3), 212–223 (1999)
8. Liu, F.T., Ting, K.M., Zhou, Z.H.: Isolation forest. In: 2008 Eighth IEEE International Conference on Data Mining, pp. 413–422. IEEE (2008)
9. Liu, F.T., Ting, K.M., Zhou, Z.H.: Isolation-based anomaly detection. ACM Trans. Knowl. Discov. Data (TKDD) **6**(1), 3.1–3.39 (2012)
10. Wilson, D.L.: Asymptotic properties of nearest neighbor rules using edited data. IEEE Trans. Syst. Man Cybern. **SMC-2**(3), 408–421 (1972)
11. Wilson, D.R., Martinez, T.R.: Reduction techniques for instance-based learning algorithms. Mach. Learn. **38**(3), 257–286 (2000). https://doi.org/10.1023/A:1007626913721
12. Kalegele, K., Takahashi, H., Sveholm, J., Sasai, K., Kitagata, G., Kinoshita, T.: On-demand data numerosity reduction for learning artifacts. In: 2012 IEEE 26th International Conference on Advanced Information Networking and Applications (AINA), pp. 152–159. IEEE (2012)
13. Liu, C., Wang, W., Wang, M., Lv, F., Konan, M.: An efficient instance selection algorithm to reconstruct training set for support vector machine. Knowl. Based Syst. **116**, 58–73 (2017)
14. Ghojogh, B.: Principal sample analysis for data ranking. In: Meurs, M.-J., Rudzicz, F. (eds.) Canadian AI 2019. LNCS (LNAI), vol. 11489, pp. 579–583. Springer, Cham (2019). https://doi.org/10.1007/978-3-030-18305-9_62
15. Ghojogh, B., Crowley, M.: Instance ranking and numerosity reduction using matrix decomposition and subspace learning. In: Meurs, M.-J., Rudzicz, F. (eds.) Canadian AI 2019. LNCS (LNAI), vol. 11489, pp. 160–172. Springer, Cham (2019). https://doi.org/10.1007/978-3-030-18305-9_13
16. Coxeter, H.S.M.: Regular Polytopes. Courier Corporation, North Chelmsford (1973)
17. Markvorsen, S.: Curvature and shape. In: Yugoslav Geometrical Seminar, Fall School of Differential Geometry, Yugoslavia, pp. 55–75 (1996)
18. Descartes, R.: Progymnasmata de solidorum elementis. Oeuvres de Descartes X, pp. 265–276 (1890)
19. Richeson, D.S.: Euler's Gem: The Polyhedron Formula and the Birth of Topology, vol. 64. Princeton University Press, Princeton (2019)
20. Hilton, P., Pedersen, J.: Descartes, Euler, Poincare, Polya and polyhedra. Séminaire de Philosophie et Mathématiques **8**, 1–17 (1982)
21. Pimentel, M.A., Clifton, D.A., Clifton, L., Tarassenko, L.: A review of novelty detection. Signal Process. **99**, 215–249 (2014)
22. Hofmann, T., Schölkopf, B., Smola, A.J.: Kernel methods in machine learning. Ann. Stat. **36**, 1171–1220 (2008)
23. Schölkopf, B.: The kernel trick for distances. In: Advances in Neural Information Processing Systems, pp. 301–307 (2001)
24. Ah-Pine, J.: Normalized kernels as similarity indices. In: Zaki, M.J., Yu, J.X., Ravindran, B., Pudi, V. (eds.) PAKDD 2010. LNCS (LNAI), vol. 6119, pp. 362–373. Springer, Heidelberg (2010). https://doi.org/10.1007/978-3-642-13672-6_36
25. Rayana, S.: Outlier detection data sets (2019). http://odds.cs.stonybrook.edu/

Ethical Requirements for AI Systems

Renata Guizzardi[1]([⊠]), Glenda Amaral[2], Giancarlo Guizzardi[2],
and John Mylopoulos[3]

[1] NEMO/UFES, Espírito Santo (UFES), Vitoria, Brazil
`rguizzardi@inf.ufes.br`
[2] CORE/KRDB, Free University of Bozen-Bolzano, Bolzano, Italy
{`gmouraamaral,giancarlo.guizzardi`}`@unibz.it`
[3] University of Ottawa, Ottawa, Canada
`jm@cs.toronto.edu`

Abstract. AI systems that offer social services, such as healthcare services for patients, driving for travellers and war services for the military need to abide by ethical and professional principles and codes that apply for the services being offered. We propose to adopt Requirements Engineering (RE) techniques developed over decades for software systems in order to elicit and analyze ethical requirements to derive functional and quality requirements that together make the system-to-be compliant with ethical principles and codes. We illustrate our proposal by sketching the process of requirements elicitation and analysis for driverless cars.

Keywords: Requirements Engineering · Ethical requirements · AI systems

1 Introduction

The advent of Artificial Intelligence (AI) technologies, including machine learning, computer vision and natural language processing, has made it possible to build autonomous cyber-physical systems (CPSs), systems consisting of software and physical components, for example robots. Some CPSs being developed, including driverless cars and autonomous weapons, have raised ethical questions and even calls for their banning altogether [2]. Since AI is often built to stand in situations where human decision-making would otherwise be required, a big aspect one takes into account in decision-making processes is one's own ethics. Thus, systems should likewise be built based on ethical principles. But ethical questions about CPSs that socially interact with humans are not limited to AI systems and apply to all CPSs, including car cruise control systems, drones and photo cameras. It seems that the publicity surrounding AI systems has focused the limelight on a neglected dark corner of Software Engineering (SE): *Ethical Requirements*.

© Springer Nature Switzerland AG 2020
C. Goutte and X. Zhu (Eds.): Canadian AI 2020, LNAI 12109, pp. 251–256, 2020.
https://doi.org/10.1007/978-3-030-47358-7_24

Requirements Engineering (RE) is the area of research within SE concerned with the elicitation and analysis of requirements for a system-to-be (for our purposes, an AI system). Requirements are elicited from stakeholders: persons, groups, or organizations that are actively involved in the design of the system-to-be, may be affected by its outcomes, or can influence its outcomes. Analysis of stakeholder requirements leads to a specification for the system-to-be, consisting of functional and quality constraints the system-to-be must satisfy in order to meet the needs of its stakeholders.

Ethical requirements are requirements for AI systems derived from ethical principles or ethical codes (norms). They are akin to *Legal Requirements* [8], i.e., requirements derived from laws and regulations[1]. We are interested in characterizing the sources of ethical requirements, ethical principles and ethical codes, also sketching a systematic process for deriving requirements from such sources. The AI systems built on the basis of our proposal are not ethical agents who can reason and act on the basis of ethical principles. Rather, they are software systems that have the functionality and qualities to meet ethical requirements, in addition to other requirements they are meant to fulfill. We illustrate our initial proposal with a case study involving a driverless car. The main thesis of this paper is that techniques developed in RE that have been practiced for decades can also be used for making AI systems compliant with ethical principles and codes.

Defining ethical requirements allows ethical issues to be considered from the beginning in the CPSs development process. Hence, first of all, developers and stakeholders (e.g. those paying for the development of the system or the actual users of the system) shall include these issues during requirements elicitation, aiming at achieving a consensual agreement in their regard. Moreover, during requirements validation activities, i.e., when it is time to evaluate if each requirement is met by the system, a focus on ethical aspects is assured.

The remainder of the paper is structured as follows. Section 2 introduces ethical principles and codes, while Sect. 3 sketches a systematic process for identifying ethical requirements. By leveraging on this process, Sect. 4 briefly discusses the case of driverless cars, discussing their compliance to ethical considerations. Section 5 discusses related work, and Sect. 6 presents some final considerations.

2   Ethical Principles and Codes

Ethical principles are general principles of conduct towards others. For example, The European Commission's draft ethical guidelines for trustworthy AI [5] lists five such principles: *Autonomy* (respect for human dignity), *Beneficience* (doing good to others), *Nonmaleficence* (doing no harm to others), *Justice* (treating others fairly), *Explicability* (behaving transparently towards others). For example, from the Principle of Autonomy one may derive "Respect for a person's privacy", and from that an ethical requirement "Take a photo of someone only after her consent"

---

[1] But note, there are ethical requirements that are not legal, and legal ones that are not ethical.

for a phone camera. As another example, from Nonmaleficence, we may derive a functional requirement "Do not drive fast past a bystander" for a driverless car.

Ethical principles are generally domain-independent and rather abstract, so they require some analysis to fit them to a particular domain so as to derive ethical requirements. Ethical codes specialize ethical principles into particular domains, such as codes of conduct for employees of an organization, and codes of professional conduct for members of a professional society. The medical profession has adopted elaborate rules for an ethical code of practicing doctors, and so have research organizations for the conduct of research. There are codes of conduct for the military, by national jurisdiction, and numerous ethical codes for drivers in regional or municipal jurisdictions depending on driver responsibilities (such as taxi/track/school bus driving). Notably, Germany is the first country to adopt an ethical code for driverless cars [10]. Finally, and perhaps most importantly for autonomous weapons, there are international conventions for the conduct of war, the use of weapons, the treatment of civilians and prisoners, etc.

3   Deriving Ethical Requirements

The key concept to deriving ethical requirements is that of *Runtime Stakeholders*. These include those stakeholders that are using, affected by, or influencing the outcomes of a system as it is operating. Traditional RE often limits runtime stakeholders to just users of the system-to-be. However, for AI systems this needs to be extended to other parties. For example, for a driverless car, runtime stakeholders include passengers – i.e., the users of the car – but also pedestrians, whose path may cross that of the car and shouldn't be hit; bystanders, who shouldn't be scared or splashed as the car drives by; nearby drivers, who as a courtesy, should be allowed to cut in front in the car's lane; and fellow drivers in general, who might benefit from information about an accident that just happened in the vicinity of the car.

Runtime stakeholders are often ignored in classical RE as they are perceived to lack a concrete "stake" in the system-to-be. But the intrusion of AI systems in social settings is dictating a shift in the theory and practice of RE to include also these somehow indirect stakeholders into the RE process. Considerations such as the examples given above may seem trivial in the dawn of a new technological era. But they aren't! Think of ten thousand driverless cars added to a local setting, say Ottawa (population approximately 1,000,000), who are aggressive and inconsiderate in their driving in the sense that they don't fulfill simple ethical requirements, such as the above. Wouldn't this constitute an act of maleficence towards local drivers and pedestrians alike? Manufacturers of driverless cars should produce cars that can do more than meet legal, safety, security and other requirements: the cars they produce must be *good* drivers. And what constitutes good driving is defined in terms of ethical requirements, to be derived from ethical principles and codes.

We could categorize Ethical Requirements for an artificial system as types of *Ecological Requirements*, in the sense that they are necessarily requirements that are derived from the whole ecosystem in which the system is included. From an ontological perspective, there is a fundamental reason why this is the case, namely, given that these requirements are derived from assessments of *value* and *risk*. In a nutshell, value can be seen as a relational property, emerging from a set of relations between the intrinsic properties of a *value object* (or a value experience) and the goals of a *Value Subject* [9]. Roughly speaking, the value of an object (or experience) amounts to the degree to which the properties (*affordances*) of that object positively contribute (help, make) to the achievement of the value subject goals. Mutatis Mutandis, risk can also be seen as a relational property, emerging from a set of relations between the intrinsic properties of an *Object-at-Risk* (vulnerabilities), as well as *Threat Objects* and *Risk Enablers* (capacities, intentions) and the goals of a *Risk Subject* [9][2]. Again, roughly speaking, the risk of an object-at-risk given threat objects and risk enablers amounts to the degree to which the properties of those entities can be enacted to negatively contribute to denting (hurt, break) the risk subject goals. Now, ontologically speaking, affordances, vulnerabilities, capacities, intentions are all types of *dispositions*, which are themselves ecological properties, i.e., those that essentially depend on their environment (context) for their manifestation [9].

For example, given that we (as a society) value life, we would of course like to reduce as much as possible the risk of serious accidents with threats to human life (humans being the object at risk). For this, we must both consider the vulnerabilities of cars and their passengers, as well as the possible threats posed by other entities (e.g., other cars, road conditions). We must also endow driveless cars with a number of security features, but we must also do that for the entire platform in which driveless cars operate, including the consideration of features for roads, coordination points (the digital equivalent of traffic lights and road signs).

Given a set of runtime stakeholder types with their associated value and risk assessments, the next step is to introduce functional requirements that ensure that the car-to-be can actually recognize with adequate accuracy when it encounters instances of each type, under different weather and lighting conditions. In addition, we need functional requirements for recognizing notable events in the traffic environment of a car, such as accidents, slow/fast/very fast moving vehicles. Reports from different driverless car projects suggest that this is a step that has been recognized and adopted by driverless car manufacturers. Ethical requirements are functional and quality requirements elicited from runtime stakeholders in accordance with the five ethical principles discussed above.

4   The Case of Driveless Cars

We can now conduct an analysis of how to apply ethical principles, such as those listed above, to the case-at-hand. Explicability towards passengers may lead to a functional requirement for the driverless car to engage in conversations to

---

[2] In [9], the focus is on *use value* as opposed to *ethical value*. However, we believe the analysis still holds, in particular, regarding the connection between value and risk.

explain the route it is following and why. Explicability towards nearby drivers, pedestrians and bystanders leads to a functional requirement for the car to signal on turns and changes of lane. Explicability towards society in general benefits from the type of analysis aforementioned in which requirements can be traced back to the explicit identification of stakeholders, and an explicit and semantically transparent analysis of their values and risk[3]. Respect for human dignity calls for the car to stop in case it encounters a runtime stakeholder in need of assistance. Beneficience calls for the car to let a nearby driver cut in front, also to notify traffic authorities of an accident. Nonmaleficence calls for the car to slow down in the presence of nearby pedestrians and bystanders, independently of any speed limits that might apply. And in the case of two lanes merging into one, Justice calls for treating drivers from the other lane fairly, rather than in a me-first manner.

This analysis can be made more concrete and guided if it is based on an ethical code that applies for the system-to-be. Firstly, ethical codes often identify some of the runtime stakeholders, also include concrete applications of ethical principles that make the derivation of ethical requirements more direct and less controversial.

5 Related Work

In [6], the authors offer an excellent discussion on the incorporation of ethics into AI systems in the context of driverless cars. Two approaches are considered: (a) Make the AI system an ethical agent who can reason top-down from first principles to an ethical problem-at-hand and choose a suitable action; (b) Have the AI system learn bottom-up the most suitable ethical choice in different circumstances. Both alternatives are found to be problematic and both assume that for an AI system to comply with ethical principles or codes, it must be capable of reasoning on its own about the ethical merits of alternative decisions.

The US Department of Defense directive on autonomous weapon system [1] adopts a human-in-the-loop approach to such weapons. It also proposes policies that emphasize thorough testing and Verification & Validation for all semi-autonomous weapons to ensure that they function as designed. Arkin [4] discusses the merits and pitfalls of autonomous weapons, emphasizing that they could end up saving civilian lives. On the other hand, O'Connell [7] considers the politics of banning autonomous killing altogether.

---

[3] Notice that transparency w.r.t. the entities that compose an ecosystem regarding their capabilities, intentions, vulnerabilities, and goals strongly connects also to the notion of *trust*. In a nutshell, trust amounts to a set of relations connecting the beliefs of a (trustor) agent regarding the capabilities, vulnerabilities and intentions of a trustee insomuch as they can affect that agent's goals [3]. From this we directly have that: (1) trustworthiness assessment can and should be grounded in the explicit assessment of these aspects; (2) trustworthiness is not an absolute property of a system, but one that depends on all these aspects. To put it bluntly, it is meaningless to speak of trustworthy systems in an unqualified manner.

6   Conclusions

We have argued that RE techniques can be applied in the design of AI systems, such as driverless cars and autonomous weapons, to ensure that they comply to ethical principles and codes. It is important to emphasize that the solution we propose doesn't render such systems autonomous in ethical decision-making, since ethical matters are dealt with by their designers and built into the systems. Our proposal, however, does suggest a way to go forward with AI systems where technology is available, but we don't know how to deal with the ethical implications of their outcomes.

As to the implementation of functional and quality requirements derived from ethical requirements, it is important to emphasize that the system-to-be should be able to perform as well as well-trained humans performing the same task. For instance, similarly to a medical doctor, who writes a detailed report explaining her findings, AI systems should explain their reasoning rather than only providing results and taking decisions. This has important implications because some of the most successful AI technologies, notably Machine Learning ones, cannot currently deal well with explainability and other transparency-related requirements.

Acknowledgments. This research is supported by the Strategic Partnership Grant "Middleware Framework and Programming Infrastructure for IoT Services".

References

1. Autonomy in weapon system, DoD directive. Technical report, Department of Defence (2012). https://tinyurl.com/vh2qhej
2. Artificial intelligence: Mankind's last invention (2019). https://tinyurl.com/y9moo26c
3. Amaral, G., Sales, T.P., Guizzardi, G., Porello, D.: Towards a reference ontology of trust. In: Panetto, H., Debruyne, C., Hepp, M., Lewis, D., Ardagna, C.A., Meersman, R. (eds.) OTM 2019. LNCS, vol. 11877, pp. 3–21. Springer, Cham (2019). https://doi.org/10.1007/978-3-030-33246-4_1
4. Arkin, R.: Lethal autonomous systems and the plight of the non-combatant. AISB Q. **137**, 1–9 (2013)
5. High-Level Expert Group on Artificial Intelligence, E.C: Draft ethics guidelines for trustworthy AI, draft document (2018)
6. Etzioni, A., Etzioni, O.: Incorporating ethics into artificial intelligence. J. Ethics **21**(4), 403–418 (2017). https://doi.org/10.1007/s10892-017-9252-2
7. O'Connell, M.E.: Banning autonomous killing. In: Evangelista, M., Shue, H. (eds.) The American Way of Bombing: How Legal and Ethical Norms Change. Cornell University Press, Ithaca (2013)
8. Otto, P.N., Antón, A.I.: Addressing legal requirements in requirements engineering. In: Proceedings of the 15th IEEE RE 2007, New Delhi, 15–19 October 2007, pp. 5–14 (2007)
9. Sales, T.P., Baião, F., Guizzardi, G., Almeida, J.P.A., Guarino, N., Mylopoulos, J.: The common ontology of value and risk. In: Trujillo, J.C., et al. (eds.) ER 2018. LNCS, vol. 11157, pp. 121–135. Springer, Cham (2018). https://doi.org/10.1007/978-3-030-00847-5_11
10. Tuffley, D.: At last! the world's first ethical guidelines for driverless cars. The Conversation, September 2017. https://tinyurl.com/u4gbskh

A Deep Neural Network for Counting Vessels in Sonar Signals

Hamed H. Aghdam[1($\boxtimes$)], Martin Bouchard[1], Robert Laganiere[1],
Emil M. Petriu[1], and Philip Wort[2]

[1] Ottawa University, Ottawa, ON, Canada
{h.aghdam,bouchm,laganier,petriu}@uottawa.ca
[2] General Dynamics Mission Systems–Canada, Ottawa, ON, Canada
philip.wort@gd-ms.ca

Abstract. Monitoring the oceanographic activity of ships in restricted areas is an important task that can be done using sonar signals. To this end, a human expert may regularly analyze passive sonar signals to count the number of vessels in the region. To automate this process, we propose a deep neural network for counting the number of vessels using sonar signals. Our model is different from common approaches for acoustic signal processing in the sense that it has a rectangular receptive field and utilizes temporal feature integration to perform this task. Moreover, we create a dataset including $117K$ samples where each sample resembles a scenario with at most 3 vessels. Our results show that the proposed network outperforms traditional methods substantially and classifies 96% of test samples correctly. Also, we extensively analyze the behavior of our network through various experiments. Our codes and the database are available at https://gitlab.com/haghdam/deep_vessel_counting.

Keywords: Deep neural networks · Sonar signal · Vessel counting

1 Introduction

Sonar is an important technology in oceanographic monitoring that might be used for mapping the ocean floor or studying the population of certain species [1,2,10,14]. These applications usually depend on active sonar that is also used for creating sonar images. Active sonar emits a signal and listens to the signal that is reflected after hitting an object or the ocean floor. However, active sonar is not suitable for applications such as detecting the presence of vessels or counting them in restricted areas since vessels can acquire the signal that is sent by the active sonar.

In contrast to active sonar, passive sonar does not emit a signal and, fundamentally, it listens to ambient noise and the signal generated by marine objects. Human-made objects such as cargo vessels produce a certain pattern that could

Supported by General Dynamics Mission Systems–Canada.

be acquired and processed for detecting their presence or counting them. Passive sonar offers a powerful technology for detecting and counting illegal vessels since it does not send any signal and it will not be detectable by other vessels. Understanding the raw signal is not trivial and it must be processed before analyzing it by a human expert. Assigning experts to regularly check the data might not be effective assuming the scarcity of illegal activities.

An alternative method is to analyze these signals automatically. However, signals vary substantially depending on the ambient noise, the number of vessels, and the bearing, heading, speed, frequency and the distance of each vessel. As it turns outs, it is unlikely that a hand-engineered method counts the number of vessels correctly for a diverse set of signals that are produced by different combinations of these factors. In this paper, we propose a method for automatically counting the number of vessels using deep neural networks that accepts pre-processed signals (*ie*. spectrograms) and outputs the number of vessels.

Contribution: The raw signal is processed in overlapping time windows in the frequency domain to obtain the spectrogram of the signal. The sonar signal could be very long that in turn generates a large spectrogram. Our contribution has three folds. First, we devise a statistical method to determine if a spectrogram contains a vessel. Second, we propose a network with a rectangular receptive field and temporal feature aggregation to count the number of vessels in the input spectrogram. Third, we synthesize a large dataset of signals containing $117K$ spectrograms. Our experiments show that the proposed network performs significantly better than traditional approaches and networks designed for processing the spectrogram holistically. Our dataset and networks are available in https:// gitlab.com/haghdam/deep_vessel_counting. In this paper, we directly work on *sonar signals* that is different from *sonar images*.

2   Materials

We generated raw sonar signals using noise levels $\bar{n} \in \{0.2, 0.4, 0.8, 1\}$. For each noise level, 7000, 500 and 2250 samples were generated as training, validation and test samples, respectively. Overall, there are 84K training samples, 6K validation samples, and 27K test samples in the dataset. The generated signal depends on variables such as the initial position of vessels, heading, bearing, speed, frequencies of each target and power of each frequency, which are set randomly for each signal. Raw signals are further processed to obtain their spectrograms and feed them to vessel detection and vessel counting methods.

Each spectrogram is computed using different segment lengths (window sizes, integration times) where 500 overlapping segments are used with a shift of 2.5 s between two consecutive segments. Therefore, spectrograms represent different physical time and frequency resolutions. However, for ease of processing, the spectrograms are computed such that the number of frequency bins in each row of the spectrogram (i.e. resolution) is constant. Each spectrogram shows if there are 0, 1, 2 or 3 ships in the scanned region. Our goal is to train a model to count the number of submarines depicted by the spectrogram.

3   Proposed Methods

In this section, we propose two different methods for vessel detection (*ie.* presence or absence of targets) and counting problems.

3.1   Vessel Detection

Figure 1 illustrates sample log-spectrograms with various number of targets. All log-spectrograms have been normalized for consistent visualization by setting the minimum value to -40 and the maximum value to 20. Each row in these spectrograms indicates the spectrum of the signal at a specific time frame. Comparing one row from the spectrogram without any vessels with other spectrograms, we realize that there is a significant difference between them. An effective idea to perform target detection might be computing the maximum and mean of each spectrogram and detecting the presence/absence of vessels by passing them through a decision stump. Figure 2 shows the scatter plot of these quantities for the training set.

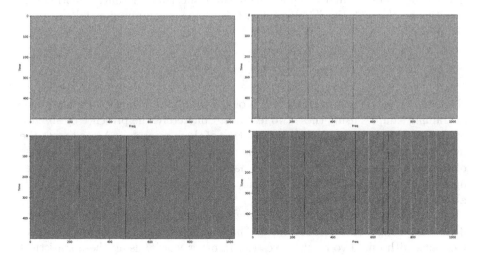

Fig. 1. Samples of spectrograms without any vessels (upper left), one vessel (upper right), two vessels (bottom left) and three vessels (bottom right). The horizontal and vertical axis shows the frequency and time, respectively.

First, depending on the noise level, the mean of spectrograms is scattered in four regions. Nonetheless, the maximum of spectrogram varies between 0 and 33 regardless of the noise level. Second, while it is possible to perform target detection accurately when the noise level is 0.2 and lower, it does not work accurately with higher noise levels. Consequently, the mean and maximum of spectrograms are not good features to perform vessel detection. Below, we propose a method that computes higher-order statistics for this purpose.

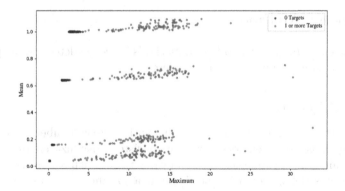

Fig. 2. Presence and absence of targets is detectable with reasonable accuracy using the mean and maximum of spectrograms.

To be more specific, given a spectrogram of size $S = \mathbb{R}^{N_f \times N_t}$ where N_t denotes the number of time frames and N_f shows the number of frequency bins in each spectrum, we first compute the mean of spectra over time. Denoting this quantity by $\hat{S} = [\hat{s}_i]$, the ith element in this vector is computed as follows:

$$\hat{s}_i = \frac{1}{N_t} \sum_{j=0}^{N_t} S_{ij} \quad \forall i = 0 \ldots N_f \tag{1}$$

Next, the *skewness* and *kurtosis* of $\hat{S}$ is obtained. Figure 3 shows the scatter plot of training samples using these two features. Concretely, the presence or absence of targets is perfectly detectable using either of these measures. According to the figure, by comparing the kurtosis of the mean spectrum with 2 or the skewness of the mean spectrum with 1, we can detect the presence or absence of target on training samples with 100% accuracy. As we will show in experiments, this also holds on validation and test sets.

In contrast to vessel detection, counting the number of vessels cannot be done using simple statistical features. The blue cluster in Fig. 3 represents spectrograms with one, two or three vessels. In other words, using these statistical features, spectrograms with a different number of vessels fall into the same cluster in the feature space. Consequently, we need a more expressive mapping to be able to form a separate cluster for each category in the feature space. To achieve this goal, we propose a neural network in the next section.

3.2 Counting Targets

Processing spectrograms with *only* convolutional neural networks is commonly done using square filters and processing the entire inputs to extract a feature vector for classification [3,13,17]. This approach is different from using recurrent neural networks for analyzing spectrograms [4,8] since the latter approach

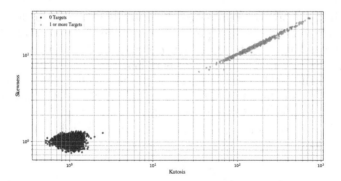

Fig. 3. The presence and absence of targets is perfectly detectable using either the skewness or kurtosis of the mean spectrum. All points are shifted by one along both axis to prevent negative values for log-scale plots. (Color figure online)

considers the spectrogram as a sequence of inputs and learns the temporal dependency between these inputs.

Our proposed network is inspired by the method in the previous section for detecting vessels. We started by aggregating spectra over time to compress the information into a single vector. Then, two features were computed from the aggregated spectrum. Depending on how spectra are aggregated over time, we might lose the temporal information in the spectrogram. Each spectrogram from a class represents a sequence of spectra. By comparing sequences of two different spectrograms of the same class, we will realize that they might be potentially different from each other. Therefore, strictly encoding long-term temporal information in the aggregated spectrum might not be essential.

In contrast, there might be a strong dependency between consecutive frequencies and spectra in the short-term. As a result, it is important to analyze frequencies in each time frame, locally. Based on these observations, we propose the approach that is illustrated in Fig. 4 for counting targets. Given a spectrogram of size 1024 × 500, we design a neural network (NN) that operates over entire frequencies, but it covers a small fraction of time frames (*ie.* spectra). In other words, we aim to design a neural network whose receptive field is $1024 \times H$ where H is much smaller than 500.

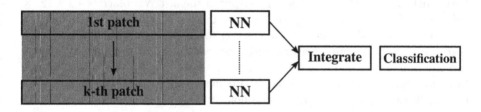

Fig. 4. Our network performs feature extraction in small time frames in a sliding fashion and integrates them to form the final feature vector.

Then, the input spectrogram is *hypothetically* divided into smaller patches of size $1024 \times H$, and each patch is processed using the neural network. Assuming there are K overlapping patches on the spectrogram, K feature vectors will be produced in the output. The next stage is to integrate these feature vectors into one vector. Finally, the integrated feature vector is classified using a linear classifier. Our network is different from both above approaches in the sense that it has a rectangular receptive field and it aggregates features that are extracted over a short time. In practice, this network can be implemented using all-convolutional neural networks [11].

Neural Network. Our network is composed of an initial layer, five blocks of wide residual (WR) modules [16], and an integration layer and a classification layer that are collectively 33 layers. Table 1 shows the architecture of the network. Each row inside brackets represents one WR module. Each WR module contains two convolution layers. In contrast to the original WR module, we use vertical convolution kernels instead of square kernels and one-sided strides as opposed to square strides in the first convolution layer. The size of the second convolution layers is always 3×3 except in the last block where the second convolution uses 1×1 kernels. Moreover, we apply spatial dropout [12] after the first convolution and batch normalization [5] after all convolution layers in each WR module. Figure 5 shows the architecture of our WR module. Note that the 1×1 convolution layer in this module is applied if the depth of the input is different from the depth of the second convolution. According to the output size in the last WR block, K is equal to 125 using our network.

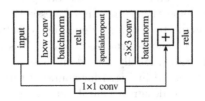

Fig. 5. The wide residual module in our proposed network.

Integration. Table 1 shows that the output of the last WR block is a $1 \times 125 \times 256$ tensor. This means, there are 125 time-frames in which entire frequencies have been processed to generate 256 features for each time-frame. The next step is to integrate time-frame features into one feature vector encoding the whole spectrogram in one vector. In this work, we integrate time-frame specific features by computing the maximum of each feature across all time-frames. This method is computationally efficient and it does not add new parameters to the network.

Table 1. The architecture of our proposed network.

Block name	Architecture	Input size	Output size
Init	$3 \times 3@16,\ st = (2, 2)$	1024×500	512×250
Block 1	$\begin{cases} 7 \times 3@32,\ st = (5, 2) \\ 3 \times 3@32,\ st = (1, 1) \\ 3 \times 3@32,\ st = (1, 1) \end{cases}$	$512 \times 250 \times 16$	$103 \times 125 \times 32$
Block 2	$\begin{cases} 7 \times 1@64,\ st = (5, 1) \\ 3 \times 3@64,\ st = (1, 1) \\ 3 \times 3@64,\ st = (1, 1) \end{cases}$	$103 \times 125 \times 32$	$21 \times 125 \times 64$
Block 3	$\begin{cases} 7 \times 1@128,\ st = (5, 1) \\ 3 \times 3@128,\ st = (1, 1) \\ 3 \times 3@128,\ st = (1, 1) \end{cases}$	$21 \times 125 \times 64$	$5 \times 125 \times 128$
Block 4	$\begin{cases} 7 \times 1@256,\ st = (5, 1) \\ 3 \times 3@256,\ st = (1, 1) \\ 3 \times 3@256,\ st = (1, 1) \end{cases}$	$5 \times 125 \times 128$	$1 \times 125 \times 256$
Block 5	$\begin{cases} 1 \times 1@256,\ st = (1, 1) \\ 3 \times 1@256,\ st = (1, 1) \\ 3 \times 1@256,\ st = (1, 1) \end{cases}$	$1 \times 125 \times 256$	$1 \times 125 \times 256$
Integrate	Max	$1 \times 125 \times 256$	$1 \times 1 \times 256$
Logits	3	$1 \times 1 \times 256$	$1 \times 1 \times 3$

Receptive Field. The *effective* value of H depends on the weights of the network and it is related to the effective size of the receptive field (ERF) [7]. We computed the ERF for 1000 samples *after training the proposed network*.

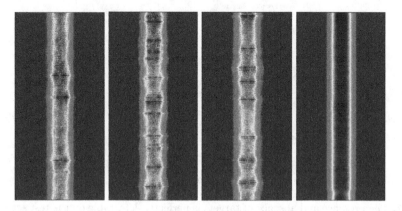

Fig. 6. Effective receptive field of the trained network computed for the middle neuron in the last convolution layer. Left three images show the results for three random samples and the right most image illustrates the accumulated results of 1000 samples. Dark blue color regions do not have any impact on the output of the neuron and red regions have the highest impact on the neuron. (Color figure online)

Figure 6 shows that the ERF in our trained network is roughly $H = 160$ time frames from the input spectrogram. Compared to Fig. 4, the axis of input in Fig. 6 are transposed because of the space limitation in the paper. That means the horizontal axis indicates the time and the vertical axis corresponds to frequencies.

4   Experiments

In this section, we first present the results of vessel detection. Then, we will analyze the results of target counting from different perspectives. Since the accuracy of target detection is 100% and it is a fairly simple problem, we do not perform extensive analysis for this problem.

4.1   Vessel Detection

We computed the kurtosis margin and the skewness margin between spectrograms without any vessels and spectrograms with one or more than one vessel on the training set. The minimum and maximum margins are (1.25, 5.63) for kurthosis and (0.32, 1.17) for skewness, respectively. As it turns out, there is a clear space between two classes using any of these features. We fixed the threshold T to the middle point in each feature and classified the given spectrogram X using a decision stump with the threshold T. In both cases, the accuracy of the classification on the test set is 100%. Figure 7 shows the plot of two classes using these features on the test set.

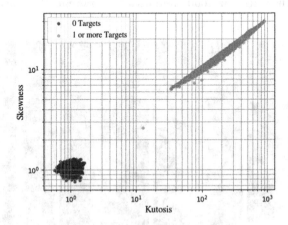

Fig. 7. The presence and absence of targets is perfectly detectable on the test set using either the skewness or kurtosis of the mean spectrum. All points are shifted by one along both axis to prevent negative values for log-scale plots.

4.2   Counting Targets

We train the network using the Momentum Gradient Descend with batch size 32 for 100 epochs. The initial learning rate is set to 0.001, and it is multiplied with 0.1 at the 70th and 90th epochs. Moreover, the weight of the L2 regularization term in loss function is fixed to $1e-4$. The network is evaluated using 27K test samples. We repeated the training procedure three times starting from random initialization at each trial. Table 2 shows the mean and standard deviation of the three runs. First, the precision and recall of the test set deviate infinitesimally at each run. This shows that the training procedure is stable, and we will obtain highly similar results if the network is trained several times. Second, the network has high precision and recall on the spectrograms with only one target. Nonetheless, the precision of the second class and the recall of the third class are lower. Overall, the accuracy of the network is greater than 96%.

To study the reason, we embedded the integrated feature into the 2D space using the t-SNE [9] technique. Figure 8 illustrates the features in the embedded space. Spectrograms with only one target are distinguishable from spectrograms with two or three targets. In contrast, some of the samples with three targets fall into the region where samples with two targets are located. As a result, these samples are classified as the second class (*ie.* two targets). This increases the number of false-positives in the second class which in sequel reduces the precision. The same argument holds for the reduction in the recall of the third class. The results could be improved by creating an ensemble of models and using the ensemble to make predictions. In this work, we compute the mean of probability scores for each sample and take the class with the maximum score as the class of the sample. Table 3 illustrates the precision and recall of the ensemble.

Comparison. The main idea behind deep neural networks is to learn a set of feature functions to map the input into a feature space. Then, the extracted feature is classified using a linear classifier. Traditionally, this would be done by designing feature functions manually and training a nonlinear classifier. To compare the results using traditional approaches with our network, we extracted feature vectors using different methods from spectrograms and trained a random forest to perform classification. Table 4 shows the results.

As it turns out, the mfcc [15] and spectral centroid [6] features are not able to accurately classify spectrograms in this problem. Moreover, there is a improvement compared to holistic network where we use square kernels and process the

Table 2. Precision and recall of the proposed network on the test set.

# of targets	Precision (%)	Recall (%)	Support
1 target	98.4 ± 0.001	99.9 ± 0.0002	9003
2 targets	91.9 ± 0.004	97.0 ± 0.003	9004
3 targets	98.5 ± 0.003	91.5 ± 0.004	9003

Table 3. Precision and recall of the proposed network ensemble on the test set.

# of targets	Precision (%)	Recall (%)	Support
1 target	98.5	1.00	9003
2 targets	92.9	97.9	9004
3 targets	99.4	92.5	9003

Fig. 8. Embedding the integrated feature vectors into the 2D space using the t-SNE technique. Red, blue and pink points show spectrograms with 1, 2 and 3 targets, respectively. (Color figure online)

entire input as opposed to integrating features temporally. In contrast, we use rectangular field-of-view and perform feature integration directly on the spectrogram. Second, our network has a high capacity that makes it possible to learn complex mappings.

Table 4. Comparing our network with hand-crafted features.

Method	1 target (%)		2 targets (%)		3 target (%)	
	Precision	Recall	Precision	Recall	Precision	Recall
Our network	98.5	1.00	92.9	97.9	99.4	92.5
Holistic network	80.0	70.0	50.0	50.0	67.0	77.0
mfcc	70.0	77.0	47.0	41.0	66.0	67.0
Spectral centroid	47.0	51.0	30.0	27.0	39.0	39.0

Temporal Dependency. In this section, we empirically study how important is temporal information in this problem. First, we start by flipping the spectrogram such that the first spectrum becomes the last spectrum and the last spectrum become the first spectrum after flipping. Second, spectra are dropped out every x frames where $x \in \{2, 3, 5, 10\}$. Fourth, The last half of the spectra are zeroed out. Fifth, the first half of the spectra are zeroed out. Sixth, half of the spectra in the middle are zeroed out. Finally, spectra are randomly shuffled. In this experiment, we use only one of our trained networks rather than using the ensemble. The results are summarized in Table 5.

Table 5. Precision and recall of the proposed network on the test set.

	1 target (%)		2 targets (%)		3 target (%)	
	Precision	Recall	Precision	Recall	Precision	Recall
Original network	98.2	99.9	91.5	96.6	98.2	91.0
Flip	98.3	98.4	90.7	94.8	96.3	91.9
Drop 10	97.8	99.9	89.3	96.6	98.7	88.5
Drop 5	97.6	99.9	90.5	95.8	98.1	90.0
Drop 3	97.7	99.8	90.5	94.8	96.9	90.2
Drop 2	98.4	74.8	70.6	73.4	78.9	94.6
Zero upper	81.7	99.9	53.6	79.7	1.0	29.0
Zero lower	48.4	1.00	22.3	19.7	1.00	5.1
Drop middle	84.8	98.9	52.8	83.8	99.9	24.7
Shuffle	44.7	98.3	28.5	22.6	94.0	0.5

Interestingly, flipping the spectrogram along spectra produces comparable results. In addition, dropping out spectra every x frames does not affect the output significantly as long as there are at least two consecutive spectra that are not dropped out. In contrast, zeroing out last spectra reduces the accuracy dramatically compared to zeroing out first or middle spectra. In all cases, zeroing out spectra affects the performance adversely. The results in this table suggest that temporal information play an important role in this problem.

5   Conclusion

In this work, we first propose a statistical technique to identify the presence or absence of a vessel in spectrograms. Then, we proposed a new network architecture to extract features from the spectrogram of sonar signals and count the number of targets. Our network has a rectangular field-of-view and it works by extracting features in short time-frames and integrating all of them into one vector. Our experiments showed that the proposed network is able to classify 96% of samples correctly and compared to hand-crafted features, our network produces considerably higher accuracy. Visualizing the feature space showed that while there is a negligible overlap between the first class and the other two classes, the second class overlaps with the thirds class causing a reduction in the overall performance. Then, we analyzed the histogram of entropy and probability scores individually. The histograms reveal that there are some samples that are incorrectly classified with high confidence due to the class overlap. There are also samples that are classified correctly with very low confidence since they lie close to the decision boundary in the feature space.

References

1. French, G., et al.: JellyMonitor: automated detection of jellyfish in sonar images using neural networks. In: 2018 14th IEEE International Conference on Signal Processing (ICSP), pp. 406–412, August 2018. https://doi.org/10.1109/ICSP.2018. 8652268
2. Galusha, A., Dale, J., Keller, J.M., Zare, A.: Deep convolutional neural network target classification for underwater synthetic aperture sonar imagery. In: Bishop, S.S., Isaacs, J.C. (eds.) Detection and Sensing of Mines, Explosive Objects, and Obscured Targets XXIV. vol. 11012, pp. 18–28. International Society for Optics and Photonics, SPIE (2019). https://doi.org/10.1117/12.2519521
3. Han, Y., Lee, K.: Acoustic scene classification using convolutional neural network and multiple-width frequency-delta data augmentation. CoRR abs/1607.02383 (2016)
4. Hannun, A.Y., et al.: Deep speech: scaling up end-to-end speech recognition. CoRR abs/1412.5567 (2014)
5. Ioffe, S., Szegedy, C.: Batch normalization: accelerating deep network training by reducing internal covariate shift. CoRR abs/1502.03167 (2015)
6. Klapuri, A., Davy, M.: Signal Processing Methods for Music Transcription, 1st edn. Springer, Boston (2010). https://doi.org/10.1007/0-387-32845-9
7. Luo, W., Li, Y., Urtasun, R., Zemel, R.S.: Understanding the effective receptive field in deep convolutional neural networks. CoRR abs/1701.04128 (2017)
8. Maas, A.L., Hannun, A.Y., Jurafsky, D., Ng, A.Y.: First-pass large vocabulary continuous speech recognition using bi-directional recurrent DNNs. CoRR abs/1408.2873 (2014)
9. van der Maaten, L., Hinton, G.: Visualizing data using t-SNE. J. Mach. Learn. Res. **9**, 2579–2605 (2008)
10. Neves, G., Ruiz, M., Fontinele, J., Oliveira, L.: Rotated object detection with forward-looking sonar in underwater applications. Expert Syst. Appl. **140**, 112870 (2020)
11. Springenberg, J.T., Dosovitskiy, A., Brox, T., Riedmiller, M.A.: Striving for simplicity: the all convolutional net. CoRR abs/1412.6806 (2014)
12. Tompson, J., Goroshin, R., Jain, A., LeCun, Y., Bregler, C.: Efficient object localization using convolutional networks. CoRR abs/1411.4280 (2014)
13. Valenti, M., Squartini, S., Diment, A., Parascandolo, G., Virtanen, T.: A convolutional neural network approach for acoustic scene classification. In: 2017 International Joint Conference on Neural Networks (IJCNN), pp. 1547–1554, May 2017. https://doi.org/10.1109/IJCNN.2017.7966035

14. Wang, X., Jiao, J., Yin, J., Zhao, W., Han, X., Sun, B.: Underwater sonar image classification using adaptive weights convolutional neural network. Appl. Acoust. **146**, 145–154 (2019)
15. Xu, M., Duan, L.-Y., Cai, J., Chia, L.-T., Xu, C., Tian, Q.: HMM-based audio keyword generation. In: Aizawa, K., Nakamura, Y., Satoh, S. (eds.) PCM 2004. LNCS, vol. 3333, pp. 566–574. Springer, Heidelberg (2004). https://doi.org/10.1007/978-3-540-30543-9_71
16. Zagoruyko, S., Komodakis, N.: Wide residual networks. CoRR abs/1605.07146 (2016)
17. Zhang, Y., et al.: Towards end-to-end speech recognition with deep convolutional neural networks. CoRR abs/1701.02720 (2017)

Partial Label Learning by Entropy Minimization

Xuejun Han[✉]

Carleton University, Ottawa, ON, Canada
xuejunhan@cmail.carleton.ca

Abstract. Partial label learning deals with the problem where each training example is associated with a set of candidate labels, only one of which is assumed to be valid. To learn from such ambiguous labeling information, the critical point is to disambiguate the set of candidate labels, thereby targeting the ground-truth label. By utilizing the nature that only one of the candidate labels is correct, we employ the entropy minimization strategy to force the model making confident predictions of the training data. By doing this, the ground-truth labels are likely to make more contributions to the model training. Finally, comparative experiments on a number of real-world datasets are conducted, demonstrating the effectiveness of the proposed approach.

Keywords: Weak supervision · Partial label learning · Entropy minimization

1 Introduction

Partial label learning is regarded as a weakly supervised learning framework. It is also named as *ambiguous label learning* [1–3] or *superset label learning* [4,5]. The partial label learning deals with the problem where each training example is associated with a set of candidate labels, only one of which is assumed to be valid [6,7]. Due to the difficulty in collecting the unambiguous or clean labeled data, partial label learning has been widely applied to many real-world scenarios, such as web mining [8], automatic image annotation [3,9], ecoinformatics [10], etc.

Formally speaking, let $\mathcal{X} = \mathbb{R}^n$ be the n-dimensional feature space and $\mathcal{Y} = \{1, 2, \cdots, l\}$ be the label space including l possible labels. Suppose there are m training examples and each example is assigned with several labels. Denote $\mathcal{D} = \{(\mathbf{x}_i, S_i)_{i=1}^m\}$ the partial training set, where $\mathbf{x}_i \in \mathcal{X}$ is the feature vector and $S_i \subseteq \mathcal{Y}$ is the candidate label set of $\mathbf{x}_i$. The ground-truth label y_i of $\mathbf{x}_i$ is assumed to reside in S_i, i.e. $y_i \in S_i$. The task of partial label learning is to learn a multi-class classifier $f : \mathcal{X} \mapsto \mathcal{Y}$ from training set $\mathcal{D}$, which aims to correctly predict the ground-truth label of a test example.

Due to the inaccessibility of the ground-truth labels, the key to successful partial label learning is trying to disambiguate the set of candidate labels, namely

© Springer Nature Switzerland AG 2020
C. Goutte and X. Zhu (Eds.): Canadian AI 2020, LNAI 12109, pp. 270–275, 2020.
https://doi.org/10.1007/978-3-030-47358-7_26

targeting the ground-truth label. To achieve this, there are normally two general strategies, which are disambiguation by averaging [2, 6, 7] and disambiguation by identification [10–13]. The average-based strategy treats every candidate label equally, and the final prediction is made by averaging the modelling outputs of all the candidate labels. It is worth pointing out the potential drawback of this strategy, that is the significant output yielded by the ground-truth label is likely to be overwhelmed by the undesired outputs yielded by the false positive labels, so that the model could be misled into making wrong predictions. The identification-based strategy ties to regard the ground-truth label as a latent variable and employ an iterative refining process to gradually identify the confidence of each candidate label. However, these approaches fail to consider the mutually exclusive relationship among the candidate labels.

Motivated by the entropy minimization [14], a strategy widely-used in the semi-supervised learning forcing the confident prediction of the unlabelled data, we follow the similar idea to try to make confident predictions of the partial training data. Extensive experiments on a number of real-world datasets are conducted, which clearly demonstrates the effectiveness of the proposed approach.

2 Related Work

As a result of the difficulty in learning from ambiguous labeling information of partial training examples, there exist two general strategies to disambiguate the candidate labels, which are the average-based strategy and the identification-based strategy.

The average-based strategy makes the final predictions by averaging the modelling outputs of all the candidate labels, as we mentioned above. Following this strategy, some instance-based approaches [2, 5] predict the label of a test example by averaging the candidate labels of its nearest neighbours in the partial training set. Moreover, some parametric approaches train a parametric model [6, 7] trying to distinguish the average modelling output of candidate labels from that of non-candidate labels.

Different from the average-based strategy dealing with each candidate label equally, the identification-based strategy attempts to treats candidate labels with discrimination, and refine different confidences of candidate labels. Following this strategy, conventional approaches try to optimize the objective function in term of the maximum likelihood criterion [11] or the maximum margin criterion [12]. Furthermore, some approaches [5, 7] in leveraging the topological information of the feature space are recently proposed, to derive the confidence of each candidate label.

In this paper, a simple but effective partial label learning approach called *Partial Label Learning by Entropy Minimization* is proposed, and the performance is demonstrated by the detailed results of comparative experiments.

3   Proposed Method

Following the notations stated in Sect. 1, we denote the partial training set by
$\mathcal{D} = \{(\mathbf{x}_i, S_i)_{i=1}^m\}$. Let $\mathbf{X} = [\mathbf{x}_1, \cdots, \mathbf{x}_m]^\top \in \mathbb{R}^{m \times n}$ be the training data matrix
and $\mathbf{Y} = [\mathbf{y}_1, \cdots, \mathbf{y}_m]^\top \in \{0, 1\}^{m \times l}$ be the corresponding partial label matrix,
where $y_{ij} = 1$ means the j-th label is a candidate label of $\mathbf{x}_i$, i.e. $y_{ij} \in S_i$,
otherwise $y_{ij} \notin S_i$.

Assuming a probabilistic classifier $f(\mathbf{x}; \theta)$ such that $f(\mathbf{x}; \theta) \in [0, 1]^l$ and
$f(\mathbf{x}_i)^\top \mathbf{1} = 1$, we propose a unified partial label learning formulation with proper
constraints as follows:

$$\min_\theta \sum_{i=1}^m (L(f(\mathbf{x}_i, \theta), \mathbf{y}_i) + \lambda H(f(\mathbf{x}_i, \theta))) + \beta \Omega(\theta) \tag{1}$$

$$\text{s.t.} \quad f(\mathbf{x}_i) \leq \mathbf{y}_i + \epsilon, \quad \forall i \in [m]$$

where $[m] := \{1, 2, \cdots, m\}$. L is the loss function, here we employ the mean-
squared error loss that is $L(f(\mathbf{x}_i), \mathbf{y}_i) = (f(\mathbf{x}_i; \theta) - \mathbf{y}_i)^2$. H is the entropy function, i.e. $H(\mathbf{p}) = -\mathbf{p}^\top \log \mathbf{p}$. $\epsilon > 0$ is a very small constant value. $\Omega(\cdot)$ is a
regularization function to control the complexity of model parameters, here we
define $\Omega(f) = \|\theta\|^2$. λ and β are the tradeoff parameters.

By incorporating the constraints as a hinge loss term into the objective function, the final optimization problem is presented as follows:

$$\min_\theta \sum_{i=1}^m ((f(\mathbf{x}_i; \theta) - \mathbf{y}_i)^2 - \lambda f(\mathbf{x}_i; \theta)^\top \log f(\mathbf{x}_i; \theta) + \alpha[f(\mathbf{x}_i; \theta) - \mathbf{y}_i - \epsilon]_+) + \beta \Omega(\theta) \tag{2}$$

where $[\cdot]_+ = \max(\cdot, 0)$. This optimization problem can be solved by stochastic
gradient descent. The entropy term encourages the model to make confident
predictions for training data, so that the ground-truth labels are likely to contribute more to the model training. Furthermore, the hinge loss term enhances
this target.

4   Experiment

To demonstrate the effectiveness of our proposed method, we conduct comparative experiments against five partial label learning algorithms, each configured
with parameters suggested in respective literature:

- CLPL [6]: a convex formulation for the partial label setting via averaging-
  based disambiguation [suggested configuration: SVM with squared hinge loss];
- PALOC [15]: an approach that adapts one-vs-one decomposition strategy for
  partial label learning [default configuration for partial label learning: $\mu = 10$]
- PLKNN [2]: a k-nearest neighbour approach to make predictions by averaging
  the labeling information of neighbouring examples [suggested configuration:
  $k \in \{5, 6, \cdots, 10\}$];

- PLSVM [12]: a maximum margin approach learning from partial examples by optimizing margin-based objective function [suggested configuration: $\lambda \in \{10^{-3}, 10^{-2}, \cdots, 10^3\}$];
- LSBCMM [10]: a maximum likelihood approach learning from partial label examples via mixture models [suggested configuration: $L = 10 \log 2(l)$].

For our model, we employ a three-layer neural network as the classifier and apply softmax function to the final layer to get the probabilistic outputs. For each dataset, the ten-fold cross-validation is conducted and the mean prediction accuracies and the standard deviations are reported. Moreover, we deploy pairwise t-test at 0.05 significance level to investigate whether our model significantly outperforms the comparing algorithms for all the experiments.

Table 1. Characteristics of the real-world partial label datasets.

Data set	#Examples	#Features	#Labels	Avg. CLs	Task domain
BirdSong	4998	38	13	2.18	Bird song classification [16]
Lost	1122	108	16	2.23	Automatic face naming [17]
FG-NET	1002	262	78	7.48	Facial age estimation [17]
Soccer Player	17472	279	171	2.09	Automatic face naming [9]
Yahoo! News	22991	163	219	1.91	Automatic face naming [18]

Table 1 summarizes the characteristics of real-world partial label datasets, including BirdSong [16], Lost [17], FG-NET [17], Soccer Player [9], and Yahoo! News [18]. These real-world partial label datasets are from various application domains, such as bird song classification (BirdSong), automatic face naming (Lost, Soccer Player, and Yahoo! News) and facial age estimation (FG-NET). The average number of candidate labels (Avg. CLs) for each partial label instance is also reported in Table 1.

Table 2. Classification accuracy of each algorithm on the real-world datasets. The symbols •/∘ indicate whether our approach is statistically superior/inferior to the comparing algorithms (t-test at 0.05 significance level for two independent samples).

Data	Ours	CLPL	PALOC	PL-KNN	PL-SVM	LSBCMM
Bir.	0.698 ± 0.018	0.632 ± 0.019•	0.711 ± 0.016	0.614 ± 0.021•	0.660 ± 0.037•	0.672 ± 0.056
Los.	0.649 ± 0.043	0.742 ± 0.038∘	0.629 ± 0.056	0.424 ± 0.036•	0.729 ± 0.042∘	0.693 ± 0.035∘
FG.	0.099 ± 0.020	0.063 ± 0.027•	0.065 ± 0.019•	0.038 ± 0.025•	0.063 ± 0.029•	0.059 ± 0.025•
Soc.	0.501 ± 0.010	0.368 ± 0.010•	0.537 ± 0.015∘	0.497 ± 0.015	0.464 ± 0.011•	0.494 ± 0.017
Yah.	0.490 ± 0.007	0.462 ± 0.009•	0.625 ± 0.005∘	0.457 ± 0.004•	0.629 ± 0.010∘	0.645 ± 0.005∘

Table 2 reports the mean classification accuracies of our approach as well as comparing algorithms. As shown in Table 2, it is worth to point out that: 1)

On all data sets, our proposed method significantly outperforms or at least is comparable to PLKNN; 2) On the FG-NET dataset, our approach achieves superior performance compared to all the approaches; 3) our approach statistically outperforms CLPL on all datasets except Lost; 4) our approach is comparable to PALOC on BirdSong & Lost and to LSBCMM on BirdSong & Soccer Player.

5   Conclusion

In this paper, a new partial label learning approach with entropy minimization is presented. An experimental study was conducted to evaluate the performance and demonstrate the effectiveness of the models. If we have a number of unlabeled data at hand, how to utilize unlabelled data to help partial label learning would be an interesting direction.

References

1. Chen, Y., Patel, V.M., Chellappa, R., Phillips, P.J.: Ambiguously labeled learning using dictionaries. IEEE Trans. Inf. Forensics Secur. **9**(12), 2076–2088 (2014)
2. Hüllermeier, E., Beringer, J.: Learning from ambiguously labeled examples. In: Famili, A.F., Kok, J.N., Peña, J.M., Siebes, A., Feelders, A. (eds.) IDA 2005. LNCS, vol. 3646, pp. 168–179. Springer, Heidelberg (2005). https://doi.org/10.1007/11552253_16
3. Cour, T., Sapp, B., Jordan, C., Taskar, B.: Learning from ambiguously labeled images. In: 2009 IEEE Conference on Computer Vision and Pattern Recognition, pp. 919–926, June 2009
4. Liu, L.P., Dietterich, T.G.: Learnability of the superset label learning problem. In: Proceedings of the 31st International Conference on International Conference on Machine Learning - Volume 32. ICML 2014, JMLR.org II-1629-II-1637 (2014)
5. Gong, C., Liu, T., Tang, Y., Yang, J., Yang, J., Tao, D.: A regularization approach for instance-based superset label learning. IEEE Trans. Cybern. **48**(3), 967–978 (2018)
6. Cour, T., Sapp, B., Taskar, B.: Learning from partial labels. J. Mach. Learn. Res. **12**(null), 1501–1536 (2011)
7. Zhang, M.L., Zhou, B.B., Liu, X.Y.: Partial label learning via feature-aware disambiguation. In: Proceedings of the 22nd ACM SIGKDD International Conference on Knowledge Discovery and Data Mining. KDD 2016, New York, NY, USA, Association for Computing Machinery, pp. 1335–1344 (2016)
8. Luo, J., Orabona, F.: Learning from candidate labeling sets. In: Lafferty, J.D., Williams, C.K.I., Shawe-Taylor, J., Zemel, R.S., Culotta, A. (eds.) Advances in Neural Information Processing Systems 23. Curran Associates, Inc., pp. 1504–1512 (2010)
9. Zeng, Z., et al.: Learning by associating ambiguously labeled images. In: 2013 IEEE Conference on Computer Vision and Pattern Recognition, pp. 708–715, June 2013
10. Liu, L., Dietterich, T.G.: A conditional multinomial mixture model for superset label learning. In Pereira, F., Burges, C.J.C., Bottou, L., Weinberger, K.Q. (eds.) Advances in Neural Information Processing Systems 25. Curran Associates, Inc., pp. 548–556 (2012)

11. Jin, R., Ghahramani, Z.: Learning with multiple labels. Advances in Neural Information Processing Systems, July 2003

12. Nguyen, N., Caruana, R.: Classification with partial labels, vol. 08, pp. 551–559 (2008)

13. Yu, F., Zhang, M.L.: Maximum margin partial label learning. Mach. Learn. **106**(4), 573–593 (2017)

14. Grandvalet, Y., Bengio, Y.: Semi-supervised learning by entropy minimization. In: Saul, L.K., Weiss, Y., Bottou, L., (eds.) Advances in Neural Information Processing Systems 17, pp. 529–536. MIT Press (2005)

15. Wu, X., Zhang, M.L.: Towards enabling binary decomposition for partial label learning. In: Proceedings of the Twenty-Seventh International Joint Conference on Artificial Intelligence, IJCAI-18, International Joint Conferences on Artificial Intelligence Organization, pp. 2868–2874, July 2018

16. Briggs, F., Fern, X.Z., Raich, R.: Rank-loss support instance machines for MIML instance annotation. In: Proceedings of the 18th ACM SIGKDD International Conference on Knowledge Discovery and Data Mining. KDD 2012, New York, NY, USA. Association for Computing Machinery, pp. 534–542 (2012)

17. Panis, G., Lanitis, A., Tsapatsoulis, N., Cootes, T.: An overview of research on facial aging using the FG-NET aging database. IET Biom. 5 (2015)

18. Guillaumin, M., Verbeek, J., Schmid, C.: Multiple instance metric learning from automatically labeled bags of faces. In: Daniilidis, K., Maragos, P., Paragios, N. (eds.) ECCV 2010. LNCS, vol. 6311, pp. 634–647. Springer, Heidelberg (2010). https://doi.org/10.1007/978-3-642-15549-9_46

Low-Dimensional Dynamics of Encoding and Learning in Recurrent Neural Networks

Stefan Horoi[1,2(✉)], Victor Geadah[1,2], Guy Wolf[1,2], and Guillaume Lajoie[1,2]

[1] Department of Mathematics and Statistics, Université de Montréal,
Montreal, QC, Canada
`stefan.horoi@umontreal.ca`
[2] Mila - Quebec Artificial Intelligence Institute, Montreal, QC, Canada

Abstract. In this paper, we use dimensionality reduction techniques to study how a recurrent neural network (RNN) processes and encodes information in the context of a classification task, and we explain our findings using tools from dynamical systems theory. We observe that internal representations develop a task-relevant structure as soon as significant information is provided as input and this structure remains for some time even if we let the dynamics drift. However, the structure is only interpretable by the final classifying layer at the fixed time step for which the network was trained. We measure that throughout the training, the recurrent weights matrix is modified so that the resulting dynamical system associated with the network's neural activations evolves into a non-trivial attractor, reminiscent of neural oscillations in the brain. Our findings suggest that RNNs change their internal dynamics throughout training so that information is stored in low-dimensional cycles, rather than in high-dimensional clusters.

Keywords: Recurrent neural networks · Internal representations geometry · Dynamical systems

1 Introduction

Recurrent neural networks (RNNs) contain intra-layer connections which allow the networks to preserve past information across multiple time steps. Because of this feedback mechanism RNNs are well-suited for the analysis of temporal or sequential data [1]. Furthermore, this temporal dependence suggests that progress in understanding RNNs resides in analysing them as nonlinear dynamical systems [1,8,10]. At every time step, the network state, which is governed by past inputs, is moved in the high dimensional internal state space by the network dynamics, which depend in part of the present input.

While research has been done studying the geometry of internal representations and its effect on the performance of deep neural networks [2], this was not

---

G. Wolf, G. Lajoie—Equal senior authorship contributions

© Springer Nature Switzerland AG 2020
C. Goutte and X. Zhu (Eds.): Canadian AI 2020, LNAI 12109, pp. 276–282, 2020.
https://doi.org/10.1007/978-3-030-47358-7_27

studied in RNNs where network dynamics largely dictate the geometry [5,10]. Here, we analyse the geometric properties of internal representations and the network dynamics in parallel. This allows us to shed some light onto how RNNs process information at every time step and encode it over long periods of time.

2 Methods

2.1 Experimental Setup

Task: We trained the network on Sequential MNIST classification, a dynamical task where an RNN has to classify a digit by being shown one row of pixels at a time. Each image being 28×28, the inputs to the network were vectors of dimension 28 and the sequence length was equally 28, corresponding to the number of rows. This toy task requires the network to integrate past inputs quickly into its computations and its decision making process hence making the task relevant and representative of more general RNN tasks.

Model: The recurrent neural network used contained a single recurrent layer of 200 tanh neurons and a linear output layer. Both layers had only weight parameters and no bias to affect the recurrent dynamics. The vector equation for the recurrent unit activations h_t in response to input x_t is:

$$h_t = \tanh(W_{in} \, x_t + W_{rec} \, h_{t-1}) \tag{1}$$

The weights matrices for each layer, W_{in}, W_{rec} and W_{out} are sampled i.i.d from a Gaussian distribution, i.e. $W_{ij} \sim \mathcal{N}(0, 1/\text{dim}W))$. The output y is generated by a linear readout at the last time step $t = T$, via the equation:

$$y = W_{out} \, h_T \tag{2}$$

Training and Implementation: The network is trained using the Adam optimization algorithm with a crossentropy loss function. After training the network for 30 epochs we achieve a testing accuracy of a little over 93%. While this accuracy is far from the state-of-the-art, our purpose here is to study the dynamics of network computations, which should not change qualitatively in circumstances where the network architecture, task and training method are similar. The code was implemented in Python using PyTorch and all the experiments reported below were conducted using the trained networks and the validation dataset, which the network did not see during training.

2.2 Experimental Datasets

Our experiments were conducted on the MNIST validation dataset as well as on three modified versions of it. **Dataset 1: Hidden images** - The last n pixel rows of each image are blank (value 0), for $1 \leq n \leq 27$ (e.g., Fig. 1b). This aims to determine how the recurrent dynamics alone help the classification task. **Dataset 2: Cut images** - Only the first n lines of the images are shown to the networks, for $1 \leq n \leq 27$ (e.g., Fig. 1c). This allows to evaluate the importance of sequence length and the interpretability of early internal representations by the last linear layer. **Dataset 3: Extended images** - Blank pixel rows are added at the end of the full input sequences, increasing their length (e.g., Fig. 1d). This is in order to see how network dynamics affect the internal representations after the network was provided with all the available information.

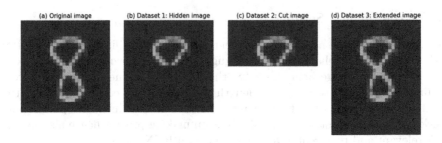

Fig. 1. One MNIST image across all tested validation datasets

2.3   The Spectrum of Lyapunov Exponents

To analyse and understand the behaviour of an autonomous dynamical system such as the recurrent internal dynamics associated with (1)

$$h_t = \sigma(W_{rec}h_{t-1}) \tag{3}$$

we use its *spectrum of Lyapunov exponents*, a powerful classification tool for the geometry of the attractor associated with the dynamics (e.g. [3,5]). For such a system, or in a more compact form $h_t = f(h_{t-1})$, the following limit exists and is constant for almost-all values of the initial starting point $h_0 \in \mathbb{R}^m$:

$$\lambda_0 = \lim_{n \to \infty} \frac{1}{n} \ln \left| \mu_0 \left(F_n^T F_n \right) \right| \tag{4}$$

where $\mu_0(A)$ denotes the first eigenvalue of matrix A, $F_n = f'(h_n) \cdot \ldots \cdot f'(h_0)$ and $f'(h)$ denotes the Jacobian of f at h. To compute λ_0, we decompose the product of Jacobians in (4) using its QR-decomposition.

The scalar $\lambda_0 \in \mathbb{R}$ is referred to as the *maximal Lyapunov exponent* (MLE), and corresponds to the rate of growth of a volume element, governed by the dynamics. Its value can be interpreted as follows :

- $\lambda_0 > 0$: exponential volume expansion, meaning initially close trajectories will exponentially differ ;
- $\lambda_0 = 0$: volume preservation, such as trajectories in a limit cycle ;
- $\lambda_0 < 0$: exponential volume compression, meaning converging trajectories.

3   Results

3.1   Development of a Task Relevant Structure

In this section, we expose the impact of input sequence length on classification accuracy and relate these findings to the developed internal representations associated with network responses to inputs. The results were generated by training ten networks and evaluating them on the modified datasets from Sect. 2.1. Mean and standard deviation across networks are plotted in Fig. 2.

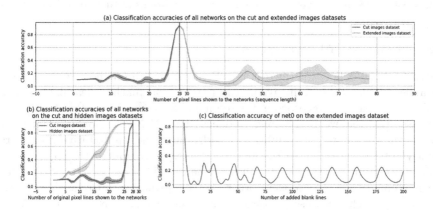

Fig. 2. Accuracy of the trained networks on the modified validation datasets

Figure 2a shows the importance of sequence length for the classification task. The networks only reach over 90% classification accuracy when the sequence length is of 28 pixel lines – the full training image length. For almost all other sequence lengths the classification is, on average, no better than random.

Figure 2b further emphasises this idea by showing that despite being provided with the same amount of relevant information (i.e., nonblank pixel lines), the networks classify better when input sequences are of length 28. This also indicates that up to time step 28, the recurrent dynamics driven by W_{rec} significantly help classification. However, after time step 28 the effect is opposite: the hidden dynamics hinder classification as shown by the orange line of Fig. 2a and by Fig. 2c. The recurrent pattern in Fig. 2c suggests that the network relies on oscillatory dynamics in the representation space to implement the task. We further investigate this idea in Sect. 3.2.

If we run unsupervised clustering on the internal representations at different time steps using t-SNE [4], we observe that early in the classification process the networks identify the class of certain inputs and separates their internal representations from those of the rest of the inputs (Fig. 3a). By the time entire images are shown, the clustering is highly correlated with the true classes (Fig. 3b). As we let the recurrent dynamics drift, we see a slow but steady degradation of the task-relevant geometric structure (Fig. 3c).

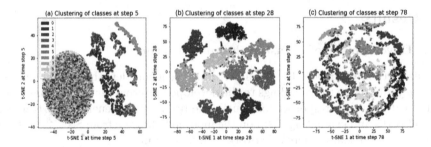

Fig. 3. Class coloured clustering of internal representations in three different settings

3.2 Formation of Limit Cycles

To establish a link between the results in Sect. 3.1 and autonomous RNN internal dynamics we use tools from dynamical systems and machine learning. Neural oscillations have been recorded in the brain and are central to different cognitive processes [7], and we find strong evidence that RNNs develop internal limit cycles over the course of training. To understand and quantitatively measure the asymptotic behaviour of the internal states, we computed the spectrum of Lyapunov exponents associated with the dynamical system in (3) (see Sect. 2.3). The maximal Lyapunov exponent (MLE) for the system tends to 0 as the number of training epochs increases, regardless of initialisation. This value for a MLE indicates that the volume carried by the asymptotic ergodic measure of the system is preserved along a direction and all other directions compress; i.e., the attractor is composed of one or several limit cycles. Since the MLE was negative before training, as the networks we consider are initialised in a convergent regime [9], this shows that RNNs change their internal dynamics during training so that the attractor associated with their learned autonomous internal states settles into limit cycles (Fig. 4).

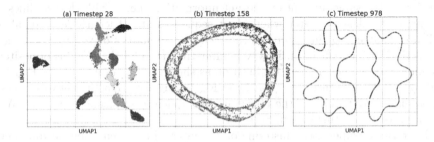

Fig. 4. Network internal representations in phase space at three different time steps. Each subplot contains 10^4 points corresponding to MNIST test image. Time step 28 corresponds to the final line of the images; further time steps correspond to blank lines added to the images where we let the neural dynamics drift.

To analyse the global geometrical characteristics of the internal representations as they vary with time, we use the UMAP visualisation [6], as shown in Fig. 4. We determine that not one, but in fact two limit cycles are used by the network to complete the task. Furthermore, the cycles are of low-dimensional nature despite residing in a high-dimensional space.

4 Summary and Conclusions

In this paper, we provide different perspectives on how information is processed in a standard RNN in the context of the sMNIST classification task, using dimensionality reduction techniques and dynamical systems theory. While the information is initially kept as multi-dimensional clusters, our work suggests that when the recurrent dynamics drift, learned internal representations are compressed into a non-trivial attractor. This attractor is composed of two limit cycles of intrinsic dimension far smaller than that of the space they reside in.

While the setup used is unconventional and highly artificial for RNNs, our analysis offers interesting insight into the network's underlying dynamics. Considering the fact that recurrent models such as RNNs, LSTMs and GRUs are defined by their dynamics, we expect the same analytical framework to be effective in other scenarios as well. A clear description of the internal dynamical system can then help us better understand the way information is encoded and decoded by these models. In this scenario we hypothesise that information about the classes associated with internal representations might be encoded in their respective phases on these limit cycles. Representations corresponding to different classes would have different phases on the limit cycles. Further work is required to verify this hypothesis and to extend our analysis to less artificial setups.

Acknowledgements. We would like to thank Aude Forcione-Lambert and Giancarlo Kerg for useful discussions. This work was partially funded by: CRM-ISM scholarships [*S.H.*, *V.G.*]; IVADO (l'institut de valorisation des données), NIH grant R01GM135929 [*G.W.*]; NSERC Discovery Grant (RGPIN-2018-04821), FRQNT Young Investigator Startup Program (2019-NC-253251), and FRQS Research Scholar Award, Junior 1 (LAJGU0401-253188) [*G.L.*] The content is solely the responsibility of the authors and does not necessarily represent the official views of the funding agencies.

References

1. Bengio, Y., Mikolov, T., Pascanu, R.: On the difficulty of training Recurrent Neural Networks. arXiv e-prints, November 2012
2. Cohen, U., Chung, S., Lee, D.D., Sompolinsky, H.: Separability and geometry of object manifolds in deep neural networks. bioRxiv (2019)
3. Crisanti, A., Sompolinsky, H.: Path integral approach to random neural networks. Phys. Rev. E **98**(6), 062120 (2018)
4. van der Maaten, L., Hinton, G.: Visualizing data using t-SNE. J. Mach. Learn. Res. **9**, 2579–2605 (2008)

5. Marquez, B.A., Larger, L., Jacquot, M., Chembo, Y.K., Brunner, D.: Dynamical complexity and computation in recurrent neural networks beyond their fixed point. Sci. Rep. **8**(1), 3319 (2018)
6. McInnes, L., Healy, J.: UMAP: Uniform manifold approximation and projection for dimension reduction. ArXiv 1802.03426 (2018)
7. Pina, J., Bodner, M., Ermentrout, B.: Oscillations in working memory and neural binding: a mechanism for multiple memories and their interactions. PLoS Comput. Biol. **14**(11), 1–30 (2018)
8. Poole, B., Lahiri, S., Raghu, M., Sohl-Dickstein, J., Ganguli, S.: Exponential expressivity in deep neural networks through transient chaos. arXiv e-prints (2016)
9. Sompolinsky, H., Crisanti, A., Sommers, H.J.: Chaos in random neural networks. Phys. Rev. Lett. **61**, 259–262 (1988)
10. Sussillo, D., Barak, O.: Opening the black box: low-dimensional dynamics in high-dimensional recurrent neural networks. Neural Comput. **25**, 626–649 (2012)

From Explicit to Implicit Entity Linking: A Learn to Rank Framework

Hawre Hosseini[(✉)] and Ebrahim Bagheri[(✉)]

LS3 Laboratory, Ryerson University, Toronto, ON, Canada
{hawre.hosseini,bagheri}@ryerson.ca

Abstract. Implicit entity linking is the task of identifying an appropriate entity whose surface form is not explicitly mentioned in the text. Unlike explicit entity linking where an entity is linked to an observed phrase within the input text, implicit entity linking is concerned with determining specific yet implied entities. Existing work in the literature have already identified appropriate features that can be used for ranking relevant entities for explicit entity linking. In this paper, we (1) consider the applicability of such features for implicit entity linking, (2) introduce features that are suited for this task, (3) compare our work with the state of the art in implicit entity linking, and (4) and report on feature importance values and present their interpretations.

1 Introduction

When producing content on social media, such as Twitter, users often refer to people, places and things without explicitly mentioning them [2–4,8]. For instance, in the tweet 'and thats why he is the King of Pop, Duke of Dance, Master of The Moonwalk...but mostly the King of our Hearts for all eternity.' Michael Jackson is the main person who is being referred to but without being mentioned. For such cases, traditional entity linking methods and named entity taggers cannot identify or link the content to an appropriate entity. According to [4], on average, 15% of tweets contain implicit mentions and according to [8], 21% of tweets in the domain of movies and 40% of tweets in the domain of books, contain implicit references to entities. This translates into a large number of information-rich content that cannot be readily processed by existing entity linking techniques. The task of *implicit entity linking* is concerned with identifying and linking such implied mentions. In this paper, we adopt a learning to rank approach for performing implicit entity linking. Our work's main motivation is that existing work on explicit entity linking have successfully adopted the learning to rank approach for identifying suitable entities [6]. These approaches rely on a collection of features that link the content space to the entity space. We systematically categorize and present features for the context of implicit entity linking. We will show that features showing strong performance on explicit linking do not necessarily have the same linking power for implicit linking. We identify most suitable features for implicit linking and show that when used in

© Springer Nature Switzerland AG 2020
C. Goutte and X. Zhu (Eds.): Canadian AI 2020, LNAI 12109, pp. 283–289, 2020.
https://doi.org/10.1007/978-3-030-47358-7_28

the context of a learning to rank approach, they can provide significantly better performance compared to the state of the art. Summarily, the contributions of our work are as follows: 1) We provide a framework for systematically categorizing features that can be used for explicit and implicit entity linking within the context of a learning to rank approach; and, 2) We examine the suitability of these features for both of the entity linking tasks and show their importance.

2   Proposed Approach

This paper rests on the foundation offered by work in the learning to rank literature, which rely on the definition of effective features for ranking relevant items for an input query. In our case, we are interested in ranking relevant entities from the knowledge graph for an input text, e.g., a tweet. Here, we introduce our different feature types and explain how they can be extracted. As seen in Table 1, the features are structured based on four main categories: term-based, string-based, graph-based (network properties) and graph-based (popularity properties). In the table, we additionally specify whether the feature is extracted based on input text (specified as T) or the target entity representation (specified as E) or both (denoted as TE). We further denote the *unit* of each feature as to the form by which the feature is extracted, which can include Unigrams (u), Bigrams (b), Unordered Bigrams (ub), entities identified using an explicit entity tagger (e), or anchor texts (a). The final column of the table indicates which features, and if so which variation, is not applicable for the task of explicit entity linking. In the following, we provide the details of each feature category.

2.1   Term-Based and String-Based Features

This category encompasses features that extract information from textual content of tweets and/or entities' textual representations. Features in this category are mainly Information Retrieval (IR)-based, such as, term frequency also discounted with inverse document frequency, sequential dependence model (SDM) [7] as well as textual similarity through cosine similarity. All of these features can be applied to both the tweet content and the entities' representations. We additionally extract three other term-based features, which are not based on IR methods, yet commonly used in entity linking: (1) *PARC* considers presence of anchors within the tweet. Presence of an entity anchor in the tweet can be an indicator for relevance to the anchor's pertinent entity. Also, it might serve as a textual reference to the target implicit mention; (2) *TitleContainsTweet*, investigates the presence of a substring of the tweet in the title of the candidate entity. This can be an effective feature in the case of explicit entity linking since surface forms of entities can appear in the text. We study the effectiveness of this feature in implicit entity linking and we hypothesize that this feature will not perform as well, since implicit references do not contain surface forms of entities; and, (3) *URLEntityCount* depends on the URLs found within the tweet. This feature extracts and additionally considers the webpage content of URLs found within a tweet and counts the number of times a candidate entity appears in them.

Table 1. Description and categorization of features proposed for entity linking.

Category	Feature name	Description	Type	Unit					Not applicable in explicit linking
				u	b	ub	e	a	
Term-based	TF	Considers frequency of tweet term in the content of entity	TE	✓	✓	✓	✓	✓	TF(e)
	TF-IDF	Considers the inverse frequency of tweet term in entities' content	TE	✓	✓	✓	✓	✓	TF-IDF(e)
	SDM	SDM model with different feature functions	TE	✓	✓	✓	✓	✓	SDM(e)
	Cosine similarity	Cosine similarity of tweet text and entity representation	TE	✓	✓	✓	✓	✓	Cosine similarity(e)
	PARC	Presence of an Anchor Referring to a Candidate Entity inside tweet	TE	–	–	–	–	✓	–
	TitleContains tweet	If title of entity contains substring of the tweet	TE	–	–	–	–	–	–
	URLEntity count	Number of times entity appears on a webpage whose URL is in the tweet	T	–	–	–	–	–	–
	ECoocKB	Co-occurrence of tweet Explicit entities with Candidate entities on KB	E	–	–	–	✓	–	ECoocKB(e)
String-based	TitleCharLength	Character length of title of the entity	E	–	–	–	–	–	–
	TitleTermCount	Number of terms in title of entity	E	–	–	–	–	–	–
Graph-based (network properties)	InkinksKB	Number of entities on a KB linking to e	E	–	–	–	✓	–	–
	OutlinksKB	Number of KB articles linking from e	E	–	–	–	✓	–	–
	CatKB	Number of categories associated with e on a KB hierarchy	E	–	–	–	✓	–	–
	Redirect	Number of redirect pages linking to e on Wikipedia	E	–	–	–	✓	–	–
	Betweenness	Betweenness measure of each candidate entity in a constructed graph	E	–	–	–	✓	–	–
	PageRank	PageRank measure of each candidate entity in Wikipedia graph	E	–	–	–	✓	–	–
	EmbedEntSimilarity	Embedding-based Similarity Measure between the candidate entity and the entities in the tweet (Cosine similarity of embeddings)	TE	–	–	–	✓	–	EmbedEnt similarity(e)
Graph-based (popularity properties)	ViewCount	Number of times {e} was visited in a specific time frame	E	–	–	–	✓	–	–
	ClickStream	The number of times Wikipedia users have navigated from a tweet explicit entity to a candidate entity	TE	–	–	–	✓	–	ClickStream(e)

On the other hand, string-based features consists of two features, namely *TitleCharLength* and *TitleTermCount*, both of which are primarily syntactic features. The first feature calculates the number of characters in the entity title and the second counts the number of terms in the entity title. These features can be important for the explicit entity linking task given significantly longer entity titles have a lower likelihood of appearing within a text.

2.2 Graph-Based Features

Network Properties. The most common form of network measures focus on centrality of nodes. In the context of a knowledge graph, node centrality can indicate the importance and/or relevance of the content represented by that node within the graph. As such, we adopt two widely used centrality measures, namely Betweenness Centrality and PageRank. Furthermore, we introduce additional locally defined features based on the neighborhood of an entity within the knowledge graph. For instance, we measure how many Wikipedia categories are associated with the entity, or how many inbound and outbound links are connected to the entity of interest. We consider such features as an indication of the extent to which a given entity is involved in relationships with other entities. We also employ entity representations that have been trained based on neural networks on the structure of the knowledge graph [9] to compute the similarity between the input tweet and the target entity from the knowledge graph. The neural representations of entities capture geometric relations between entities given their proximity and position to each other on the knowledge graph and can hence be an indication for the relevance of the tweet and the target entity.

Popularity Properties. Features in this category depend on the meta-data associated with the knowledge graph that are collected from external sources. We introduce two features in this category. *ViewCount* takes into account the number of times a specific entity was visited by viewers during a certain time frame. This feature aims at capturing hotness of entities in the real world at different times. Moreover, we introduce a novel feature denoted *ClickStream*, which captures the way users navigate on Wikipedia. This feature is extracted from the Wikipedia metadata and records how many times each linked entity on a specific entity's Wikipedia page has been clicked. This stream of clicks is hypothesized to show the different levels of relevance between an entity and other reachable entities. The feature takes explicit entities within the tweet and the target entity and calculates the click frequency between them.

3 Datasets and Experimental Setup

For implicit entity linking, we exploit the dataset introduced in [4], which contains $1,345$ tweets with implicit mentions. The dataset's taxonomy contains 6 coarse-grained entity types, namely Person, Organization, Location, Product, Event, and Work. In this dataset, every tweet is labelled with one target entity.

For explicit entity linking, we use the dataset provided by [6] consistinng 318 available tweets, with an average of 2.22 mentions per tweet. In our ranking problem, we consider each tweet-entity pair as one training instance resulting in a total of 707 samples. Here, for the sake of reproduciblity, we clearly describe the process for extracting the introduced features. For features requiring identification of explicit entities within the tweet, we employ TagMe entity tagger. For Wikipedia textual content, we extract entities by processing Wikipedia dumps. In order to extract entity inlinks, outlinks, and redirects and the number of categories associated with each entity, we exploit the Wikipedia API. For EmbedEntSimilarity, we use embeddings trained by Li et al. [5]. We extract Betweenness Centrality and PageRank from a graph that we construct from the DBpedia RDF dump. Finally, in order to extract PARC, we build a mapping from anchors on Wikipedia to entities, extracted by processing Wikipedia dump's textual content. To build the rankers, we exploit SVM^{rank} model trained with features described in Table 1. The choice of SVM^{rank} is motivated by the fact that SVM^{rank} has been shown to perform well in ranking problems similar to ours [6]. The specification of our model, identical to the baseline, is as follows: linear kernel, 0.01 as the trade-off between training error and margin, and the loss function is the number of swapped pairs summed over all inputs.

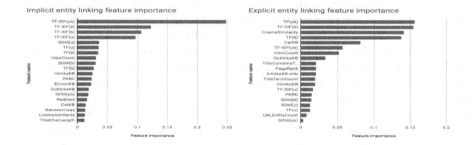

Fig. 1. Feature importance ranked by value for implicit and explicit entity linking.

Table 2. P@1 of this work for implicit entity linking as compared to baselines. The average is calculated with weighting of ratio of queries in each domain.

	Person	Organization	Location	Event	Product/Device	WrittenWork	Film	Average
This work (Implicit entity linking)	66.37	62.6	62.02	77.77	70.58	74	76.78	67.53
Hosseini et al. 2019-a	59.82	61.23	58.25	54.09	67.63	72.97	76.43	64.34
Perera et al. 2016	49.6	49	49.8	50.4	48.9	61.05	60.97	52.81

4   Results and Discussion

We report the overall results for the implicit entity linking task in Table 2 and compare performance of the introduced features against two state of the art baselines. As seen, the proposed features when employed within the context of a learning to rank approach show improved performance compared to the two baselines over all entity domains. To evaluate the importance of features, we draw upon their Gini scores, as reported in Fig. 1. Literature on explicit entity linking report that graph-based features such as ViewCount alone are enough for successfully performing linking [1]. This point is reassured in our experiments as well; as seen in Fig. 1, CatKB, ViewCount, and OutlinksKB are among the top 10 best performing features for explicit entity linking. In case of implicit entity linking, however, only one of the graph-based features, i.e., ViewCount, is to be found among the top 10. A lower feature importance for Graph-based features for implicit entity linking as compared to explicit linking shows the difference between the two tasks. We further perform experiments with different groups of features as categorized in our work. We run our systems with features of the following four groups: Term-based and String-based (we combine these two categories as string-based features alone do not produce any noticeable results), Graph-based (popularity-based), Graph-based (network-based), and Graph-based (combined), i.e., the combination of popularity-based and network-based features; results are reported in Table 3. As seen, there are significant differences between feature performance for implicit and explicit entity linking tasks. For implicit linking there is significant difference between the performance of term-based features and that of the graph-based features. However, such a difference is not noticeable for explicit linking. Most specifically, we find that: (1) Term-based features are the most discriminative features for performing implicit entity linking. This is because those terms that appear in the input text, e.g. tweet, have close resemblance to the textual representation of the target entity. While strong features, these term-based features are not as effective for explicit entity linking; (2) Graph-based popularity features are quite effective for implicit entity linking. This can be in part due to the fact that users often use implicit mentions when they believe their audience can understand the implicit reference. Such identifiable entities are often those which have become 'hot' in the social sphere or widely mentioned by the community. As such, graph-based popularity features that capture these characteristics are effective. On the other hand, these features are not useful for explicit entity linking at all. (3) On the contrary, graph-based network features are quite effective for explicit entity linking. This can be explained by the fact that network measures determine the importance of entities that form effective priors for the likelihood of that entity being mentioned in text. When explicitly mentioned, these priors accurately

Table 3. P@1 of implicit and explicit entity linking with subsets of features.

	Term and string-based	Graph-based (popularity)	Graph-based (network)	KB (combined)
Implicit linking	70.58	41.17	5.88	47.05
Explicit linking	31.62	7.26	25.21	27.35

estimate the likelihood of the entity to be mentioned. However, when discussing implicit mentions, these priors are not accurate but rather priors based on popularity of entities are more accurate; and, (4) Finally, we find that popularity and network features have reinforcing effect on each other for implicit entity linking and as such, it is helpful to include features from both categories when building an implicit entity linker. On the other hand, these features have an overlapping effect on each other for explicit entity linking and as such the inclusion of only network-based features seems to be a better strategy.

References

1. Guo, Y., Che, W., Liu, T., Li, S.: A graph-based method for entity linking. In: IJCNLP, pp. 1010–1018 (2011)
2. Hosseini, H.: Implicit entity recognition, classification and linking in tweets. In: The 42nd International ACM SIGIR, pp. 1448–1448 (2019)
3. Hosseini, H., Nguyen, T.T., Bagheri, E.: Implicit entity linking through ad-hoc retrieval. In: ASONAM, pp. 326–329. IEEE (2018)
4. Hosseini, H., Nguyen, T.T., Wu, J., Bagheri, E.: Implicit entity linking in tweets: an ad-hoc retrieval approach. Appl. Ontol. (Preprint) 14, 1–27 (2019)
5. Li, Y., Zheng, R., Tian, T., Hu, Z., Iyer, R., Sycara, K.: Joint embedding of hierarchical categories and entities for concept categorization and dataless classification. In: Proceedings of COLING (2016)
6. Meij, E., Weerkamp, W., De Rijke, M.: Adding semantics to microblog posts. In: Proceedings of WSDM, pp. 563–572. ACM (2012)
7. Metzler, D., Croft, W.B.: Latent concept expansion using markov random fields. ACM SIGIR 2007, 311–318 (2007)
8. Perera, S., Mendes, P.N., Alex, A., Sheth, A.P., Thirunarayan, K.: Implicit entity linking in tweets. In: European Semantic Web Conference, pp. 118–132 (2016)
9. Wang, Q., Mao, Z., Wang, B., Guo, L.: Knowledge graph embedding: a survey of approaches and applications. TKDE 29(12), 2724–2743 (2017)

Automatic Polyp Segmentation
Using Convolutional Neural Networks

Sara Hosseinzadeh Kassani[1]([✉])[ID], Peyman Hosseinzadeh Kassani[2],
Michal J. Wesolowski[3], Kevin A. Schneider[1], and Ralph Deters[1]

[1] Department of Computer Science, University of Saskatchewan, Saskatoon, Canada
sara.kassani@usask.ca
[2] Department of Neurology and Neurological,
University of Stanford, Stanford, CA, USA
[3] Department of Medical Imaging, University of Saskatchewan, Saskatoon, Canada

Abstract. Colorectal cancer is the third most common cancer-related death after lung cancer and breast cancer worldwide. The risk of developing colorectal cancer could be reduced by early diagnosis of polyps during a colonoscopy. Computer-aided diagnosis systems have the potential to be applied for polyp screening and reduce the number of missing polyps. In this paper, we compare the performance of different deep learning architectures as feature extractors, i.e. ResNet, DenseNet, InceptionV3, InceptionResNetV2 and SE-ResNeXt in the encoder part of a U-Net architecture. We validated the performance of presented ensemble models on the CVC-Clinic (GIANA 2018) dataset. The DenseNet169 feature extractor combined with U-Net architecture outperformed the other counterparts and achieved an accuracy of 99.15%, Dice similarity coefficient of 90.87%, and Jaccard index of 83.82%.

Keywords: Convolutional neural networks · Polyp segmentation · Colonoscopy images · Computer-aided diagnosis · Encoder-decoder

1 Introduction

Colorectal cancer is the third most common cancer-related death in the United States in both men and women. According to the annual report provided by American cancer society [3], approximately 101,420 new cases of colon cancer and 44,180 new cases of rectal cancer will be diagnosed in 2019. Additionally, 51,020 patients are expected to die from colorectal cancer during 2019 in the United States. Most colorectal cancers start as benign polyps in the inner linings of the colon or rectum. Removal of these polyps can decrease the risk of developing cancer. Colonoscopy is the gold standard for screening and detecting polyps [5]. Screening and analysis of polyps in colonoscopy images is dependent on experienced endoscopists [21]. Polyp detection is considered as a challenging task due to the variations in size and shape of polyps among different patients. This is illustrated in Fig. 1, where the segmented regions vary in size, shape and position.

C. Goutte and X. Zhu (Eds.): Canadian AI 2020, LNAI 12109, pp. 290–301, 2020.
https://doi.org/10.1007/978-3-030-47358-7_29

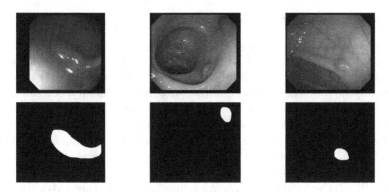

Fig. 1. Some examples of polyps from colonoscopy images (first row) and their corresponding manual segmentations provided by expert endoscopists (second row).

The miss rates of smaller polyps during the colonoscopy is also another issue that needs to be addressed. Developing computer-aided diagnosis (CAD) systems can assist physicians in the early detection of polyps. CAD systems using convolutional neural networks (CNN) is an active research area and has the potential to reduce polyp miss rate [20]. Recent developments based on the application of deep learning-based techniques achieved promising results for the segmentation and extraction of polyps and improved the detection rate, despite the complexity of the case during colonoscopy [12,17,19,26]. The presence of visual occlusions such as shadows, reflections, blurriness and illumination conditions, as shown in Fig. 2 can adversely affect the performance of CNN and the quality of the segmented polyp region.

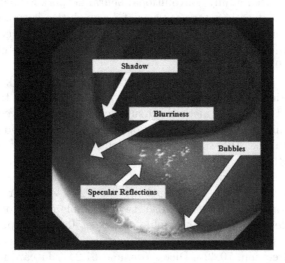

Fig. 2. Examples of different noises exist in colonoscopy images.

1.1   Motivation and Contributions

The main motivation of this paper is to compare the performance of different CNN modules, i.e. Squeeze-and-Excitation (SE) blocks, inception blocks, residual blocks and dense blocks for building automatic polyp segmentation systems. Considering the problem of intra- and inter-observer variability, designing hand-crafted features with limited representation capability requires expert knowledge and extensive application-specific fine-tuning. Also, employing very deep networks for small data samples suffers from gradient vanishing and poor local minima issues. In this study, we evaluate the performance of different CNN architectures (e.g., ResNet [10], DenseNet [14], InceptionV3 [24], InceptionResNet [23], SE-ResNeXt [11]) with various modules as feature extractor to the encoder part of a U-Net architecture to investigate the impact of incorporating modules in extracting high-level contextual information from the input image. In this way, we provide better insights on how different convolutional pathways efficiently incorporate both local and contextual image information for training producers and cope with the inherent variability of medical data. We validated the performance of presented ensemble models using the CVC-ClinicDB (GIANA 2018) dataset.

The rest of this study is organized as follows. Section 2 provides the related work in the literature on polyp segmentation approaches. Section 3 presents a detailed description of materials and the methodology. Section 4 describes experimental analysis and discussion of the performance of the segmentation models. Finally, Sect. 5 concludes the paper and provides future directions.

2   Related Work

Li et al. [17] presented a fully convolutional neural network for polyp segmentation. The feature extraction stage consists of 8 convolution layers, and 5 pooling layers. The presented method evaluated on CVC-ClinicDB. Li et al. approach obtained an accuracy of 96.98%, f1-score of 83.01%, sensitivity of 77.32% and specificity of 99.05%.

Akbari et al. [4] applied a fully convolutional neural network (FCN-8S) for polyp segmentation. An image patch selection method used for training procedure. Also, a post-processing method (Otsu thresholding) employed on the probability map to improve the performance of the proposed method on the CVC-ColonDB [1] database. Akbari et al. method achieved an accuracy of 97.70% and a Dice score of 81.00%.

Qadir et al. [21] trained a Mask R-CNN with different CNN architectures (Resnet50, Resnet101 and InceptionResnetV2) as a feature extractor for polyp detection and segmentation. Also, two ensemble models of (ensemble of Resnet50 and Resnet101) and (ensemble of Resnet50 and InceptionResnetV2) were employed on CVC-ColonDB dataset. Qadir's approach achieved 72.59% recall, 80.00% precision, 70.42% Dice score, and 61.24% Jaccard index.

Nguyen and Lee [19] used a deep encoder-decoder network method for polyp segmentation from colonoscopy images. The presented encoder-decoder structure consists of atrous convolution and depthwise separable convolution. To improve the performance, the proposed model pre-trained with the VOC 2012 dataset and achieved 88.9% of Dice score and 89.35% of Jaccard index on the CVC-ColonDB database.

Kang and Gwak [15] employed Mask R-CNN to segment polyp regions in colonoscopy images. Also, an ensemble Mask R-CNN model with different backbone structures (ResNet50 and ResNet101) was adopted to further improve the model performance. The Mask R-CNN was first trained on the COCO dataset and then fine-tuned for polyp segmentation. Three datasets, i.e. CVC-ClinicDB, ETIS-Larib, and CVC-ColonDB, used to measure the performance of the proposed model. The best result achieved on the CVC-ColonDB dataset with 77.92% mean pixel precision, 76.25% mean pixel recall and 69.4% intersection over the union.

3   Methods and Materials

3.1   Experimental Dataset

In this paper, CVC-ClinicDB [7,8,25] database, publicly available at [2], is used to validate the performance of the presented method. The database consists of 300 Standard Definition (SD) colonoscopy images with a resolution of 574×500 pixels, and each image contains one polyp. Each frame has a corresponding ground truth of the region covered by the polyp.

3.2   Data Pre-processing

Resizing: Regarding to the black margin of each image as illustrated in Fig. 3, we center-cropped all images of SD-CVC-ClinicDB from the original size of 574×500 pixels to the appropriate size 500×500 pixels using bicubic interpolation to reduce the non-informative adjacent background regions.

Data Augmentation: Recent works have demonstrated the advantages of data augmentation methods in extending the size of training data to cover all of the data variances. In this regard, various data augmentation techniques such as horizontal and vertical flipping, rotating and zooming are applied to enlarge the dataset and aid to successfully accomplish segmentation task. Figure 3 shows the examples of the original polyp image (Fig. 3.a) after applying different data augmentation methods. The used methods of augmentation are vertical flipping (Fig. 3.b), horizontal flipping (Fig. 3.c), random filter such as blur, sharpen (Fig. 3.d), random contrast by a factor of 0.5 (Fig. 3.e), and finally, random brightness by a factor of 0.5 (Fig. 3.f).

Z-Score Normalization: To have a uniform distribution from input images and remove bias from input features, we re-scaled the intensity values of the input images to have a zero mean and a standard deviation of one to standardize the input images.

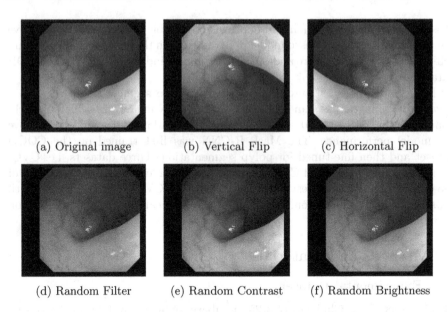

(a) Original image   (b) Vertical Flip   (c) Horizontal Flip

(d) Random Filter   (e) Random Contrast   (f) Random Brightness

Fig. 3. Examples of data augmentation methods.

Image Normalization: Before feeding images into the CNN models, we also normalize the intensity values of input images using ImageNet mean subtraction [16]. The ImageNet mean is a pre-computed constant derived from ImageNet [9] database.

3.3   Feature Extraction Using Transfer Learning Strategy

The intuition behind transfer learning is that knowledge learned by a cross-domain dataset transfer into the new dataset in another domain. The main advantages of transfer learning are the improvement of the network performance, reducing the issue of over-fitting, reducing the computational cost, and also the acceleration of the convergence of the network [18]. In this approach, instead of training a model from scratch, the weights trained on ImageNet dataset or other similar cross-domain dataset is used to initialize weights for the current task. Providing training data large enough to sufficiently train a CNN model is limited due to privacy concerns, which is a common issue in the medical domain. To address the issue of insufficient training samples, transfer learning strategy has also been widely used for accurate and automatic feature extraction in developing various CAD systems.

3.4   Ensemble Method

U-Net Architecture. U-Net, proposed by Ronneberger et al. [22] in 2015, is an encoder-decoder convolutional network that won ISBI cell tracking challenge.

The encoder or down-sampling layers of U-Net architecture learn the feature maps and the decoder or up-sampling layers provide precise segmentation. The encoder part has alternating convolutional filters and max-pooling layers with ReLU activation function to down-sample the data. When the input image is fed into the network, representative features are produced by convolutions at each layer.

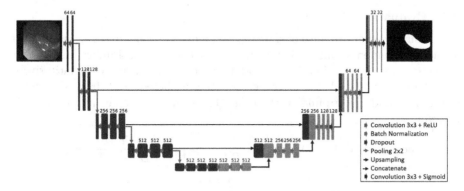

Fig. 4. Proposed Approach for polyp segmentation. The CNN network is based on encoder-decoder of U-Net architecture with an encoder of pre-trained VGG16 as an example.

Pre-trained CNN Feature Extractors. For the down-sampling part of the U-Net architecture, different deep CNN-based feature extractors were selected to extract high-level features from the input image. The choice of the feature extractor is based on different modules incorporated into the associated CNN models that successfully achieve the best segmentation performance in the literature. In this study, we selected five Deep CNN architectures as feature extractors, namely ResNet, DenseNet, InceptionV3, InceptionResNetV2 and Squeeze-and-Excitation Networks (SE-ResNeXt), to compare their performance in polyp segmentation task. Residual blocks in ResNet architecture consists of two or three sequential convolutional layers and a supplementary shortcut connection. This shortcut connection adds the output of the previous layer to the output of the next layer, enabling to pass the signal without modification. This architecture helps reduce the degradation of the gradient in deep networks. The inception module creates wider networks rather than deeper by adding filters of three different sizes (1×1, 3×3, and 5×5) and an additional max-pooling layer. The output is then concatenated together and is sent to the next inception module. Also, before 3×3 and 5×5 convolutions, an extra 1×1 convolution is added to limit the number of input channels. In Dense modules, the previous layer is merged into the future layer by concatenation, instead of using shortcut connections as in ResNet modules. In the Dense module, all feature maps from a layer are connected to all other subsequent layers. SE-ResNeXt introduced an operation that can adaptively recalibrate channel-wise feature responses of each feature map. SE-ResNeXt is the integration of ResNet into squeeze-and-excitation blocks to further improve the accuracy of the network. Inception-ResNet is a hybrid of the Inception architecture with residual connections to boost the representational power of the network. The proposed CNN network based on U-Net architecture with a pre-trained VGG16 feature extractor is illustrated in Fig. 4.

3.5   Evaluation Criteria

To measure the performance of the proposed method for polyp segmentation, we employed common segmentation evaluation metrics: Jaccard index, also known as intersection over union (IoU), and Dice similarity score to quantitatively measure similarity and difference between the predicted mask from segmentation model and the ground-truth mask. These metrics are computed by the following:

$$Jaccard\ index\,(A, B) = \frac{|\ A \cap B\ |}{|\ A \cup B\ |} = \frac{|\ A \cap B\ |}{|\ A\ | + |\ B\ | - |\ A \cap B\ |} \tag{1}$$

$$Dice\,(A, B) = \frac{2 \times |\ A \cap B\ |}{|\ A\ | + |\ B\ |} \tag{2}$$

Where A represents the output binary mask, produced from the segmentation method and B represents the ground-truth mask, $\cup$ represents union set between A and B, and $\cap$ represents the intersection set between A and B.

We also used accuracy to measure the overall accuracy of the segmentation models (binary classification). A high accuracy demonstrates that most of the polyp pixels were classified correctly.

$$Accuracy = \frac{TP + TN}{TP + TN + FP + FN} \tag{3}$$

True Positive (TP) represents the number of correctly predicted pixels as polyp. False Positive (FP) represents misclassified background pixels as polyp. False Negative (FN) represents misclassified polyp pixels that misclassified as background, and True Negative (TN) represents the background pixels that are correctly classified as background.

4   Experiments and Results

4.1   Experimental Setup

For this study, we randomly selected 80% of the CVC-ClinicDB images as the training and validation set and the remaining 20% for the test set. There is no intersection between the training and test images. To update the weight, we used Adam optimizer with a learning rate, $\beta 1$ and $\beta 2$ of 10-5, 0.9, 0.999, respectively. The batch size was set to 2, and all models were trained for 50 epochs. Our experiment is implemented in Python using Keras package with Tensorflow as backend and run on Nvidia GeForce GTX 1080 Ti GPU with 11 GB RAM.

4.2 Results and Discussion

The accuracy, Dice score and Jaccard index of the obtained results are summarized in Table 1. There is a level of variation in the performance of all models. Analyzing Table 1, U-Net with DenseNet169 backbone feature extractor outperformed the other approaches, where the U-Net with InceptionResNetV2 backbone feature extractor achieved the second-best results with a slightly lower performance rate. We believe that dense modules, inception modules and also residual blocks as part of U-Net encoder provide an efficient segmentation process and overcome the issue of over-segmentation [13]. U-Net with DenseNet169 achieved an accuracy of 99.15% in comparison to 99.10% for InceptionResNetV2 architecture. Also, Dice score for DenseNet169 model was 90.87% compared to 90.42% for InceptionResNetV2. DenseNet169 also had better results for Jaccard index, 83.82% compared to 83.16% for InceptionResNetV2 architecture.

Table 1. Evaluation of the segmentation results from different combinations of the pre-trained feature extractors and U-Net architecture.

	Accuracy (%)	DICE (%)	Jaccard index (%)
Baseline U-Net [22]	97.92	75.86	63.53
SegNet [6]	95.12	68.39	61.57
U-Net+ResNet34	98.09	88.08	79.22
U-Net+ResNet50	98.77	86.06	77.62
U-Net+ResNet152	98.9	87.67	79.22
U-Net+DenseNet121	98.72	85.42	77.35
U-Net+DenseNet169	**99.15**	**90.87**	**83.82**
U-Net+DenseNet201	98.85	87.54	80.2
U-Net+InceptionV3	99.08	89.63	81.84
U-Net+InceptionResNetV2	99.1	90.42	83.16
U-Net+SE-ResNeXt50	98.79	86.61	79.05
U-Net+SE-ResNeXt101	98.9	87.63	80.09

To justify the performance of the ensemble architectures, the performance of baseline U-Net and SegNet architectures are also evaluated and compared with the presented approach. The worst performance is for SegNet with a Jaccard index of 61.57%, Dice score of 68.39%, and accuracy of 95.12%. U-Net with DenseNet169 significantly improves the baseline U-Net up to 15.01%, and the baseline SegNet architecture up to 22.48% in terms of Dice score. Moreover, U-Net with DenseNet169 improves baseline U-Net up to 20.29% and the SegNet architecture up to 22.25% in terms of Jaccard index. Similar conclusions can be drawn for accuracy metrics. The experimental results indicate the important role of incorporating modules in encoder part of a convolutional segmentation architecture in extracting hierarchical information from input images.

Table 2. Comparison of performance of polyp segmentation models on the CVC-ClinicDB dataset.

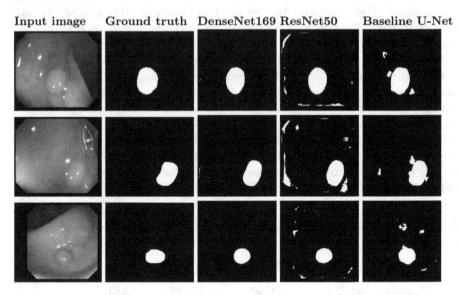

| Input image | Ground truth | DenseNet169 | ResNet50 | Baseline U-Net |

In Table 2, we illustrate three segmentation output results produced by the DenseNet169, ResNet50 and the baseline U-Net model. As the results indicate, the examples selected in the column of DenseNet169 can accurately segment polyps from the background. Also, DensNet169 feature extractor can adequately address different noises present in the input images, including shadows, reflection and blurriness, etc. It should be noted that feature extractors such as ResNet50 and baseline U-Net suffer from over-segmentation. Over-segmentation affects the Dice score and Jaccard index adversely. The main cause of over-segmentation is the low-intensity variations between the foreground and the background and also the lack of enough spatial information. Dense and InceptionResNet modules can eliminate the over-segmentation and effectively segment out polyps with a better performance rate than other models, as demonstrated in Table 2. Table 3 compares the performance of the proposed methods with that of [4,15,17,19, 21]. The obtained results were comparable with prior CNN-based methods in the literature, as shown in Table 3. Both DenseNet169 and InceptionResNetV2 methods show better performance when compared with the existing methods. However, Nguyen and Lee's approach achieved better results in terms of Jaccard index while the Dice score and accuracy of DenseNet169 and InceptionResNetV2 models outperformed those of Nguyen and Lee's approach.

Table 3. Quantitative comparison of the segmentation results with prior CNN-based works on the polyp segmentation task.

Model	Jaccard index (%)	Dice score (%)	Accuracy (%)
Li et al. [17]	-	-	96.98
Akbari et al. [4]	-	81	97.7
Qadir et al. [21]	61.24	70.42	
Nguyen and Lee [19]	89.35	88.9	-
Kang and Gwak [15]	69.46	-	-
U-Net+DenseNet169	**83.82**	**90.87**	**99.15**
U-Net+InceptionResNetV2	83.16	90.42	99.1

5 Conclusion

In this work, we presented a transfer learning-based encoder-decoder architecture for automated polyp segmentation. The proposed framework consists of a U-Net architecture with different backbone feature extractors, i.e. ResNet, DenseNet, InceptionV3, InceptionResNetV2 and SE-ResNeXt. Our method is validated using a dataset from the CVC-ClinicDB polyp segmentation challenge. The experimental results showed that the proposed ensemble method using DenseNet169 and InceptionResNetV2 feature extractors achieved good results and significantly outperformed the baseline U-Net, and SegNet approaches for polyp segmentation. The main limitation of this work is the limited number of polyp shapes and structures present in the provided dataset, which is a focus of future work. By adding more training samples from external datasets, the deep learning-based segmentation models could gain a better performance and further improve the generalization ability of the network. Our future work will also be dedicated to the investigation of the post-processing methods to reduce the over-segmentation issue.

References

1. CVC-ColonDB. http://mv.cvc.uab.es/projects/colon-qa/cvccolondb
2. GianaDataset. https://giana.grand-challenge.org/PolypSegmentation/
3. Key Statistics for Colorectal Cancer. https://www.cancer.org/cancer/colon-rectal-cancer/about/key-statistics.html
4. Akbari, M., et al.: Polyp segmentation in colonoscopy images using fully convolutional network. In: 2018 40th Annual International Conference of the IEEE Engineering in Medicine and Biology Society (EMBC), pp. 69–72. IEEE (2018)
5. Akbari, M., et al.: Classification of informative frames in colonoscopy videos using convolutional neural networks with binarized weights. In: 2018 40th Annual International Conference of the IEEE Engineering in Medicine and Biology Society (EMBC), pp. 65–68. IEEE (2018)

6. Badrinarayanan, V., Kendall, A., Cipolla, R.: SegNet: a deep convolutional encoder-decoder architecture for image segmentation. IEEE Trans. Pattern Anal. Mach. Intell. **39**(12), 2481–2495 (2017)
7. Bernal, J., Sánchez, F.J., Fernández-Esparrach, G., Gil, D., Rodríguez, C., Vilariño, F.: WM-DOVA maps for accurate polyp highlighting in colonoscopy: validation vs. saliency maps from physicians. Comput. Med. Imag. Graph. **43**, 99–111 (2015)
8. Bernal, J., Sánchez, J., Vilarino, F.: Towards automatic polyp detection with a polyp appearance model. Pattern Recogn. **45**(9), 3166–3182 (2012)
9. Deng, J., Dong, W., Socher, R., Li, L.J., Li, K., Fei-Fei, L.: ImageNet: a large-scale hierarchical image database. In: 2009 IEEE Conference on Computer Vision and Pattern Recognition, pp. 248–255. IEEE (2009)
10. He, K., Zhang, X., Ren, S., Sun, J.: Deep residual learning for image recognition. In: Proceedings of the IEEE Conference on Computer Vision and Pattern Recognition, pp. 770–778 (2016)
11. Hu, J., Shen, L., Sun, G.: Squeeze-and-excitation networks. In: Proceedings of the IEEE Conference on Computer Vision and Pattern Recognition, pp. 7132–7141 (2018)
12. Huang, C.H., Xiao, W.T., Chang, L.J., Tsai, W.T., Liu, W.M.: Automatic tissue segmentation by deep learning: from colorectal polyps in colonoscopy to abdominal organs in CT exam. In: 2018 IEEE Visual Communications and Image Processing (VCIP), pp. 1–4. IEEE (2018)
13. Huang, D., Lai, J.H., Wang, C.D., Yuen, P.C.: Ensembling over-segmentations: from weak evidence to strong segmentation. Neurocomputing **207**, 416–427 (2016)
14. Huang, G., Liu, Z., Van Der Maaten, L., Weinberger, K.Q.: Densely connected convolutional networks. In: Proceedings of the IEEE Conference on Computer Vision and Pattern Recognition, pp. 4700–4708 (2017)
15. Kang, J., Gwak, J.: Ensemble of instance segmentation models for polyp segmentation in colonoscopy images. IEEE Access **7**, 26440–26447 (2019)
16. Krizhevsky, A., Sutskever, I., Hinton, G.E.: ImageNet classification with deep convolutional neural networks. In: Advances in Neural Information Processing Systems, pp. 1097–1105 (2012)
17. Li, Q., et al.: Colorectal polyp segmentation using a fully convolutional neural network. In: 2017 10th International Congress on Image and Signal Processing, BioMedical Engineering and Informatics (CISP-BMEI), pp. 1–5. IEEE (2017)
18. Lu, S., Lu, Z., Zhang, Y.D.: Pathological brain detection based on alexnet and transfer learning. J. Comput. Sci. **30**, 41–47 (2019)
19. Nguyen, Q., Lee, S.W.: Colorectal segmentation using multiple encoder-decoder network in colonoscopy images. In: 2018 IEEE First International Conference on Artificial Intelligence and Knowledge Engineering (AIKE), pp. 208–211. IEEE (2018)
20. Qadir, H.A., Balasingham, I., Solhusvik, J., Bergsland, J., Aabakken, L., Shin, Y.: Improving automatic polyp detection using CNN by exploiting temporal dependency in colonoscopy video. IEEE J. Biomed. Health Inform. **24**, 180–193 (2019)
21. Qadir, H.A., Shin, Y., Solhusvik, J., Bergsland, J., Aabakken, L., Balasingham, I.: Polyp detection and segmentation using mask R-CNN: does a deeper feature extractor cnn always perform better? In: 2019 13th International Symposium on Medical Information and Communication Technology (ISMICT), pp. 1–6. IEEE (2019)

22. Ronneberger, O., Fischer, P., Brox, T.: U-Net: convolutional networks for biomedical image segmentation. In: Navab, N., Hornegger, J., Wells, W.M., Frangi, A.F. (eds.) MICCAI 2015. LNCS, vol. 9351, pp. 234–241. Springer, Cham (2015). https://doi.org/10.1007/978-3-319-24574-4_28

23. Szegedy, C., Ioffe, S., Vanhoucke, V., Alemi, A.A.: Inception-v4, inception-ResNet and the impact of residual connections on learning. In: Thirty-First AAAI Conference on Artificial Intelligence (2017)

24. Szegedy, C., Vanhoucke, V., Ioffe, S., Shlens, J., Wojna, Z.: Rethinking the inception architecture for computer vision. In: Proceedings of the IEEE Conference on Computer Vision and Pattern Recognition, pp. 2818–2826 (2016)

25. Vázquez, D., et al.: A benchmark for endoluminal scene segmentation of colonoscopy images. J. Healthcare Eng. **2017** (2017)

26. Wickstrøm, K., Kampffmeyer, M., Jenssen, R.: Uncertainty modeling and interpretability in convolutional neural networks for polyp segmentation. In: 2018 IEEE 28th International Workshop on Machine Learning for Signal Processing (MLSP), pp. 1–6. IEEE (2018)

Augmented Out-of-Sample Comparison Method for Time Series Forecasting Techniques

Igor Ilic[1(✉)], Berk Gorgulu[2], and Mucahit Cevik[1]

[1] Ryerson University, Toronto, Canada
{iilic,mcevik}@ryerson.ca
[2] University of Toronto, Toronto, Canada
bgorgulu@mie.utoronto.ca

Abstract. Time series data consists of high dimensional sets of observations with strong spatio-temporal relations. Accordingly, conventional methods for comparing different regression methods, such as random train-test splits, do not sufficiently evaluate time series forecasting tasks. In this work, we introduce a robust technique for out-of-sample forecasting that takes the spatio-temporal nature of time series into account. We compare well-known auto-regressive integrated moving average (ARIMA) models with recurrent neural network (RNN) based models using Turkish electricity data. We observe that RNN-based models outperform ARIMA models. Moreover, as the length of forecast interval increases, the performance gap widens between these two approaches.

Keywords: Time series · Performance measures · Deep learning · ARIMA

1 Introduction

Time series, $\{y_t\}$, are high dimensional sets of observations with strong spatio-temporal relations. They can be broken down into time-dependent deterministic component, Ω_t, and random components, ϵ_t, and are represented as follows:

$$y_t = \Omega_t + \epsilon_t$$

The goal of prediction models is to identify as much of the deterministic component, Ω_t, without overfitting and capturing random noise. With the amount of available modern tools, there is not enough analysis on the accuracy of a given prediction model. Due to strong spatio-temporal relations between observations, time series forecasting can not be treated as a regular regression problem. Learning the underlying deterministic component (Ω_t) requires the capabilities of modeling these spatio-temporal relations. Therefore, evaluation of different time series forecasting algorithms requires special attention. The de facto evaluation methods in regular regression tasks, such as random train-test splits, are

© Springer Nature Switzerland AG 2020
C. Goutte and X. Zhu (Eds.): Canadian AI 2020, LNAI 12109, pp. 302–308, 2020.
https://doi.org/10.1007/978-3-030-47358-7_30

no longer suitable for time series forecasting. In this work, we introduce a new approach to compare models, known as *an augmented out-of-sample model* comparison method. This method is able to accurately compare different models on a dataset, with a higher degree of certainty.

Current comparison techniques for time series models such as one shot comparison and random interval testing lack robustness and the testing usually does not correctly reflect real-world situations. Our augmented out-of-sample model comparison approach alleviates these issues by providing a more flexible and robust technique. To demonstrate the effectiveness of the proposed approach, we compared numerous models using highly seasonal Turkish electricity consumption data. We find that recurrent neural network (RNN) architectures outperform classical algorithms in the associated forecasting task. Moreover, the gap widens as the forecast interval grows.

2 Background

2.1 Time-Series Prediction

Earlier studies on time series forecasting focus on linear prediction models such as auto regressive (AR), moving average (MA) and auto-regressive integrated moving average (ARIMA) which predict future values based on a linear function of past observations [2]. Recent advances in artificial neural networks and deep learning allowed researchers to utilize deep structures in time series forecasting. Several recent studies focus on Long Short Term Memory Neural Networks (LSTM) and Gated Reccurent Units (GRU) architecture for forecasting problems in various fields including electricity demand prediction [3,8]. Further, a recently developed GluonTS package provides a comprehensive deep-learning-based time series modeling environment [1].

2.2 Time-Series Model Comparison Techniques

Commonly used methods for comparing time-series prediction models include single forecast testing [7], multiple dataset testing [9] and random test interval sampling [6]. The strengths to single interval tests is the simplicity of implementation as well as the translation to real-world tests. However, this approach allows for lucky one-shot tests to determine the most accurate method. This issue might be averted by expanding the algorithm to predict the future (data points) for multiple datasets. One pitfall to multiple dataset testing is that it is computationally expensive because of the effort to find numerous datasets and training each model on each dataset. As an alternative, random test interval sampling allows all tests to be done on the same dataset, hence there is no need for multiple datasets to have multiple tests. However, disadvantage is that we are no longer predicting the immediate future, as required in most practical applications. As well, we lose the ability to compare against algorithms that are data dependent such as ARIMA, which can not make a prediction at some arbitrary point in the future without the preceding residuals.

To remedy aforementioned issues, Tashman et al. [10] proposed an augmented training approach. The idea is to first train over the dataset, $train_1$, then test the trained model on a varying interval length of the next time period $test_1$. Next, the model's hyper parameters are tuned, the entire model is retrained on the initial test dataset $train_1 + test_1 = train_2$, and then the model is tuned on the next test interval $test_2$. This process is repeated until some stopping condition is met. This method has shown to be a fast ad hoc way to train a model with the best hyper parameters, as well as a way to determine the performance of a single model. Figure 1 provides a visualization of the augmented training approach.

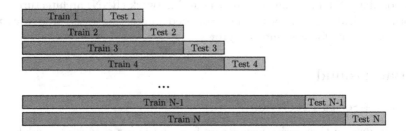

Fig. 1. Train-test rolling visualization

3   Augmented Out-of-Sample Comparison Technique

Augmented training method can be used to evaluate numerous models on test datasets. Specifically, the entire dataset is used to obtain the train-test sets, a forecast interval, and the number of tests. After each test, the model is updated to include the test data. In cases where old data should be discarded, a sliding

---

Algorithm 1: Augmented Out-Of-Sample Testing

Input   : Dataset sorted by ascending date as $\mathcal{D}$, algorithm as f, test interval length as ℓ, number of tests as n

Output: Array of predicted values and real values

1 $\mathcal{T}, \mathcal{U} \leftarrow$ **TrainTestSplit**$(\mathcal{D})$;

2 $model \leftarrow$ **TrainUsing**$(f, \mathcal{T})$;

3 $\{C_j\}_{j=1}^{n} \leftarrow$ **Split**$(\mathcal{U}, n)$;

4 $results \leftarrow \emptyset$;

5 **for** $i = 1 \ldots n$ **do**

6 $\quad$ $testingData \leftarrow$ **RetrieveFirst**(ℓ, C_i);

7 $\quad$ $testResults \leftarrow$ **TestUsing**$(model, testingData)$;

8 $\quad$ $results \leftarrow results \cup testResults$;

9 $\quad$ $model \leftarrow$ **UpdateUsing**$(model, C_i)$;

10 **end**

11 **return** $results$;

---

window approach should be used. This would require model retraining, which is a laborious task. Using the proposed expanding horizon approach, the model is instead updated on a small portion of data which prevents full retraining. The proposed approach is summarized in Algorithm 1.

By using fixed forecast intervals, numerous models can all be compared using the same testing points. The test values can be then amalgamated using any preferred metric, e.g. mean absolute error and mean squared error. Comparison plots can be made to ensure that outlier data points do not impact the results. If the dataset has seasonality, s, it is important to check that the test interval length, ℓ, is chosen such that $gcd(s, \ell) = 1$. This ensures that the test intervals capture a wide variety of tests, instead of a particular segment of the seasonality.

4 Experiments

To demonstrate the power of the proposed technique, we evaluate numerous models on the Turkish electricity dataset[1]. Specifically, we consider two RNN algorithms: LSTM [5] and GRU [4]. In addition, we implement a variant of the ARIMA model, called SARIMAX (Seasonal ARIMA with Regressors), and a naive baseline model which uses last week's data to predict the current week.

The electricity dataset is broken down into five years worth of hourly data. Each day, on average, follows a typical trend where the electric usage starts to rise in the morning hours, peaks in the middle of the day, and then tapers off in the evening. There is weekly seasonality as well, with more electricity usage in the weekdays compared to the weekends.

4.1 Experimental Settings

In our analysis, each model was fit using an 80–20 train-test split, with 25 individual tests embedded in the test data. Then, three forecast intervals have been chosen: 6 h, 24 h, and 48 h, which allows for 150, 600, and 1200 test points, respectively. We only considered multi-step forecasts, but the same principles translate to single-step forecasting as well.

Parameters of the SARIMAX model were determined by optimization in order to ensure the best performance. The resulting SARIMAX model has an order of $(1, 1, 2) \times (1, 0, 1)24$ for the $(p, d, q) \times (P, D, Q)m$ parameters, respectively. Both LSTM and GRU models contain two layers with 32 and 16 hidden units, which is useful for making a multi-step forecast. In addition, the final 15% of the training data was placed into a validation set. This was done in order to perform cross validation during training. Once the loss in the validation plateaued, training is stopped and the trained models were saved. Using a set of test values $\mathcal{T}$ of length N, we use Mean Absolute Percentage Error (MAPE) as our aggregation metric, which can be defined as $\text{MAPE} = 100 \times \frac{1}{N} \sum_{t \in \mathcal{T}} \frac{|\hat{y}_t - y_t|}{y_t}$

---

[1] https://seffaflik.epias.com.tr/transparency/tuketim/gerceklesen-tuketim/gercek-zamanli-tuketim.xhtml.

where, for a given time $t \in \mathcal{T}$, the predicted value is represented as $\hat{y}_t$ and the observed value is y_t.

Additional regressors considered were cyclic day-of week and hour-of-day regressors, which are incorporated using sinusoidal transformation. Due to the inherent seasonality, instead of letting the model pick up this information, we pass it to the model explicitly. We observed that addition of these regressors improved the MAPE value by 10–30% depending on the model. Several other potential regressors such as holidays and weather information were not included.

4.2   Benefits of Augmented Out-of-Sample Testing

Since the subtests were evaluated on many different points, our testing captured many cases. In Fig. 2, each point on the x-axis is an individual subtest, which could be a lucky-one shot test. We see that the subtest lines overlap, and the performance of the subtest is heavily dependent on the subtest interval. If we had used only one test, the choice could have been unlucky. For example, subtest 3 ranks algorithms from best to worst as SARIMAX, Baseline, LSTM, and GRU, which is different from the actual raking of GRU, LSTM, Baseline, SARIMAX. This comparison would have taken 25 times as long using a sliding window approach. This highlights the benefits of updating models instead of retraining.

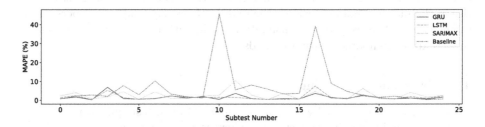

Fig. 2. Subtest MAPEs for 24-h Forecast

4.3   Results

Table 1 provides a numerical comparison between different models. LSTM and GRU models produce similar results for a 6-h and a 48-h forecast, and the GRU performs best for the 24-h forecast. Both RNN models perform fairly consistently across all the tests with low deviations. In fact, while our objective was not to develop the best possible prediction model, we note that predictive performance of GRU is better than recently published studies with the same dataset (e.g. see [11]). We note that each trained model predicts less accurately as the time horizon increased. Interestingly, SARIMAX perform worse than the naive baseline model for the long forecast since the SARIMAX model is not well-suited to predict 48 intervals into the future with its relatively low number of parameters. The SARIMAX model error fluctuates a lot more and it misses many segments in the testing, which can be seen by the larger error bounds on its predictions. If augmented out-of-sample testing had not been used, this pattern in the predictive nature of SARIMAX would not have been evident.

Table 1. MAPE by prediction forecast interval with 95% error bounds

Hours	Baseline (%)	SARIMAX (%)	LSTM (%)	GRU (%)
6	6.8 ± 1.8	2.6 ± 0.5	1.8 ± 0.4	1.6 ± 0.3
24	6.8 ± 0.9	5.4 ± 0.5	2.6 ± 0.2	1.9 ± 0.1
48	6.2 ± 0.6	7.9 ± 0.5	3.2 ± 0.2	3.3 ± 0.2

5   Conclusion

The augmented out-of-sample method alleviates many shortcomings of the standard approaches that are currently used to compare different time series models. By allowing for more testing on the same dataset, in a realistic manner to real-world training, augmented out-of-sample comparison is able to determine the best algorithm. In our numerical analysis, we found neural networks to outperform classical models to predict electricity consumption rates in Turkey using the augmented out-of-sample model comparison. Beyond the scope of the electricity dataset, this comparison method is flexible to be used in comparing many different time series models.

References

1. Alexandrov, A., Benidis, K., Bohlke-Schneider, M., Flunkert, V., Gasthaus, J., Januschowski, T., et al.: Gluonts: probabilistic time series models in python. arXiv preprint arXiv:1906.05264 (2019)
2. Box, G.E., Jenkins, G.M., Reinsel, G.C., Ljung, G.M.: Time Series Analysis: Forecasting and Control. Wiley, Hoboken (2015)
3. Cheng, Y., Xu, C., Mashima, D., Thing, V.L.L., Wu, Y.: PowerLSTM: power demand forecasting using long short-term memory neural network. In: Cong, G., Peng, W.-C., Zhang, W.E., Li, C., Sun, A. (eds.) ADMA 2017. LNCS (LNAI), vol. 10604, pp. 727–740. Springer, Cham (2017). https://doi.org/10.1007/978-3-319-69179-4_51
4. Cho, K., et al.: Learning phrase representations using RNN encoder-decoder for statistical machine translation. arXiv preprint arXiv:1406.1078 (2014)
5. Hochreiter, S., Schmidhuber, J.: Long short-term memory. Neural Comput. **9**(8), 1735–1780 (1997)
6. Huang, Z., Xu, W., Yu, K.: Bidirectional LSTM-CRF models for sequence tagging. arXiv preprint arXiv:1508.01991 (2015)
7. Kane, M.J., Price, N., Scotch, M., Rabinowitz, P.: Comparison of arima and random forest time series models for prediction of avian influenza h5n1 outbreaks. BMC Bioinform. **15**(1), 276 (2014)

8. Kuan, L., et al.: Short-term electricity load forecasting method based on multilayered self-normalizing GRU network. In: 2017 IEEE Conference on Energy Internet and Energy System Integration (EI2), pp. 1–5. IEEE (2017)

9. Merh, N., Saxena, V.P., Pardasani, K.R.: A comparison between hybrid approaches of ann and arima for indian stock trend forecasting. Bus. Intell. J. **3**(2), 23–43 (2010)

10. Tashman, L.: Out-of sample tests of forecasting accuracy: a tutorial and review. Int. J. Forecast. **16**, 423–450 (2000)

11. Yukseltan, E., Yucekaya, A., Bilge, A.H.: Forecasting electricity demand for turkey: modeling periodic variations and demand segregation. Appl. Energy **193**, 287–296 (2017)

Predicting the Number of Reported Bugs in a Software Repository

Hadi Jahanshahi[✉], Mucahit Cevik, and Ayşe Başar

Data Science Lab, Ryerson University, Toronto, Canada
{hadi.jahanshahi,mcevik,ayse.bener}@ryerson.ca

Abstract. The bug growth pattern prediction is a complicated, unrelieved task, which needs considerable attention. Advance knowledge of the likely number of bugs discovered in the software system helps software developers in designating sufficient resources at a convenient time. The developers may also use such information to take necessary actions to increase the quality of the system and in turn customer satisfaction. In this study, we examine eight different time series forecasting models, including Long Short Term Memory Neural Networks (LSTM), auto-regressive integrated moving average (ARIMA), and Random Forest Regressor. Further, we assess the impact of exogenous variables such as software release dates by incorporating those to the prediction models. We analyze the quality of long-term prediction for each model based on different performance metrics. The assessment is conducted on Mozilla, which is a large open-source software application. The dataset is originally mined from Bugzilla and contains the number of bugs for the project between Jan 2010 and Dec 2019. Our numerical analysis provides insights on evaluating the trends in a bug repository. We observe that LSTM is effective when considering long-run predictions whereas Random Forest Regressor enriched by exogenous variables performs better for predicting the number of bugs in the short term.

Keywords: Time series prediction · Software quality · Bug number prediction

1 Introduction

Bug prediction is a crucial task in software engineering practice as it lends insight for practitioners to prepare their resources before their system becomes overwhelmed with defects. Predicting the number of reported bugs to a software system enables developers, managers, and product owners to distribute limited resources, take timely decisions towards effort reduction, and maintain a high-level of software quality. Therefore, there is a need for an automated bug number estimation to facilitate decision making in software development.

The bug growth pattern is a complex and tedious task, and there is uncertainty in the reporting time, assigning time, and fixing time of a bug [7]. Despite

© Springer Nature Switzerland AG 2020
C. Goutte and X. Zhu (Eds.): Canadian AI 2020, LNAI 12109, pp. 309–320, 2020.
https://doi.org/10.1007/978-3-030-47358-7_31

the random bug introduction pattern, there are certain rules and patterns in those interactions which can be a valuable source of information [16]. In this paper, we have extracted the number of bugs from the Mozilla bug repository from Jan 2010 to Dec 2019. The data are split to the weekly bug number, and there exist 522 weeks in total. As an exogenous factor, we extracted the release times of Mozilla updates to see whether multivariate modelling can enhance the performance of the prediction models.

Previous studies on modelling the bug growth patterns mostly use generic time series models, e.g. auto-regressive integrated moving average (ARIMA), X12 enhanced ARIMA, exponential smoothing, and polynomial regression [12,13,17,21,23]. We note that the previous studies lack a rational baseline to compare our methods against. In most cases, they compared different algorithms with each other without having a concrete baseline. To alleviate the issue, we defined lag(1) prediction as a naive baseline - that is, predicted value for the target step is exactly equal to the last observed value. The worst-case scenario is that the prediction model should outperform the naive baseline, otherwise the model can be considered useless.

Time series prediction models have been used in the software engineering domain for the past 20 years. Choraś et al. [2] exploit time series methods such as ARIMA, Holt-Winters, and random walk to forecast various project-related characteristics. Pati and Shukla [15] compared ARIMA and Artificial Neural Network models for Debian Bug Number Prediction. Their comparative analysis indicates that the combination of ANN and ARIMA improves prediction accuracy. Destefanis et al. [3] used time series analysis to determine seasonality and trends of Affective Metrics for Software Development. They consider the evolution of human aspects in software engineering. Our study is different than these studies in that we focused on the reported number of bugs in software as a metric which helps developers maintain the software quality.

Wang and Zhang [20] use a different approach to predict Defect Numbers. They design Defect State Transition models and apply the Markovian method to predict the number of defects at each state in the future. There are also studies that consider software defect number prediction in method-level and file-level [1,5,6]. Our work differs from these two studies as we consider the bug reported to the system regardless of whether it is valid or not rather than the number of bugs existing in different granularities of the system. Hence, we set out to study different approaches that are not well investigated to identify the number of bugs reported to the Mozilla project. We structure our study along with the following two research questions:

RQ1: How accurately the number of bugs in a project can be predicted using time series analysis?

> Time series prediction assumes that there are some patterns in the time series, making it feasible for prediction. Nonetheless, it is not always the case. We tried multiple time series models, including Long Short Term Memory (LSTM), ARIMA, Exponential Smoothing (EXP), Weighted Moving Average (WMA), and Random Forest (RF) regressor with or

without exogenous features. Surprisingly, the performance of a one-step prediction for all models is not significantly different. Furthermore, the baseline seems as good as the others, a new finding which was not considered in previous studies.

RQ2: How feasible is long-term bug number prediction?
Not all models are able to predict more than one step ahead. Hence, we investigate the feasibility and sensitivity of different models to long-term prediction. Specifically, we consider a 3-month prediction (or equivalently 13-week prediction). For the Mozilla project, LSTM shows a significant improvement compared to traditional time series models. The performance of the model is almost of the same quality as it was for a one-step prediction.

The rest of the paper is organised as follows. Section 2 discusses the experimental setting of the models, including the preprocessing phase, the datasets, and a brief discussion of forecasting models. Section 3 presents the performance results of different models for predictions over a long and short horizon. Section 4 discloses the threats to the validity of our findings. Finally, Sect. 5 concludes the paper.

2   Methodology

To predict the number of introduced bugs in the system, we used both statistical models and machine learning techniques. Before applying time series models, we first check certain requirements to ensure that data is stationary [10]. We conduct Augmented Dickey-Fuller (ADF) for this purpose. The ADF test determines the number of lags by the Akaike information criterion (AIC). The null hypothesis (at the significance level $\alpha = 0.05$) of the test is that the data are non-stationary [4]. The p-value of the test is 0.012 thus rejecting the null hypothesis. This suggests that the time series does not have a unit root, and in turn it is stationary. Hence, the data can be modelled directly and there is no need to have supplementary preprocessing or transformation. To further investigate the result of the test, we conduct the auto-correlation function and partial auto-correlation function as well (see Sect. 2.2).

In order to train time series models on the data, we applied a rolling method for training a time series dataset [18]. The idea is to train the dataset, $train_{(0-t)}$, including time series from time 0 to t, and test the model on time $t+1$, $test_{t+1}$. In the next step, the ground truth at time $t+1$ is added to the training set and tested on $test_{t+2}$. Figure 1 demonstrates a visual representation of this approach.

We used the weekly reported bug dataset from January 2010 to January 2017 as the training set and February 2017 to December 2019 as the test set (see Sect. 2.1). After the training-test split based on the rolling approach, different models have been evaluated for the given dataset (see Sect. 2.3).

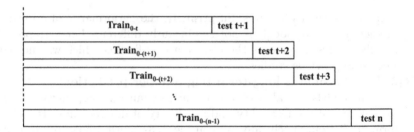

Fig. 1. Train-test rolling visualisation

2.1 Data

We have extracted the number of reported bugs from the Mozilla bug repository[1]. Mozilla is established in 1998 by Netscape as an open-source community. The bug-related information for Mozilla and its products is tracked in the Bugzilla system. We gather reported number of bugs data for Mozilla for the past decade (from January 2010 to the end of December 2019) with a total count of 100,450. We divided data into weekly bug number- that is 522 weeks in total. Mozilla suite weekly bug arrival ranges from 50 to 558.

Figure 2 demonstrates the arrival and resolved bugs in the Mozilla project. The clear outlier in the number of resolved bugs related to the end of 2013 is manually removed from the dataset. Here, our aim is to predict the number of bugs arriving to the system (arrival bugs) to help practitioners efficiently allocate resources.

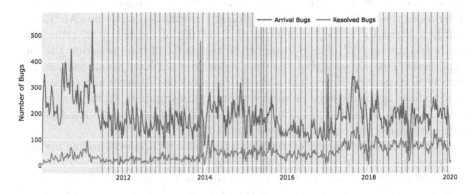

Fig. 2. The number of bugs introduced and resolved in Mozilla during the past decade. Horizontal black lines indicate release dates of different versions of the Mozilla project.

---

[1] Mozilla Bug Tracking System. https://bugzilla.mozilla.org/.

2.2   ACF and PACF

Auto-correlation function (ACF) defines the correlation of the series between different lags. In other words, it specifies how well the current value of the series is related to its past values. Therefore, this characteristic enables it to depict seasonality, cycles, and trends in data. On the other hand, partial auto-correlation function (PACF) shows the correlation of the residuals with the next lag value. The sharp cut-off in PACF represents the number of Auto-Regressive (AR) terms, p, while a sharp drop in ACF is associated with Moving Average (MA) terms, q. Figure 3 illustrates the general tendency in ACFs and PACFs calculated on our training set. The PACF drops sharply when the lag is 2, indicating the number of AR terms, p, is two. ACF does not have such behaviour, so we assume q is equal to zero.

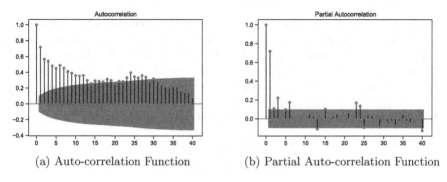

(a) Auto-correlation Function   (b) Partial Auto-correlation Function

Fig. 3. ACF and PACF plots. ACF Long short term memory gradual decline whereas PACF cuts off sharply.

2.3   Forecasting Models

In order to forecast the issues reported to the Mozilla project, we consider different algorithms, including LSTM, ARIMA, Exponential Smoothing, Weighted Moving Average, and RF regressor. We also defined exogenous variables (covariates) to examine the effect of external and seasonal variables on the prediction performance. The exogenous variables include *Branch dates* (from version history), *week of the month, month of the year*, and *year*. For the models which incorporate exogenous variables, we add "x" at the end of their name; for instance, RFx is a random forest containing exogenous variables. The brief explanations for each model are provided below.

1. **Naive Baseline:** It assumes the number of bugs at time t is equal to that at time $t-1$. Note that we expect to see a better result for the proposed models compared to naive baseline defined in this manner.

2. **EXP:** It considers two factors in its prediction: the forecast value at the previous timestamp and its actual value. Therefore, it is defined as

$$X_t = \alpha X_{t-1} + (1 - \alpha)\hat{X}_{t-1}$$

where X_t and $\hat{X}_t$ are the actual and predicted values at time t, respectively, and α is the smoothing level.

3. **WMA:** Weighted Moving Average simply forecasts based on a weighted average of the previous steps.

4. **ARIMA:** It is one of the most popular models used for time series prediction [2,14,15,24]. The general ARIMA model (p, q, d) is formulated as

$$W_t = \sum_{i=1}^{p} \alpha_i W_{t-i} + \sum_{j=0}^{q} \beta_j e_{t-j}$$

where $W_t = X_t - X_{t-d}$, α_i and β_i represent the linear coefficients of the model, $e(t)$ is the error concerning the mean, and d is the degree of non-stationary homogeneity. The value associated with each parameter has been discussed in Sect. 2.2.

5. **LSTM:** Long short-term memory is widely used for time series analysis. They capture both long temporal dependencies and short term patterns in data. We use the LSTM cell architecture defined by [8] as follows:

$$f_t = \sigma\left(W_f.[h_{t-1}, x_t] + b_f\right)$$
$$i_t = \sigma\left(W_i.[h_{t-1}, x_t] + b_i\right)$$
$$\hat{C}_t = \tanh\left(W_c.[h_{t-1}, x_t] + b_c\right)$$
$$C_t = f_t \times C_{t-1} + i_t \times \hat{C}_t$$
$$o_t = \sigma\left(W_o.[h_{t-1}, x_t] + b_o\right)$$
$$h_t = o_t \times \tanh(C_t)$$

where b is the bias term in all equations. f_t is the forget gate deciding which information coming from hidden state h_{t-1} and the new input x should be discarded. Update layer, i_t, selects critical information to be stored and multiplied with candidate vector, $\hat{C}_t$, generating C_t vector as the result. Finally, the model decides which information should be reported, o_t, and which one should be passed to the next cell, h_t.

6. **RF:** Unlike the greedy nature of uni-variate trees, Random Forest is less prone to overfitting on the data as it is an amalgam of DTs which takes the random subset of features in each tree. We applied RF Regressor as a new method that has not been used in this domain.

The parameters of the models are shown in Table 1.

Table 1. Models' parameter settings

Model	Settings
EXP	$\alpha = 0.5$
WMA	Number of previous steps = 2
	Weights = [0.66, 0.33]
ARIMA	$[p, q, d] = [2, 0, 0]$
LSTM	Number of units = 100
	Number of epochs = 50
	with log and difference transformation
RF	Number of trees = 100
	Number of feature = sqrt (total features)

3   Results

After designing the experiment, Root Mean Square Error (RMSE), R-squared (R^2), Error percentage, Median Absolute Error (MAE), and Error Stand Deviation (Std) are used to contrast the algorithms' performance.

Table 2 shows the performance of each method. Random Forest with exogenous variables (RFx) has the best performance in terms of RMSE, error percentage and standard deviation of the errors. Surprisingly, the simple exponential smoothing generates reliable predictions in terms of MAE and R^2. We observe that adding exogenous variables does not necessarily augment the performance of the algorithms, including ARIMA and LSTM. One of the reasons could be the sensitivity of those algorithms to the new features. They cannot differentiate between the time series and exogenous variables; thus, the effect of the values of previous time steps will be eclipsed by the newly defined variables. On the other hand, in the software engineering domain, the Random Forest is proved to be robust, highly accurate, and resilient to noisy data [9,11]. Therefore, it can deal with new features that might be unimportant or have a negligible effect on the output. RFx has a better performance than a simple RF Regressor as more exogenous features will reduce its bias. Figure 4 shows the general overview of the training-test split and predicted time series using Random Forest Regressor with covariates.

The answer to the RQ1 is that due to random fluctuation in the number of bugs introduced to the Mozilla project, the performance of the proposed models remains almost the same. However, Random Forest with exogenous features outperforms the others in most cases with the least variance (see Fig. 5).

In RQ1, we investigated the prediction performance of the models for the next step (next week) whereas in the RQ2 we want to see the effect of long-term prediction. As for the long-term prediction, the error of the models is accumulated, the future predictions would not be as accurate as the closer ones. Here, we analyse the performance of the models for a 3-month bug number prediction. Figure 6 shows that LSTM with the capability of having both Long and Short

Table 2. The result of different time series prediction models in terms of RMSE, R^2, Error percentage, MAE, and Std. The best value for each metric is bold-faced.

Method	RMSE	R-squared	Error (%)	MAE	Std
EXP	39.36	**0.114**	0.178	**22.09**	39.36
WMA	39.26	0.270	0.181	25.17	39.25
ARIMA	37.46	−0.131	0.170	22.62	37.43
ARIMAx	42.97	−0.448	0.199	29.24	42.92
LSTM	39.86	0.127	0.185	25.07	39.78
LSTMx	42.29	0.383	0.194	25.81	42.29
RF	40.18	0.167	0.171	23.30	39.87
RFx	**36.06**	0.179	**0.160**	24.39	**35.88**
Base	41.04	0.346	0.183	27.00	41.03

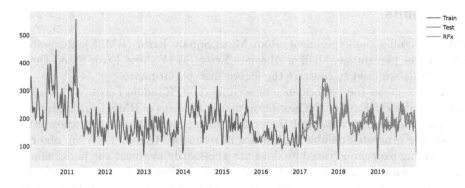

Fig. 4. A sample output of the bug number prediction for the last 3 years

Term Memory performs at the same level in long-run without compromising accuracy. This merit makes it the most suitable model for long-term prediction.

4   Threats to Validity

In this section, we disclose threats to the validity of our empirical study.

4.1   Construct Validity

We estimated the performance of the models using a train-test rolling approach. However, the rolling approach can be used when we have enough history of data. As cross-validation can not be directly applied in time series prediction, the rolling strategy maintains the chronological order of the series and increases its reliability. Other experiment designs, such as nested rolling Cross-validation [19], may yield different results.

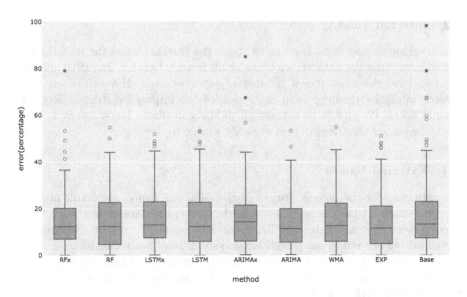

Fig. 5. Prediction performance of different algorithms for one-step prediction (based on error percentage)

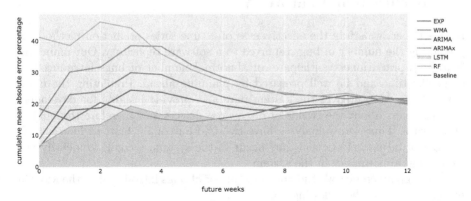

Fig. 6. 3-month prediction performance of the models, reported as cumulative mean absolute error percentage

Although we study four different exogenous variables, there are likely more external features (e.g., major release or social media rumours about a defect in a project) that may impact the trends in the dataset. We plan to expand our exogenous variables to include additional external factors since some previous works that the correct implementation of exogenous terms will improve the performance of time series prediction [22].

4.2   Internal Validity

The tracking dataset data is extracted from the Bugzilla using the REST API[2]. While extracting the dataset, we consider all records between Jan 2010 and Dec 2019 to have the most recent number of reported bugs. However, some issue records might be removed from the repository or imposed restricted access to normal users. We ensure to extract all publicly available bugs; however, there might be some other defects that are only visible to developers.

4.3   External Validity

We only consider the biggest project in Bugzilla, and hence, our result may not be generalizable to all software systems. However, the dataset is a large, long-lived system that alleviates the likelihood of bias in our report. Nonetheless, replication of our study using additional systems may prove fruitful[3].

We used 8 different models to evaluate the feasibility of bug number prediction across the projects whereas using other forecasting techniques may boost the prediction performance.

5   Conclusion and Future Work

In this paper, we study the effectiveness of a time series prediction methods in predicting the number of bugs reported to a software repository. Our main aim is to help practitioners anticipate an abnormal number of bugs introduced in a specific time and be well prepared by planning ahead. Predicting the number of bugs discovered in the system may also provide insights for developers and managers about the trends of discovered defects and therefore the quality of software. Time series analyses have two-fold outcomes: first, to forecast the number of future defects that may occur in the system; second, to identify the trends and abnormality in the system.

Through an empirical study on the number of bugs introduced to the Mozilla project, we made the following observations:

- The number of bugs introduced to the aforementioned system does not have a unit root and therefore is stationary. Furthermore, no specific trend has been observed in the dataset.
- Using eight different forecasting methods, we conclude that there are some improvements in the prediction performance compared to the baseline. Considering five different metrics, Random Forest with exogenous variables exceeds other methods. Nevertheless, we expected to observe a significantly better performance while all models have almost the same error distribution.

---

[2] https://wiki.mozilla.org/Bugzilla:REST_API.
[3] https://github.com/HadiJahanshahi/Bug-Number-Prediction.

- Using a neural network, especially LSTM, significantly improve the long-term prediction. LSTM is not affected by the residual error of the prediction when applied on the long-term test dataset whereas other methods construct their prediction on the error of the last period; hence they are unable to challenge the robustness of LSTM.

References

1. Chen, X., Zhang, D., Zhao, Y., Cui, Z., Ni, C.: Software defect number prediction: unsupervised vs supervised methods. Inf. Softw. Technol. **106**, 161–181 (2019)
2. Choraś, M., Kozik, R., Pawlicki, M., Hołubowicz, W., Franch, X.: Software development metrics prediction using time series methods. In: Saeed, K., Chaki, R., Janev, V. (eds.) CISIM 2019. LNCS, vol. 11703, pp. 311–323. Springer, Cham (2019). https://doi.org/10.1007/978-3-030-28957-7_26
3. Destefanis, G., Ortu, M., Counsell, S., Swift, S., Tonelli, R., Marchesi, M.: On the randomness and seasonality of affective metrics for software development. In: Proceedings of the Symposium on Applied Computing, SAC 2017, pp. 1266–1271. Association for Computing Machinery, New York (2017)
4. Dickey, D.A., Fuller, W.A.: Likelihood ratio statistics for autoregressive time series with a unit root. Econometrica **49**(4), 1057–1072 (1981)
5. Gao, K., Khoshgoftaar, T.M.: A comprehensive empirical study of count models for software fault prediction. IEEE Trans. Reliab. **56**(2), 223–236 (2007)
6. Graves, T.L., Karr, A.F., Marron, J.S., Siy, H.: Predicting fault incidence using software change history. IEEE Trans. Softw. Eng. **26**(7), 653–661 (2000)
7. Herraiz, I., Gonzalez-Barahona, J.M., Robles, G.: Forecasting the number of changes in eclipse using time series analysis. In: Fourth International Workshop on Mining Software Repositories (MSR 2007: ICSE Workshops 2007), p. 32, May 2007
8. Hochreiter, S., Schmidhuber, J.: Long short-term memory. Neural Comput. **9**(8), 1735–1780 (1997). https://doi.org/10.1162/neco.1997.9.8.1735
9. Jiang, Y., Cukic, B., Menzies, T.: Can data transformation help in the detection of fault-prone modules? In: Proceedings of the 2008 Workshop on Defects in Large Software Systems, DEFECTS 2008, pp. 16–20. Association for Computing Machinery, New York (2008)
10. Jothimani, D., Yadav, S.S.: Stock trading decisions using ensemble-based forecasting models: a study of the Indian stock market. J. Bank. Financial Technol. **3**(2), 113–129 (2019). https://doi.org/10.1007/s42786-019-00009-7
11. Kamei, Y., Fukushima, T., McIntosh, S., Yamashita, K., Ubayashi, N., Hassan, A.E.: Studying just-in-time defect prediction using cross-project models. Empir. Softw. Eng. **21**(5), 2072–2106 (2015). https://doi.org/10.1007/s10664-015-9400-x
12. Kenmei, B., Antoniol, G., di Penta, M.: Trend analysis and issue prediction in large-scale open source systems. In: 2008 12th European Conference on Software Maintenance and Reengineering, pp. 73–82, April 2008
13. Krishna, R., Agrawal, A., Rahman, A., Sobran, A., Menzies, T.: What is the connection between issues, bugs, and enhancements? Lessons learned from 800+ software projects. In: Proceedings of the 40th International Conference on Software Engineering: Software Engineering in Practice, ICSE-SEIP 2018, pp. 306–315. Association for Computing Machinery, New York (2018)

14. Pati, J., Kumar, B., Manjhi, D., Shukla, K.K.: A comparison among ARIMA, BP-NN, and MOGA-NN for software clone evolution prediction. IEEE Access **5**, 11841–11851 (2017). https://doi.org/10.1109/ACCESS.2017.2707539

15. Pati, J., Shukla, K.K.: A comparison of ARIMA, neural network and a hybrid technique for Debian bug number prediction. In: 2014 International Conference on Computer and Communication Technology (ICCCT), pp. 47–53, September 2014. https://doi.org/10.1109/ICCCT.2014.7001468

16. Pati, J., Swarnkar, K., Dhakad, G., Shukla, K.K.: Temporal modelling of bug numbers of open source software applications using LSTM. In: Thampi, S.M., Mitra, S., Mukhopadhyay, J., Li, K.-C., James, A.P., Berretti, S. (eds.) ISTA 2017. AISC, vol. 683, pp. 189–203. Springer, Cham (2018). https://doi.org/10.1007/978-3-319-68385-0_16

17. Rastogi, A., Sureka, A.: What community contribution pattern says about stability of software project? In: 2014 21st Asia-Pacific Software Engineering Conference, vol. 2, pp. 31–34, December 2014

18. Tashman, L.J.: Out-of-sample tests of forecasting accuracy: an analysis and review. Int. J. Forecast. **16**(4), 437–450 (2000). The M3- Competition

19. Varma, S., Simon, R.: Bias in error estimation when using cross-validation for model selection. BMC Bioinform. **7**(1), 91 (2006)

20. Wang, J., Zhang, H.: Predicting defect numbers based on defect state transition models. In: Proceedings of the 2012 ACM-IEEE International Symposium on Empirical Software Engineering and Measurement, pp. 191–200, September 2012

21. Wu, W., Zhang, W., Yang, Y., Wang, Q.: Time series analysis for bug number prediction. In: The 2nd International Conference on Software Engineering and Data Mining, pp. 589–596, June 2010

22. Xiao, S., Yan, J., Farajtabar, M., Song, L., Yang, X., Zha, H.: Learning time series associated event sequences with recurrent point process networks. IEEE Trans. Neural Netw. Learn. Syst. **30**(10), 3124–3136 (2019)

23. Yazdi, H.S., Angelis, L., Kehrer, T., Kelter, U.: A framework for capturing, statistically modeling and analyzing the evolution of software models. J. Syst. Softw. **118**, 176–207 (2016)

24. Zhang, W.: SamEn-SVR: using sample entropy and support vector regression for bug number prediction. IET Softw. **12**, 183–189 (2018)

Evaluation of a Failure Prediction Model for Large Scale Cloud Applications

Mohammad S. Jassas$^{(\boxtimes)}$ and Qusay H. Mahmoud

Department of Electrical, Computer and Software Engineering,
Ontario Tech University, Oshawa, ON L1G 0C5, Canada
{mohammad.jassas,qusay.mahmoud}@ontariotechu.net

Abstract. Modern cloud-based applications, including smart homes and cities require high levels of reliability and availability. All cloud services, including hardware and software experience failures because of their large scale and heterogeneity nature. In this paper, the main objective is to develop a failure prediction model that can early detect failed jobs. The advantage of the proposed model is to enhance resource utilization and to increase the efficiency of cloud applications. The proposed model is evaluated based on three public available traces, which are the Google cluster, Mustang, and Trinity. Moreover, four different machine learning algorithms have been applied to the traces in order to select the best accurate model. Furthermore, we have improved the prediction accuracy using different feature selection techniques. The evaluation results show that the proposed model has achieved a high rate of precision, recall, and f1-score.

Keywords: Failure prediction · Fault tolerance · Google cluster trace · Trinity trace · Mustang trace

1 Introduction

Fault tolerance can be defined as the capability of cloud computing that includes all cloud layers to deliver uninterrupted services even though one or more cloud components can be failed for any reason. The cloud architectures have become more complicated due to their large scale and heterogeneity nature. Thus, there is a significant need to design and implement a dependable cloud computing environment. The reliability and availability concerns have become one of the most significant challenges facing cloud computing [8]. Failed jobs consume a notable amount of computational resources and memory. Hence, wherever the number of failed tasks increases, cloud resources such as CPU, memory, and disk space will be wasted. This paper extends our previous work in [5] and establishes a new generalized version of our failure prediction model that can be applied to different workload traces. To the best of our knowledge, no study has focused on the design and implement of the failure prediction model based on comparing four different classification algorithms: Decision Trees (DTs), Random

© Springer Nature Switzerland AG 2020
C. Goutte and X. Zhu (Eds.): Canadian AI 2020, LNAI 12109, pp. 321–327, 2020.
https://doi.org/10.1007/978-3-030-47358-7_32

Forest (RF), Naive Bayes (NB), and Quadratic Discriminant Analysis (QDA), and three different datasets: Google, Mustang, and Trinity. In addition, we evaluated and compared the model performance using different metrics to ensure that the proposed model provides high accuracy of prediction.

To this end, the main contribution of this paper is to design and implement a failure prediction model based on machine learning methods and provide a comprehensive review of recent advances in this field. The remainder of this paper is organized as follows: Sect. 2 discusses the related work, while the proposed model is presented in Sect. 3. The experiments and evaluation results of the proposed model are presented in Sect. 4, and Sect. 5 concludes the paper.

2   Related Work

Failure analysis and characterization have been studied widely in grid computing, cloud cluster and supercomputer [2]. In [4], we have studied the workload features such as memory usage, CPU speed, disk space. We find that there is a clear correlation between the failed jobs and workload attributes. The Google traces [6] are used in different research studies, including workload trace characterization [2] and applying statistical methods in order to compare Google datacentres, which consider as a cloud environment, to Grid or HPC systems. Sun et al. [9] have applied deep learning model to predict a software fault. Also, they have utilized a technique for generating new samples to produce failure data. Most studies have focused mainly on failure analysis and characterization while there is limited research has been done on failure prediction [2,3,7]. El-Sayed et al. [3] have designed a job failure prediction model using a RF classifier. Amvrosiadis et al. [1] have introduced four new traces which are: two from the private cloud of Two Sigma, and two from HPC clusters located at the Los Alamos National Laboratory (LANL).

3   Failure Prediction Model

We improved and extended our framework of failure analysis and prediction of previous studies that were presented in [4,5]. The framework phases are presented in Fig. 1.

The five phases of the proposed model are summarized as follows: (1) The data are loaded from different datasets, and the reason for using different traces is to ensure that the proposed model is generic and can be applied to any Cloud workload trace. (2) The data prepossessing and filtration techniques are applied to the traces. The data preprocessing steps for Google trace have been presented in detail in [4]. (3) Different feature selection algorithms are applied to improve the model accuracy, then we can select the most important features that can be used as input for the proposed model. (4) Four different classification algorithms will be applied to the three different traces. (5) Finally, based on the best prediction results, the failure prediction model will be selected, then the cloud management system will make the appropriate decision. If the job is predicted

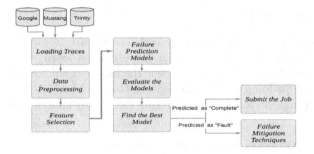

Fig. 1. Proposed evaluation process

as a completed job, the job continues to be submitted and normally scheduled to the available nodes. Otherwise, if the incoming job is predicted as a failed job, failure mitigation techniques will be applied.

4 Experiments and Evaluation Results

4.1 Trace Description

Our experiments utilize three different datasets of large scale traces that were collected from different organizations which are Google and LANL. The general description of each trace is presented in Table 1.

Table 1. Basic description of each trace

Dataset	Num of nodes	Sample size	Features	Failed sample ratio (%)
Google	12550	28,546,501	11	36.2
Mustang	1600	2,113,175	9	7.2
Trinity	9408	20,277	14	16.5

Google cluster traces [6] contains many files of monitored data that are generated from a large system includes more than 12,500 nodes. The Google traces contain 672,074 jobs and more than 28 million tasks, and these jobs are submitted from May 1st to May 29th. Mustang is one of HPC clusters that used for capacity computing at LANL from October 2011 to November 2016. Thus, we can consider that Mustang trace is the longest publicly available trace to date. Mustang contained of 1600 compute nodes, with 102 TB RAM and an overall of 38,400 AMD Opteron 6176 2.3 GHz cores. The trace contain of 2.1 million jobs, and these jobs are submitted by 565 users [1]. Trinity is the largest supercomputer at LANL. Trinity consisted of 9408 compute nodes with a total of 301,056 Intel Xeon E5-2698v3 2.3 GHz cores and 1.2 PB RAM. Thus, Trinity trace is the largest cluster with a publicly available trace by the number of CPU cores. This dataset covers three months, from February to April 2017 [1].

4.2 Experimental Setup

The traces size of Google, Mustang, and Trinity are approximately 15 GB, 280 MB, and 14 MB, respectively. Then, we use scikit-learn, which is machine learning packages in python, to implement the failure prediction model. We run this experiment on Microsoft Azure because Google trace has large volumes of data requiring high performance computing (HPC) for analysis and prediction.

4.3 Failure Analysis and Characterization

For Google trace, we used all jobs that were submitted in 29 days of the trace (500 files). Our goal is to study the failure behaviour by focusing on the most important events ("Fail" and "End"), which are presented in the Google trace as "3" and "4", respectively. Figure 2 shows the distribution of Google trace. Approximately 46% of jobs are finished while a very high percentage of approximately 54% for failed jobs in the first ten days of the trace. However, the percentage of completed jobs has increased to be approximately 93% while the number of failed jobs has sharp decreased to be only 7% in the last ten days of the trace. Figure 3 and 4 present the distribution of Mustang and Trinity traces. Note: in Fig. 2, 3, and 4, the overlap colors present both classes: failed and completed tasks.

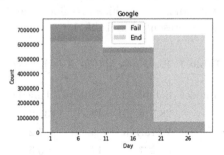

Fig. 2. Distribution of job status for Google Trace (Color figure online)

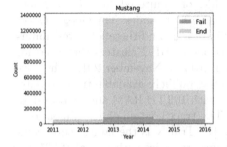

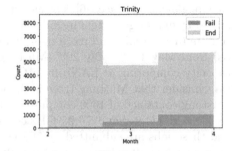

Fig. 3. Distribution of job status for Mustang (Color figure online)

Fig. 4. Distribution of job status for Trinity (Color figure online)

4.4   Classifiers and Prediction Techniques

After the datasets were preprocessed, the goal is to build a prediction model that can predict the value of the target variables. The most accurate classifiers can learn from a relatively small quantity of training data.

As shown in Fig. 5, it is clear that after we increase the number of trace observations (Google trace in 29 days), the accuracy of Precision, Recall, and F1-score for the RF and DTs classifiers have increased to be 98%, 95%, and 97% respectively. The RF-based model has the longest time at 1023 s using Google trace for 29 days compared to DTs, NB, and QDA, which have training time of 174, 3, and 9.2 s respectively.

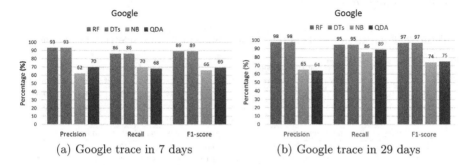

(a) Google trace in 7 days   (b) Google trace in 29 days

Fig. 5. Performance evaluation of different algorithms applied to the Google trace

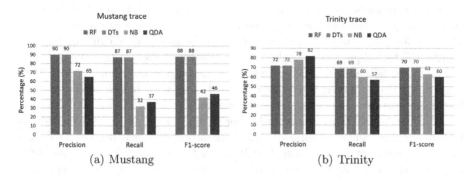

(a) Mustang   (b) Trinity

Fig. 6. Performance evaluation of different algorithms applied to the Mustang and Trinity traces

As depicted in Fig. 6, the same classifiers were applied to Mustang and Trinity traces. The RF and DTs have achieved the best accuracy compared to other classifiers. For Mustang trace, the accuracy of precision, recall, and f1-score for the RF and DTs are 90%, 87%, and 88%, respectively. However, the RF and DTs have achieved lower accuracy when they are applied to Trinity, the accuracy of precision, recall, and fl-score for the RF and DTs algorithms are 72%, 69%, and 70%, respectively. As a result, it is clear that DTs and RF classification are the best two models. However, the DTs have less complexity than the RF because the classifier does not require long training time.

4.5   Feature Selection Algorithms

Features selection is one of the most important methods that can be used to increase the accuracy of our model based on automatically select a number of features that have a high contribution to the model output. Many benefits of applying features selection techniques, such as reducing overfitting and also reducing training time.

Table 2. Evaluation results of performing different feature selection techniques

| | | Decision tree classifier | | | | | Random forest classifier | | | | |
		Prec.	Rec.	F1-score	Train. (t)	Test. (t)	Prec.	Rec.	F1-score	Train. (t)	Test. (t)
Google	SelectKBest	97%	97%	97%	491.3	4.17	98%	97%	97%	2683	39.5
	Feature importance	93%	93%	93%	518.3	7.9	95%	93%	94%	2924	60.40
	RFE	99%	99%	99%	542.18	3.39	99%	99%	99%	2812	36.43
Mustang	SelectKBest	92%	93%	93%	0.9	0.03	94%	94%	94%	9.66	0.32
	Feature importance	94%	94%	93%	1.2	0.09	95%	94%	95%	10.3	0.56
	RFE	93%	93%	93%	1.4	0.12	94%	93%	93%	11.1	0.73
Trinity	SelectKBest	72%	69%	70%	0.05	0.0008	72%	65%	69%	0.27	0.008
	Feature importance	88%	85%	87%	0.07	0.001	89%	85%	89%	0.38	0.02
	RFE	84%	83%	84%	0.09	0.008	85%	81%	83%	0.58	0.09

Table 2 presents the evaluation results of performing different feature selection techniques. For Google trace, the DTs and RF have achieved the highest accuracy for predicting failed class to be 99% for precision, recall, and fl-score using RFE technique. As a result, applying feature selection techniques to the proposed model has a significant impact on increasing the accuracy. For Mustang, the RF classifier can predict the failed class with higher accuracy using feature importance algorithm. The accuracy of Precision, Recall, and F1-score are 95%, 94%, and 95%, respectively. For Trinity trace, we have achieved the highest accuracy when we applied the feature importance algorithm and RF classifier. The accuracy of Precision, Recall, and F1-score are 89%, 85%, and 89%, respectively. RF achieves higher accuracy compare to DTs when it is applied to Trinity and Mustang. As a result, we selected RF as a failure prediction model which can be considered as a generic model that can be applied to different traces to achieve the best accurate results. El-Sayed et al. [3] have designed a job failure prediction model using Random Forests (RF). The results show that they can successfully recall up to 94% of all failed jobs with at least 95% precision. A comparison with the previous studies, we have achieved the highest precision, recall, fl-score. For Google trace, the DTs and RF classifiers have achieved the highest accuracy for predicting failed class.

5   Conclusion and Future Work

We have developed a prediction model for failed jobs based on machine learning methods. The proposed model can be applied to the large datacenters in aim to detect the failed jobs before the cloud management system schedules them. Moreover, we have applied different classification algorithms to various workload traces to come up with a new general model that can provide a high rate of accuracy of predicting failed jobs. In future work, we will develop the proposed model using a deep learning approach to improve the accuracy. Besides, future research will consider mitigation policies and techniques.

Acknowledgement. The first author would like to thank Umm Al-Qura University, Saudi Arabia for funding this work as part of his graduate scholarship.

References

1. Amvrosiadis, G., et al.: The Atlas cluster trace repository. ;login **43**(4), 29–35 (2018)
2. Chen, X., Lu, C.D., Pattabiraman, K.: Failure analysis of jobs in compute clouds: a Google cluster case study. In: 2014 IEEE 25th International Symposium on Software Reliability Engineering, pp. 167–177. IEEE (2014)
3. El-Sayed, N., Zhu, H., Schroeder, B.: Learning from failure across multiple clusters: a trace-driven approach to understanding, predicting, and mitigating job terminations. In: 2017 IEEE 37th International Conference on Distributed Computing Systems (ICDCS), pp. 1333–1344. IEEE (2017)
4. Jassas, M., Mahmoud, Q.H.: Failure analysis and characterization of scheduling jobs in Google cluster trace. In: IECON 2018-44th Annual Conference of the IEEE Industrial Electronics Society, pp. 3102–3107. IEEE (2018)
5. Jassas, M., Mahmoud, Q.H.: Failure characterization and prediction of scheduling jobs in Google cluster traces. In: 2019 10th IEEE-GCC Conference and Exhibition (GCCCE). IEEE (2019)
6. Reiss, C., Wilkes, J., Hellerstein, J.L.: Google cluster-usage traces: format+ schema. Google Inc., White Paper, pp. 1–14 (2011)
7. Ros, A., Chen, L.Y., Binder, W.: Predicting and mitigating jobs failures in big data clusters. In: 2015 15th IEEE/ACM International Symposium on Cluster, Cloud and Grid Computing, pp. 221–230 (2015)
8. Snir, M., et al.: Addressing failures in exascale computing. Int. J. High Perform. Comput. Appl. **28**(2), 129–173 (2014)
9. Sun, Y., Xu, L., Li, Y., Guo, L., Ma, Z., Wang, Y.: Utilizing deep architecture networks of VAE in software fault prediction. In: 2018 IEEE International Conference on Parallel & Distributed Processing with Applications, pp. 870–877. IEEE (2018)

Customer Segmentation and Churn Prediction in Online Retail

Nilay Jha, Dhruv Parekh, Malek Mouhoub$^{(\boxtimes)}$, and Varun Makkar

Department of Computer Science, University of Regina, Regina, Canada
{nhj123,djp657,mouhoubm,vmq810}@uregina.ca

Abstract. The online retail industry has changed the way customers shop as everything is available online. In order to build a loyal customer base, a company needs to deploy various marketing strategies focused on the diverse nature of its customers. We propose a model, abbreviated as RFMOC, based on extension of recency frequency, monetary (RFM) analysis with two new variables to segment customers. The model also studies the segmentation performance for the k-means clustering algorithm. Moreover, customer lifetime value (CLV) is calculated for the weighted RFMOC with weights for variables calculated by the analytic hierarchy process (AHP) and customer segments are then ranked accordingly which helps to create targeted marketing strategies. At last, the customer churn prediction is performed using logistic regression by further extending the RFMOC with one more variable, abbreviated as RFMOCD, in order to predict the churning behaviour of the customers. The proposed approach is helpful to assess customer loyalty and to manage customer relationships in an effective manner.

Keywords: Online shopping · E-commerce · Customer segmentation · Churn prediction

1 Introduction

To retain existing customers and attract new ones is challenging for online retailers. [1]. A possible solution is to segment customers and make targeted marketing strategies for which historical data of customers is required. RFM analysis is a technique that helps in extracting insights from the records and can be used for segmentation of customers as well. However, sometimes RFM analysis alone is not sufficiently insightful. In such situations, it is extended with other variables [1]. This paper focuses on improving customer segmentation by extended RFM model which is named RFMOC (Recency, Frequency, Monetary, Offer Factor and Category Variance). K-means clustering is used to access RFMOC for better segmentation in comparison to RFM. The values are standardized before clustering using min-max normalization as it increases clustering performance [6]. Thus, the study proves better feasibility of RFMOC as compared to classical RFM. Moreover, the RFMOC model is used to calculate CLV which is used to identify the loyalty of customers across segments. In order to calculate weights used by CLV for

C. Goutte and X. Zhu (Eds.): Canadian AI 2020, LNAI 12109, pp. 328–334, 2020.
https://doi.org/10.1007/978-3-030-47358-7_33

RFMOC, AHP is used with recency and frequency having the highest importance [4]. Apart from customer segmentation, the study also discusses the use churning behavior of the customers and suggests a way to predict customer churn.

2   Background

2.1   Analytic Hierarchy Process (AHP)

AHP is used to calculate the relative importance of the desired variables. It is a structured method that helps in studying and organizing complex decisions. AHP helps decision-makers to set priorities and make the best decision by reducing complex decisions to a series of pairwise comparisons [8]. AHP helps to capture both the objective aspect (price, weight, etc.) and subjective (feelings, preferences, etc.) aspects of the problem [8].

2.2   RFM Analysis

It is a process of studying the purchase behaviour of customers for making the process of customer segmentation efficient. RFM analysis helps to recognize the customers which are most likely to respond to marketing strategies. The interpretation for R, F and M variables is as follows:

- **Recency (R):** It indicates number of days from the last purchase [3].
- **Frequency (F):** It indicated how frequently a customer makes purchases. Basically, it is the number times a customer makes transactions over specific period of time [3].
- **Monetary (M):** It indicates the sum of the amount spent by the customer within a particular period of time [3].

The below-average values of all these parameters indicates that the company may lose customers if any actions are not taken. The above-average value would indicate that such customers require attention, which can be achieved by rewarding in an appropriate manner.

Once these values are known for each customer, the customers can be segmented into 5 equal quantiles based on the distribution of recency (R), frequency (F) and the monetary (M) values of a customer [1]. Based on the quintile values for R, F, and M an RFM score is assigned which is obtained by concatenation of R-F-M values [3]. For eg: If the customer has a value of R-Quartile = 4, F-Quartile = 1 and M-Quartile = 2, his/her RFM score will be 412.

3   The Proposed Model

The main steps of the model, as shown in Fig. 1, are described below in detail.

Fig. 1. The proposed architecture.

3.1   Data Pre-processing

In initial phase of the process, irrelevant fields are removed and the format of some fields are changed to relevant ones. The customer data is aggregated based on orders and additional variables are derived which contains Recency (R), Frequency (F), Monetary (M), Offer Factor (O), Categorical Variance (C) and Distribution Delay (D) are calculated. Then, the remaining process of RFM analysis is carried out where R, F and M values are assigned respective quartiles and RFM score is calculated. In proposed model, two more attributes are considered as for extending RFM. The attributes derived from the data are as follows:

- **The Offer Factor (O):** This variable refers to how often a customer buys a product during the offer periods. The offer factor can be obtained by the following formula:

$$O = PIO/TP \tag{1}$$

  Here, PIO is total products purchased during offers and TP is total products purchased. The value of O closer to one indicates that the customer generally purchases more during the offer periods whereas a value closer to zero indicates that the customer is not much concerned about offers.
- **Category Variance (C):** This variable indicates how much variation is there in the customer's purchase behavior with respect to categories. The category variance can be obtained by the formula:

$$C = PC/TC \tag{2}$$

  Here, PC is total number of categories from which the customer has purchased products and TC is total number of available categories. The value of C closer to 1 shows that the customer is interested in products from different categories whereas the value closer to 0 means the customer considers the products from a specific set of categories.
- **Delay in Distribution (D):** This variable will be used for the purpose of churn prediction. It is borrowed from the RFMITSDP model and it the indicates average time difference between the date on which the order is placed and the date on which the order was shipped [1].

3.2   Standardization

According to various studies, clustering with standardized attributes results into better cluster outputs compared to non standardized values [6]. The standardization technique used here is Min-Max normalization. The equation for Min-Max normalization is same as mentioned in [2]. Here values of R, F and M are normalized to match with range of variables O and C.

3.3   Clustering

In order to group the customers on the basis of recency, frequency and monetary value, k-means clustering is used [5]. The possible number of groups (k) is calculated according to Davies–Bouldin Index [5], the Average Silhouette score and the Elbow test. The appropriate number of clusters obtained by all three of them is 4. Once the appropriate number of clusters are finalized, the clustering is applied. The performance is evaluated on the basis of silhouette index and the cluster size ratio (the ratio of the largest cluster size to smallest cluster size). The configurations on which the k-means clustering was performed are K-means with RFM attributes, K-means with normalized RFM attributes and K-means with normalized RFMOC attributes.

3.4   Calculating Weights

AHP is used to calculate weights for the RFMOC attributes. These weights are used to calculate the CLV for the clusters obtained by applying k-means clustering on normalized RFMOC attributes. The final calculated weights for R, F, M, O and C variables using AHP are 0.398067, 0.287158, 0.156691, 0.0950614, 0.0630227 respectively. The precedence of the importance of the variables is:

$$W_F > W_R > W_M > W_O > W_C$$

3.5   Calculating the CLV Score for Clusters

CLV is calculated for all the clusters to identify which cluster has customers with the highest level of loyalty using the equation below mentioned:

$$CLV_i = W_R * a(R_i) + W_F * a(F_i) + W_M * a(M_i) + W_O * a(O_i) + W_C * a(C_i) \quad (3)$$

Here i = ith cluster, a(X) stands for mean value of parameter X and Wx stands for the weight of parameter X, computed using AHP. Once CLV, for each cluster is calculated, the clusters are ranked accordingly. The cluster with the highest rank represents the customers with the highest loyalty and likewise the loyalty decreases with the ranking. The following are the results of calculating CLV with clustering on normalized values of RFMOC:

Table 1. Clusters with calculated CLV and assigned ranks.

	Cluster-1	Cluster-2	Cluster-3	Cluster-4
CLV score	0.0883	0.1300	0.1309	0.0738
CLV rank	3	2	1	4

From Table 1, one can depict that cluster-3 has the highest number of loyal customers whereas the cluster-2 comes second, cluster-1 third and then cluster-4. The outcome of the CLV was verified by manually inspecting the values of the RFM quartiles for different clusters.

3.6 Churn Prediction

Churn Prediction is an approach used to predict the churning behavior a customer. A churned customer is one who is no longer making purchases. Here, churn prediction is carried out using Logistic Regression with L1 penalty [7]. Here, RFMOC with derived varied D (discussed earlier) is used to predict customer churn. For the purpose of this study, customer belonging to the 4th quartile for recency and 3rd and 4th quartile for frequency are considered as churned. The data used for this purpose is a subset of the dataset of 798 customers.

4 Experimentation

4.1 Dataset Description

The dataset used for this study is a combination of two datasets of the online sale of Superstore with one dataset having data from 2011-to-2015 and other from 2015–2018. The dataset set has historical data of 1590 customers and 30039 orders.

4.2 Customer Segmentation

To perform and evaluate the clustering a tool called SPSS Modeler (Statistical Package for Social Sciences) was used. The tool provides the silhouette score and the cluster ratio as discussed earlier in this paper are the metrics to access the clustering performance. On evaluation of the outputs for the clustering, normalized (represented by prefix 'n') RFMOC model performed better, the results are as follows (Table 2):

Table 2. Output of k-means clustering with different configurations.

	RFM	n-RFM	n-RFMOC
Silhouette score	0.6	0.6	0.7
Cluster ratio	5.96	4.17	4.08

4.3   Churn Prediction

On accessing the performance of RFMOCD and RFM for churn prediction using Logistic regression, It was found that RFMOCD performed better. As presented in [7], the regression model was evaluated based on: (1) Precision, (2) Recall and (3) Area under Reverse Operating Characteristics (ROC) Curve. The f1- score is another parameter which is used to present balance the trade off between recall and precision [4]. The results are shown in Table 3 are average of 4 tests.

Table 3. Churn prediction results.

	Area under ROC	Precision	Recall	F1-score
RFM	0.88	0.84	0.79	0.81
RFMOCD	0.91	0.93	0.85	0.88

5   Conclusions

A model abbreviated as RFMOC was proposed in the paper which is an extension of RFM analysis. The performance of RFMOC model for customer segmentation against RFM was evaluated on normalized values for attributes. The dataset used contained customer data of transactions from 2011–2018 of a Superstore. It was found that RFMOC performs better at segmentation. The CLV score for each obtained segment was calculated to measure loyalty of each. The outcome of CLV was verified manually by inspecting the data. Finally, one more variable D was used with RFMOC for churn prediction and it was found that the RFMOCD is better as compared to RFM. The proposed model thus can be effectively used for customer segmentation and churn prediction in order planning the targeted marketing strategies.

References

1. Khodabandehlou, S., Zivari Rahman, M.: Comparison of supervised machine learning techniques for customer churn prediction based on analysis of customer behavior. J. Syst. Inf. Technol. **19**(1/2), 65–93 (2017)
2. Khajvand, M., Zolfaghar, K., Ashoori, S., Alizadeh, S.: Estimating customer lifetime value based on RFM analysis of customer purchase behavior: case study. Procedia Comput. Sci. **3**, 57–63 (2011)
3. Wei, J.T., Lin, S.Y., Wu, H.H.: A review of the application of RFM model. Afr. J. Bus. Manage. **4**(19), 4199–4206 (2010)
4. Sharda, R., Delen, D., Turban, E.: Business Intelligence and Analytics: Systems for Decision Support, 10th edn. Pearson Education, Harlow (2014)
5. Hosseini, S.M.S., Maleki, A., Gholamian, M.R.: Cluster analysis using data mining approach to develop CRM methodology to assess the customer loyalty. Expert Syst. Appl. **37**(7), 5259–5264 (2010)

6. Davies, D.L., Bouldin, D.W.: A cluster separation measure. IEEE Trans. Pattern Anal. Mach. Intell. **2**, 224–227 (1979)
7. Subramanya, K. B., Somani, A.: Enhanced feature mining and classifier models to predict customer churn for an E-retailer. In 2017 7th International Conference on Cloud Computing, Data Science and Engineering-Confluence, pp. 531–536. IEEE, January 2017
8. Saaty, T.L.: Decision making with the analytic hierarchy process. Int. J. Serv. Sci. **1**(1), 83–98 (2008)

Detection and Diagnosis of Breast Cancer Using a Bayesian Approach

Nathina Krishnakumar[(✉)] and Tamer Abdou[(✉)]

Ryerson University, Toronto, ON, Canada
{nathina.krishnakumar,tamer.abdou}@ryerson.ca

Abstract. This paper aims to build a reliable decision support detection system to help physicians give more accurate risk profiles for breast cancer patients. In this quantitative study, we use Bayesian network to uncover hidden insights from the Wisconsin Breast Cancer Diagnostic dataset. A Bayesian network was learned from the data and conditional probability queries were performed. Lastly, a Bayesian network classifier was built. We found diagnosis was conditionally dependent upon two features: worst concave points and worst radius. Both the highest probability for malignant cancer and lowest probability for benign diagnosis were detected by Very High, (0.161, 0.291] for worst concave points and Very High, (18.8, 36] μm for worst radius. The highest probability for benign diagnosis and lowest probability for malignant cancer were detected by Low, [0, 0.0649] for worst concave points and Low, [7.93, 13] μm for worst radius. Our proposed Bayesian network classifier had: 96.31% accuracy, 92.92% sensitivity, 98.32% specificity, 97.04% positive predictive value, and 95.90% negative predictive value, making it a robust model in terms of accuracy, sensitivity, and specificity.

Keywords: Bayesian network · Inference · Decision support system · Healthcare · Breast cancer

1 Introduction

Breast cancer is the most common cancer found among women around the world. In the year 2018 alone, there was a report of over two million cases accounting for 25.4% of all new cases of cancer in women [2]. As breast cancer rates increase, early detection and diagnosis are very important for early intervention and treatments. False positive is misdiagnosing a healthy patient to have cancer and false negative is misdiagnosing a cancer patient as healthy, which is detrimental. Additionally, in areas with limited resources, a reliable decision support system can help clinicians detect cancer quickly and with greater confidence.

A Bayesian network is a probabilistic graphical model that represents conditional dependencies between random variables. It is a directed acyclic graph

This paper was written as part of the Certificate in Data Analytics, Big Data, and Predictive Analytics at Ryerson University.

© Springer Nature Switzerland AG 2020
C. Goutte and X. Zhu (Eds.): Canadian AI 2020, LNAI 12109, pp. 335–341, 2020.
https://doi.org/10.1007/978-3-030-47358-7_34

(DAG) consisting of nodes and edges. Nodes represent random variables and edges represent conditional dependencies between these variables [9]. Every node has a local probability distribution contributed by its parents [8]. In this paper, we conducted a quantitative Bayesian analysis using the Wisconsin Breast Cancer Diagnostic dataset, which is open source and publicly available[1].

Our research questions are:

RQ_1 Which features and intervals detect the highest and lowest probabilities for malignant cancer? Which features and intervals detect the highest and lowest probabilities for benign diagnosis?

RQ_2 What is the probability of a patient having a diagnosis of malignant or benign event given feature measurements as evidence?

RQ_3 How do models from literature compare to our proposed Bayesian network classifier in terms of accuracy, sensitivity, and specificity?

2   Related Work

Wolberg et al. proposed three robust classification methods using different feature sets [14]. Logistic regression used Worst Radius, Worst Texture, and Worst Concave Points features and Multisurface Method-Tree (MSM-T) used Worst Area, Worst Texture, and Worst Smoothness features, which performed the same as Xcyt program using Extreme Area, Extreme Smoothness, and Mean Texture features (Table 2) [7,13]. In 2018, Westerdijk looked at correlation, recursive feature elimination, and genetic algorithm feature selection techniques and classification models, logistic regression, random forest, support vector machine, artificial neural network, and ensemble, using 10-fold cross-validation (Table 2) [11].

In 2017, Abdou et al. quantitatively identified causal relationships for software defects by employing Bayesian belief networks [1]. Abdou et al. learned the structure of the Bayesian network using Hill-Climbing search algorithm, then estimated the Bayesian network using the Peter and Clark (PC) algorithm and conditional independence tests at significance level of 0.01 [1,6]. False positives were excluded by checking the Markov blanket [1]. Cruz-Ramírez et al. investigated seven Bayesian network (BN) classifiers for single and multiple observers using two breast cancer databases (Table 2) [3]. Fallahi and Jafari found the Bayesian network algorithm outperformed models AR+NN and NN (Table 2) [5]. Witteveen et al. compared logistic regression with various Bayesian networks and found Bayesian networks with the most links performed better and logistic regression slightly outperformed Bayesian networks [12].

3   Methods and Proposed Bayesian Network Model

All numeric attributes were discretized into four categories of equal number of observations, each corresponding to a certain level: Low - minimum to 25%,

---

[1] https://archive.ics.uci.edu/ml/machine-learning-databases/breast-cancer-wisconsin.

Medium - 25% to 50%, High - 50% to 75%, and Very High - 75% to maximum. The Bayesian network was learned using the greedy Hill-Climbing search algorithm and the Akaike information criterion (AIC) metric for the score algorithm [10]. To estimate the parameters, the PC algorithm was implemented using the conditional independent test, G^2 statistic, at a significance alpha level of 0.05. This could confirm the arc directions that are statistically significant [1,6] The Markov blanket (parents, children, and spouses) of the diagnosis node was checked for false positives [1]. Conditional probability distributions were determined to find the features conditionally dependent upon diagnosis (RQ_1). Conditional probability queries were performed using the likelihood weighting (lw) algorithm to give the probability of diagnoses as events given feature intervals as evidence for potential patients. RQ_2 was answered using the query: $cpquery\,(fitted, event, evidence, method = "lw")$. The BN classifier has a high classification accuracy and optimal Bayes's error [4]. A BN classifier was built using 10-fold cross-validation and its parameters fitted using the Bayes method. The posterior classification error "pred-lw" was used to obtain Bayesian posterior estimates [10]. The accuracy, sensitivity, specificity, positive predictive value, and negative predictive value were calculated.

4   Results

4.1   Bayesian Network

The resultant Bayesian network is shown in Fig. 1 and its corresponding conditional probability table bar chart in Fig. 2. This shows that the diagnosis is conditionally dependent upon two features: worst concave points and worst radius and the combinations of feature intervals of parent nodes give patients either a high or low probability of malignant or benign diagnosis (RQ_1).

- The highest probability for malignant cancer was detected by Very High, (0.161, 0.291] for worst concave points and Very High, (18.8, 36] μm for worst radius.
- The lowest probability for malignant cancer was detected by Low, [0, 0.0649] for worst concave points and Low, [7.93, 13] μm for worst radius.
- The highest probability for benign diagnosis was detected by Low, [0, 0.0649] for worst concave points and Low, [7.93, 13] μm for worst radius.
- The lowest probability for benign diagnosis was detected by Very High, (0.161, 0.291] for worst concave points and Very High, (18.8, 36] μm for worst radius.

The PC algorithm could not estimate the cause and effect relationships at an alpha significance level of 0.05 as the estimated arcs were undirected. From Fig. 1, the Markov blanket consists of seven nodes: two parents - worst radius and worst concave points, two children - mean texture and worst smoothness, and three spouses - compactness standard error, mean smoothness, and worst fractal dimension. There were no false positives in the Markov blanket.

4.2   Conditional Probability Querying

Conditional probability queries were performed to predict outcomes for potential patients (RQ_2). This would help physicians assess and give patients a quantitative probabilistic risk profile. Examples are:

$$P(M|wr = (18.8, 36] \, \mu m, wcp = [0, 0.0649]) = 44\%$$
$$P(B|wr = (15, 18.8] \, \mu m, wcp = (0.0649, 0.0999]) = 90\%$$
$$P(M|wr = (15, 18.8] \, \mu m, wcp = [0, 0.0649]) = 10\%$$

B, M, wr, and wcp are benign, malignant, worst radius, and worst concave points.

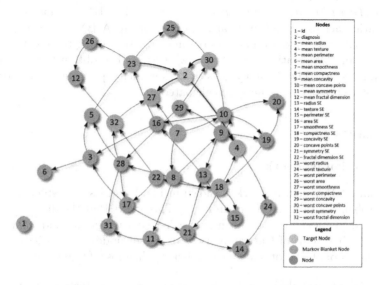

Fig. 1. The learned Bayesian network structure

4.3   Bayesian Network Classifier

Our BN classifier had 96.31% accuracy and 3.69% classification error. The sensitivity, specificity, positive predictive value, and negative predictive value were: 92.92%, 98.32%, 97.04%, and 95.90%, as determined from Table 1.

Table 2 compares performance measures of our model to models found in literature [3, 5, 11, 13]. Westerdijk's logistic regression had 97.35% accuracy compared to Wolberg et al.'s 96.2% accuracy [11, 13]. While Wolberg et al.'s highest model accuracy was 97.5% for the MSM-T model, both Westerdijk's support vector machine and ensemble was 98.23% accuracy [11, 13]. Running a logistic model on our discretized dataset with Westerdijk's features gave 94.71% accuracy and with Wolberg et al.'s features, 96.47% accuracy. The small difference for Westerdijk's model could have been due to different pre-processing techniques.

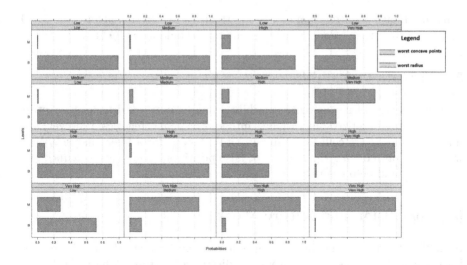

Fig. 2. Conditional probabilities for the diagnosis node

Table 2. Model performance measures

Model	Accuracy	Sensitivity	Specificity
BN classifier	0.9631	0.9292	0.9832
LR [13]	0.9620		
MSM-T [13]	0.9750		
LR [11]	0.9735	0.9524	0.9859
RF [11]	0.9735	0.9286	1
ANN [11]	0.9735	0.9286	1
SVM [11]	0.9823	0.9524	1
Ensemble [11]	0.9823	0.9524	1
BN classifier 1 [3]	0.9304		
BN classifier 2 [3]	0.8331		
BN classifier [5]	0.9815		
AR+NN [5]	0.9740		
NN [5]	0.9520		

Table 1. Confusion matrix for BN classifier

Actual	Benign	Malignant
Benign	351 (TN)	6 (FP)
Malignant	15 (FN)	197 (TP)

For RQ_3, our goal was to build a model that was explainable rather than a black box model that lacks interpretability. Our proposed model's performance was slightly lower but the difference is negligible, so we can conclude that our model is robust in terms of accuracy, sensitivity, and specificity. Compared to BN classifiers found in literature on breast cancer, our model outperforms those models in terms of accuracy except for Fallahi and Jafari's model (Table 2) [3, 5].

This could have resulted from employing the feature selection algorithm (RELI-EFF) and Synthetic Minority Over-Sampling Technique (SMOTE).

5   Threats to Validity

Continuous variables have more variability making patterns difficult to capture; therefore, we discretized our variables into four ordinal categories. However, with finer granularity, we could determine more intervals for a finer analysis. While we focused on reducing the number of false positives, techniques to reduce false negatives could help evaluate other aspects of the model.

6   Conclusion and Future Work

Assuming there were no hidden variables in the dataset, the most probable Bayesian network was learned from the discretized dataset. The highest probability for malignant cancer and lowest probability for benign diagnosis were detected by Very High, (0.161, 0.291] for worst concave points and Very High, (18.8, 36] μm for worst radius. The highest probability for benign diagnosis and lowest probability for malignant cancer were detected by Low, [0, 0.0649] for worst concave points and Low, [7.93, 13] μm for worst radius. The probabilities of a patient having diagnoses given feature measurements as evidence were determined by querying. This enables the decision support system of clinicians to give more accurate risk profiles for patients. The proposed BN classifier had 96.31% accuracy, 92.92% sensitivity, and 98.32% specificity, suggesting it was a robust model in addition to its reliability, reproducibility, and explainability. With the help of a domain expert and more available current data, we can improve the model. For future work, a finer granular quantitative analysis can be conducted by discretizing numeric attributes into more intervals for further insights into breast cancer detection and diagnosis.

References

1. Abdou, T., Soltanifar, B., Bener, A., Neal, A.: What is the cause for a defect to be re-assigned? In: Proceedings - 2016 IEEE International Conference on Software Maintenance and Evolution, ICSME 2016, pp. 502–508 (2017)
2. Bray, F., Ferlay, J., Soerjomataram, I., Siegel, R.L., Torre, L.A., Jemal, A.: Global cancer statistics 2018: GLOBOCAN estimates of incidence and mortality worldwide for 36 cancers in 185 countries. CA Cancer J. Clin. **68**(6), 394–424 (2018)
3. Cruz-Ramírez, N., Acosta-Mesa, H.G., Carrillo-Calvet, H., Alonso Nava-Fernández, L., Barrientos-Martínez, R.E.: Diagnosis of breast cancer using Bayesian networks: a case study. Comput. Biol. Med. **37**(11), 1553–1564 (2007)
4. Duda, R.O., Hart, P.E., Stork, D.G.: Maximum-likelihood and Bayesian parameter estimation. In: Pattern Classification, Chap. 3, pp. 84–159, 2nd edn. Wiley (2012)
5. Fallahi, A., Jafari, S.: An expert system for detection of breast cancer using data preprocessing and Bayesian network. Int. J. Adv. Sci. Technol. **34**, 65–70 (2011)

6. Kalisch, M., Mächler, M., Colombo, D., Maathuis, M.H., Bühlmann, P.: Causal inference using graphical models with the R package pcalg. J. Stat. Softw. **47**(11), 1–26 (2012)
7. Mangasarian, O.L., Street, W.N., Wolberg, W.H.: Breast cancer diagnosis and prognosis via linear programming. Opera. Res. **43**(4), 570–577 (1995)
8. Glymour, C., Copper, G.F.: Computation, causation and discovery. J. Am. Stat. Assoc. **95**(451), 1019 (2000)
9. Nagarajan, R., Scutari, M., Lèbre, S.: Bayesian Networks in R with Applications in Systems Biology. Springer, New York (2013). https://doi.org/10.1007/978-1-4614-6446-4
10. Scutari, M.: Learning Bayesian networks with the bnlearn R Package. J. Stat. Softw. **35**(3), 1–22 (2010)
11. Westerdijk, L.: Predicting malignant tumor cells in breasts. Ph.D. thesis, Vrije Universiteit Amsterdam, Faculty of Science (2018). https://www.math.vu.nl/sbhulai/papers/paper-westerdijk.pdf
12. Witteveen, A., Nane, G.F., Vliegen, I.M., Siesling, S., IJzerman, M.J.: Comparison of logistic regression and Bayesian networks for risk prediction of breast cancer recurrence. Med. Decis. Making **38**(7), 822–833 (2018)
13. Wolberg, W.H., Street, W.N., Heisey, D.M., Mangasarian, O.L.: Computer-derived nuclear features distinguish malignant from benign breast cytology. Hum. Pathol. **26**(7), 792–796 (1995)
14. Wolberg, W.H., Street, W.N., Mangasarian, O.L.: Machine learning techniques to diagnose breast cancer from image-processed nuclear features of fine needle aspirates. Cancer Lett. **77**(2–3), 163–171 (1994)

Query Focused Abstractive Summarization via Incorporating Query Relevance and Transfer Learning with Transformer Models

Md Tahmid Rahman Laskar[1,3]($\boxtimes$), Enamul Hoque[2], and Jimmy Huang[2,3]

[1] Department of Electrical Engineering and Computer Science,
York University, Toronto, Canada
tahmedge@cse.yorku.ca
[2] School of Information Technology, York University, Toronto, Canada
{enamulh,jhuang}@yorku.ca
[3] Information Retrieval and Knowledge Management Research Lab,
York University, Toronto, Canada

Abstract. In the Query Focused Abstractive Summarization (QFAS) task, the goal is to generate abstractive summaries from the source document that are relevant to the given query. In this paper, we propose a new transfer learning technique by utilizing the pre-trained transformer architecture for the QFAS task in the Debatepedia dataset. We find that the Diversity Driven Attention model (DDA), which was the first model applied on this dataset, only performs well when the dataset is augmented by creating more training instances. In contrast, without requiring any in-domain data augmentation, our proposed approach outperforms the DDA model as well as sets a new state-of-the-art result.

Keywords: Query · Abstractive summarization · Transformer · BERT

1 Introduction

In abstractive text summarization task, the goal is to create summaries containing novel words or phrases which have not appeared in the source document. In the query focused summarization task, a query is also given along with document(s) and the aim is to generate summaries based on the relevance between the query and the document(s) [13]. While significant research has been done on different datasets for generic abstractive summarization, the number of datasets available for query-based abstractive summarization is very small [1,13]. For the QFAS task, two of the most used datasets are the Debatepedia dataset[1] [8] and the datasets based on Document Understanding Conferences (DUC) between 2005 and 2007 [1]. However, these datasets are very small in size compared to the generic abstractive summarization datasets.

---

[1] http://www.debatepedia.org/.

© Springer Nature Switzerland AG 2020
C. Goutte and X. Zhu (Eds.): Canadian AI 2020, LNAI 12109, pp. 342–348, 2020.
https://doi.org/10.1007/978-3-030-47358-7_35

To address this problem, we introduce a transfer learning approach for the QFAS task based on the transformer architecture by first pre-training our model on a large generic abstractive summarization dataset followed by fine-tuning it for the QFAS task via incorporating query relevance. Our contributions presented in this paper are summarized as follows. First, we propose a transfer learning technique with transformer for the QFAS task. To the best of our knowledge, this is the first work where transfer learning is utilized with transformer for QFAS. Second, we perform extensive experiments on the Debatepedia dataset and observe a new state-of-the-art result with our proposed approach. Unlike the DDA model [8], which we find that fails to perform well without augmenting the training data, our approach does not require any in-domain data augmentation. Finally, we make the source code of this work publicly available[2].

2   Related Work

Recently, researchers have applied various neural models based on encoder-decoder architecture to generate abstractive summaries [7,10]. A major problem of such models is that they tend to repeat the same word multiple times and thus produce non-cohesive summaries. See et al. [11] address the problem by utilizing the Pointer Generation Network (PGN) which discourages repetition using a copy and coverage mechanism. More recently, the Bidirectional Encoder Representations from Transformers (BERT) architecture for SUMmarization (BERT-SUM) [4] model obtained state-of-the-art-results which used BERT [2] as encoder and transformer decoder [12] as decoder for abstractive summarization.

While significant progress has been made on generic abstractive summarization, applying neural encoder-decoder models for query focused summarization has been rare [1]. One notable exception is the Diversity Driven Attention (DDA) model [8], which alleviates the problem of repeating phrases and can focus on different portions of a document based on a given query at different times. However, their Debatepedia dataset is very small compared to the datasets used for generic abstractive summarization. [1,4,11]. Thus, the lack of large training data makes the QFAS task on this dataset a few-shot learning problem. To tackle this issue, the Relevance Sensitive Attention for Query Focused Summarization (RSA-QFS) [1] utilized transfer learning by first pre-training the PGN model on large generic abstractive summarization dataset and then incorporated query relevance into the pre-trained model to generate query focused summaries. However, they did not fine-tune their model on QFAS datasets and obtained a very low Precision score [1]. In contrast to the above body of work that are based on Recurrent Neural Network (RNN) models, we utilize the transformer architecture along with leveraging transfer learning and fine-tuning since the former and the latter with the transformer based models have been found to be more effective on various natural language processing tasks [2,3].

---

[2] https://github.com/tahmedge/QR-BERTSUM-TL-for-QFAS.

3   Our Proposed Approach

Let's assume that we have a query $q = q_1, q_2, ..., q_k$ containing k words and a source document $d = d_1, d_2, ...d_n$ containing n words. Our task is to generate a contextual summary $y = y_1, y_2, ...y_m$ containing m words. In other words, our goal is to find the summary y_* that maximizes the probability $p(y|q, d)$.

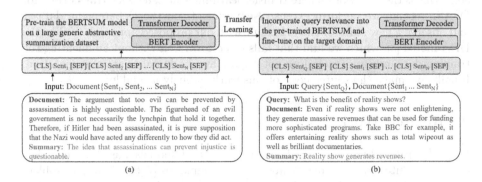

Fig. 1. Our proposed approach works in two steps: (a) Pre-train the BERTSUM model on a generic abstractive summarization corpus (e.g. XSUM) and (b) Fine-tune the pre-trained model for the QFAS task on the target domain (i.e. Debatepedia).

To achieve this goal, our proposed method adopts the BERTSUM model [4] which used transformer-based architecture for abstractive summarization. However, the BERTSUM model was designed for a generic summarization task without considering the query relevance [4]. Therefore, we incorporate query relevance (QR) and transfer learning (TL) within the BERTSUM model for the QFAS task. More specifically, our model (denoted as QR-BERTSUM-TL) performs the QFAS task in two steps as shown in Fig. 1. In the first step, we pre-train the BERTSUM model on a large training corpus of generic abstractive summarization. Then, we fine-tune the pre-trained model for the QFAS task by utilizing the query relevance. In the following, we describe these steps.

1) Pre-training the BERTSUM Model: In this step, we pre-train the BERTSUM model on a large generic abstractive summarization dataset. Among the datasets used for BERTSUM [4], the XSUM[3] dataset was the most abstractive one containing highest number of novel bi-gram. Therefore, we pre-train the BERTSUM model on this dataset. During the training process, the model utilizes the pre-trained BERT model [2] as the encoder and the randomly initialized Transformer decoder [12] as the decoder. However, the original BERT model inserted the special token [CLS] at the beginning of only the first sentence. In contrast, the BERTSUM model inserts the [CLS] token at the beginning of each sentence. Moreover, each sentence-pair in BERTSUM is also separated by the [SEP] token.

---

[3] https://github.com/EdinburghNLP/XSum/tree/master/XSum-Dataset.

2) Incorporating Query Relevance and Fine-tuning BERTSUM: In this step, we fine-tune the BERTSUM model on the Debatepedia dataset which was pre-trained on the XSUM dataset. During fine-tuning, we incorporate the query relevance by concatenating the query with the document as the input of the encoder (see Fig. 1b). We do this because we find that a similar approach of concatenating question with document works well for the question-answering task [3]. Through this process, our model is adapted in the target domain.

4   Debatepedia Dataset

Debatepedia is an encyclopedia of pro and con arguments and quotes on debate topics. Nema et al. [8] utilized Debatepedia to create a dataset for the QFAS task [8]. The average number of words per document, summary, and query in the Debatepedia dataset is 66.4, 11.16, and 9.97 respectively. They used 10-fold cross validation in their experiments with the DDA model on this dataset. The average number of instances in each fold is 10,859 for training, 1,357 for testing, and 1,357 for validation respectively. However, we find in the source code[4] of the DDA [8] model that the dataset was augmented to create new training instances. In the augmented dataset, the test and validation data were the same as the original, but the average training instances in each fold were 95,843. Since the original paper [8] did not address the data augmentation approach, we briefly explain this process below based on the source code (see Footnote 4) of DDA.

Augmenting the Debatepedia Dataset: For data augmentation, a pre-defined vocabulary of 24,822 words was used where each word had been associated with a synonym. Then for each training instance, N ($10 \leq N \leq 17$) words in each document and M ($1 \leq M \leq 3$) words in each query were randomly selected (except stop words and numerical values) and then replaced with their synonyms found in the vocabulary. If a selected word was not found in the vocabulary, it was added there with the most similar word found based on cosine similarity in the GloVe [9] vocabulary. For each training instance, this process is repeated 8 times to create 8 new document and query instances. But the same summary of the original instance was used in the newly generated instances.

In this work, we did not leverage any data augmentation. We used the original Debatepedia dataset for evaluation and pre-processed it by removing the start token <s> and the end token <eos> to evaluate our QR-BERTSUM-TL model.

5   Experimental Setup

In this section, we describe the baselines used to evaluate the effectiveness of our proposed approach followed by the training parameters used in our experiment.

---

[4] https://git.io/JeBZX.

5.1   Baselines

We consider the following models as baselines:

QR-BERTSUM: This model adopted the BERTSUM architecture [4] and incorporated Query Relevance (QR) by concatenating the query with the document. We trained it end-to-end on the original version of Debatepedia Dataset.

BERTSUM$_{XSUM}$: This baseline used the BERTSUM model pre-trained on the XSUM dataset and did not do any fine-tuning on the Debetepedia dataset.

We also compared our proposed model with the current state-of-the-art in the Debatepedia dataset RSA-QFS + PGN model [1] as well as the first model proposed for this dataset, the DDA model [8]. In addition, we experimented with DDA on both the original and augmented versions of the Debatepedia dataset.

Table 1. Performance of different models. Here, '*' denotes 'our implementation of DDA'. 'R', 'P', and 'F' denote 'Recall', 'Precision', and 'F1' respectively. The 'Original' and 'Augmented' versions of DDA are denoted by 'ORG' and 'AUG' respectively.

MODEL	ROUGE-1			ROUGE-2			ROUGE-L		
	R	P	F	R	P	F	R	P	F
QR-BERTSUM	22.31	35.68	26.42	9.94	16.73	11.90	21.22	33.85	25.09
BERTSUM$_{XSUM}$	17.36	11.48	13.32	3.03	2.47	2.75	14.96	9.88	11.46
DDA*(ORG)	7.52	7.67	7.35	2.83	2.88	2.84	7.13	7.54	7.24
DDA*(AUG)	37.80	47.38	40.49	27.55	33.74	29.37	37.27	46.68	39.90
DDA [8]	41.26	-	-	18.75	-	-	40.43	-	-
RSA-QFS + PGN [1]	53.09	-	-	16.10	-	-	46.18	-	-
QR-BERTSUM-TL	**57.96**	**60.44**	**58.50**	**45.20**	**46.11**	**45.47**	**57.05**	**59.33**	**57.73**

5.2   Training Parameters

To pre-train the BERTSUM model on the XSUM dataset, we kept the parameters similar to the original work [4]. To fine-tune it on Debatepedia, we set new values to the following parameters: batch size = 500, warmup_steps_encoder = 6000, warmup_steps_decoder = 2000, and total training steps = 60000.

6   Results and Discussions

Experimental results of our proposed approach and other models are shown in Table 1. We report the ROUGE-1, ROUGE-2, and ROUGE-L scores[5] which were calculated based on the average across 10-folds. Among the baselines, both QR-BERTSUM and BERTSUM$_{XSUM}$ models outperform the DDA*(ORG) model

---

[5] We used the following package for calculation: https://pypi.org/project/pyrouge/.

for few-shot learning. Moreover, the transformer based QR-BERTSUM out-performing the RNN based DDA*(ORG) model suggests the effectiveness of using transformer instead of RNN. We find that data augmentation significantly improves the performance of DDA, with the DDA*(AUG) model outperforming all baselines. As our result with DDA*(AUG) could not fully reproduce the result in [8], we assume that different pre-processing settings could be the possible reason behind this since Nema et al. [8] did not mention their pre-processing approach.

When we compare our proposed QR-BERTSUM-TL model with the baselines, we find that our model significantly improved performance over the QR-BERTSUM model as well as the BERTSUM$_{XSUM}$ model, which suggests the effectiveness of utilizing both transfer learning and fine-tuning in QR-BERTSUM-TL. In comparison to the recent progress, we observe that the proposed QR-BERTSUM-TL model sets a new state-of-the-art result with an improvement of 9.17%, and 23.54% in terms of ROUGE-1, and ROUGE-L respectively over the RSA-QFS + PGN model [1]. As mentioned in [1], the RSA-QFS + PGN model generated summaries 10 times longer than the required length. In contrast, our proposed model shows high precision score by effectively generating summaries according to the required length. We also observe a huge gain in terms of ROUGE-2 compared to the previous models, with an improvement of 141.07% from DDA [8] and an improvement of 180.75% over RSA-QFS + PGN.

7 Conclusions and Future Work

In this work, we presented a transfer learning technique with the transformer-based BERTSUM model and utilized it for the QFAS task via incorporating query relevance. Our approach shows state-of-the-art result in the Debatepedia dataset without the leverage of any data augmentation. This suggests that our model can overcome the lack of availability of large training data for QFAS. Our experimental result also suggests the effectiveness of using the transformer model instead of RNN for such tasks. In future, we will investigate the performance of our proposed approach on more datasets in other applications [5,6].

Acknowledgements. This research is supported by the Natural Sciences & Engineering Research Council (NSERC) of Canada and an ORF-RE (Ontario Research Fund-Research Excellence) award in BRAIN Alliance.

References

1. Baumel, T., et al.: Query focused abstractive summarization: Incorporating query relevance, multi-document coverage, and summary length constraints into seq2seq models. arXiv preprint arXiv:1801.07704 (2018)
2. Devlin, J., et al.: BERT: pre-training of deep bidirectional transformers for language understanding. In: Proceedings of NAACL-HLT, pp. 4171–4186 (2019)

3. Lewis, M., et al.: Bart: denoising sequence-to-sequence pre-training for natural language generation, translation, and comprehension. arXiv preprint arXiv:1910.13461 (2019)
4. Liu, Y., Lapata, M.: Text summarization with pretrained encoders. In: Proceedings of EMNLP-IJCNLP, pp. 3721–3731 (2019)
5. Liu, Y., et al.: ARSA: a sentiment-aware model for predicting sales performance using blogs. In: Proceedings of SIGIR, pp. 607–614 (2007)
6. Liu, Y., et al.: Modeling and predicting the helpfulness of online reviews. In: Proceedings of ICDM, pp. 443–452 (2008)
7. Nallapati, R., et al.: Abstractive text summarization using sequence-to-sequence RNNs and beyond. In: Proceedings of ACL, pp. 280–290 (2016)
8. Nema, P., et al.: Diversity driven attention model for query-based abstractive summarization. In: Proceedings of ACL, pp. 1063–1072 (2017)
9. Pennington, J., et al.: GloVe: global vectors for word representation. In: Proceedings of EMNLP, pp. 1532–1543 (2014)
10. Rush, A.M., et al.: A neural attention model for abstractive sentence summarization. In: Proceedings of EMNLP, pp. 379–389 (2015)
11. See, A., et al.: Get to the point: Summarization with pointer-generator networks. In: Proceedings of ACL, pp. 1073–1083 (2017)
12. Vaswani, A., et al.: Attention is all you need. In: Proceedings of NIPS, pp. 5998–6008 (2017)
13. Yao, J., Wan, X., Xiao, J.: Recent advances in document summarization. Knowl. Inf. Syst. **53**(2), 297–336 (2017). https://doi.org/10.1007/s10115-017-1042-4

Word Representations, Seed Lexicons, Mapping Procedures, and Reference Lists: What Matters in Bilingual Lexicon Induction from Comparable Corpora?

Martin Laville[1]([✉]), Mérième Bouhandi[1], Emmanuel Morin[1],
and Philippe Langlais[2]

[1] LS2N, UMR CNRS 6004, Universite de Nantes, Nantes, France
{martin.laville,merieme.bouhandi,emmanuel.morin}@ls2n.fr
[2] Université de Montréal, Montreal, QC H3C 3J7, Canada
felipe@iro.umontreal.ca

Abstract. Methods for bilingual lexicon induction are often based on word embeddings (WE) similarity. These methods must be able to project the WE to the same space. Uncontextualized WE proved to be useful for this task. We compare them to contextualized WE and Bag of Words, using specialized and general datasets. We also evaluate the impact of seed lexicons and check the existing reference lists validity, claiming that extracting the translation of some words in those lists is not useful and confirming the need to have more fine-grained reference lists.

1 Introduction

Bilingual lexicons are mainly made of word pairs considered to be word-level translations of each other. They are an essential resource for several bilingual tasks, such as machine translation, cross-lingual information retrieval, and their automatic extraction, from parallel and comparable corpora, is a very active research topic. With word embeddings (WE) [3,7], being greatly in fashion these past few years and with the emergence of various mapping methods to project different languages in the same embedding space [2,8], several solutions to compare word meaning across languages have been implemented.

The recent surge of contextual embedding models [5,9] allows an interesting extension of previous work on uncontextualized WE, and various solutions have been built [11,12] to adapt these WE to actual mapping methods.

In this work, we challenge the current evaluation protocol by studying the reference lists used for bilingual lexicon induction (BLI) from comparable corpora. These lists are often used as-is, and there is a general tendency to think

This research has received funding from the French National Research Agency under grant ANR-17-CE23-0001 and the Canadian Institute for Data Valorisation.

that the larger the list size, the more significant the results, even if they are filled with proper names, perfect cognates or even incorrect words in the language of interest. We examine these issues by filtering down general and specialized reference lists into sublists, examining the resulting differences when using supervised and unsupervised methods and the Bag of Words (BoW) method.

Our main contributions seek to observe and understand the difference between BLI techniques when using specialized and general corpora and to take a more critical and precise look at these lists.

2   Methods

We compare three word representations (BoW, uncontextualized and contextualized WE methods) and various mapping methods (unsupervised and supervised).

2.1   Bag of Words

The Distributional Standard Approach [10] is the historical method for BLI from comparable corpora. Based on the idea that a word is defined by its context, the semantic proximity of two words is determined by the degree of overlap of their contexts. For each source language word, its context vector is translated into the target language using a bilingual seed lexicon, allowing the source word and its translation to appear in similar contexts, enabling their alignment.

For specialized corpora, [6] shows that adding general data to the corpus improves the representations of general words, allowing for an increase in results.

2.2   Embedding Methods

The introduction of deep distributed representations [7] renewed this historical method. In [8] the authors proposed an approach to learn a linear transformation from the source to the target embedding spaces.

We use fastText [3] as our uncontextualized WE. For contextualized WE, we use ELMo [9] and pre-trained models from [11]. To make contextualized embeddings suitable for classic mapping methods, we follow [11], creating anchors for each word by averaging the embeddings of each occurrences of this word.

After extracting our embeddings separately from the source and target corpora, and in order to be able to compare them, we map the obtained matrices into the same space.

We use two different approaches to map both fastText and anchored ELMo embeddings. The unsupervised one [2] creates and refines an initial seed lexicon based on the idea that source (X) and target (Y) embeddings space are perfectly isometric. The similarity matrices $(M_X = XX^T$ and $M_Y = YY^T)$ are, then, just a permutation of their rows and columns. We also experiment with a supervised [1] method, with seed lexicon of different sizes.

2.3 Cross-domain Similarity Local Scaling (CSLS)

After mapping the word vectors in a shared space, we measure the similarity between each source word of the reference list and the target words. Since the usage of cosine similarity measure suffers from the hubness problem (some points tend to be nearest neighbours to many others), we reduce the similarity for word vectors in dense areas and increase it for isolated ones by using CSLS [4].

3 Data and Analysis

In this section, we describe the corpora, seed lexicons and reference lists used in most evaluations conducted recently [11,12].

We use two English/French comparable corpora. The Breast Cancer (BC) corpus represents our specialized domain and contains 500,000 words for each language. It is composed of scientific documents, available in open access on the ScienceDirect portal, where the title or the keywords contain the term *breast cancer* in English (and their French translation). Our general corpus is a fraction of the same Wikipedia (WIKI) dumps than [11] (100M words for each language).

For the supervised mapping and the BoW approach, we use a first seed lexicon (10,872 pairs) from MUSE [4] and a second one from the general domain ELRA-M0033 French/English dictionary (243,539 pairs).

As for evaluation (see Table 1), we use one general domain reference list from [4], and a specialized one from the UMLS. The reference list in the specialized domain is smaller than the general one because it is harder to find many words in the same specialized domain if we do not want to incorporate less specific words. However, the specialized reference list represents a more significant part of its corpus vocabulary than the general one does (6% versus 0.5%).

Table 1. Size of the reference lists and their sublists

Domain	Original	In-dictionary	Lev. ≥ 3	Freq. ≤ 100
General	1,446	1,139 (79%)	783 (54%)	146 (10%)
Specialized	248	216 (87%)	85 (34%)	18 (8%)

In the general domain list, we found many words that do not belong to the language of interest (i.e. *garrison* or *enjoy* being in the French part). Translating such entities is not of much interest and pollutes the conducted evaluation. In order to verify this claim, we filter our lists with monolingual general dictionaries[1,2], removing also proper names in the process, but isolating a subset of pairs that makes more sense to translate. This cuts down the general domain list by almost a quarter of its words. We also apply the dictionary filtering on

---

[1] English dictionary: github.com/dwyl/english-words.
[2] French dictionary: infolingu.univ-mlv.fr/DonneesLinguistiques/.

the specialized domain reference. Here, unlike for the general domain, we see that most of the words found to be out-of-dictionary are acronyms (e.g., *DNA*, *AIDS*) or in-domain words which are not part of a standard dictionary that we still want to translate because they are part of the specialized domain.

Furthermore, in the remaining pairs for both general and specialized lists, we found many pairs with nearly identical words. Even if usage of monolingual dictionaries already solved part of this with the deletion of proper names and city or country names, we use the Levenshtein distance (the number of deletions, insertions, or substitutions required to transform a string to another) and add a third reference list trying to study words with no shared morphology.

We create one last sublist to study rare words, only keeping the ones appearing less than 100 times, dropping the size of our specialized domain list to only 18 pairs, making it quite hard to draw meaningful conclusions.

4   Results

Table 2 shows the results obtained using the different reference lists, word representations models, and mapping approaches on the general and specialized domains. For the general domain, we extract the vectors from WIKI. For the specialized domain, we enrich BC with general data from WIKI.

We can observe that, even if the results of the unsupervised methods are close to the supervised ones, the latter is still the way to go, being at least two points higher in most configurations. We can also see that using a more substantial dictionary does not mean getting better results for ELMo and fastText.

Table 2. P@1 (%) on multiple lists using different word representations.

Domain	Mapping	Embeddings	Original	In-dictionary	Lev. ≥ 3	Freq. ≤ 100
General	Unsupervised	fastText	**68.9**	60.2	38.4	30.8
		ELMo	62.1	**72.2**	**57.2**	**68.0**
	Supervised (MUSE)	fastText	<u>**70.4**</u>	64.6	44.1	44.9
		ELMo	63.4	<u>**72.7**</u>	<u>**59.0**</u>	<u>**70.1**</u>
		BoW	53.4	49.9	35.7	4.3
	Supervised (ELRA)	fastText	63.8	63.2	44.2	41.7
		ELMo	57.4	70.1	55.9	58.5
		BoW	43.8	46.2	34.7	3.8
Specialized	Unsupervised	fastText	**80.6**	**81.4**	60.0	<u>**94.4**</u>
		ELMo	70.4	77.7	**61.2**	61.1
	Supervised (MUSE)	fastText	<u>**81.8**</u>	<u>**82.3**</u>	<u>**63.5**</u>	83.3
		ELMo	68.4	75.3	62.4	50.0
		BoW	59.5	65.3	53.8	16.7
	Supervised (ELRA)	fastText	80.2	81.9	62.4	77.8
		ELMo	68.8	75.8	61.2	50.0
		BoW	67.6	73.5	61.2	27.8

The results obtained with the three mapping methods all have the same trends for the different lists, but the list-based variations are more significant. On the original list, fastText always gets the most interesting results. However, when we filter the list down, its results degrade notably, while ELMo is way less affected, showing that fastText is better at predicting graphically close words since it works with character n-grams. BoW is left behind, especially with the filtered lists (less than 5% for general domain and with frequency ≤ 100).

5 Analysis

In this section, we provide a more qualitative analysis of the results obtained for both general and specialized domains to illustrate the trends mentioned above from studied lists. To do this, we show in Table 3 some word pairs with their frequencies and their n-best translations as found by the different approaches.

For the general domain, we observe that fastText mostly finds graphically close words, without really grasping the concept behind the words ("napoléone" is a plant, and "rings" while "wrestlers" are not French words). Conversely, ELMo seems to capture their meaning, finding war-related concepts for "napoleon", or geometric shapes for "rings". BoW seems really affected by word occurrences.

In the specialized domain, since words are supposed to have only one specific meaning, they are less likely to be found in varying contexts. FastText and

Table 3. 4 best translations obtained for pairs on Supervised (MUSE).

Domain	Method	Word translation	Top 1	Top 2	Top 3	Top 4
General	fastText	napoleon: 2.1k napoléon: 5.2k	napoléon	napoléone	napoléonienne	napoléonnien
	ELMo		bélisaire	napoléon	guerry	salinator
	BoW		napoléon	bonaparte	xiv	prussien
	fastText	rings: 710 anneaux: 117	anneaux	rings	ring	anneau
	ELMo		anneaux	ceintures	sphères	balls
	BoW		anneaux	rouhault	penon	mémère
	fastText	wrestlers: 27 lutteurs: 10	catches	catchers	wrestlers	catch
	ELMo		lutteurs	joueurs	joueuses	joueuses
	BoW		grandidieri	bergroth	committeer	shinjitsu
Specialized	fastText	birth: 9k naissance: 14k	naissance	décès	âge	deuil
	ELMo		naissance	baptême	éclosion	décès
	BoW		naissance	enfant	mère	femme
	fastText	keratin: 66 kératine: 52	kératine	fibroblaste	adipocyte	prolactine
	ELMo		kératine	collagène	mélanine	tanin
	BoW		kératine	luminales	fibrine	hyperdensité
	fastText	vincristine: 23 vincristine: 15	vincristine	dominique	monique	colette
	ELMo		vincristine	raloxifène	fusarium	doxorubicine
	BoW		vinorelbine	herceptin	rechuter	docétaxel

BoW do a better job at understanding these words, even if they can have some problems for infrequent words like "vincristine".

6   Conclusion

This work sought to study the different BLI methods and their evaluation when using specialized and general comparable corpora. Comparing mapping methods, we observe that the results follow the same trends. The choice of seed lexicon, however, is more impactful as bigger lexicons cause the performances to decrease in the general domain and increase in the specialized one. Supervised mapping is still the way to go without needing large lexicons. FastText gives the best results on the original lists, as they are composed of a lot of graphically close words. ELMo gets better results with the sublists, as it is better at capturing concepts.

Reference lists for these approaches are often used as-is. For both domains, we challenge the validity of these lists, arguing that not all the words are worth translating. To verify this claim, we broke down our lists into sublists, isolating subsets that make more sense to translate. When comparing the results for the original list and the sublists, we see clear differences in performances, indicating the necessity of having more fine-grained reference lists.

References

1. Artetxe, M., Labaka, G., Agirre, E.: Generalizing and improving bilingual word embedding mappings with a multi-step framework of linear transformations. In: AAAI, New Orleans, LA, USA, pp. 5012–5019 (2018)
2. Artetxe, M., Labaka, G., Agirre, E.: A robust self-learning method for fully unsupervised cross-lingual mappings of word embeddings. In: ACL, Melbourne, Australia, pp. 789–798 (2018)
3. Bojanowski, P., Grave, E., Joulin, A., Mikolov, T.: Enriching word vectors with subword information. CoRR abs/1607.04606 (2016)
4. Conneau, A., Lample, G., Ranzato, M., Denoyer, L., Jégou, H.: Word translation without parallel data. CoRR abs/1710.04087 (2017)
5. Devlin, J., Chang, M.W., Lee, K., Toutanova, K.: BERT: pre-training of deep bidirectional transformers for language understanding. In: NAACL-HLT, Minneapolis, MN, USA, pp. 4171–4186 (2019)
6. Hazem, A., Morin, E.: Leveraging meta-embeddings for bilingual lexicon extraction from specialized comparable corpora. In: COLING, Santa Fe, NM, USA, pp. 937–949 (2018)
7. Mikolov, T., Chen, K., Corrado, G., Dean, J.: Efficient estimation of word representations in vector space. arXiv:1301.3781 (2013)
8. Mikolov, T., Le, Q.V., Sutskever, I.: Exploiting similarities among languages for machine translation. CoRR abs/1309.4168 (2013)
9. Peters, M.E., et al.: Deep contextualized word representations. CoRR abs/1802.05365 (2018)
10. Rapp, R.: Automatic identification of word translations from unrelated English and German corpora. In: ACL, College Park, MD, USA, pp. 519–526 (1999)

11. Schuster, T., Ram, O., Barzilay, R., Globerson, A.: Cross-lingual alignment of contextual word embeddings, with applications to zero-shot dependency parsing. In: NAACL-HLT, Minneapolis, MN, USA, pp. 1599–1613 (2019)
12. Zhang, Z., Yin, R., Zhu, J., Zweigenbaum, P.: Cross-lingual contextual word embeddings mapping with multi-sense words in mind. arXiv:1909.08681 (2019)

Attending Knowledge Facts
with BERT-like Models
in Question-Answering: Disappointing
Results and Some Explanations

Guillaume Le Berre[✉] and Philippe Langlais

Université de Montréal, Montréal, QC, Canada
`guillaume.le.berre@umontreal.ca`, `felipe@iro.umontreal.ca`

Abstract. Since the first appearance of BERT, pretrained BERT inspired models (XLNet, Roberta, ...) have delivered state-of-the-art results in a large number of Natural Language Processing tasks. This includes question-answering where previous models performed relatively poorly particularly on datasets with a limited amount of data. In this paper we perform experiments with BERT on two such datasets that are OpenBookQA and ARC. Our aim is to understand why, in our experiments, using BERT sentence representations inside an attention mechanism on a set of facts tends to give poor results. We demonstrate that in some cases, the sentence representations proposed by BERT are limited in terms of semantic and that BERT often answers the questions in a meaningless way.

1 Introduction

Question answering has long been a core task in Natural Language Processing (NLP). Due to the booming of deep learning, there has been recently a resurgence of work on question-answering, leading to multiplications of benchmarks. Despite the amazing progress made by deep learning methods, current models fail to achieve human performance on a lot of these benchmarks. Similarly to other NLP tasks, question answering features a wide range of sub-tasks. Some extractive question answering datasets provide an open question on a short text and require the models to select a chunk in the text (often corresponding to an entity) that answers the question. SQuAD 1 and 2 [1,2] are two popular exemples of this type of benchmark. On the other hand, datasets like CoQA [3] rather ask models to generate an answer that is typically not a span of the input text or like in RACE [4] provide multiple answer choices from which to choose while still using a provided text as reference. In our work, we will focus on two datasets, OpenBookQA [5] and ARC [6], representative of another type of question answering task. OpenBookQA and ARC feature questions adjoined with 4 possible answers (similar to RACE), but unlike the datasets presented above do not provide a reference text with each question. Instead, both datasets

C. Goutte and X. Zhu (Eds.): Canadian AI 2020, LNAI 12109, pp. 356–367, 2020.
https://doi.org/10.1007/978-3-030-47358-7_37

are adjoined with a set of common sense sentences supposedly containing all the knowledge needed to answer the questions but not directly linked to any question in particular. The competing models thus can retrieve information from this set of sentences in order to answer the proposed question. However, most state of the art models rather choose to ignore this additional information and instead rely on learned world knowledge. We believe that learning to use this knowledge can lead to improved performances and higher generalization capability.

Recently, [7] introduced BERT, a pretrained deep learning model which showed huge improvements in a large number of NLP tasks including question answering. Models inspired from BERT are currently widely used on a lot of question answering datasets and often hold the first places in the leaderboards. Notably, on SQuAD, most recent models even beat the human level performance.

These pretrained models can take advantage of a massive quantity of unlabelled data and are thus particularly useful for tasks that require common sense knowledge since they already embed some semantic knowledge from the pretraining. Furthermore, pretraining allows for shorter training time on specific data and is particularly advantageous for relatively small datasets like ARC and OpenBookQA that each only gathers a few thousands of questions. Nearly all current state-of-the-art models on ARC and OpenBookQA are pretrained models, but they still do not reach human level performances.

In this work, we report the results of multiple experiments we conducted with BERT-like models on OpenBookQA and ARC. More precisely, our objective is to evaluate the possibility of using external common sense knowledge to enhance the current models. Although OpenBookQA and ARC do not provide a reference text for each question, both datasets are adjoined with a list of common sense items written in natural language, that is, short sentences such as "A bee is a pollinating animal". To the best of our knowledge, among all the models proposed to address this tasks, most of them only use the common sense knowledge acquired during training (including pretraining) and only a few models really used this dataset of common knowledge. A model able to use this extra data may allow at test time to add sentences into the common sense database and thus adapt to some extent to new domains without retraining. In addition, using this common sense database is a first step toward building a model able to reason using the short sentences (facts) in the database and combining them to assess the rightness of an answer.

2   Related Work

In this section, we first sketch how BERT works. Then we introduce other models and techniques we implemented in this work.

BERT [7] is a deep language model pretrained on the BooksCorpus [8] and English Wikipedia. The model itself is composed of a multi-layer transformer [9]. Once pretrained, it provides a context dependant embedding of all the words in a sentence and thus can provide a sentence embedding as well by either taking the first token's embedding ('CLS' special symbol) or the mean of the words embeddings. BERT is pretrained on two unsupervised tasks: "Masked LM" and "Next Sentence Prediction". In Mask LM, the model is fed with a sentence where some words are randomly masked and has to predict the missing words. While in Next Sentence Prediction, the model is fed with two sentences and has to determine if the last is the actual sentence following the first. Two versions of BERT have been released: the "base" version with 12 layers and representation vectors of dimension 768 and the "large" version with 24 layers and vectors of dimension 1024.

Following the release of the initial paper, multiple adaptations of BERT have been proposed (XLNet [10], Roberta [11]), each one improving the model or the pretraining procedure. Sentence-BERT (SBERT) [12] is one of these models that aim to increase the semantic meaningfulness of the sentence representation provided by BERT. The authors propose to add to the standard BERT pretraining an additional pretraining step on the SNLI dataset [13]. SNLI is a dataset containing sentence pairs labelled as entailment, contradiction or neutral. They fine-tune BERT using a Siamese neural architecture so that two sentences marked as entailment have a representation close to one another (cosine distance) while two sentences marked as contradiction have representations that are far apart. Their claim is that the resulting sentence representations are more semantically relevant than the representations obtain by a vanilla BERT model.

In this work, we also experiment with MAC Cells [14]. This architecture was first introduced on the CLEVR dataset [15] (question-answering using an image) in order to improve the reasoning capability of attention-based neural networks. This model is build as a recurrent network maintaining two state vectors (memory and control). Control vector is used to determined which reasoning action must be performed at each step while memory vector is a representation of all the information the model has obtained. The MAC cell is composed of 3 modules (see Fig. 1): at each step the "control" module updates the control vector using the previous control vector. A "read" module then uses this new control vector and the previous memory vector to perform an attention on a database (attention over the image in the case of CLEVR) and thus creates a proposed new memory vector. The final new memory is a linear combination of the previous and proposed memory decided by the "write" module as a function of the control state. MAC Cells obtained good results on CLEVR and showed they were capable of basic reasoning.

3   Experimental Protocol

For this study, we use 2 datasets: OpenBookQA [5] and ARC [6]. Both are multiple-choice question datasets with 4 choices for each question. The questions are about a broad variety of subjects related to everyday life logic or general knowledge.

The OpenBookQA dataset is composed of 3 parts: train, validation and test with respectively 4958, 500 and 500 questions in each. It is adjoined with two small datasets of common sense sentences. They are similar in their content

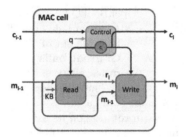

Fig. 1. Inner working of a MAC cell. The cell is a recurrent unit maintaining 2 state vectors c_i and m_i. c_i is the control state and determine which action is to be done at a given time step and m_i is the memory state that stores the information from previous steps. Image from [14]

but one of them is composed of all the sentences provided to the annotators as an inspiration during the creation of the dataset (each question is linked to 1 sentence but each sentence may have been used for multiple questions) and thus these sentences most of the time contain the necessary information to answer the questions. The correspondence between the questions and these sentences is known. The second one is composed of the same kind of sentences but these ones are not directly related to any question in particular. The first and second datasets are composed of 1327 and 5168 sentences respectively. See Fig. 2 for examples of a question and fact sentences.

The ARC dataset is split into 2 parts. The "Easy" part corresponds to questions well answered by classical machine learning models while more complicated questions belong to the "Challenge" part. The train sets for "Easy" and "Challenge" contain 2252 and 1120 questions respectively. Similarly to the fact dataset in OpenBookQA, we have access in ARC to a 1.4 GB dataset of common sense sentences data-mined from the web but no information about which ones can help a given question.

The metric used for evaluation is accuracy defined as the percentage of questions correctly answered. Current state-of-the-art models on OpenBookQA reach 78% accuracy using Roberta [11] and an additional pretraining on RACE dataset [4] while BERT Large with no additional pretraining achieves 60% accuracy. On ARC challenge, the best model scores 68% accuracy. The scores of BERT Base and Large on ARC challenge are around 36% and 40% respectively.

4   Models

In this section we present different model architectures we implemented. In all of these models, we use BERT or SBERT as a sentence embedding technique. To obtain the embedding of a sentence we use a mean pooling over the word representations provided by (S)BERT. We observed no significant differences in performance between the mean pooling and other methods of pooling (e.g. take the start token representation vector as sentence embedding) as long as

OpenBookQA	
Question	Stars are ... - A: warm lights that float B: made out of nitrate **C: great balls of gas burning billions of miles away** D: lights in the sky
Related fact	A star is made of gases
Other interesting facts	▷The Earth rotating on its axis causes the sun to appear to move across the sky at night ▷The Earth rotating on its axis causes stars to appear to move across the sky at night ▷The north star does not move in the sky in the Northern ▷Hemisphere each night ▷Burning wood is used to produce heat
ARC	
Question	Which is a nonrenewable resource? - **A: oil** B: trees C: solar energy D: food crops

Fig. 2. An example of 2 questions in ARC and OpenBookQA with the fact given to the annotator and 4 additional facts automatically selected by word co-occurrence with the question and all possible choices.

the weights of BERT are fine-tuned on the end task. Although we only refer to BERT in the following model description, BERT and SBERT embeddings are commutable, and we tested both.

4.1 Model A

This model is the vanilla question-answering setting for BERT used by the large majority of the proposed models on ARC and OpenBookQA. The answer choices are concatenated to the question thus obtaining 4 "question + choice" sequences. From there, we use BERT to get a sentence embedding vector and we send this embedding vector through a 2 layer perceptron with a ReLU activation function in between to obtain a single scalar score for each sequence. The scores corresponding to the 4 choices are then gathered and a softmax is applied on them. During training, we use a cross-entropy loss and at test time, the choice with the highest score is selected as the predicted answer. See Fig. 3A.

4.2 Model B

In addition to the "question + choice" sequence embedding we provide the model with an additional sentence also embedded with (S)BERT. Note that this model and the following ones are trained on OpenBookQA only since they require the link between the common sense database and questions. We use the common knowledge sentence associated to the question in OpenBookQA. This has been provided to annotators as an inspiration to write the question and thus, this sentence is supposed to give enough information to the model to answer the question. The idea behind this model is to evaluate the semantic quality of the sentence embedding. Since the additional sentence is supposed to have a meaning closer to the answer than the other choices, their embedding should also be closer. The new sentence embedding is concatenated to the "question + choice" embedding and fed to the final perceptron. See Fig. 3B.

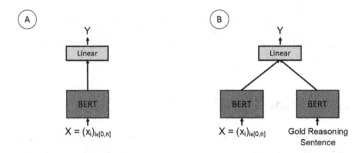

Fig. 3. Description of model A (left) and B (right). X represents the input sequence consisting of the question plus one answer choice. Y represents the output scalar score for this particular answer choice. The scores for all answer choices are then gathered and passed through a softmax function (omitted here).

4.3 Model C

Model B is not applicable to any real case scenario since it presupposes having access to the sentence that inspired the question. In this third architecture, we add an attention mechanism to B. Instead of a unique sentence, the model now has to select the right sentence from a set of 10 sentences extracted from the common knowledge dataset of OpenBookQA. We select the 9 sentences with the highest number of words in common with the question and add the "target" sentence if not already selected. See Fig. 4C.

4.4 Model D

Finally, we experimented with MAC Cells [14]. Usually, multiple facts are relevant for a given question. The reasoning capability of MAC Cells can thus be useful in order to assemble multiple pieces of information from different facts. We replace the simple attention of model C by a MAC Cell in the hope, this would help the model to extract more information from multiple sentences. See Fig. 4D.

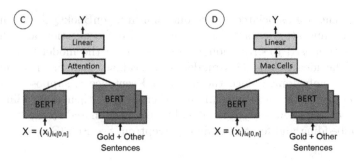

Fig. 4. Description of model C (left) and D (right). The only difference with model B is that instead of a single help sentence, the model has access to multiple sentences including the help and needs to select the right one with an attention (model C) or MAC cells (model D).

5   Implementation Details

We use the Hugging Face Pytorch implementation of BERT [16] and the implementation of SBERT provided with the original paper. We use the base version of both BERT and SBERT. This means that the word embeddings are of dimension 768 and we kept the same vector size along all the steps of the models. We implemented the rest of the code using Pytorch.

The training of all the models is made with a mini-batch of size 8. The shorter sentences in the mini-batch are padded using a special token and we cropped the sequences longer than 40 tokens by removing the first tokens thus ensuring that we keep the answer choice untouched (Table 3 provides some length statistics for OpenBookQA). We train the models for a maximum of 5 epochs after which the models always starts to overfit since the datasets are small and BERT has a huge capacity. To compensate for overfitting, BERT weights are frozen during the first 2 epochs to let a chance to the other parts of the network to converge. On models that require an attention mechanism, we added an additional cross-entropy loss directly on the attention distribution to help the models to quickly identify what is the sentence that gives the required information. A weighting coefficient is applied to this part of the loss. It starts to 1.0 and decreases to 0.2 after the first 2 epochs. All the hyper-parameters above are chosen to maximize accuracy on validation set.

When training on the ARC dataset, we use both "Easy" and "Challenge" training sets and report separated results for test and validation sets.

6   Experiments

In this section, we present the experiments we made with BERT on Open-BookQA and ARC. All the experiments shown below were made in an attempt to explain why in our preliminary experiments we failed to apply an attention mechanism (model C) over a knowledge base of sentences embedded with BERT.

First, in order to appreciate what parts of a question BERT is using while answering it, we decrease the number of tokens available to the model. To do this, we use the A setup described in Sect. 4. We provide the results (in terms of accuracy) of a model that has access to the complete question and a model with access to only the last 4 tokens of the question. Eventually, we completely removed the question thus feeding only the answer choice to BERT. We of course expect the model to reach around 25% accuracy (OpenBookQA has 4 answer choices for each question) with this setup since without the question, there is supposedly no way to differentiate the right answer from the other choices. We verified that the dataset (train, validation and test) is balanced in the sense that all answers choices (A, B, C and D) appear approximately 25% each.

Table 1. Accuracy on OpenBookQA when the model has access to the full question (full) or only the 4 last tokens of the question (4 tokens) or no question at all (none). model A + help refers to model A in which the input is the concatenation of the help sentence, question and answer choice instead of just question + answer choice.

	Full	4 tokens	None
BERT (model A)	**55.8**	52.0	51.2
SBERT (model A)	53.2	**53.0**	**53.6**
BERT (model A + help)	**64.5**	64.9	–
SBERT (model A + help)	63.6	**65.2**	–
BERT (model B)	53.0	53.4	–
SBERT (model B)	55.0	**61.3**	—
SBERT (model C)	54.8	56.8	–
SBERT (model D)	**56.8**	**56.8**	–

Table 2. Accuracy on ARC (Easy and Challenge) when model A has access to the full question (full), only the 4 last tokens of the question (4 tokens) or no question at all (none).

	Full	4 tokens	None
BERT (Easy)	51.7	46.9	36.5
BERT (Challenge)	36.5	36.2	34.2

Table 1 shows the resulting accuracies on the validation and test sets on OpenBookQA (first 2 lines) for both BERT and SBERT (for comparison purposes since we use SBERT in later experiments). What comes out of this first experiment is that, surprisingly, giving only the last 4 tokens of the question to BERT only slightly decreases the accuracy. Furthermore, if we completely remove the question, BERT is still managing to get an accuracy greater than 50%, not so far from the original accuracy and even, in the case of SBERT, a better accuracy.

We ran the same experiment on both parts (Easy and Challenge) of ARC. The results are reported in Table 2. The challenge set results are quite similar to what we observed for OpenBookQA: the accuracy for 4 tokens and no token are close to the accuracy obtained by the vanilla BERT model. On the Easy set however, we see a clear degradation of the accuracy when removing tokens but we still obtain at worse 36% accuracy which is significantly better than the expected 25%. This shows that in any case, a significant part of the questions can be answered without even looking at the question itself.

Table 3. Mean/min/max length of the questions and answers in OpenBookQA. Correct answers tend to be slightly longer than wrong ones.

Length in words (train set/test set)	Mean	Min	Max
Questions	10.7/10.3	1/1	68/61
Wrong answers	2.7/3.0	1/1	20/16
Correct answers	3.0/3.3	1/1	21/15

Table 4. Mean/min/max frequencies of tokens in an answer choice averaged over the dataset in OpenBookQA. Correct answers tend to contain at least a token that is more specific (less frequent) than the other answers.

Word frequency (train set/test set)	Mean	Min	Max
Wrong answers	0.0054/0.0056	0.00029/0.00033	0.016/0.017
Correct answers	0.0050/0.0058	0.00014/0.00019	0.016/0.019

Our models are thus mainly learning what the characteristics of a good answer are instead of learning the logical link between a question and the corresponding answer. We found some pieces of explanation for this. First, we compute statistics about the length of the questions and answers. These statistics are reported in Table 3 and show that the right answers are on average longer than their counterparts. In practice, a dummy model that selects the longest answers among the 4 proposed achieves a 33% accuracy. However, this does not explain the entirety of the 51.2% of BERT on OpenBookQA.

Question	What impacts an objects ability to reflect light?
Answer choices	**A: color pallete** B: weights C: height D: smell
4 tokens	...ability to reflect light?

Fig. 5. An example of question in OpenBookQA (the spelling error is from the dataset) for which it is relatively easy to answer using only the last 4 tokens. Note that the correct answer is more complex than the other three and so it might be possible to guess the answer without reading the question.

In Table 4, we further provide statistics about the relative frequencies of the tokens in correct vs. incorrect answers. What is interesting here is that the least frequent token in the correct answers is on average less frequent than the least frequent token in incorrect answers. This means that the correct answers tends to include more specific words than incorrect ones. This could be a bias linked to the annotation method in which annotators are required to invent incorrect answers. It is possible that one tends to be more generic when trying to write a wrong answer without inspiration. A dummy model that selects the answer with the least frequent token achieves 36.8% accuracy on OpenBookQA. Finally, our experiments shows that 54.6% of the correct answers are either the longest sequences or the one with the least frequent token in it. These two experiments although they do not explain the entirety of the phenomenon, show that the models (BERT and SBERT) can be heavily biased by factors unrelated to the question-answering logic (Fig. 5).

As explained in Sect. 3, OpenBookQA provides along each question a common knowledge short sentence that was provided to the annotator as an inspiration for the question. Although using this sentence directly causes the results to be incomparable with results showed above and state-of-the-art, we can use it to evaluate how much accuracy could be gained if we could make a decent selection of related knowledge in a database. We refer at this particular sentence as the help sentence in this section.

In the following, we still use BERT/SBERT in the A setting on OpenBookQA but instead of the simple question + answer sequence, we concatenate at the beginning of the sequence the common sense sentence related to the question. The idea for now on is to see if the representations from BERT/SBERT are usable in a configuration where the solution is directly given to the model. Table 1 shows that as expected, we obtain a significant increase in performance with both BERT and SBERT. Now, what happens if the new sentence is not given prior to BERT embedding but rather posterior to it? To test this, we use the B setting. In this configuration, the model now has to rely on a good semantically significant sentence embedding to answer since it has to evaluate the similarity between the help sentence and each of the answer choices. The accuracy is again described in Table 1 (lines 5 & 6). We observe that BERT performs relatively poorly at this task and shows no significant change of its accuracy between this configuration and no help sentence at all. SBERT however obtains a gain similarly to what happens when concatenating the help sentence to the question. Thus, the representations of SBERT seem to be more adapted to find semantic similarity links between sentences.

To confirm this observation we ran yet another dummy model. This model compute the similarity (cosine distance) between the BERT/SBERT representation of the help sentence and the representations of the question + answers. The process is done without any training using only the pretrained BERT and SBERT. Using this configuration, SBERT achieves 57.6% accuracy on Open-BookQA test set while BERT only reaches 37.4% once again demonstrating that SBERT is more adapted to the task.

Finally, we tested more realistic settings. Instead of giving the help sentence directly, we "hide" it among other similar sentences thus simulating a scenario in which we use a selector able to accurately select a pool of fact sentences useful for answering the question. As explained in Sect. 4, we select 9 sentences from the fact dataset of OpenBookQA according to the maximum word overlap with the question and we add the help sentence to ensure a good selection. We report the results of SBERT for configuration C and D in Table 1. BERT performs poorly for both configurations during our experiments (likely due to the reasons previously exposed) and so only the accuracy of SBERT are shown here. Overall, the models C and D are weaker than the configuration B with direct help but still perform better than model A where no help is given. This tends to show that there is value to be earned by adding additional common sense information to the inputs of the model and in the future, we intend to continue working on new ways to achieve this objective. We would like for example to try to re-balance correct and incorrect answers by picking correct answers from another question as new incorrect answers. Like this, we could potentially improve the training of BERT by removing the local minimum created by the length and frequencies bias.

7   Conclusion

In this work, we compared a number of models based on BERT-like models for question-answering. We report disappointing but informative results. Our experiments show that in the case of OpenBookQA more than 50% of the questions can be answered without even looking at them, which represents a big bias that has to be taken into account when considering state-of-the-art results. This observation also transposes to some extent to other question-answering benchmarks such as the ARC dataset. In addition we present a comparison of BERT and SBERT representations in term of semantic usefulness and show that SBERT is more apt to combine sentence representations together in order to answer the question.

References

1. Rajpurkar, P., Jia, R., Liang, P.: Know what you don't know: unanswerable questions for squad. CoRR, abs/1806.03822 (2018)
2. Rajpurkar, P., Zhang, J., Lopyrev, K., Liang, P.: Squad: 100, 000+ questions for machine comprehension of text. CoRR, abs/1606.05250 (2016)
3. Reddy, S., Chen, D., Manning, C.D.: CoQA: a conversational question answering challenge. CoRR, abs/1808.07042 (2018)
4. Lai, G., Xie, Q., Liu, H., Yang, Y., Hovy, E.H.: RACE: large-scale reading comprehension dataset from examinations. CoRR, abs/1704.04683 (2017)
5. Mihaylov, T., Clark, P., Khot, T., Sabharwal, A.: Can a suit of armor conduct electricity? A new dataset for open book question answering. CoRR, abs/1809.02789 (2018)
6. Clark, P., et al.: Think you have solved question answering? Try ARC, the AI2 reasoning challenge. CoRR, abs/1803.05457 (2018)
7. Devlin, J., Chang, M.-W., Lee, K., Toutanova, K.: BERT: pre-training of deep bidirectional transformers for language understanding. CoRR, abs/1810.04805 (2018)
8. Zhu, Y.: Aligning books and movies: towards story-like visual explanations by watching movies and reading books. arXiv e-prints, page arXiv:1506.06724, June 2015
9. Vaswani, A., et al.: Attention is all you need. arXiv e-prints, page arXiv:1706.03762, June 2017
10. Yang, Z., Dai, Z., Yang, Y., Carbonell, J.G., Salakhutdinov, R., Le, Q.V.: XLNET: generalized autoregressive pretraining for language understanding. CoRR, abs/1906.08237 (2019)
11. Liu, Y.: RoBERTa: a robustly optimized BERT pretraining approach. CoRR, abs/1907.11692 (2019)
12. Reimers, N., Gurevych, I.: Sentence-BERT: sentence embeddings using Siamese BERT-networks. arXiv e-prints, page arXiv:1908.10084, August 2019
13. Bowman, S.R., Angeli, G., Potts, C., Manning, C.D.: A large annotated corpus for learning natural language inference. In: Proceedings of the 2015 Conference on Empirical Methods in Natural Language Processing, Lisbon, Portugal, pp. 632–642. Association for Computational Linguistics, September 2015
14. Hudson, D.A., Manning, C.D.: Compositional attention networks for machine reasoning. CoRR, abs/1803.03067 (2018)
15. Johnson, J., Hariharan, B., van der Maaten, L., Fei-Fei, L., Lawrence Zitnick, C., Girshick, R.B.: CLEVR: a diagnostic dataset for compositional language and elementary visual reasoning. CoRR, abs/1612.06890 (2016)
16. Wolf, T.: HuggingFace's transformers: state-of-the-art natural language processing (2019)

Machine Learning the Donor Journey

Greg Lee[1][✉], Ajith Kumar Veera Raghavan[1], and Mark Hobbs[2]

[1] Acadia University, Wolfville, Canada
{glee,146771v}@acadiau.ca
[2] Fundmetric, Halifax, Canada
mark@fundmetric.com

Abstract. A fundamental question for charities is what action should they take next to maximize the chance of receiving a donation from a particular individual. We solve this problem using time-series data, showing that based on the previous five actions in which a charitable constituent was involved, their likely donation amount after a sixth action can be predicted. Once an accurate model is learned, the best next action can be selected to maximize donations. We show that Recurrent Neural Networks can learn accurate models of how much a constituent will donate, and we use these models to suggest actions for charities to take on an individual basis.

Keywords: Machine learning · Recurrent Neural Networks · Charity

1 Introduction

When charities make data-driven decisions, they generally rely heavily on the donation history of a donor, and not on their engagement behavior, such as the actions taken by the constituent with respect to appeals sent their way. Thus, while the intention is to give each donor a personalized experience, this personalization is done on small amounts of (largely) dated donation data, ignoring what is generally the most recent donor information – the constituent's interaction with charity communications. In terms of the message being sent to these constituents, most charities use the same sequence of communications and receive similar results. Donor actions include opening an appeal email and visiting a charity donation portal.

We define the donor journey as the content and mediums of communication that a donor engages with prior to making a gift. Our goal involves modeling this journey using engagement data – how the donor interacts with the charity. Most of the data to change the sequence (and collect more money) is readily available to be collected and modelled, but the charitable world generally ignores it.

In this paper, we use deep learning, specifically Recurrent Neural Networks (RNNs), to model the donor journey using the ordered actions of the constituent and charity as data, and then suggest best actions for a charity to take, on an *individual basis*. That is, for a particular constituent, we suggest what action a charity should take to increase the amount the constituent will donate. As

© Springer Nature Switzerland AG 2020
C. Goutte and X. Zhu (Eds.): Canadian AI 2020, LNAI 12109, pp. 368–374, 2020.
https://doi.org/10.1007/978-3-030-47358-7_38

most actions are actually performed by the constituent and out of control of the charity, in our experiments we frame this question as "what are the parameters of the next email the charity sends that will maximize expected donation?" This is a first step towards the complete personalization of charitable emails. In order to have the donor journey model trusted by real-world charities, we sought an error of $30 or lower in our models.

2   Related Research

To the best of our knowledge, most fundraising decisions about emails are not based on any empirical evidence. Instead, anecdotal "best practices" are often shared on the web [1]. In [2] the authors evaluate email campaigns from perspectives such as demographics, interest, and social network influence of their constituents, as well as external time-related factors. Basing email decisions on previous actions of the charity or the constituent has not previously been considered. In [3], direct email content is explored, through experiments with Red Cross mailings. The authors found that membership cards lead to repeat donations, while providing donors with gifts actually hurt retention.

While charities have been hesitant to adopt machine learning techniques, for-profit organizations have long been using them, and the literature is extensive in terms of their strategies [4]. Some of the research in machine learning for targeted marketing in for-profits can be translated to charities, but how people spend their money on products is not necessarily related to how they donate their money to charities. Charitable giving is not as simple as one might assume [5].

The use of RNNs to help predict how people will spend money, rather than donate, is common. In [6], RNNs are used to predict the customer influx and outflux at a location based on historical parking data. This is analogous to our work, as we are using the activity of constituents to predict whether they will be spending their money to help a particular charity.

3   Problem Formulation

Email campaigns involve a charity sending multiple appeal emails to constituents in order to try to raise money for a designation within a charity's mandate. Once a constituent donates to a cause, they receive no further solicitation emails, and instead receive a thank you message. This demonstrates two actions that can take place during a campaign, the *delivery* of an email, and a *donation*. The full set of 10 actions in our experiments was: no action (a filler when an insufficient number of actions are available), delivered, opened, complained, bounced, unsubscribed, dropped (all corresponding to an email), pageview (donation portal was viewed), donated, and clicked (a link or donation amount was clicked).

Charities must choose which action to take next given the previous actions taken as part of a campaign, or even beyond a campaign (in the past). If a constituent has *opened* an email three times and then *clicked* a link, but not *donated*, what is the best action for the charity to take? Should the charity send an email? If so, should it be brief as to not annoy the constituent? Should it be long to provide more information to the constituent about the cause for which the charity is currently seeking funds?

All of these are valid questions encountered by fundraisers on a daily basis. In the next section, we demonstrate how RNNs can be used to automatically learn a model of a sequence of actions that leads to donation, and use the model to select actions and email parameters.

4   Our Approach

Our goal was to both model the donor journey and predict the next best action to be taken for a specific constituent. This would allow us to suggest actions and email parameters to charities that they should take or use with their constituents, to increase donations to their causes. We sought to accomplish this using RNNs.

RNNs using Long Short-Term Memory (LSTM) and rectified linear units (ReLU) are described in [7]. RNNs allow information to persist over time to model situations where past actions have an influence on future actions. Thus, they are an appropriate machine learning algorithm to use since we are investigating how a constituent's past actions influence their likelihood to donate to a charity in the future.

4.1   Preliminary Experimental Setup

We chose to consider the previous six actions related to a constituent since campaigns typically involve this number of constituent and charity actions. A larger window would lead to most donor journeys being padded with one or more *no action* actions, since there is no interaction between the charity and the constituent beyond these six actions.

Note that the only action that is controlled by the charity is *delivered*. Thus, the preliminary experiment tested not necessarily what action the charity should take next, but tested whether the RNN could learn a sequence of actions that leads to donations. One-hot encoding of the action label was used to eliminate any false sense of ordering over the actions. The RNN chose only the following actions as the next best action: *pageview, donated, delivered, clicked* and *opened*. This showed that the network had learned which actions lead to donations.

4.2   Email Optimization

In email appeals the only action available to charities is actually sending an email. All other actions are taken by the constituent. We thus considered parameters of an appeal email that could be changed in order to optimize donation amounts. The email parameters we considered are the number of each of the following in the email: words, paragraphs, images, links, blocks, HTML divs, variables, and editable blocks. These were chosen as they are easy to measure, and we aim to extend this list in the future.

These parameters apply only to emails, but every action in the donor journey is either directly related to an email, or one step removed from an email – except for *no action*. Thus, in addition to the one-hot encoding of the action taken, 8 additional values are added to each action, being the parameters of the associated email. If no email is associated with an action (for us, only *no action*), the 8 additional values are set to 0. As an example of an action represented in this format, consider an opened email with 235 words, 4 paragraphs, 2 images, 1 link, 2 blocks, 3 divs, 4 variables and 4 editable content sections. This action is represented as $\{0, 0, 0, 1, 0, 0, 0, 0, 0, 0, 235, 4, 2, 1, 2, 3, 4, 4\}$.

For almost all campaigns the set of donor data is much smaller than the set of non-donor data. Thus we balance the two sets by limiting the larger (non-donor) set to be the size of the smaller (donor) set. Several iterations of this process are performed to ensure the error of this method is accurate. We experimented with both types of training data and explain this in the next section.

5 Empirical Evaluation

Given our positive preliminary experiment results (Sect. 4), we determined that the basic structure of the donor journey should be able to be modeled with RNNs, as the goal of this journey is to receive donations. For the eight email parameters, the values used were similar to parameter values of emails sent by actual charities. Table 1 lists all values tried in various experiments.

Table 1. Variable email parameter values used in actual emails and experiments.

Parameter	Description
Words	{150, 200, 393, 445, 457, 474, 500, 1000}
Paragraphs	{1, 2, 3}
Images	{2, 3, 6}
Links	{28, 29, 35}
Blocks	{8, 9}
Divs	{40, 41, 44}
Variables	{2, 3}
Editable content	{8, 9, 15}

5.1 Training the RNN

Preliminary experiments were performed with data sets from three charities – a wildlife conservation charity (C1), an Alzheimer's charity (C2), and a children's charity (C3). Data was gathered from C1's Giving Tuesday appeal in 2018, C2's Spring Appeal in 2019. and C3's Year End appeal in 2018. Actions were recorded as described in Sect. 3. The data for C1, C2 and C3 are shown in Table 2.

Table 2. Training data for RNN from three charities.

	C1	C2	C3
Donors	640	229	316
Non-donors	195669	195688	50811
Total raised	$54,387	$55,952	$130,034
Mean donation	$85	$245	$409
Median donation	$50	$100	$100
Standard deviation	$105	$514	$1520
Minimum donation	$5	$1	$1
Maximum donation	$1,000	$5,000	$10,000

For each charity, 60% of the data was used to train the RNN and 40% was used to test the learned model, with no overlap between the sets. Other training and testing splits tried were 75/25 and 80/20, with inferior results. Balancing the donors with the non-donors obviously dramatically reduced the training and testing data, and we ran 10 iterations to check the accuracy and error of this setup, changing the non-donors selected for training and testing in each iteration.

All architectures for the RNN tried for each charity used 18 inputs (10 actions and 8 email parameters). The best performing architectures in terms of mean absolute error (MAE) had two LSTM layers with 64 and 32 nodes, a dropout rate of either 0.2 and 0.5, learning rates of 0.001, momentum of 0.9 and the ReLU activation function. Deeper networks did not result in more accurate predictions, nor did using a sigmoid activation function. The MAE testing error these architectures for C1, C2, and C3 were $25, $139, and $359, respectively. For C1, The error was within our margin of error to trust the model ($30). Having accomplished our first goal, we decided to query this model for the best email parameters for non-donors to each charity.

5.2   Full Email Experiments

After gaining confidence in the RNN model through some preliminary experiments, we experimented with parameters shown in Table 1, with the exception of *variables*, which we held at 2 after some analysis of the training data. Results were gathered on 4,261 constituents for C1, 1914 constituents for C2 and 1763 constituents for C3. While the goal of this project is to provide *personalized* email recommendations, several lessons were learned on the group as a whole. For parameters that showed variance, the mean, median, mode values for C1 are shown in Table 3. Results for C2 and C3 are not shown as the model did not meet the error threshold of $30, and the model always chose the same values.

For C1, emails with fewer words (150) and paragraphs (1) were chosen most often, but for some constituents, 445 words and 3 paragraphs were selected. These constituents were those who had fewer interactions with the campaign

Table 3. Mean, median and mode email parameter values chosen by RNN for C1

	Mode	Median	Mean	St. Dev
Words	150	150	155.94	41.57
Paragraphs	1	1	1.1	0.414
Images	6	6	4.65	1.89
Links	35	29	31.7	2.99
Editable content	9	9	9.24	1.18

(thus had one or more *no action* actions in their sequence). We hypothesize that the model has learned that for constituents who have received less messaging, more content in the next email will increase the donation amount. This is a first step towards our goal of personalizing email appeals.

6 Discussion and Future Work

In this paper, we showed that RNNs could learn a model of the donor journey that was within \$25 of the true donation value of a constituent for a given campaign. We used this model to suggest actions for a charity to take next to elicit donations from non-donors, and found consistent and reasonable patterns in these predictions. We would like to test the email parameter selection in an actual campaign. We have agreements with some charities to do so, given the accuracy of the models. This will involve feeding current constituents into the learned RNN model and choosing the email parameters based on the model's predictions. In this way, the true performance of the predictions can be measured. The content of the messages will be determined by outside means - namely natural language processing research that is ongoing.

To strengthen the accuracy of the RNN, we will feed features of the constituents to the model, in addition to the action and email features. This will allow for more personalization, as well as give the RNN more data to learn on.

References

1. Jepson, T.: 11 Fundraising Email Best Practices To Drive High Response Rates (2020). https://www.causevox.com/blog/11-fundraising-email-best-practices-drive-high-response-rates/
2. Mejova, Y., Garimella, V.R.K., Weber, I., Dougal, M.: Giving is caring: understanding donation behavior through email. In: The 17th ACM Conference on Computer Supported Cooperative Work and Social Computing, CSCW 2014, pp. 1297–1307 (2014)
3. Ryzhov, I., Han, B., Bradic, J.: Cultivating disaster donors using data analytics. Manag. Sci. **62**, 849–866 (2015)
4. Bose, I., Mahapatrab, R.: Business data mining — a machine learning perspective. Inf. Manag. **39**, 211–225 (2001)

5. Andreoni, J.: Giving with impure altruism: applications to charity and ricardian equivalence. J. Polit. Econ. **97**(6), 447–458 (1989)
6. Mudassar, L., Byun, Y.C.: Customer prediction using parking logs with RNNs. Int. J. Netw. Distrib. Comput. **6**, 133–142 (2018)
7. Goodfellow, I., Bengio, Y., Courville, A.: Deep Learning. MIT Press, Cambridge (2016). http://www.deeplearningbook.org

Exploring Deep Anomaly Detection Methods Based on Capsule Net

Xiaoyan Li$^{1(\boxtimes)}$ ⓘ, Iluju Kiringa^1, Tet Yeap^1, Xiaodan Zhu^2 ⓘ, and Yifeng Li^3 ⓘ

^{1} Electrical and Computer Engineering, University of Ottawa, Ottawa, Canada
{xli343,Iluju.Kiringa,tyeap}@uottawa.ca
^{2} Electrical and Computer Engineering, Queen's University, Kingston, Canada
xiaodan.zhu@queensu.ca
^{3} Computer Science, Brock University, St. Catharines, Canada
yli2@brocku.ca

Abstract. In this paper, we develop and explore deep anomaly detection techniques based on capsule network (named AnoCapsNet) for image data. Being able to encode intrinsic spatial relationship between parts and a whole, CapsNet has been applied as both a classifier and deep autoencoder. This inspires us to design three normality score functions: prediction-probability-based (PP-based), reconstruction-error-based (RE-based), and combination of both (PP+RE-based) for evaluating the "outlierness" of unseen images. Our results on four datasets demonstrate that PP-based and RE-based methods outperform the principled benchmark methods in many cases and the pp-based method performs consistently well, while the RE-based approach is relatively sensitive to the similarity between labeled and unlabeled images. The PP+RE-based approach effectively takes advantages of both methods and achieves state-of-the-art results.

Keywords: Anomaly detection · Capsule net · Normality score

1 Introduction

As real-time tracking & diagnosis systems and autonomous controlling devices are strongly demanded in various domains in the current era of Internet of Things (IoTs), smart cities, big data, and deep learning, anomaly detection (also known as *outlier detection*) is becoming increasingly critical. It aims at uncovering abnormal data points which may stand for novel or alarming events. Anomaly detection is of key importance in IoT systems, data centres, security platforms, and life science to diagnose system failure, detect intruders or attackers, and discover novel knowledge.

Prior to the use of deep learning approaches, statistical and heuristic methods were main tools for anomaly detection in restricted application domains. Kernel methods, such as kernel density estimation (KDE) [15, 25] and support vector domain description (SVDD) [31], were the most successful ones to deal with non-linearity of the input feature space through kernel trick. However, nowadays the

© Springer Nature Switzerland AG 2020
C. Goutte and X. Zhu (Eds.): Canadian AI 2020, LNAI 12109, pp. 375–387, 2020.
https://doi.org/10.1007/978-3-030-47358-7_39

tremendous amount of data of various types (such as images, texts, omics data, etc.) have been collected, posing new challenges in handling the scalability and complexity of such data. With billions of samples in modern datasets, traditional methods become less effective. For example, conventional feature encoding and extraction methods are incompetent to capture informative factors from the input feature space of complex data.

Embracing the wealth of data, deep learning models have achieved significant successes in various discriminative and generative modellings of modern data [4,8,9,11,16–18], encouraging to explore deep anomaly detection solutions. The core of deep learning is learning complex representations for the data at different levels in the latent space [3]. For example, convolutions on sequence data and embedding technologies on discrete or symbolic data allow to encode visible examples to low-dimensional continuous dense vectors in a latent space, showing the advantage of distributed representation learning over alternative approaches. Furthermore, stochastic gradient descent using mini-batches makes learning of deep networks very scalable to data of big size. The past few years have witnessed progresses made by the work reviewed in Sect. 2. Accordingly, comparative studies, e.g., [30], unconsidering these two advantages of deep learning over classic methods, could be biased.

The technique in [7] is based on the insightful observation: learning to discriminate between many types of geometrically transformed images encourages learning of features that are useful for detecting novelties. Among all geometric transformations, they only considered compositions of horizontal flipping, translations, and rotations, resulting in 72 distinct transformations. Their main focus was tackling the problem of identifying anomalous images in pure single class setting, even though it was mentioned their method may also be effective at distinguishing out-of-distribution samples from multiple-class data.

In fact, learning transformation is very challenging in computer vision tasks. Convolutional neural network (CNN) [19], a hierarchy of convolution operations, has been widely used as a highly effective technique in classifying images. However, one arguable key limitation of CNN is that the neurons do not sufficiently capture the properties of entities such as position, orientation, and sizes, as well as their part-whole relationship. The capsule network (CapsNet) [10,12,27] has been proposed and shown advantages in maintaining such information, which is a novel and promising structure that may be more closely related to biological neural organization. A capsule is a group of neurons whose activity vector represents the instantiation parameters of a specific type of entity such as an object or part. It has been demonstrated that CapsNet is capable of preserving hierarchical pose (position, size, and orientation) relationships between image features. For a given image, CapsNet can automatically and dynamically model affine transformations and part-whole relationships using an iterative routing-by-agreement mechanism.

Inspired by these developments, we propose that, instead of using geometric transformations to learn distinct features of images, CapsNet can be employed to automatically learn transformations from the training examples such that a test example that cannot be explained by the network should be viewed as anomaly. Unlike [7], our work concentrates on the cases where normal samples come from multiple classes. Our contributions are three-fold: (1) based on unique characteristics of CapsNet, we propose three normality score functions that work well; to the best of our knowledge, this is the first attempt to explore and test CapsNet for deep anomaly detection; (2) we provide insights and categorize existing ideas for deep anomaly detection into boundary-based and distribution-based families, paving the road for future studies; (3) we compared our methods with principled benchmark methods and three advanced deep generative techniques and assessed their capacities for deep anomaly detection.

2   Insights into Existing Work

Instead of simply enumerating all existing work for deep anomaly detection, we categorize these solutions into two families, and provide insights into their characteristics, strengths and challenges.

A classical viewpoint regarding anomaly detection is that, learning the boundary of data mass is more effective and straightforward than learning the density distribution of the data, because (1) available data were too few to cover the distribution in many cases, and (2) it was much more complicated and difficult to model data distribution using a generative model. However, in the big data era, massive amount of data become available in many domains; many of the data are *unstructured* (such as images, graphs, time-series, text, etc.); and deep generative models have achieved promising successes in modelling such modern data, offering a new avenue for exploring distribution-based methods for anomaly detection. These two categories, boundary-based and distribution-based, are respectively discussed below.

2.1   Boundary-Based Methods

Kernel based support vector domain description (SVDD) or one-class support vector machine (OCSVM) methods, including hypersphere [31] and hyperplane [29] models, is a successful family for anomaly detection in the pre-deep-learning era. The idea of the hypersphere-based SVDD is to map data points from input space to high-dimensional space and learn a hypersphere that capture the core mass of the data distribution. Any data point outside the hypersphere is viewed as an abnormal sample. Please see [21,22] for a systematic discussion of SVDD methods. It is quite natural to think of deep extension of these methods to continue their success in the deep learning age. In pursuit of this aim, there are two efforts: deep hybrid methods and one-class neural network models. In deep hybrid methods (e.g., VAE+OCSVM [2] and DBN+OCSVM [6]), a supervised (e.g., CNN or recurrent net) or unsupervised (e.g., deep belief net (DBN) or variational autoencoder (VAE)) neural network is first employed to learn embedding representations of samples in the hidden space, then using these latent representations as inputs an SVDD method is employed to detect abnormal data points.

One-class neural network models (e.g., one-class deep SVDD [26]) learn a neural network and SVDD together by maximising an adapted SVDD objective in the prime form. Both classes of methods have pros and cons. Deep hybrid methods are pipelines that are very flexible in choosing and combining different representation learning (or pretrained embeddings) and SVDD models. However, these methods face the challenge of scalability due to size of kernel matrices in dual form of SVDD. Deep SVDD models explicitly use deep neural networks as feature extractors in replacement of implicit kernel tricks, and are scalable to large data due to use of SVDD's prime forms and stochastic gradient descent. Nevertheless, this strategy is lack of flexibility in practice, that is a specific model needs to be built for each specific type of data.

2.2   Distribution-Based Methods

As mentioned above, deep generative models (DGMs), such as deep belief net (DBN) [13,23] and variational autoencoder (VAE) [16], can be applied as unsupervised feature learning techniques. More importantly, since DGMs aim at modelling the joint distribution of visible and latent variables (that is $p(x, h)$), their likelihood $p(x)$ by marginalising out h may serve as an anomaly describer. However, exact likelihood can only be obtained in quite few generative models, such as exponential family restricted Boltzmann machines (RBMs) [24]. In many cases when a recognition component is used for approximate inference, only the variational/evidence lower bound (ELBO) of log-likelihood is available. Unfortunately, ELBO may be too loose to be an normality indicator. Oftentimes, the ELBOs of normal and abnormal samples indistinguishably fall into the same range. Luckily, this is not the end of the story. In an architecture with encoder (recognition) component and decoder (generative) component, reconstruction error could serve as anomaly measure based on the intuition that out-of-distribution samples can be reconstructed badly [1]. But in the case of generative adversarial net (GAN) based methods [28], an encoder is unavailable. For an inquiry sample x, a supervised learning process has to be executed to search for a hidden representation h' such that the corresponding generated sample x' best approximates x. The approximation error and probability from GAN's discriminator can together indicate the extent of anomaly.

2.3   Benchmarks

Our CapsNet based anomaly detection methods were compared with three principled methods from the two family of methods discussed above. These benchmarks are described as below. The method presented in [7] is unavailable for multi-class normal data, thus we were unable to compare with it. (1) We implemented a deep hybrid method named CNN+OCSVM, obviously CNN was used to learn the latent representations and OCSVM was used for anomaly detection. The CNN component has two convolutional layers (3×3 receptive fields in both layers, 32 and 64 feature maps respectively for the first and second layers, max-pooling with pooling size of 2×2 after the second layer, ReLU activation function), one fully connected layer (128 units), and a softmax layer for class labels of normal samples. The outputs of the fully connected layer were latent representations extracted for searching the hyperparameters and optimizing the model parameters of OCSVM. (2) We employed a three-layer DBN to model data distribution and then measured reconstruction error to score abnormality. There are 500 units in each hidden layer. The model was layer-wise pretrained by RBMs. Bernoulli distribution was assumed for both visible and hidden layers. The pixel values were scaled to interval [0, 1] as input to DBN. Reconstruction error of an inquiry sample was used to detect anomalous samples. (3) Similarly, a convolutional VAE was also applied to capture data distribution. The inference component (encoder) has the same architecture as the CNN component (disregarding the output layer) in CNN+OCSVM. The latent space size was set to 64. The structure of the inference component was mirrored for the structure of the generative component. Reconstruction error was used in determination of anomaly.

3   CapsNet-Based Normality Score Functions

Unlike [7], we actually do not need to manually transform each training image. Using CapsNet, transformations ought to be automatically learned via an iterative routing-by-agreement mechanism. We can thus avoid the stage of labeling each transformed image. After a CapsNet classifier is trained, unseen images (either normal or out-of-distribution) could be directly fed into the learned model. We present two normality score function to determine the *outlierness* of these images.

3.1   Prediction-Probability-Based Normality Score

The activation probabilities of digital capsules at the last layer of a CapsNet [27] indicate the probabilities of the input sample belonging to the classes. However, unlike softmax probabilities in CNN, the activation probabilities of all digit capsules do not necessarily sum to one. Assuming the network is trained sufficiently well, for a normal test image, there should be one and only one probability being close to "1", representing the possibility of this image belonging to its true class. However, when an anomalous sample cannot be explained by the network, all activation probabilities of digital capsules would be very low. Therefore, this unique characteristic inspires us to define a normality score function $n_s(s)$ based on prediction probabilities (PP):

$$n_{PP}(x) \triangleq \max_{c=1,\cdots,C}(\|h_c\|_2), \tag{1}$$

where x is an input image, h_c represents the c-th digit capsule, $\|h_c\|_2$ denotes the probability of x belonging to the c-th class. Hereafter, we simply call Eq. (1) *PP score function*. Since the threshold, dividing the normal and the anomalous, is hard to decide, as per convention, we use the area under the receiver operating characteristic curve (auROC) to measure its performance.

3.2 Reconstruction-Error-Based Normality Score

In CapsNet [27], reconstruction error is used as a regularization term through a decoder component. The classifier is also an encoder for disentangled representation learning. In this perspective, a CapsNet thus meanwhile functions as a deep autoencoder, offering an idea to score normality based on reconstruction error.

In this method, we use normalized squared error (NSE) to measure the quality of reconstructed images. The reason of using NSE instead of MSE (mean squared error) is that, different objects in different images (e.g., MNIST digits) can have different numbers of nonzero pixels in contrast with pure background. Using MSE will significantly weaken the actual difference between the input image and the reconstructed image. For example, as digit image "1" takes much less numbers of pixels than digit "8", using MSE the reconstruction loss of image "1" will be reduced at a greater extent than that of image "8", even though image "1" may be reconstructed worse than image "8". The reconstruction error (RE) based normality score can thus be defined as:

$$n_{RE}(\boldsymbol{x}) \triangleq -\text{NSE}(\boldsymbol{x}') = -\frac{\|\boldsymbol{x} - \boldsymbol{x}'\|_2^2}{\|\boldsymbol{x}\|_2} \tag{2}$$

where $\boldsymbol{x}$ represents an actual image and $\boldsymbol{x}'$ the reconstructed image. When the background in an image takes a large portion of the space, and the values of background pixels are near zeros, the advantage of NSE will be more obvious. Hereafter, we refer to Eq. (1) as *RE score function*.

3.3 Combined Normality Score

Since prediction-probability-based and reconstruction-error-based normality score functions demonstrate their own strengths on a range of datasets, why don't we combine them together so that they can complement each other? Therefore, the combined normality score function (denoted as $\mathcal{N}_{PP+RE}$) is introduced and defined as below:

$$\mathcal{N}_{PP+RE}(\boldsymbol{x}) = \alpha \mathcal{N}_{RE}(\boldsymbol{x}) + (1 - \alpha)\mathcal{N}_{PP}(\boldsymbol{x}), \tag{3}$$

where $\alpha \in [0, 1]$ is the combination hyperparameter such that the two terms can effectively complement each other. The experimental results to be presented in Sect. 4.2 verify the effectiveness of this normality score function.

4 Experiments

4.1 Comparison on MNIST, Fashion-MNIST, and Small-Norb Data

Both PP and RE methods and 3 benchmarks: CNN+OCSVM, DBN, and VAE were compared on MNIST by respectively treating each digit class as anomalous class and the rest as normal. Their performance in terms of auROC is displayed in Fig. 1a. Evidently the CapsNet-based PP method works consistently the best, while CapsNet-based RE method in general is slightly inferior to the PP method, but has comparable results as CNN+OCSVM. The two DGMs (DBN and VAE) are not competitive to the CapsNet-based and deep hybrid methods.

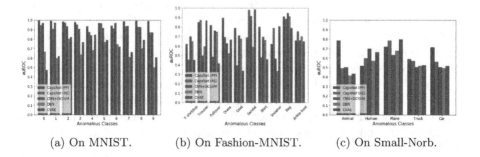

(a) On MNIST.   (b) On Fashion-MNIST.   (c) On Small-Norb.

Fig. 1. Performance on PP, RE, and benchmark methods.

When comparing all five methods on the Fashion-MNIST data (see Fig. 1b), generally speaking all methods tended to get lower results, which is reasonable because Fashion-MNIST samples is more complicated than MNIST samples. When using footwear (Sandals, Sneakers, or Ankle boots) as anomalous samples, the CapsNet(RE) method outperformed the CapsNet(PP) method. It is because the CapsNet classifier could carry some wrong confident information of classifying a footwear anomalous sample (e.g., Sneaker) to the normal footwear classes (e.g., Sandal or Ankle boots), while reconstruction errors can pick up differences in details. In the case of using data from a topwear class as anomalous samples, CapsNet(PP) worked better than CapsNet(RE). When trousers and bags were viewed as abnormal samples, both methods worked quite well without big differences in performance. The performances of CNN+OCSVM and VAE vary largely. In few cases, they can obtain similar results as CapsNet-based methods. DBN did not behave impressively on the data.

From Fig. 1c, one can see that Small-Norb is a challenging data for all five methods. When using animals and cars as anomalies, only CapsNet(PP) performed reasonably good, the other methods behaved randomly. When trucks were used as anomalous samples, only CapsNet(PP) and CapsNet(RE) were able to behave non-randomly. In the case of planes as abnormal data, both CapsNet(RE) and VAE worked the best. Only in the case of humans as anomalies, CNN-OCSVM reached 0.7 auROC.

4.2 Case Studies in Normality Score Functions

The performance of three normality score functions is evaluated through two case studies on the MNIST data.

Multi-class Training Data and Single-Class Anomalous Digits

PP-Based Normality Score VS. RE-Based Normality Score: Figures 2a and b shows the performance of two methods when selecting digits "2" and "9" as abnormal samples, respectively, and the rest digits as normal samples. When digit "2" was treated as anomaly class, both PP and RE score functions achieved near perfect auROCs (0.9841 and 0.9699). When "9" was viewed as anomaly class, however, the PP score function outperformed the RE score function by auROC of 0.12. The reason for this can be found from Figs. 2c and d, which depicts 50 real digits and their reconstructed ones for both digits. From Figs. 2c and d, one can easily notice that anomalous digit "9" was mostly reconstructed as digit "4", while anomalous digit "2" was reconstructed as several different digits such as "1", "3", "6", and "7". As images "4" and images "9" are quite similar, the normality scores of images "9" become very high, even though the true digits "9" and their reconstructed versions "4" are two different numbers. Similar situations can be observed when considering other digits as anomaly, like "4" and "1".

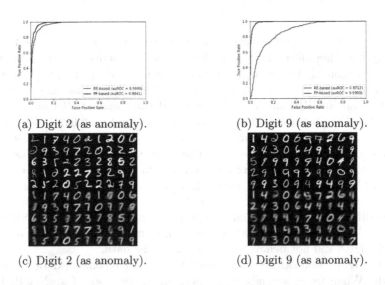

(a) Digit 2 (as anomaly).

(b) Digit 9 (as anomaly).

(c) Digit 2 (as anomaly).

(d) Digit 9 (as anomaly).

Fig. 2. ROC curves and images (upper half: original digits, lower half: reconstructed digits), when detecting anomalous digits: "2" and "9" using AnoCapsNet.

Performance of PP+RE-Based Normality Score: From the aforementioned experiments, one can be convinced that both CapsNet-based PP and RE methods displayed their own strengths on specific anomalous classes. Not surprisingly, the combined method (PP+RE) performed the best among all three normality score functions by making optimum use of advantages of PP and RE methods. Table 1 depicts the results of three normality score functions on four image datasets. From Table 1, one can easily notice that outstanding averaged

Table 1. Performance of AnoCapsNet in comparison with other methods in terms of auROC on image data. In the column for β-VAE, only reconstruction error is used as anomaly score. Results for AnoGAN and ADGAN are from [5].

Dataset	Class	PP	RE	PP+RE	AnoGAN	ADGAN	β-VAE	AnoDM
MNIST	0	0.997	0.947	0.997	0.990	**0.999**	0.890	0.985
	1	0.990	0.907	0.990	**0.998**	0.992	0.841	0.987
	2	0.984	0.970	**0.991**	0.888	0.968	0.967	**0.991**
	3	0.976	0.949	**0.984**	0.913	0.953	0.947	0.969
	4	0.935	0.872	0.939	0.944	0.960	0.968	**0.975**
	5	0.970	0.966	**0.983**	0.912	0.955	0.966	0.976
	6	0.942	0.909	0.959	0.925	0.980	0.907	**0.983**
	7	**0.987**	0.934	**0.987**	0.964	0.950	0.899	0.977
	8	**0.993**	0.929	**0.993**	0.883	0.959	0.946	0.982
	9	**0.990**	0.871	**0.990**	0.958	0.965	0.794	0.928
	Avg	0.977	0.925	**0.981**	0.937	0.968	0.913	0.975
Fashion-MNIST	0	0.620	0.454	0.620	–	–	0.500	**0.844**
	1	0.851	0.871	0.915	–	–	0.860	**0.978**
	2	**0.818**	0.486	**0.818**	–	–	0.459	0.783
	3	0.895	0.693	**0.898**	–	–	0.730	0.886
	4	**0.790**	0.394	**0.790**	–	–	0.379	0.763
	5	0.691	0.982	0.982	–	–	0.985	**0.990**
	6	**0.801**	0.480	**0.801**	–	–	0.501	0.713
	7	0.619	0.787	0.787	–	–	0.842	**0.952**
	8	0.912	0.885	0.960	–	–	0.876	**0.980**
	9	0.656	0.754	0.754	–	–	0.701	**0.944**
	Avg	0.765	0.679	0.833	–	–	0.683	**0.883**
Small-Norb	0	**0.785**	0.491	**0.785**	–	–	0.346	0.520
	1	0.519	0.598	0.598	–	–	0.393	**0.656**
	2	0.718	0.785	**0.812**	–	–	0.772	0.772
	3	0.589	0.570	**0.605**	–	–	0.581	0.581
	4	0.713	0.562	**0.714**	–	–	0.564	0.564
	Avg	0.665	0.601	**0.703**	–	–	0.531	0.619
CIFAR-10	0	0.645	0.377	0.645	0.610	**0.661**	0.368	0.635
	1	0.452	0.736	0.736	0.565	0.435	0.746	**0.754**
	2	0.646	0.413	0.646	**0.648**	0.636	0.397	0.589
	3	0.666	0.597	**0.686**	0.528	0.488	0.604	0.608
	4	0.670	0.390	0.670	0.670	**0.794**	0.387	0.564
	5	0.645	0.590	**0.669**	0.592	0.640	0.611	0.638
	6	**0.723**	0.486	**0.723**	0.625	0.685	0.500	0.600
	7	0.704	0.628	**0.733**	0.576	0.559	0.614	0.648
	8	0.477	0.403	0.477	0.723	**0.798**	0.399	0.642
	9	0.504	0.688	**0.788**	0.582	0.643	0.698	0.718
	Avg	0.613	0.531	**0.677**	0.612	0.634	0.532	0.640

0.981 auROC was achieved by PP+RE-based method when detecting anomalous data on MNIST. In cases of Fashion-MNIST, Small-Norb and CIFAR-10, the averaged auROCs were also improved by around 7%, 4% and 6% of the highest auROCs respectively. It can be concluded that (1) PP-based method had consistently better performance than RE-based method when evaluating AnoCapsNet on MNIST. And it outperformed RE method mostly on Fashion-MNIST, Small-Norb and CIFAR-10. (2) PP+RE method sometimes overlapped with PP method (when PP outperformed RE at a great degree) and occasionally overlapped with PP method (considering RE had much better performance than PP). In other cases (PP and RE had close performance), PP+RE method improved both methods at some degrees. Three normality scores of AnoCapsNet are also compared with other advanced models (AnoGAN [28], ADGAN [5] and AnoDM [20]). The results displayed in Table 1, prove that the proposed PP+RE-based AnoCapsNet framework improved current successful image anomaly detection techniques at a great degree. It accomplished the state-of-art performance on a range of image datasets, even for some challenging datasets such as Small-Norb and CIFAR-10.

Multi-class Training Data and Multiple Anomalous Digits
In this case, the three CapsNet-based normality score functions are tested by considering digits "0", "3" and "5" as abnormal digits, and the rest as normal. The results is displayed in Fig. 3. Both RE and PP methods achieved similar auROCs. By combining them together (using PP+RE normality score function), the state-of-the-art result (0.9919 auROC) is achieved by AnoCapsNet framework.

5   Conclusion

Many modern data-driven intelligent systems require more accurate anomaly detection techniques. In this paper, we explore novel solutions in consideration of CapsNet's distinct characteristics. We devise three normality score functions based on CapsNet's activation probability and reconstruction error respectively. Experiments on four image data sets show that these CapsNet-based methods outperform existing solutions in many setups.

In this paper, we did not discuss classification-based anomaly detection methods, as they essentially treat an anomaly detection task as a two-category classification problem by using both normal and abnormal samples in the training process. While all out-of-distribution anomaly detection methods as discussed in this paper only take normal samples for training, because an anomalous data point could unpredictably come from

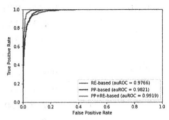

Fig. 3. ROC curves of detecting abnormal digits "0", "3", and "5" using three CapsNet-based score functions.

anywhere outside the normal data distribution, which is true in many application domains. We did not discuss deep reinforcement learning methods for anomaly detection in this paper as well. But this is a new area worthy of future investigation. The performance of CapsNet-based normality score functions depends on CapsNet's learning capacity on certain data. Prediction-probability-based and reconstruction-error-based normality scores possess complementary strengths, and their combination presents the state-of-art performance on a range of image data sets by making optimum use of their advantages. Even though Sabour et al. [27] mentioned that same as deep generative models, current CapsNets do not perform very well when the backgrounds of images vary too much (such as CIFAR-10) to be modelled, the AnoCapsNet framework using combined (PP+RE) normality score technique still accomplishes quite impressive results.

Considering the impressive performance of AnoCapsNet, it might be fruitful to continue deeper research for CapsNet-based anomaly detection problem. As CapsNets acquire the capability of handling complicated data, CapsNet-based anomaly detection methods could become more accurate and robust. Since CapsNet is famous with its state-of-the-art performance on some public image datasets, currently there are quite few works for sequence (text) modeling based on CapsNets [14,32]. This research could be a promising and productive area considering CapsNet is the theoretical improvement of convolution neural network.

Expectation-Maximization routing algorithm in Matrix CapsNet is essentially an inference technique. Therefore, CapsNet is actually not purely discriminative (or supervised) model, having a potential to improve and develop into a fully generative model in the near future. Although CapsNet learns some affine transformations as a supervised method, it is worth exploring unsupervised learning of affine transformations as a future topic.

References

1. An, J., Cho, S.: Variational autoencoder based anomaly detection using reconstruction probability. Technical report, Data Mining Center, Seoul National University, Seoul, South Korea (2015)
2. Andrews, J.T.A., Morton, E.J., Griffin, L.D.: Detecting anomalous data using autoencoders. Int. J. Mach. Learn. Comput. **6**(1), 21–26 (2016)
3. Bengio, Y., Courville, A., Vincent, P.: Representation learning: a review and new perspectives. IEEE Trans. Pattern Anal. Mach. Intell. **35**(8), 1798–1828 (2013)
4. Bengio, Y., Ducharme, R., Vincent, P., Jauvin, C.: A neural probabilistic language model. J. Mach. Learn. Res. **2**, 1137–1155 (2003)
5. Deecke, L., Vandermeulen, R., Ruff, L., Mandt, S., Kloft, M.: Image anomaly detection with generative adversarial networks. In: Berlingerio, M., Bonchi, F., Gärtner, T., Hurley, N., Ifrim, G. (eds.) ECML PKDD 2018. LNCS (LNAI), vol. 11051, pp. 3–17. Springer, Cham (2019). https://doi.org/10.1007/978-3-030-10925-7_1
6. Erfani, S.M., Rajasegarar, S., Karunasekera, S., Leckie, C.: High-dimensional and large-scale anomaly detection using a linear one-class SVM with deep learning. Pattern Recognit. **58**, 121–134 (2016)

7. Golan, I., EI-Yaniv, R.: Deep anomaly detection using geometric transformations. arXiv preprint arXiv:1805.10917 (2018)
8. Goodfellow, I., et al.: Generative adversarial networks. In: Advances in Neural Information Processing Systems, pp. 2672–2680 (2014)
9. Graves, A., Mohamed, A., Hinton, G.: Speech recognition with deep recurrent neural networks. In: IEEE International Conference on Acoustics, Speech and Signal Processing, pp. 6645–6649 (2013)
10. Hinton, G., Sabour, S., Frosst, N.: Matrix capsules with EM routing. In: International Conferences on Learning Representations (2018)
11. Hinton, G., Salakhutdinov, R.: Reducing the dimensionality of data with neural networks. Science **313**, 504–507 (2006)
12. Hinton, G.E., Krizhevsky, A., Wang, S.D.: Transforming auto-encoders. In: Honkela, T., Duch, W., Girolami, M., Kaski, S. (eds.) ICANN 2011. LNCS, vol. 6791, pp. 44–51. Springer, Heidelberg (2011). https://doi.org/10.1007/978-3-642-21735-7_6
13. Hinton, G., Osindero, S., Teh, Y.: A fast learning algorithm for deep belief nets. Neural Comput. **18**, 1527–1554 (2006)
14. Kim, J., Jang, S., Choi, S., Park, E.: Text classification using capsules. ArXiv p. arXiv:1808.03976v2 (2018)
15. Kim, J., Scott, C.: Robust kernel density estimation. J. Mach. Learn. Res. **13**, 3381–3384 (2008)
16. Kingma, D., Welling, M.: Auto-encoding variational Bayes. In: International Conference on Learning Representations (2014)
17. Krizhevsky, A., Sutskever, I., Hinton, G.E.: ImageNet classification with deep convolutional neural networks. In: Pereira, F., Burges, C.J.C., Bottou, L., Weinberger, K.Q. (eds.) Advances in Neural Information Processing Systems 25, pp. 1097–1105. Curran Associates, Inc. (2012)
18. LeCun, Y., Bengio, Y., Hinton, G.: Deep learning. Nature **521**, 436–444 (2015)
19. LeCun, Y., Bottou, L., Bengio, Y., Haffner, P.: Gradient-based learning applied to document recognition. Proc. IEEE **86**(11), 2278–2324 (1998)
20. Li, X., Kiringa, I., Yeap, T., Zhu, X., Li, Y.: Anomaly detection based on unsupervised disentangled representation learning in combination with manifold learning. In: NeurIPS 2019 Workshop on Learning with Rich Experience: Integration of Learning Paradigms (2019)
21. Li, Y.: Sparse machine learning models in bioinformatics. Ph.D. thesis, School of Computer Science, University of Windsor, Windsor, Ontario, Canada (2013)
22. Li, Y., Oommen, B., Ngom, A., Rueda, L.: Pattern classification using a new border identification paradigm: the nearest border technique. Neurocomptuing **157**, 105–117 (2015)
23. Li, Y., Zhu, X.: Exploring Helmholtz machine and deep belief net in the exponential family perspective. In: ICML 2018 Workshop on Theoretical Foundations and Applications of Deep Generative Models, July 2018
24. Li, Y., Zhu, X.: Exponential family restricted Boltzmann machines and annealed importance sampling. In: IJCNN, pp. 39–48, July 2018
25. Parzen, E.: On estimation of a probability density function and mode. Ann. Math. Stat. **33**(3), 1065–1076 (1962)
26. Ruff, L., et al.: Deep one-class classification. In: ICML, vol. 80, pp. 4393–4402. PMLR (2018)
27. Sabour, S., Frosst, N., Hinton, G.: Dynamic routing between capsules. In: Neural Information Processing Systems, pp. 3856–3866 (2017)

28. Schlegl, T., Seeböck, P., Waldstein, S.M., Schmidt-Erfurth, U., Langs, G.: Unsupervised anomaly detection with generative adversarial networks to guide marker discovery. In: Niethammer, M., Styner, M., Aylward, S., Zhu, H., Oguz, I., Yap, P.-T., Shen, D. (eds.) IPMI 2017. LNCS, vol. 10265, pp. 146–157. Springer, Cham (2017). https://doi.org/10.1007/978-3-319-59050-9_12

29. Scholkopf, B., Platt, J., Shawe-Taylor, J., Smola, A., Williamson, B.: Estimating the support of a high-dimensional distribution. Neural Comput. **13**, 1443–1471 (2001)

30. Skvara, V., Pevny, T., Smidl, V.: Are generative deep models for novelty detection truly better? In: ACM SIGKDD 2018 Workshop on Outlier Detection Deconstructed. ACM (2018)

31. Tax, D., Duin, R.: Support vector domain description. Pattern Recognit. Lett. **20**, 1191–1199 (1999)

32. Zhao, W., Ye, J., Yang, M., Lei, Z., Zhang, S., Zhao, Z.: Investigating capsule networks with dynamic routing for text classification. ArXiv p. arXiv:1804.00538v4 (2018)

Question-Worthy Sentence Selection for Question Generation

Sedigheh Mahdavi[1(✉)], Aijun An[1], Heidar Davoudi[2], Marjan Delpisheh[1], and Emad Gohari[3]

[1] York University, Toronto, Canada
{smahdavi,aan,mdelpishe}@eecs.yorku.ca
[2] Ontario Tech University, Oshawa, Canada
heidar.davoudi@uoit.ca
[3] iNAGO Inc., Toronto, Canada
emadg@inago.com

Abstract. The problem of automatic question generation from text is of increasing importance due to many useful applications. While deep neural networks achieved success in generating questions from text paragraphs, they mainly focused on a whole paragraph in generating questions, assuming all sentences are question-worthy sentences. However, a text paragraph often contains only a few important sentences that are worthy of asking questions. To that end, we present a feature-based sentence selection method for identifying question-worthy sentences. Such sentences are then used by a sequence-to-sequence (i.e., *seq2seq*) model to generate questions. Our experiments show that these features significantly improves the question generated by *seq2seq* models.

Keywords: Question Generation (QG) · Sentence selection

1 Introduction

In recent years, automatic question generation (QG) has attracted a considerable attention in both machine reading comprehension [6,34] and educational settings [5,33]. Automatic question generation aims to generate natural questions from a given text passage (e.g., a sentence, a paragraph). There are two main categories of QG methods: *rule-based* approaches [17,18] and *deep neural network* approaches based on sequence-to-sequence (*seq2seq*) models [6,29,36,37]. Rule-based approaches mainly use rigid heuristic rules to transform the source sentence into the corresponding question. However, rule-based methods heavily depend on hand-crafted templates or linguistic rules. Therefore, these methods are not able to capture the diversity of human-generated questions [35], and also may not be transformed to other domains [33]. Recently, *seq2seq* neural network models [6,29,36,37] have shown good performance to generate better-quality questions when a huge amount of labeled data is available.

© Springer Nature Switzerland AG 2020
C. Goutte and X. Zhu (Eds.): Canadian AI 2020, LNAI 12109, pp. 388–400, 2020.
https://doi.org/10.1007/978-3-030-47358-7_40

This section provides a brief overview about some of the important features that may or may not be on your specific vehicle. for more detailed information, refer to each of the features which can be found later in this owner's manual. *The remote keyless entry (rke) transmitter is used to remotely lock and unlock the doors from up to 60m (197ft) away from the vehicle. press to unlock the driver door. press unlock-symbol again within three seconds to unlock all remaining doors. press to lock all doors. lock and unlock feedback can be personalized.* see vehicle personalization.

Fig. 1. Sample paragraph from car manuals. Green sentences are question-worthy. (Color figure online)

Moreover, it has been shown that utilizing the paragraph-level context can improve the performance of seq2seq models in the question generation task [6, 36].

Most existing *seq2seq* methods generate questions by considering all sentences in a paragraph as question-worthy sentences [6, 29, 36, 37]. However, not all the sentences in a text passage (a paragraph or an article) contain important concepts or relevant information, making them suitable for generating useful questions. For example, in Fig. 1 only the underlined sentences in a sample paragraph from a car manual dataset (one of datasets used to evaluate the proposed method) are question-worthy (i.e., human may ask questions about them), and other sentences are irrelevant. Therefore, extracting question-worthy sentences from a text passage is a crucial step in question generation for generating high-quality questions.

Sentence selection has been investigated for the purpose of text summarization [9, 11, 26], where sentences in a document are ranked based on sentence-level and/or contextual features. However, few works exist for sentence selection for the task of question generation (QG). Recently, question-worthy sentence selection strategies using different textual features were compared for educational question generation [4]. However, these strategies identify question-worthy sentences by considering features individually, which may not be powerful enough to distinguish between irrelevant and question-worthy sentences.

In this paper, we use two types of features: *context-based* and *sentence-based* features to identify question-worthy sentences for the QG task. Given a passage (e.g., a paragraph), our goal is to investigate the effectiveness of using these features for extracting question-worthy sentences from the passage on the QG performance. In addition, we consider using only the question-worthy sentences in a passage as the context for question generation instead of using the whole passage. We incorporate the context into a *seq2seq* question generation model with a 2-layer *attention* mechanism. We conduct comprehensive experiments on two datasets: Car Manuals and SQuAD [24] and show that the proposed question-worthy sentence selection method significantly improves the performance of the current state-of-the-art QG approaches in terms of different criteria.

2   Related Work

2.1   Question Generation

Question Generation (QG) can be classified into two categories. (1) rule-based approach [12,19,21] and (2) neural network approach [6,29,37]. Rule-based methods rely on human-designed transformation or template-based approaches that may not be transferable to other domains. Alternatively, end-to-end trainable neural networks are applied to the QG task to address the problem of designing hand-crafted rules, which is hard and time-consuming. Du et al. [6] utilized a sequence-to-sequence neural model based on the attention mechanism [1] for the QG task and achieved better results in contrast to the rule-based approach [12]. Zhou et al. [37] further modified the attention-based model by augmenting each input word vector with the answer position-aware encoding, and lexical features such as part-of-speech and named-entity recognition tag information. They also employed a copy mechanism [10], which enables the network to copy words from the input passage and produce better questions. Both works take an answer as the input sentence and generate the question from the sentence accordingly.

Yuan et al. [34] introduced a recurrent neural model that considers the paragraph-level context of the answer sentence in the QG task. Sun et al. [29] additionally improved the performance of the pointer-generator network [27] modified by features proposed in [37]. Based on the answer position in the paragraph, a question word distribution is generated which helps to model the question words. Furthermore, they argued that context words closer to the answer are more relevant and accurate to be copied and therefore deserve more attention. They modified the attention distribution by incorporating trainable positional word embedding of each word in the sentence w.r.t its relative distance to the answer. Zhao et al. [36] improved the QG by utilizing paragraph-level information with a gated self-attention encoder. However, these methods commonly use the whole paragraph as the context. Our method uses only question-worthy sentences in a paragraph as the context.

2.2   Feature and Graph-Based Sentence Ranking and Selection

A variety of rich features have been used to score sentences in a text passage for summarization purposes [9,11,15,26]. In [26], the authors summarized these features in two general categories: importance features and sentence relation features. Importance features (e.g, length of a sentence, average term frequency (*Tf–idf*) for words in a sentence, average word embedding of words in the sentence, average document frequency, position of a sentence, and Stop words ratio of a sentence) are considered to measure importance of a sentence individually. Sentence relation features determine the content overlap between two sentences.

In [9], the number of named entities in a sentence was considered as one of sentence importance features. In [23], three types of features: statistical, linguistic, and cohesion, were applied to score sentences for selecting important sentences. Statistical features assign weights to a sentence according to several features: keyword feature, sentence position, term frequency, the length of the word, and parts of speech tag. Linguistic features: noun and pronouns give higher chances for sentences with more nouns and pronouns to be include in the summary. Cohesion features consider two kinds of features: grammatical and lexical. In order to score and extract sentences that best describe the paragraphs, a graph-based model, TextRank [20] is used. In this approach, a graph is formed by representing sentences as nodes and the similarity scores between them as vertices. By using the PageRank algorithm [3], nodes with higher scores are chosen as the significant sentences of a given paragraph. Another popular method for deriving useful sentences is LexRank [7], which is a graph-based method capturing the sentences of great importance based on the eigenvector centrality of their corresponding nodes in the graph. SumBasic [31] is another algorithm in which the frequency of words occurring across documents determines sentence significance.

To select sentences for question generation, in [4], different textual features, such as sentence length, sentence position, the total number of entity types, the total number of entities, hardness, novelty, and LexRank measure [7] are individually used to extract question-worthy sentences for a comparison purpose. Here, we train a sentence selection classifier by using multiple features including both context-based and sentence-based features.

3   Methodology

Given a text passage (e.g., a paragraph, a section or an article), our task is to select question-worthy sentences from the passage that capture the main theme of the passage, and use the selected sentences to generate questions. In this section, we first introduce a question-worthy sentence extraction method that extracts all question-worthy sentences from a paragraph. Then, we describe how the question-worthy sentences of a paragraph are incorporated into a *seq2seq* model that uses an attention strategy to generate questions. Figure 2 shows the general view of the proposed method.

3.1   Feature-Based Question-Worthy Sentence Extraction

Inspired by text summarization methods that extract rich features from a text passage (a paragraph or an article) for identifying summary-worthy sentences [9,11,15,26], we develop a new question-worthy sentence selection method. We consider question-worthy sentence selection as a classification task that evaluates each sentence in the passage utilizing context-based and sentence-based features of the sentence.

Given a training data set that contains a set of passages where each passage consists of a sequence of sentences and each sentence is labelled as question-worthy or not, our task is to learn a classifier from the training data that predicts the question-worthiness of a sentence in a passage. To learn such a classifier, we first extract features of sentences in the training data and then train a

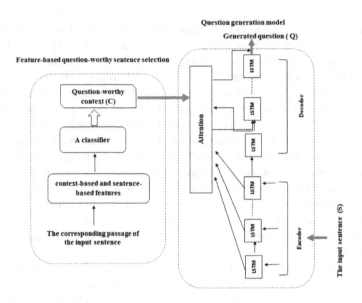

Fig. 2. Proposed framework for question generation

classifier based on the extracted features. The training data are represented as $D = \{(x_1, y_1), \ldots, (x_n, y_n)\}$, where x_i, y_i, n are the feature vector of the sentence i, its label, and the number of sentences in D, respectively. The classifier finds a mapping function $F : X \rightarrow Y$, where X is the domain of input sentences and Y is the set of labels or classes (i.e., question-worthy or not). In our experiment, a Random Forest classifier [2,13] is trained to identify question-worthy sentences due to its solid performance in text classification tasks, although other classification methods can be used.

We use two groups of features to represent a sentence: context-based and sentence-based features. Context-based features consider the passage which the sequence is in and contain rank features and the *tf–df* feature. The rank features of a sequence are the ranks of the sentence in its passage obtained from different text summarization methods. The intuition of using rank features is that sentences with important and valuable information contents are ranked higher. Therefore, high rank sentences are more suitable to ask question about. We employ four text summarization methods: TextRank [20], SumBasic [31], LexRank [7], and Reduction [8]. We use four different ranking methods because different ranking methods consider different sets of factors in sentence ranking and all these factors can be considered when incorporating all of them in our sentence representation. The sentence ranks generated by these summarization methods are used as four rank features. To compute the *tf-idf* feature of a sequence, we first compute the *tf-idf* value of each word in the sequence in the context of the passage the sentence is in. That is, the term frequence of a word is the frequency of the word in the sentence and the *inverted document frequency* of the word is the number of sentences containing the word in the passage.

We then use the average *tf–idf* value of the words in a sentence as the *tf–idf* feature of the sequence. Intuitively, the *tf-idf* value of a sentence measures the importance of a sentence in its passage.

We also use sentence-based features, which consider only the sentence without its context. Sentence-based features are of two different types: POS-tag (Parts of speech tag) and sentence importance features. Part-of-speech tagging is a basic NLP task that classifies words into their parts of speech and labeling them accordingly. We use six POS-tag features: (1) Number of verbs in a sentence, (2) Number of nouns in a sentence, (3) Number of adjectives in a sentence, (4) Number of adverbs in a sentence (5) Number of pronouns in a sentence, and (6) Number of connection words in a sentence. Our sentence importance features are the length of a sentence and the stop words ratio in a sentence [26].

3.2 Context-Aware Question Generation

We use a *seq2seq* model to generate questions from question-worthy sentences given a passage. In a *seq2seq* question generation model, the objective is to generate a question Q for a text sequence S (e.g., a sentence that answers the question). More formally, the main objective is to learn a model with parameter θ^* given a set of S and Q pairs by solving the following:

$$\theta^* = \arg\max_{\theta} \sum_{Q,S} \log P(Q|S; \theta), \tag{1}$$

Here, we also consider the context of the input sentence S when generating a question from S. We use the question-worthy sentences in the paragraph of sentence S as the context C of S. Thus, our problem is to learn a model with parameter θ^* given a set of tuples $\langle S, C, Q \rangle$, such that:

$$\theta^* = \arg\max_{\theta} \sum_{Q,S,C} \log P(Q|S, C; \theta), \tag{2}$$

To incorporate contexts into the *seq2seq* model, we use the same strategy proposed in [25] for context-aware query reformulation, where a new attention strategy (two-layer attentions) was introduced for incorporating the context of a query into a *seq2seq* model. The model proposed in [25] is called *Pair Sequences to Sequence* (*Pair S2S*) due to the fact that two input sequences are used to generate one output sequence. In the encoder stage of *Pair S2S* model, both the input sequence $S = \{w_t^S\}_{t=1}^M$ and its context $C = \{w_t^C\}_{t=1}^N$ (where w_t^S and w_t^C represent the tth word in S and C, respectively, and M and N are the number of words in S and C, respectively) are separately encoded as follows:

$$u_t^S = RNN^S(u_{t-1}^S, e_t^S) \tag{3}$$

$$u_t^C = RNN^C(u_{t-1}^C, e_t^C) \tag{4}$$

where e_t^S and e_t^C are the word embeddings for the context and the input sentence, respectively. In the decoder stage, the traditional attention mechanism is separately applied on the context and input sequence as follows:

$$c_t^C = \sum_{k=1}^{N} \alpha_{t,k}^C u_k^C \qquad c_t^S = \sum_{k=1}^{M} \alpha_{t,k}^S u_k^S \tag{5}$$

$$\alpha_{t,k}^C = \frac{e^{f(s_t, u_k^C)}}{\sum_{k_i} e^{f(s_t, u_{k_i}^C)}} \qquad \alpha_{t,k}^S = \frac{e^{f(s_t, u_k^S)}}{\sum_{k_i} e^{f(s_t, u_{k_i}^S)}} \tag{6}$$

where s_t, c_t^C, c_t^S, $\alpha_{t,k}^C$, $\alpha_{t,k}^S$, and f are represents the internal state of recurrent neural network (RNN) at time t, the attention vector for the context, the attention vector for the input sentence, the attention strength for the context, the attention strength for the input sentence, and the attention function, respectively. Then, another attention layer is applied to combine the attention vectors of the input sequence and the context:

$$c_t^{C+S} = \beta_C c_t^C + \beta_S c_t^S \tag{7}$$

$$\beta_C = \frac{e^{f(s_t, c_t^C)}}{e^{f(s_t, c_t^S)} + e^{f(s_t, c_t^C)}} \tag{8}$$

$$\beta_S = \frac{e^{f(s_t, c_t^S)}}{e^{f(s_t, c_t^S)} + e^{f(s_t, c_t^C)}} \tag{9}$$

We apply the above two-layer attentions in [25]. For each input sentence, question-worthy sentences extracted by the feature-based sentence selection method from its corresponding paragraph are considered as the question-worthy context.

Table 1. Evaluation results for important sentence selection on SQuAD. The best results is highlighted in **boldface**.

Method (SQuAD)	Precision	Recall	Accuracy	Macro-F1	Micro-F1
ConceptTypeMax	0.6021	0.3827	0.4679	0.4680	0.4682
ConceptMax	0.6021	0.3827	0.4679	0.4678	0.4681
LexRank	0.7610	0.4836	0.5915	0.5913	0.5916
Emb	0.7000	0.0002	0.3885	0.2801	0.3887
Longest	0.7235	0.4600	0.5620	0.5622	0.5624
FS-SM-IM	**0.8273**	0.6813	0.6405	0.5623	0.6407
FS-SM-Pos	0.6938	0.6920	0.6047	0.5695	0.6049
FS-SM-Rank	0.7283	0.7287	0.6510	0.6206	0.6513
FS-SM	0.7626	**0.7606**	**0.6932**	**0.6658**	**0.6932**

4   Experimental Setup and Results

4.1   Dataset and Implementation Details

We conduct our experiments on the following datasets.

– Car Manual dataset: This dataset (provided by iNAGO Inc.[1]) consists of 4672
  QAs created by human annotators from two car manuals (Ford and GM). We
  randomly divided 80% of the dataset into training, 10% validation and 10%
  test. In this dataset, sentences can be divided into two different classes with
  label '0' and '1'. Label '1' for a sentence means that humans identify it as a
  worthy sentence. Sentences with label '0' are irrelevant sentences.
– Processed SQuAD dataset: We use the Stanford Question Answering Dataset
  (SQuAD) [24], a machine reading comprehension dataset, which offers a large
  number of questions and their answers extracted from Wikipedia through
  crowdsourcing. Each example consists of a sentence from an article with its
  associated question generated by human and its corresponding paragraph.
  We use this dataset with the same setting as (Du et al. [6]). The data has
  been split into training set (70,484 question-answer pairs), dev set (10,570
  question-answer pairs) and test set (11,877 question-answer pairs).

We train our models with stochastic gradient descent using OpenNMT-py [14],
an open source neural machine translation system, with the same hyperparameters as in [6]. The learning rate starts at 1 and is halved at $8th$ epoch. We train
a two-layer LSTMs with hidden unit size 600 for 15 epochs.

Table 2. Evaluation results for sentence selection on Car Manuals dataset. The best
results are highlighted in **boldface**.

Method (Car manuals)	Precision	Recall	Accuracy	Macro-F1	Micro-F1
ConceptTypeMax	0.6689	0.3679	0.4740	0.4744	0.4747
ConceptMax	0.6690	0.3680	0.4746	0.4747	0.4750
LexRank	0.7508	0.4129	0.5318	0.5328	0.5330
Emb	0.39	0.0002	0.3548	0.2619	0.3548
Longest	0.5436	0.2990	0.3850	0.3855	0.3858
FS-SM-IM	0.5511	0.5706	0.5805	0.5798	0.5808
FS-SM-Pos	0.6576	0.6531	0.6569	0.6572	0.6574
FS-SM-Rank	0.6094	0.6077	0.6189	0.6196	0.6200
FS-SM	**0.7641**	**0.6896**	**0.7150**	**0.7152**	**0.7155**

---

[1] http://www.inago.com/.

4.2   Evaluation Metrics

To evaluate sentence selection methods, we use precision, recall, accuracy, and F1 scores. For question generation, we report BLEU-1, BLEU-2, BLEU-3, BLEU-4 [22] and ROUGE-L [16] scores based on the package in [28] for evaluating natural language generation. BLEU-n is a modified precision of n-grams between the reference and generated sentences, while ROUGE-L compares the longest matching sequence of words between system-generated and reference counterparts.

4.3   Question-Worthy Context Results

We compare our feature-based question-worthy sentence extraction method (FS-SM) with a number of baselines, including LexRank, ConceptTypeMax, ConceptMax, and Longes proposed in [4]. In [4], it was shown that LexRank is the best question-worthy sentence identification strategy on most datasets. This strategy is based on summary scores of the LexRank [7] summarization method. The ConceptMax and ConceptTypeMax strategies consider the total number of entities and the total number of entity types in a sentence, respectively. In addition, we examine the embedding feature (Emb method) proposed in [26] which represents the sentence content. To analyze the effect of each type of features, we evaluate three variants of FS-SM:

- FS-SM-Pos: A version of FS-SM whose classifier is trained by considering just the POS-tag features
- FS-SM-IM: A version of FS-SM whose classifier is trained by considering just the sentence importance features
- FS-SM-Rank: A version of FS-SM whose classifier is trained by considering just the rank features

Tables 1 and 2 show results on the Car Manuals and SQuAD datasets. The results show that the FS-SM method significantly outperforms the other baselines in terms of classification evaluation metrics. From Tables 1 and 2, it can be seen that all versions of the FS-SM method achieved better results than other strategies.

4.4   Question Generation Results

We compare FS-SM-seq2seq (our QG method) with some baselines for question generation. Tables 3 and 4 show the results for the following QG methods:

- Vanilla seq2seq: The basic *seq2seq* model [30] whose input is a sentence.
- Transformer: Transformer model is a neural network based *seq2seq* model based on the attention mechanism [1] and positional encoding [32]. Its input is a sentence.
- Para-seq2seq: A *seq2seq* model with the 2-layer attention strategy [25] where for each input sentence its whole paragraph is used as its context.
- ConceptMax-seq2seq: A *seq2seq* model with the 2-layer attention strategy [25] that uses the question-worthy sentences identified by ConceptMax from the paragraph of the input sentence as the question-worthy context.

- LexRank-seq2seq: A *seq2seq* model with the 2-layer attention strategy [25] that uses the question-worthy sentences identified by LexRank from the paragraph of the input sentence as the question-worthy context.
- FS-SM-seq2seq (our method): A *seq2seq* model with the 2-layer attention strategy [25] that uses the question-worthy sentences identified by our proposed sentence selection method from the paragraph of the input sentence as the question-worthy context.

We chose LexRank and ConceptMax as an alternative context selection method to compare with our method because they can identify important sentences better than other strategies evaluated in [25]. It can be seen from Tables 3 and 4, FS-SM-seq2seq outperform other compared methods on all metrics on the SQuAD data set and on most metrics on the Car Manuals data set.

Table 3. Question generation evaluation on car manuals on SQuAD

Model (SQuAD)	BLEU-1	BLEU-2	BLEU-3	BLEU-4	ROUGE-L
Vanilla seq2seq	31.34	13.79	7.36	4.26	29.75
Transformer	37.528	18.097	9.457	5.0143	26.600
ConceptMax-seq2seq	41.700	16.551	8.205	4.099	28.772
LexRank-seq2seq	41.057	17.168	8.494	4.099	28.055
Para-seq2seq	33.152	13.786	06.585	03.2867	27.6876
FS-SM-seq2seq	**43.27**	**18.86**	**9.00**	**4.48**	**30.58**

Table 4. Question generation evaluation on car manuals

Model (Car manual)	BLEU-1	BLEU-2	BLEU-3	BLEU-4	ROUGE-L
Vanilla seq2seq	34.6012	16.5057	**10.11052**	**6.598**	28.1247
Transformer	28.1243	11.5928	6.3074	3.4219	25.2176
ConceptMax-seq2seq	35.2702	14.7947	9.3679	5.2965	0.269364
LexRank-seq2seq	35.4368	0.1601	9.764	6.1662	0.2808
Para-seq2seq	35.13123	15.97419	09.2094	5.5600	28.0959
FS-SM-seq2seq	**36.9870**	**17.6561**	09.7696	5.41238	**29.5423**

5 Conclusion and Future Work

We presented a method for selecting question-worthy sentences from a text passage and using these sentences as contexts for question generation. For identifying question-worthy sentences, a feature-based method is designed based on context-based and sentence-based features. A 2-layer attention strategy is applied to incorporate the question-worthy context into a *seq2seq* model. Experimental results showed that using the question-worthy context for question generation *seq2seq* models have achieved better results than baselines on both Car Manuals and SQuAD datasets.

Acknowledgement. This work is funded by the Big Data Research Analytics and Information Network (BRAIN) Alliance established by Ontario Research Fund - Research Excellence Program (ORF-RE), and iNAGO Inc.

References

1. Bahdanau, D., Cho, K., Bengio, Y.: Neural machine translation by jointly learning to align and translate. In: 3rd International Conference on Learning Representations, ICLR 2015. Conference Track Proceedings, San Diego, CA, USA, 7–9 May 2015 (2015). http://arxiv.org/abs/1409.0473
2. Barandiaran, I.: The random subspace method for constructing decision forests. IEEE Trans. Pattern Anal. Mach. Intell. **20**(8), 1–22 (1998)
3. Brin, S., Page, L.: The anatomy of a large-scale hypertextual web search engine. Comput. Netw. ISDN Syst. **30**, 107 (1998)
4. Chen, G., Yang, J., Gasevic, D.: A comparative study on question-worthy sentence selection strategies for educational question generation. In: Isotani, S., Millán, E., Ogan, A., Hastings, P., McLaren, B., Luckin, R. (eds.) AIED 2019. LNCS (LNAI), vol. 11625, pp. 59–70. Springer, Cham (2019). https://doi.org/10.1007/978-3-030-23204-7_6
5. Danon, G., Last, M.: A syntactic approach to domain-specific automatic question generation. arXiv preprint arXiv:1712.09827 (2017)
6. Du, X., Shao, J., Cardie, C.: Learning to ask: neural question generation for reading comprehension. In: Proceedings of the 55th Annual Meeting of the Association for Computational Linguistics (Volume 1: Long Papers), pp. 1342–1352 (2017)
7. Erkan, G., Radev, D.R.: LexRank: graph-based lexical centrality as salience in text summarization. J. Artif. Intell. Res. **22**, 457–479 (2004)
8. Fabish, A.: MS Windows NT kernel description. https://github.com/adamfabish/Reduction
9. Galanis, D., Lampouras, G., Androutsopoulos, I.: Extractive multi-document summarization with integer linear programming and support vector regression. In: Proceedings of COLING 2012, pp. 911–926 (2012)
10. Gülçehre, Ç., Ahn, S., Nallapati, R., Zhou, B., Bengio, Y.: Pointing the unknown words. CoRR abs/1603.08148 (2016). http://arxiv.org/abs/1603.08148
11. Gupta, S., Nenkova, A., Jurafsky, D.: Measuring importance and query relevance in topic-focused multi-document summarization. In: Proceedings of the 45th Annual Meeting of the ACL on Interactive Poster and Demonstration Sessions, pp. 193–196. Association for Computational Linguistics (2007)
12. Heilman, M., Smith, N.A.: Good question! Statistical ranking for question generation. In: Human Language Technologies: The 2010 Annual Conference of the North American Chapter of the Association for Computational Linguistics, Los Angeles, California, pp. 609–617. Association for Computational Linguistics, June 2010. https://www.aclweb.org/anthology/N10-1086
13. Ho, T.K.: Random decision forests. In: Proceedings of 3rd International Conference on Document Analysis and Recognition, vol. 1, pp. 278–282. IEEE (1995)

14. Klein, G., Kim, Y., Deng, Y., Crego, J.M., Senellart, J., Rush, A.M.: OpenNMT: open-source toolkit for neural machine translation. CoRR abs/1709.03815 (2017). http://arxiv.org/abs/1709.03815
15. Li, S., Ouyang, Y., Wang, W., Sun, B.: Multi-document summarization using support vector regression. In: Proceedings of DUC. Citeseer (2007)
16. Lin, C.Y.: ROUGE: A package for automatic evaluation of summaries. In: Text Summarization Branches Out, Barcelona, Spain. Association for Computational Linguistics, July 2004
17. Lindberg, D., Popowich, F., Nesbit, J., Winne, P.: Generating natural language questions to support learning on-line. In: Proceedings of the 14th European Workshop on Natural Language Generation, pp. 105–114 (2013)
18. Mazidi, K., Nielsen, R.D.: Linguistic considerations in automatic question generation. In: Proceedings of the 52nd Annual Meeting of the Association for Computational Linguistics (Volume 2: Short Papers), pp. 321–326 (2014)
19. Mazidi, K., Nielsen, R.D.: Leveraging multiple views of text for automatic question generation. In: Conati, C., Heffernan, N., Mitrovic, A., Verdejo, M.F. (eds.) AIED 2015. LNCS (LNAI), vol. 9112, pp. 257–266. Springer, Cham (2015). https://doi.org/10.1007/978-3-319-19773-9_26
20. Mihalcea, R., Tarau, P.: TextRank: bringing order into text. In: Proceedings of the 2004 Conference on Empirical Methods in Natural Language Processing, pp. 404–411 (2004)
21. Mitkov, R., Ha, L.A.: Computer-aided generation of multiple-choice tests. In: Proceedings of the HLT-NAACL 03 Workshop on Building Educational Applications Using Natural Language Processing (2003)
22. Papineni, K., Roukos, S., Ward, T., Zhu, W.J.: BLEU: a method for automatic evaluation of machine translation. In: Proceedings of the 40th Annual Meeting of the Association for Computational Linguistics, Philadelphia, Pennsylvania, USA. Association for Computational Linguistics, July 2002
23. Patil, N.R., Patnaik, G.K.: Automatic text summarization with statistical, linguistic and cohesion features. In: International Journal of Computer Science and Information Technologies (2017)
24. Rajpurkar, P., Zhang, J., Lopyrev, K., Liang, P.: SQuAD: 100, 000+ questions for machine comprehension of text. CoRR abs/1606.05250 (2016). http://arxiv.org/abs/1606.05250
25. Ren, G., Ni, X., Malik, M., Ke, Q.: Conversational query understanding using sequence to sequence modeling. In: Proceedings of the 2018 World Wide Web Conference, pp. 1715–1724. International World Wide Web Conferences Steering Committee (2018)
26. Ren, P., Wei, F., Zhumin, C., Jun, M., Zhou, M.: A redundancy-aware sentence regression framework for extractive summarization. In: Proceedings of COLING 2016, the 26th International Conference on Computational Linguistics: Technical Papers, pp. 33–43 (2016)
27. See, A., Liu, P.J., Manning, C.D.: Get to the point: summarization with pointer-generator networks. CoRR abs/1704.04368 (2017). http://arxiv.org/abs/1704.04368
28. Sharma, S., El Asri, L., Schulz, H., Zumer, J.: Relevance of unsupervised metrics in task-oriented dialogue for evaluating natural language generation. CoRR abs/1706.09799 (2017). http://arxiv.org/abs/1706.09799

29. Sun, X., Liu, J., Lyu, Y., He, W., Ma, Y., Wang, S.: Answer-focused and position-aware neural question generation. In: Proceedings of the 2018 Conference on Empirical Methods in Natural Language Processing, Brussels, Belgium. Association for Computational Linguistics, October-November 2018. https://www.aclweb.org/anthology/D18-1427

30. Sutskever, I., Vinyals, O., Le, Q.: Sequence to sequence learning with neural networks. In: Advances in NIPS (2014)

31. Vanderwende, L., Suzuki, H., Brockett, C., Nenkova, A.: Beyond SumBasic: task-focused summarization with sentence simplification and lexical expansion. Inf. Process. Manag. **43**(6), 1606–1618 (2007)

32. Vaswani, A., et al.: Attention is all you need. In: Advances in Neural Information Processing Systems, pp. 5998–6008 (2017)

33. Yao, K., Zhang, L., Luo, T., Tao, L., Wu, Y.: Teaching machines to ask questions. In: IJCAI, pp. 4546–4552 (2018)

34. Yuan, X., et al.: Machine comprehension by text-to-text neural question generation. In: Proceedings of the 2nd Workshop on Representation Learning for NLP, Vancouver, Canada, pp. 15–25. Association for Computational Linguistics, August 2017. https://doi.org/10.18653/v1/W17-2603. https://www.aclweb.org/anthology/W17-2603

35. Yuan, X., Wang, T., Trischler, A.P., Subramanian, S.: Neural models for key phrase detection and question generation, 7 February 2019. US Patent App. 15/667,911

36. Zhao, Y., Ni, X., Ding, Y., Ke, Q.: Paragraph-level neural question generation with maxout pointer and gated self-attention networks. In: Proceedings of the 2018 Conference on Empirical Methods in Natural Language Processing, pp. 3901–3910 (2018)

37. Zhou, Q., Yang, N., Wei, F., Tan, C., Bao, H., Zhou, M.: Neural question generation from text: a preliminary study. CoRR abs/1704.01792 (2017). http://arxiv.org/abs/1704.01792

Challenges in Vessel Behavior and Anomaly Detection: From Classical Machine Learning to Deep Learning

Lucas May Petry[1]([✉]), Amilcar Soares[2], Vania Bogorny[1], Bruno Brandoli[2], and Stan Matwin[2]

[1] Universidade Federal de Santa Catarina (PPGCC/UFSC), Florianópolis, Brazil
`lucas.petry@posgrad.ufsc.br`
[2] Institute for Big Data Analytics, Dalhousie University, Halifax, Canada

Abstract. The global expansion of maritime activities and the development of the Automatic Identification System (AIS) have driven the advances in maritime monitoring systems in the last decade. Given the enormous volume of vessel data continuously being generated, real-time analysis of vessel behaviors is only possible because of decision support systems provided with event and anomaly detection methods. However, current works on vessel event detection are ad-hoc methods able to handle only a single or a few predefined types of vessel behavior. Most of the existing approaches do not learn from the data and require the definition of queries and rules for describing each behavior. In this paper, we discuss challenges and opportunities in classical machine learning and deep learning for vessel event and anomaly detection.

Keywords: Automatic Identification System · Behavior detection · Anomaly detection · Spatiotemporal data mining

1 Introduction

The worldwide growth of maritime traffic and the development of the Automatic Identification System (AIS) has led to advances in monitoring systems for preventing vessel accidents and detecting illegal activities. In addition, the integration of vessel traffic data with environmental and climatological data allows more complex analyses and a better understanding of the cause and effect of

This study was financed in part by the Brazilian agencies CNPq and CAPES - Project Big Data Analytics: Lançando Luz do Genes ao Cosmos (CAPES/PRINT process number 88887.310782/2018-00), Fundação de Amparo a Pesquisa e Inovação do Estado de Santa Catarina - Project Match (co-financing of H2020 Projects - Grant 2018TR 1266), and the European Union's Horizon 2020 research and innovation programme under Grant Agreement 777695 (MASTER). The authors also acknowledge the support of the Natural Sciences and Engineering Research Council of Canada and of Global Affairs Canada - ELAP for this research.

© Springer Nature Switzerland AG 2020
C. Goutte and X. Zhu (Eds.): Canadian AI 2020, LNAI 12109, pp. 401–407, 2020.
https://doi.org/10.1007/978-3-030-47358-7_41

maritime events [18]. While preventing vessel accidents means saving money for shipping companies, from the environmental point of view, it also protects the marine fauna and flora from irreversible damage [2]. However, real-time monitoring and analysis of vessel traffic can be overwhelming to maritime agents due to the high volume of data continuously generated. Therefore, decision support systems are fundamental to enabling efficient and effective maritime control.

The detection of events from AIS data has been the subject of study of several works in the literature [9,16,18]. In particular, some approaches have been proposed for detecting events such as changes in the speed [18,19] or in the course of vessels [19], proximity of vessels to other vessels [9], illegal fishing [16] or possibly hazardous activity [18], among others. However, to the best of our knowledge, most current works are ad-hoc approaches that do not learn from the data and are limited to detecting a restricted set of predefined events. Such methods are not able to detect unforeseen events and also require the assistance of domain specialists for defining rules that characterize each event.

Another aspect often ignored by previous research is the integration of data from different sources for analyzing vessel behavior. Even though this can be advantageous to maritime systems, only a handful of works have addressed it [18]. For example, detecting small vessels heading towards ice-infested waters and that are not equipped for handling this situation allows the decision-maker to warn the captain in advance. Such strategy avoids the deployment of a search and rescue mission, which might represent a high cost (e.g., lives, resources) to maritime authorities.

In this work, we present research gaps and challenges in machine learning for detecting different types of vessel behavior, considering several constraints imposed by real-time data streams and the maritime monitoring domain. We highlight the potential of exploiting machine learning techniques for maritime monitoring, as it has been shown to be fundamental for enabling cognitive smart cities, for instance, which is a scenario similar to ours with heterogeneous sensor data that requires real-time decision making systems [13].

2   Research Challenges and Opportunities

In this section, we explore the challenges and research gaps that are currently faced by vessel behavior monitoring techniques. We address three main tasks related to vessel behavior detection: (i) the actual detection of different vessel behaviors or behavior changes; (ii) identifying and relating recurrent behaviors; (iii) providing the user with means for interpreting and analyzing the detected vessel behaviors. In addition, we discuss about two main issues associated to the data that can be challenging for machine learning methods: (i) the data has limited or no class labels about vessel behaviors; and (ii) often, there is no knowledge whatsoever of what are the behaviors or labels present in the data.

2.1 Behavior Detection: Supervised vs. Unsupervised

Most of the existing works for behavior detection in vessel traffic monitoring do not learn from the data. Instead, algorithms are proposed to detect a restricted set of vessel activities via rules and thresholds defined beforehand. Defining such rules requires the knowledge from domain specialists, which may introduce considerable human bias in the analysis. Although the sole use of supervised learning techniques may seem appealing, they can also limit the detection of events to a predefined set, i.e., the classification labels. With that in mind, we believe that research on unsupervised learning techniques is promising for taking vessel event detection research to the next level.

A closely-related research topic that has already been explored in the literature is the detection of concept drift or change in time series data [1,17]. In streaming data, concept drift is commonly referred to as the detection of significant changes in the data distribution [8]. Several works for change detection in data streams were designed to work along with a supervised learning model, detecting drifts based on the error rate of the learner [1].

Unsupervised approaches have also been proposed for detecting concept drift [1,8,17]. However, they also exhibit some drawbacks, such as being limited to univariate data [1]; simply detecting changes based on individual feature correlation [8] instead of analyzing dependency relationships between features and how they characterize different behaviors; and even lacking direct interpretability of the detected changes [17]. In spite of their limitations, these methods can provide a solid starting point for future works on unsupervised behavior change detection and posterior event notification.

2.2 Identifying Recurrent Behavior Patterns

Detecting multiple instances of the same behavior is an essential factor in maritime monitoring. Besides avoiding multiple analyses of the same behavior by agents, it allows agents to have a higher-level picture of behavior patterns of a single vessel or even of a group of vessels in a region. In previous works for vessel event detection, identifying multiple occurrences of the same behavior was a trivial task, since the detection process consisted mostly of a query [18] or an algorithm for detecting a single behavior [9]. On the other hand, in the unsupervised setting, a pattern or behavior needs to be characterized in a way that multiple occurrences of the same pattern can be identified. Although concept drift or behavior change detection methods can indicate boundaries for different types of behavior in streaming data, they do not provide a way for detecting recurrent behavior.

To the best of our knowledge, the only unsupervised learning method proposed for detecting different and recurrent behaviors in time series data is the Toeplitz Inverse Covariance-based Clustering (TICC), introduced by Hallac et al. [6]. TICC segments multivariate sensor data into sequences of states or clusters (i.e., behavior patterns), representing each behavior as a Markov Random Field (MRF). However, the number of clusters is fixed and must be defined by the user, meaning that the number of different behaviors present in the data should be known a priori. Additionally, TICC is not directly suitable for streaming data, as it assumes that all data is available at the same time, and it requires a few iterations of the algorithm for convergence. Although TICC has some drawbacks, we believe that future research could take advantage of the method for recurrent behavior detection as, for instance, use MRFs to represent and identify the same behavior in different trajectory segments.

2.3   Towards Interpretable Behavior Patterns

Doshi and Kim [4] define interpretability in the context of machine learning as "the ability to explain or to present in understandable terms to a human." In order to enable maritime agents to make data-driven decisions about suspicious or dangerous vessel activity, discovered behavior patterns must be interpretable, i.e., understandable to the agents. Moreover, interpretability may assist in the detection of recurrent behaviors if they can be explicitly characterized based on feature observations.

Interpretability is, perhaps, the main advantage of a few existing works, since rules explicitly define events related to vessel behavior. In contrast to these approaches, detecting interpretable patterns can be challenging for unsupervised machine learning algorithms. For vessel behavior detection, we conjecture that the interpretability of the model may be negligible if guarantees are given about the characteristics of the detected behaviors. For instance, if a behavior can be represented as an MRF, and it is always defined by the same variable dependency graph, the user might not be interested in the details of the underlying model as long as this structure is guaranteed for all future occurrences of the same behavior. Thus, detection methods that are based on abstractions of the real observed features (e.g., [12,17]) can be a feasible option. Afterwards, other techniques could be exploited for correlating and providing interpretability of different behaviors from the real observed variables.

2.4   Diving into Deep Learning

Deep learning models have often been set aside in favor of linear models, because of the claimed lack of interpretability that they have [10]. We believe that for a similar reason only very few works have addressed anomaly detection in the maritime domain with deep learning [14,20]. To the best of our knowledge, no work has addressed the detection of specific vessel behaviors with deep learning techniques. More recently, however, a few works have been proposed to assist in the visual interpretation of these models (e.g. [21]), while others have even questioned previous claims over the interpretability of deep models [10].

Computer vision has experienced the most advances with deep learning, and Convolutional Neural Networks (CNNs) are nowadays widely used for image classification [7]. An intuitive approach is the use of satellite imagery data with CNNs for detecting ship anomalies. However, it may be difficult or even impossible to detect certain vessel behaviors from satellite imagery, which limits the use of modern deep learning methods exclusively based on image data [15]. Also, obtaining satellite images is generally more expensive in comparison with AIS data. On the other hand, CNNs have already been used for trajectory classification [3] and prediction [11], based on trajectory features that can be extracted from AIS data. Therefore, CNNs with visual techniques for interpretability [21] could be further exploited for behavior and anomaly detection.

2.5   Big, Yet Limited Data

The growth of maritime activity led to advances in AIS technology, and such developments resulted in large volumes of data being generated every day. Although a large amount of AIS data is available, it lacks labels. Labels are a valuable piece of information for researchers and machine learning algorithms. However, labeling data is difficult and expensive since it requires knowledge from domain experts.

The lack of labels has guided the focus of current research either to detecting a single type of vessel behavior in an unsupervised manner [9], or to proposing the detection of different events via predefined behavior rules, materialized in the form of data queries [16,18]. Future works could take advantage of the knowledge described in previous approaches for labeling data, to provide input-output examples to machine learning algorithms, as well as ground truth for evaluating novel approaches. On the other hand, research could concentrate on synthesizing new behavior data with Generative Adversarial Networks [5], for example, for enhancing the performance of other supervised models.

3   Summary and Final Remarks

Even though maritime monitoring has experienced significant progress in the last decade, most of existing works do not take full advantage of machine learning techniques for vessel behavior detection. In fact, existing research has focused on proposing queries, rules, and ad-hoc algorithms for detecting specific types of behavior. We argue that this methodology inhibits further advances to more general behavior detection approaches, constraining monitoring systems to function only under frequent human supervision. In this paper, we presented research gaps in the field, indicating points of improvement and opportunities for future works. We hope to instigate the development of new algorithms, methods, and tools for ship behavior monitoring, as many aspects of it still remain unaddressed.

References

1. Bifet, A., Gavalda, R.: Learning from time-changing data with adaptive windowing. In: Proceedings of the 2007 SIAM International Conference on Data Mining, pp. 443–448. SIAM (2007)
2. Claramunt, C., et al.: Maritime data integration and analysis: recent progress and research challenges. Adv. Database Technol. EDBT **2017**, 192–197 (2017)

3. Dabiri, S., Heaslip, K.: Inferring transportation modes from GPS trajectories using a convolutional neural network. Transp. Res. Part C Emerg. Technol. **86**, 360–371 (2018)
4. Doshi-Velez, F., Kim, B.: Towards a rigorous science of interpretable machine learning. arXiv preprint arXiv:1702.08608 (2017)
5. Goodfellow, I., et al.: Generative adversarial nets. In: Advances in Neural Information Processing Systems, pp. 2672–2680 (2014)
6. Hallac, D., Vare, S., Boyd, S., Leskovec, J.: Toeplitz inverse covariance-based clustering of multivariate time series data. In: Proceedings of the 23rd ACM SIGKDD International Conference on Knowledge Discovery and Data Mining, pp. 215–223. ACM (2017)
7. Krizhevsky, A., Sutskever, I., Hinton, G.E.: Imagenet classification with deep convolutional neural networks. In: Advances in neural Information Processing Systems, pp. 1097–1105 (2012)
8. Lee, J., Magoules, F.: Detection of concept drift for learning from stream data. In: 14th International Conference on High Performance Computing and Communication & 9th International Conference on Embedded Software and Systems, pp. 241–245. IEEE (2012)
9. Lei, P.-R.: Mining maritime traffic conflict trajectories from a massive AIS data. Knowl. Inf. Syst. **62**(1), 259–285 (2019). https://doi.org/10.1007/s10115-019-01355-0
10. Lipton, Z.C.: The mythos of model interpretability. arXiv preprint arXiv:1606.03490 (2016)
11. Lv, J., Li, Q., Sun, Q., Wang, X.: T-conv: a convolutional neural network for multi-scale taxi trajectory prediction. In: 2018 IEEE International Conference on Big data and Smart Computing (BIGCOMP), pp. 82–89. IEEE (2018)
12. Manzoor, E., Lamba, H., Akoglu, L.: xstream: outlier detection in feature-evolving data streams. In: Proceedings of the 24th ACM SIGKDD International Conference on Knowledge Discovery & Data Mining, KDD 2018, pp. 1963–1972. ACM (2018)
13. Mohammadi, M., Al-Fuqaha, A.: Enabling cognitive smart cities using big data and machine learning: approaches and challenges. IEEE Commun. Mag. **56**(2), 94–101 (2018)
14. Nguyen, D., Vadaine, R., Hajduch, G., Garello, R., Fablet, R.: A multi-task deep learning architecture for maritime surveillance using AIS data streams. In: 5th International Conference on Data Science and Advanced Analytics (DSAA), pp. 331–340. IEEE (2018)
15. Nyman, E.: Techno-optimism and ocean governance: new trends in maritime monitoring. Mar. Policy **99**, 30–33 (2019)
16. Patroumpas, K., Artikis, A., Katzouris, N., Vodas, M., Theodoridis, Y., Pelekis, N.: Event recognition for maritime surveillance. In: Advances in Database Technology - EDBT, pp. 629–640 (2015)
17. Qahtan, A.A., Alharbi, B., Wang, S., Zhang, X.: A PCA-based change detection framework for multidimensional data streams: change detection in multidimensional data streams. In: Proceedings of the 21th ACM SIGKDD International Conference on Knowledge Discovery and Data Mining, pp. 935–944. ACM (2015)
18. Soares, A., et al.: Crisis: integrating AIS and ocean data streams using semantic web standards for event detection. In: International Conference on Military Communications and Information Systems (2019)
19. Wen, Y., et al.: Semantic modelling of ship behavior in harbor based on ontology and dynamic Bayesian network. ISPRS Int. J. Geo-Inf. **8**(3), 107 (2019)

20. Zhao, L., Shi, G.: Maritime anomaly detection using density-based clustering and recurrent neural network. J. Navig. **72**(4), 894–916 (2019)
21. Zhou, B., Khosla, A., Lapedriza, A., Oliva, A., Torralba, A.: Learning deep features for discriminative localization. In: Proceedings of the IEEE Conference on Computer Vision and Pattern Recognition, pp. 2921–2929 (2016)

An Energy-Efficient Method with Dynamic GPS Sampling Rate for Transport Mode Detection and Trip Reconstruction

Jonathan Milot⬤, Jaël Champagne Gareau$^{(\boxtimes)}$⬤, and Éric Beaudry⬤

Université du Québec à Montréal, Montréal, Canada
{milot.jonathan,champagne_gareau.jael}@courrier.uqam.ca,
beaudry.eric@uqam.ca

Abstract. This paper presents a novel approach for trip reconstruction and transport mode detection. While traditional methods use a fixed GPS sampling rate, our proposed method uses a dynamic rate to avoid unnecessary sensing and waste of energy. We determine a time for each sampling that gives an interesting trade-off using a particle filter. Our approach uses as input a map, including transit network circuits and schedules, and produces as output the estimated road segments and transport modes used. The effectiveness of our approach is shown empirically using real map and transit network data. Our technique achieves an accuracy of 96.3% for a 15.0% energy consumption reduction (compared to the existing technique that has the closest accuracy) and an accuracy of 85.6% for a 56.0% energy consumption reduction.

Keywords: Particle filter · Transport mode detection · Trip reconstruction · Energy efficiency · Mobile device · GPS · Dynamic sampling

1 Introduction

The popularity of smartphones has brought out many services and applications based on their sensors (GPS, accelerometer, gyroscope, etc.) such as activity recognition [7], trip reconstruction (TR) [8,11] and transport mode detection (TMD) [3,13]. TR and TMD are both used for traffic study in urban planning to automate the process of data collection [12] instead of using the traditional survey method that is less precise and more expensive. Most GPS sensing are usually made at a fixed predetermined rate ranging from 1 s to 60 s [3,12]. However, these algorithms of reconstruction suffer from major drawbacks.

One problem is that a GPS device consumes a significant amount of energy. Using the GPS at 1 Hz consumes 143.1 mW on the HTC Dream and Google Nexus One models [5]. According to our experiments, the consumption is 439.3 mW for a Samsung Galaxy S8 and 397.44 mW for an Asus Zenphone

© Springer Nature Switzerland AG 2020
C. Goutte and X. Zhu (Eds.): Canadian AI 2020, LNAI 12109, pp. 408–419, 2020.
https://doi.org/10.1007/978-3-030-47358-7_42

4 Max. The battery life of these smartphones ranged from 8 to 10 h for normal daily use (cellular and Wi-Fi enabled, 4G web navigation when travelling in public transit, etc.) when continuous geolocation was enabled. Sensing at a rate of 1 Hz is therefore hardly acceptable for users, since it drastically reduces their smartphone's autonomy. On another side, a lower rate decreases the accuracy of TR and TMD algorithms. Hence, there is a trade-off between the quality of estimation of the algorithm and the energy consumption (EC) of the GPS.

This ideal compromise highly depends on the road network and the smartphone's state after a certain time. Indeed, in a city's downtown, the number of different possible paths and the possible change of transport mode (e.g., bus stops, metro stations, etc.) makes it harder to do TR and TMD, because many paths and transport mode combination can explain the transition between the last two GPS sensing. In comparison, on a highway there is usually only one possibility (transport by *car* and shortest path between two points). The interval Δt between sensing should thus dynamically change depending on the position to achieve the optimal trade-off between precision and EC.

This paper presents a dynamic GPS sampling rate technique for path reconstruction and transport mode detection based on a particle filter that dynamically determines the moment to use the GPS sensor in order to get a compromise, depending on preferences, between energy consumption and accuracy.

2   Related Works

Since we address two related problems at once, TR and TMD, this section presents previous works related to one or both of these problems. Furthermore, we discuss their impacts on energy consumption.

2.1   Trip Reconstruction

The simplest TR approaches use a fixed GPS sensing rate. Usually, such approaches are tested with different fixed frequencies to show the decrease in accuracy when the rate increase. One approach is iterative based [11], where each node must be within a certain Euclidean distance from the corresponding GPS reading and directly linked to the previous associated node. If no node can be associated to a reading, the maximum distance is increased and another iteration begins. Another approach [10] uses a Hidden Markov Model (HMM) [4] to determine the most likely road segment for each GPS reading. The authors limit the EC of their algorithm by updating the sampling rate according to the mobile device state (stopped, normal road, highway). However, they do not test their approach with sampling rates higher than 30 s to avoid an *arc-skipping* situation, which could greatly increase the error according to the authors. Finally, a model to generate a set of true potential paths and to associate likelihood to each of them is used in [1]. The model uses the speed, the time and the bearing of the mobile device to determine these likelihood. All these techniques highlight the same problem: an higher sampling rate to reduce EC directly leads to an increased error due to the *arc-skipping* problem. In this situation, the road segments associated with two consecutive GPS readings may not be directly connected and a shortest-path algorithm must be used to link them [11,12]. This can induce errors, since some users may not have used the shortest path for some reason (e.g., construction site, personal preference, traffic, etc.).

2.2 Transport Modes Detection

TMD algorithms are generally based on machine learning (ML), since the goal is to classify data (GPS readings) among a defined and limited set of class (transport mode). While these approaches can achieve a good accuracy on average (more than 90%), they rely on a high GPS sampling rate (from 1 s to 15 s) and do not attempt to reconstruct the smartphone's trip. Some approaches also use data from other sensors, such as the smartphone's accelerometer or field sensor, to improve their accuracy [3]. Many TMD algorithms use ML techniques such as neural network (NN) [3,15] and derivatives such as convolutional neural network (CNN) [9]. Other ML techniques such as Random Forest [13] and Support Vector Machine [2] have also been experimented.

Table 1. Transport mode detection accuracy in related works

Approach	Sampling rate (s)	Walk	Car	Bus	Average
[13]	15	98.9%	80.8%	93.0%	93.8%
[2]	60	93.8%	88.5%	58.3%	88.0%
[3]	1	95.0%	72.0%	84.0%	83.8%
[15]	1	98.5%	94.2%	88.4%	94.4%
[9]	1–5	95.7%	67.4%	81.1%	84.8%

The general accuracy of these techniques ranged from 84.8% for the CNN [9] to 94.4% for the NN with particle swarm optimization [15]. However, these average accuracies hide a deeper phenomenon that occurs in all reviewed study: distinguishing a *walk* is generally easier than a *car* or a *bus*. Table 1 presents the accuracy for the different papers previously cited. The *walk* accuracy is always the highest (more than 93.8%), while the *car* and *bus* can vary a lot (from 67.4% to 94.2% and 58.3% to 93.0% respectively). This is mainly due to the resemblance between the motorized transport modes (*car* and *bus*) regarding speed and acceleration. Thus, they are easily mixed up and wrongly classified. A note on the results of [2]: they claim to have an average accuracy of 88.0% for a sampling rate of 60 s. While this seems impressive for such a low rate, it is important to note that 52.4% of their data is labelled as *walk*. Since this is the easiest transport mode to detect, this greatly increases their average accuracy.

2.3   Combined Approach

A technique to do both TR and TMD has been proposed by [8]. They use a conventional GIS-based map-matching algorithm to reconstruct the path and a rule-based algorithm to identify transport modes (*walk, bicycle, bus* and *car*). Their average error on TR is 21% for a sample rate of 1 s, which is worse than algorithms previously cited who exclusively reconstruct path. Their accuracy for transport mode detection is 92%, which is similar to other approaches.

3   Model and Algorithm

The goal of our method is to estimate the paths and the transport modes used during a trip with the GPS sensor of a smartphone while minimizing the EC. To achieve this, we use a particle filter to estimate the smartphone's state according to GPS readings made at a dynamic rate. This approach novelty resides in the use of the GPS sensor only when really needed, unlike other methods that make GPS readings at a given fixed rate. The general outline goes as follows:

1. Make a GPS reading to estimate the smartphone's state s.
2. Simulate the evolution of s until a time t in the future.
3. Determine a time $t^\star \in \ ]0, t]$ offering an interesting compromise between accuracy and energy consumption. When $t^\star$ is reached, return to 1.

We model the space in which the smartphone evolves as a graph $G = (N, A)$. A node is a point $n = (n.\text{Lat}, n.\text{Long}) \in N$ on the map, and an arc is a road segment $a = (n_{\text{from}}, n_{\text{to}}, v_{\text{max}}, T_e) \in A$ containing respectively the start and end vertex of the segment as well as the maximum possible speed on the segment and the set of all transport modes that can cross a. Each arc thus represents a way to move between two vertices by using a transport mode (e.g., a road segment, a subway tunnel, a train rail, etc.). The smartphone's state at time t is a tuple $s_t = (p, m, v, P)$, where p, m, v and P are respectively the smartphone's current position (a point lying on an arc in A), current transport mode, current speed and actual path travelled (list of nodes and transport modes used).

The smartphone's state s is not directly observable, but is estimated from an external sensor (GPS). Since this sensor is not perfect and contains noise, $s.p$ is modelled as a Gaussian distribution $\mathcal{N}(x, \sigma^2)$, where x is the projection of the coordinate returned by the GPS on the nearest arc a and σ^2 is the accuracy of the GPS. It is important to note that the smartphone's position distribution may overlap other allowed arcs according to the current transport mode if x is near the extremity of a. The evolution of $s.p$ after a time interval Δt is given by

$$s_{t+\Delta t}.p = s_t.p + \mathcal{N}(s.v, \sigma^2) \times \Delta t. \qquad (1)$$

We need to take into account the fact that $s.v$ can vary during Δt (due to traffic, road elevation, etc.). Over a long Δt, we consider that a Gaussian distribution is a good speed approximation at which the smartphone travels.

Algorithm 1. The particle filter algorithm

1: **function** PARTICLE FILTER($\mathcal{X}_{t-1}$, u_t, z_t)
2:    $\bar{\mathcal{X}} = \mathcal{X}_t = \emptyset$
3:    **for** m = 1 to M **do**
4:        sample $x_t^{[m]} \sim p(x_t|u_t, x_{t-1}^{[m]})$
5:        $w_t^{[m]} = p(z_t|x_t^{[m]})$
6:        $\bar{\mathcal{X}}_t = \bar{\mathcal{X}}_t + \langle x_t^{[m]}, w_t^{[m]} \rangle$
7:    **for** m = 1 to M **do**
8:        draw i with probability $\propto w_t^{[i]}$
9:        add $x_t^{[i]}$ to $\mathcal{X}_t$
10:   **return** $\mathcal{X}_t$

Using this formalism, the goal of our algorithm is to determine the list of nodes and transport modes $P = \langle (n_0, m_0), (n_1, m_1), \ldots, (n_{k-1}, m_{k-1}), (n_k, m_k) \rangle$ taken during a trip, where m_i is the transport mode used to reach n_i.

Since the evolution of the state s can hardly be modelled with parametric functions (smartphones can be at different places after Δt, each with multiple different uncertainties, paths and transport modes), we use a particle filter. A generic implementation [14] is presented in Algorithm 1. Simply put, a particle filter's goal is to approximate a belief state $bel(x_t)$ by a set of particles $\mathcal{X}_t$ of size M randomly constructed by the control data u_t and the sensor data z_t. In our problem, u_t is simply the time passed since the last sampling and z_t, the data returned by the GPS sensor. Figure 1 shows a visual example of the particle filter and the evolution of $\mathcal{X}_t$ between two GPS readings. Compare the particles circled in blue (resp. red) in Fig. 1b and in Fig. 1c. We see that many instances of $s_{30}.p$ are overlapping with different path.

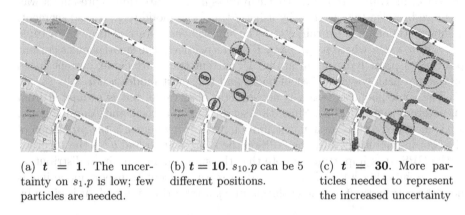

(a) $t = 1$. The uncertainty on $s_1.p$ is low; few particles are needed.

(b) $t = 10$. $s_{10}.p$ can be 5 different positions.

(c) $t = 30$. More particles needed to represent the increased uncertainty

Fig. 1. Evolution of $\mathcal{X}_t$ over 30 s for the *car* transport mode (Color figure online)

One of the challenges of the particle filter is to determine how and when to resample $\mathcal{X}_t$. After a certain amount of time, the distribution of $\mathcal{X}_t$ may

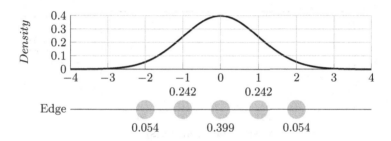

Fig. 2. Discretization of $\mathcal{N}(0,1)$ at every σ on $[-2\sigma, 2\sigma]$.

become too coarse to adequately approximate $bel(x_t)$. Therefore, a resample is eventually needed. In our case, a resampling implies a new sensor observation z_t (which incur energy consumption). We implement two methods to determine the next resampling time and avoid unnecessary sensing.

Firstly, to avoid too coarse particle distribution, we discretize all possible states with particles instead of sampling a fixed M particles from $p(x_t|u_t, x_{t-1}^{[m]})$ (Line 4 of Algorithm 1). Rather than calculating the density probability of $bel(x_t)$ with the density of particles, we associate a weight to each particle (see Fig. 2). When a smartphone reach position n_{to} of its current arc, the particle splits itself on all outgoing edges of n_{to}. An example of this splitting is shown in Fig. 1b. This ensures that all possible states are covered by the state space.

Secondly, we compute a score for $\mathcal{X}_t$ with the following formula:

$$\text{score}(\mathcal{X}_t) = \sum_{x \in \mathcal{X}_t} \left(P(x) \times x_{acc} \right) - \frac{2}{1 + e^{\lambda t}}. \tag{2}$$

This score represents the trade-off between the EC and the average estimated accuracy (TR and TMD) the algorithm would achieve if a GPS sensing was done at t, the elapsed time since the last sensing. The λ parameter in Eq. (2) controls the trade-off between EC and accuracy. A small λ saves more energy, while a bigger one gives an higher accuracy. Figure 3 shows an example of the evolution of $\text{score}(\mathcal{X}_t)$ for different values of λ. The next sensing and resampling are done at the time t that maximizes $\text{score}(\mathcal{X}_t)$. The variable x_{acc} in Eq. (2) is defined by

$$x_{acc} = \sum_{x' \in \bar{\mathcal{X}}_t} P(x') \times \frac{|x'_{Path+}| + |x'_{Path-}|}{|x_{Path}|}, \tag{3}$$

where x is the particle currently computed, $|x'_{Path+}|$ is the length of path that x' has, but not x, $|x'_{Path-}|$ is the length of path that x' doesn't have, but x has and $\bar{\mathcal{X}}_t$ is the set of particles at a lower distance from x than the GPS sensing error. x_{acc} represents the average error on the state estimated when x is the true smartphone's state and a GPS sensing is made. Due to noise induced by the GPS sensor, the true smartphone's state could be mistaken for any particles of $\bar{\mathcal{X}}_t$ (which can have different paths and transport modes).

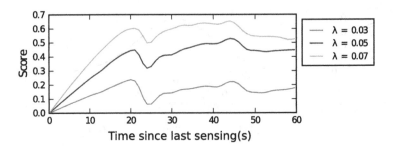

Fig. 3. score($\mathcal{X}_t$) for 3 different λ. With $\lambda = 0.03$, the next sensing is done after 21 s. With the other two λ, the next sensing is done after 44 s.

The resampling is pretty straightforward: score($\mathcal{X}_t$) is computed on $[0, T[$, where T is a time limit after which no better trade-off can be obtained. The time $t^\star$ offering the ideal compromise between accuracy and energy consumption is retained (i.e., $t^\star = argmax_t$score($\mathcal{X}_t$)). Then, a GPS sensing is done at time $t^\star$. The particles in a radius of $2\sigma^2$ are simply kept instead of generating new particles (Line 8 of Algorithm 1), where σ^2 is the radius (in meters) of 68% confidence given by the GPS sensor. In the event that no particles are in this radius, it is gradually extended by step of one σ^2. The weight of each particle is then updated according to their distance from the GPS sensing.

The change of transport mode during a trip is considered in two circumstances:

1. **Special nodes.** Some transport mode changes can only happen on predetermined nodes, such as *walk→bus*. When a particle p reaches these nodes, a copy of p is instantiated with a predefined different transport mode, changing its transition behaviour between nodes.
2. **During resampling.** Some other transport mode changes can occur practically anywhere, such as *car→walk*. Since those transport mode changes don't occur often, we only consider this change at resampling. Based on our results, the error induced by this simplification is acceptable compared to the combinatorial explosion that would arise if a transport mode change was considered every second. Therefore, after each resampling, a new state is added to $\mathcal{X}_t$ for every transport modes not already present in $\mathcal{X}_t$.

4   Experimentations

Experimentations were done in the metropolitan region of Montréal, Canada due to its range of different road network configurations (e.g., dense downtown, highways, suburbs and country roads on the peripheral region). Moreover, the presence of multiple bus and subway lines allows us to consider multiple transport modes. In this experimentation, the transport modes considered are *walk, car, bus* and *subway*. Also, score($\mathcal{X}_t$) has been evaluated on the interval $[0, 300[$.

4.1   Data for the Model

The map data comes from the OpenStreetMap (OSM) project, while the transit network schedule comes directly from the transport societies (STM, RTL, AMT, STL) that cover the region of Montréal. Bus and subway stations were integrated to the OSM data by being connected to the nearest node. To obtain the average speed on every road segment for the *car* transport mode, we used data collected by the City of Montréal. With them, we computed an average speed of 28.60 ± 9.49 km/h. The *walk* average speed has also been found to be between 3.46±0.65 km/h to 4.93 ± 0.68 km/h depending on the place he is and his gender [6].

4.2   GPS Data Collection

Table 2. GPS data collected

Transport mode	Runtime (hours)	Number of GPS readings
Walk	8.03	28514
Car	11.94	42768
Bus	5.13	17290
Subway	2.06	390
Total	27.16	88962

For two months, we recorded daily trips with an Android application on an Asus Zenphone 4 Max smartphone (model ZC554KL). All trips were collected at different times of the day and have different path. This ensures a variety regarding the traffic, weather conditions and transit network schedules. During these trips, the GPS sensor made readings at a rate of 1 Hz and the user was asked to manually specify its current transport mode (*walk, car, bus* or *subway*). A summary of the collected data is presented in Table 2. Few GPS readings have been recorded for the *subway* transport mode because GPS signal can't be received underground (but some stations are close enough to the surface to allow some signals to pass through).

Table 3. Algorithm accuracy with different λ

λ	ASR (s)	Walk	Car	Bus	Subway	Avg TMD	TR Error
0.01	20.9	0.996	0.988	0.936	0.932	0.963	0.023
0.03	32.5	0.920	0.982	0.871	0.902	0.919	0.028
0.05	48.6	0.908	0.969	0.806	0.890	0.893	0.031
0.07	59.2	0.876	0.934	0.772	0.885	0.867	0.039
0.09	66.0	0.859	0.932	0.758	0.876	0.856	0.041

Table 4. Transport mode detection matrix confusion for $\lambda = 0.05$

Detected	Real			
	Walk	Car	Bus	Subway
Walk	0.908	0.030	0.049	0.013
Car	0.001	0.969	0.021	0.009
Bus	0.048	0.055	0.806	0.091
Subway	0.048	0.018	0.044	0.890

5   Results

Using the collected data, our approach was tested with different λ values to analyze the trade-offs between accuracy and EC. Table 3 shows for every tested λ the average sampling rate (ASR), the accuracy obtained with our technique for the detection of every transport mode and the average on all of them, as well as the trip reconstruction error. As expected, a higher λ implies a lower GPS rate. With $\lambda = 0.01$, a reading is made at an average sampling rate of 20.9 s, while with $\lambda = 0.09$, the average is at 66.0 s.

5.1   Transport Mode Detection

In order to make a fair comparison between our results and those in the related works, we looked for public implementation of related works' algorithms to test them on the same datasets. However, these datasets are, to the best of our knowledge, not publicly available. Given the usage of machine learning in these techniques, a reimplementation was hardly possible, since those algorithms are very sensitive to the input dataset. Hence, we directly compare our results to those found in the related works.

The TMD accuracies we obtained are similar to those found in related works. It ranges from 96.3% with a $\lambda = 0.01$ to 85.6% with a $\lambda = 0.09$. The result for $\lambda = 0.01$ had a better average accuracy than any other approach cited (the closest average accuracy being 94.4% [15]) and was achieved with a lower sampling rate. For example, with $\lambda = 0.01$, the average sampling rate is 20.9 s while [15] had a sampling rate of 1 s. In comparison to [2], who used the highest sampling rate of 60 s and had an accuracy of 88.5% for *car* and 58.3% for *bus* [2], our technique had an accuracy of 93.2% (*car*) and 75.8% (*bus*) for an average sampling rate of 66.0 s with $\lambda = 0.09$. The TMD errors for $\lambda = 0.05$ are shown in a confusion matrix in Table 4.

The lowest accuracy is for the *bus* transport mode. It is falsely detected as *car* 5.5% of the time, which is understandable since buses and cars drive at a similar speed. However, *car* are less often mistaken for *bus*, only 2.1% of the time. This is explained by the fact that buses always share the road with cars (except some rare bus-only lanes) but cars can often be on roads with no buses route. *Bus* and *subway* transport can be confused with *walk* when the algorithm has difficulty to determine the exact bus stop or subway station a user has taken. Furthermore, *subway* are often (9.1%) confused with *bus*. This is due to the presence of a bus line running parallel to a subway line. Often, both routes can explain the transition between the last GPS location before entering a subway station and the first one after exiting the other one. Currently, our algorithm doesn't consider an absence of GPS reading as being underground. However, the presented approach's accuracy could be improved by increasing the weight of underground states when a weak GPS signal is detected.

5.2 Trip Reconstruction

Ground truth for paths taken was generated from our GPS readings and corrected by hand. We compared the path estimated by our technique to the ground truth. The path error was computed with the formula [11]:

$$E = \frac{|P^+_{Estimated}| + |P^-_{Estimated}|}{|P_{Real}|}, \qquad (4)$$

where P_{Real} is the real path, $P^+_{Estimated}$ is the part of the estimated path in extra compared to P_{Real} and $P^-_{Estimated}$ the part lacking from P_{Real}. TR error ranged from 2.3% with $\lambda = 0.01$ (ASR of 20.9 s) to 4.1% with $\lambda = 0.09$ (ASR of 66.0 s). Compared to their respective sampling rate, this is better than the accuracy found in the related works.

5.3 Energy Consumption

The energy efficiency of our approach is demonstrated by running our algorithm on smartphones and measuring the energy consumption using Android's *BatteryManager* API. Before every test, the battery was fully charged and all other applications were closed. Wifi was disabled and 4G enabled. Running the algorithm directly on the phone would consume more energy than the amount saved by making less GPS sampling. Hence, the 4G connection is required to communicate with a server running the algorithm. We also measured the EC for fixed GPS sampling rates. Table 5 shows the results obtained.

Obviously, for an equal ASR, the presented approach uses more energy than those that uses a fixed sampling rate because of the 4G usage (e.g., 286.28 mW versus 272.49 mW for a sampling rate of around 20 s). However, we can have a higher or equivalent accuracy with our approach while using less energy. Most other approaches use a sampling rate of 1 s resulting in an EC of 336.82 mW, the corresponding average accuracy ranging from 84.0% to 93.0%. In comparison, the presented approach consumes only 286.28 mW for an accuracy of 96.3% or 210.50 mW for an accuracy of 91.9%. This means a 15.0% EC reduction for a 3.3% higher accuracy and a 37.5% EC reduction for an equivalent accuracy. Furthermore, accuracy remains acceptable with greater sampling rate. With an ASR of 66.0 s, the presented approach still achieves an average accuracy of 85.6% for an EC of 148.18 mW, a 56.0% EC reduction for an equivalent accuracy compared to [15].

Table 5. Energy consumption according to the GPS sampling rate

Method	λ	ASR (s)	EC (mW)
Fixed GPS sampling rate	-	1	336.82
	-	20	272.49
	-	40	172.67
	-	60	146.74
	-	80	114.02
Dynamic GPS sampling rate	0.01	20.9	286.28
	0.03	32.5	210.50
	0.05	48.6	170.79
	0.07	59.2	151.26
	0.09	66.0	148.18

6   Conclusion

In this paper, a novel approach for trip reconstruction (TR) and transport mode detection (TMD) has been presented. It reduces significantly smartphone energy consumption by using the GPS sensor only when necessary while achieving similar or higher accuracy compared to state-of-the-art methods. This approach uses a particle filter that estimates the smartphone's state evolution and the average resulting error if a GPS sampling was made at every moment. These average estimated errors are then weighted in a smartphone energy consumption model to determine the optimal time to do the next sampling. This is to the best of our knowledge the first approach using dynamic GPS rate depending on the underlying road and transit network. Finally, field tests demonstrated the approach's accuracy and energy saving compared to other methods. In the best case, the presented approach allowed an increase of 3.3% in the average accuracy and a 15.0% energy consumption reduction compared to other approaches and a 37.5% to 56.0% energy consumption reduction for an equivalent accuracy.

Our experimentation only looked at one smartphone model (i.e. Asus Zenphone 4 Max smartphone, model ZC554KL). More tests should be done to compare the power consumption saving with different models, since each part can have a different power consumption, i.e., GPS sensor, 4G antenna, CPU, etc. This could lead to different ratios of energy consumption saving on models where other parts would consume drastically more than the GPS sensor.

Currently, our transport model uses historic data regarding travel speed on certain road segments to determine an *a priori* travel speed. However, it does not consider the possible punctual slowdowns due to traffic. Because real-time data on traffic is hard to obtain, research has already been made toward predicting and modelling traffic on a road network. Such methods would improve our transport model in order to better predict the user's state evolution when travelling by transport mode relying on road segments (e.g., *car* and *bus*), increasing the proposed approach accuracy.

References

1. Bierlaire, M., Chen, J., Newman, J.: A probabilistic map matching method for smartphone GPS data. Transp. Res. Part C: Emerg. Technol. **26**, 78–98 (2013)
2. Bolbol, A., Cheng, T., Tsapakis, I., Haworth, J.: Inferring hybrid transportation modes from sparse GPS data using a moving window SVM classification. Comput. Environ. Urban Syst. **36**(6), 526–537 (2012). https://doi.org/10.1016/j.compenvurbsys.2012.06.001
3. Byon, Y.J., Liang, S.: Real-time transportation mode detection using smartphones and artificial neural networks: performance comparisons between smartphones and conventional global positioning system sensors. J. Intell. Transp. Syst. **18**(3), 264–272 (2014). https://doi.org/10.1080/15472450.2013.824762
4. Cappé, O., Moulines, E., Rydén, T.: Inference in hidden markov models. In: Proceedings of EUSFLAT Conference, pp. 14–16 (2009)
5. Carroll, A., Heiser, G., et al.: An analysis of power consumption in a smartphone. In: USENIX Annual Technical Conference, vol. 14, pp. 21–21 (2010)
6. Chandra, S., Bharti, A.K.: Speed distribution curves for pedestrians during walking and crossing. Procedia Soc. Behav. Sci. **104**, 660–667 (2013). https://doi.org/10.1016/j.sbspro.2013.11.160
7. Cheng, W., Erfani, S.M., Zhang, R., Ramamohanarao, K.: Markov dynamic subsequence ensemble for energy-efficient activity recognition. In: Proceedings of MobiQuitous 2017, Australia, p. 10 (2017). https://doi.org/10.1145/3144457.3144470
8. Chung, E.H., Shalaby, A.: Transportation planning and technology a trip reconstruction tool for GPS-based personal travel surveys. Transp. Plan. Technol. **28**(5), 381–401 (2005)
9. Dabiri, S., Heaslip, K.: Inferring transportation modes from GPS trajectories using a convolutional neural network. Transp. Res. Part C Emerg. Technol. **86**, 360–371 (2018). https://doi.org/10.1016/j.trc.2017.11.021
10. Fang, S., Zimmermann, R.: EnAcq: energy-efficient GPS trajectory data acquisition based on improved map matching. In: Proceedings of the 19th ACM SIGSPATIAL International Conference on Advances in Geographic Information Systems, pp. 221–230 (2011). https://doi.org/10.1145/2093973.2094004
11. Li, X., Yuan, F., Lindqvist, J.: Feasibility of duty cycling gps receiver for trajectory-based services. In: 13th IEEE Annual Consumer Communications and Networking Conference (CCNC) (2016). https://doi.org/10.7282/T3VM4F56
12. Patterson, Z., Fitzsimmons, K.: DataMobile: smartphone travel survey experiment. Transp. Res. Rec. J. Transp. Res. Board **15**(2594), 35–43 (2016)
13. Stenneth, L., Wolfson, O., Yu, P.S., Xu, B.: Transportation mode detection using mobile phones and GIS information. In: Proceedings of the 19th ACM SIGSPATIAL International Conference on Advances in Geographic Information Systems, p. 54 (2011). https://doi.org/10.1145/2093973.2093982
14. Thrun, S., Burgard, W., Fox, D.: Probabilistic Robotics (Intelligent Robotics and Autonomous Agents). MIT Press, Cambridge (2005)
15. Xiao, G., Juan, Z., Gao, J.: Travel mode detection based on neural networks and particle swarm optimization. Information **6**(3), 522–535 (2015). https://doi.org/10.3390/info6030522

Similarity Matching of Temporal Event-Interval Sequences

S. Mohammad Mirbagheri[✉] and Howard J. Hamilton

Department of Computer Science, University of Regina, Regina, Canada
{mirbaghs,Howard.Hamilton}@uregina.ca

Abstract. We propose and evaluate three novel approaches to the problem of performing similarity matching on full-length event-interval sequences (e-sequences). The ERF approach represents each e-sequence as a vector of the magnitudes of the durations of the events. Euclidean distances between the vectors are then used to compare given e-sequences. The EPC approach embeds each e-sequence as a vector of position codes to capture the order of the occurrences of events explicitly and temporal relations among the events implicitly. Cosine distances between the vectors are used to infer similarity of e-sequences. Finally, the EMKL approach combines the ERF and EPC approaches using multiple kernel learning. Empirical evaluation on eight real datasets suggests that the EMKL approach outperforms existing state-of-the-art methods in terms of nearest neighbor classification accuracy.

Keywords: Event-interval sequence · Interval-based events · Temporal relation · Similarity matching · Distance measure

1 Introduction

In many application domains, sequences of events persist over intervals of time of varying lengths. Such domains include but not limited to medicine [1], sensor networks [2], and sign languages [3]. In contrast to point-based event sequences, where events occur instantaneously, we refer to a series of temporal events with duration as event-interval sequences (e-sequences). The persistence of events over intervals of time causes temporal relations, which were categorized by Allen [4].

An e-sequence dataset contains longitudinal data where instances are described by a series of event intervals over time rather than features with a single value. Such datasets are not organized appropriately for standard machine learning algorithms to build predictive or descriptive models for e-sequences. In this paper, we investigate the problem of matching full-length event interval sequences. Addressing this problem will facilitate performing data mining tasks such as classification and clustering. We propose and evaluate three novel approaches to this problem. The ERF approach uses relative frequency, as proposed by [5], to represent each e-sequence as a vector of the magnitudes of the durations of the events. Euclidean distances between the vectors are then used

© Springer Nature Switzerland AG 2020
C. Goutte and X. Zhu (Eds.): Canadian AI 2020, LNAI 12109, pp. 420–425, 2020.
https://doi.org/10.1007/978-3-030-47358-7_43

to compare the magnitude of the duration of the event intervals in the two e-sequences. The EPC approach generates a position code for the coincidence label sequence (l-sequence) [6] to embed the transformed e-sequence as a vector. Cosine similarities between the vectors are then used to measure the angles between position codes in order to explicitly compare the order of the occurrences of event labels and implicitly their temporal relations. Finally, the EMKL approach combines the ERF and EPC approaches using multiple kernel learning (MKL).

Three exisitng approaches to similarity searching and matching of event iterval sequences are DTW-based [7], IBSM [8], and Artemis [7]. The first method represents an e-sequence by a series of vectors such that a vector is created for each start- or end-point of any event-interval. The distances between the vectors are then computed using Dynamic Time Warping (DTW). IBSM uses a vector-based representation with a vector for each time point. Euclidean distance is then used to compare the e-sequences. The Artemis method measures the similarity between two e-sequences based on temporal relations shared between events by mapping the e-sequences into a bipartite graph. The main limitation of the Artemis is that it matches e-sequences only if they share common temporal relations among event intervals without taking the duration of the event intervals into account. DWT-based and IBSM on the other hand, ignore the temporal relations between event intervals.

2   Problem Statement

Definition 1. Let $\Sigma = \{A, B, ...\}$ denote a finite alphabet. A triple $e = (l, b, f)$ is called an *event interval*, where $l \in \Sigma$ is the event label and, $b, f \in \mathbb{N}$, $(b < f)$ are the beginning and finishing time, respectively. The duration of an event interval e is $d(e) = f - b$.

Definition 2. An event-interval sequence or *e-sequence* $s = \langle e_1, e_2, ..., e_m \rangle = \langle (l_1, b_1, f_1), (l_2, b_2, f_2), ..., (l_m, b_m, f_m) \rangle$ is a list of m event intervals placed in ascending order based on their beginning times. If event intervals have equal beginning times, then they are ordered lexicographically by their labels. Multiple occurrences of an event are allowed in an e-sequence if they do not happen concurrently. The duration of an e-sequence s is $d(s) = max\{f_1, f_2, ..., f_m\} - min\{b_1, b_2, ..., b_m\}$.

Definition 3. Given an e-sequence s, the multiset $T = \{b_1, f_1, b_2, f_2, ..., b_m, f_m\}$ consists of all time points corresponding to sequence s. If we sort T in ascending order and eliminate redundant elements, we can derive a sequence $T_s = \langle t_1, t_2, ..., t_{m'} \rangle$, where $t_k \in T, t_k < t_{k+1}$. T_s is called the *e-sequence sliced time* of s.

Definition 4. Given an e-sequence s, a function $\Phi_s : \mathbb{N} \times \mathbb{N} \to 2^\Sigma$ is defined as:

$$\Phi_s(t_p, t_q) = \{l_j \mid (l_j, b_j, f_j) \in s \ \wedge \ (b_j \leq t_p) \wedge (t_q \leq f_j)\} \tag{1}$$

where $1 \leq j \leq n$ and $t_p < t_q$. Given a corresponding e-sequence sliced time T_s, a *coincidence* c_k is defined as $\Phi_s(t_k, t_{k+1})$ where $t_k, t_{k+1} \in T_s$, $1 \leq k \leq m' - 1$, are two consecutive time points. The duration λ_k of coincidence c_k is $t_{k+1} - t_k$. The size of a coincidence is the number of event labels in the coincidence.

Definition 5. Given an e-sequence s and the corresponding e-sequence sliced time T_s, the *coincidence label sequence*, or l-sequence $L_s = \langle c_1, c_2, ..., c_{m'-1} \rangle$, is an ordered list of coincidences excluding *gaps* (i.e, time intervals containing no events), where each c_k, $(1 \leq k \leq m' - 1)$ is a coincidence of e-sequence s. The size of a l-sequence $|L_s|$ is the number of coincidences in the l-sequence. With this definition, all information about the order of co-occurrences of events in e-sequence s can be represented by L_s.

Problem Statement. Given an e-sequence dataset D, our goal is to specify a function f that can map any e-sequence $s_i \in D$ to an encoding that captures the main characteristics of s_i in order to facilitate the comparison of e-sequences. We intuitively view the similarity between two e-sequences s_i and s_j in terms of (1) the presence of event intervals with the same event labels, (2) the order of occurrences of these event intervals, (3) the duration of these event intervals, and (4) the temporal relations among these event intervals.

3   Similarity Matching of Interval-Based Temporal Sequences

In this section, we introduce three approaches to performing e-sequence matching. Each method defines a distance function that will be used in a nearest neighbor search. The first two methods embed e-sequences into vector-based representations so that comparing e-sequences and finding similarities among them can be performed using operations on vectors. The third method integrates these two distance functions to compute a combined distance between a given pair of e-sequences.

3.1   Matching Using Relative Frequency

The ERF approach uses the relative frequencies of event labels in e-sequenes, as defined in [5], to measure the distance between the e-sequences.

Definition 6. The *relative frequency* $R(s, l)$ of an event label $l \in \Sigma$ in an e-sequence $s \in D$, which is the duration-weighted frequency of the occurrences of l in s, is defined as the accumulated durations of all event intervals with event label l in s divided by the duration of s. Formally:

$$R(s, l) = \frac{1}{d(s)} \sum_{e \in s \, \wedge \, e.l = l} d(e) \qquad (2)$$

Suppose we want to specify a vector-based representation of an e-sequence s using relative frequency. Function f_1 is defined such that it maps an e-sequence s to a vector of the relative frequencies of event labels $l \in \Sigma$ that present in s. Formally, $f_1 : s \to \mathbb{R}^{|\Sigma|}$, $f_1(s) = \langle R(s, l_1), R(s, l_2), ..., R(s, l_{|\Sigma|}) \rangle$.

The distance between the relative frequency vectors of e-sequences s and s' can be obtained by comparing the magnitude of the relative frequencies of the event labels via Euclidean distance by computing the following:

$$\text{ERF}(s, s') = ||f_1(s) - f_1(s')||_2 = \sqrt{\sum_{j=1}^{|\Sigma|} (R(s, l_j) - R(s', l_j))^2} \qquad (3)$$

3.2   Matching Using Position Code

Encoding the temporal order of event intervals and the temporal relations among event intervals is a challenge because e-sequences have varied lengths and also event labels may occur repeatedly. The second vector-based distance method, called EPC, addresses this challenge by focusing on these two aspects of e-sequences. We use a function to generate a position code for each event label. This position code explicitly represents the positions of event labels in the e-sequence and implicitly represents the temporal relations among the event intervals, based on the coincidence label sequence (Definition 5) of the e-sequence.

Definition 7. The *position code* $P(s, l)$ of an event label $l \in \Sigma$ in an e-sequence $s \in D$, is defined as follows:

$$P(s, l) = \sum_{l \in c_k \, \wedge \, c_k \in L_s} 2^{(|L_s| - k)} \qquad (4)$$

We define the second vector-based representation of an e-sequence s based on position codes. Function f_2 maps an e-sequence s to a vector of the position codes of the event labels $l \in \Sigma$ in s. Formally, $f_2 : s \to \mathbb{N}_0^{|\Sigma|}$, $f_2(s) = \langle P(s, l_1), P(s, l_2), ..., P(s, l_{|\Sigma|}) \rangle$.

To calculate the distance between two e-sequences based on their position code vectors, we use the cosine distance measure because the angle between the vectors matters more than their magnitudes. This distance is defined as follows:

$$\text{EPC}(s, s') = 1 - \frac{f_2(s) \cdot f_2(s')}{||f_2(s)||_2 \, ||f_2(s')||_2} \qquad (5)$$

3.3   Matching Using Multiple Kernel Learning

Both the ERF and EPC approaches are able to assess the similarity between e-sequences. ERF captures aspects of the duration of the event intervals, while EPC captures aspects of the temporal order and relations of the events. Hence, it seems natural to integrate these complementary approaches.

One way to integrate these two methods is to utilize a Multiple Kernel Learning (MKL) framework, which is commonly used with SVM for classification tasks. Linear combinations of kernels can be used to solve SVM optimization problems. In this framework, the problem is solved with respect to a new kernel obtained as a weighted linear combination of a set of given kernels. Here, we use this framework to combine the ERF and EPC distance functions as kernels in order to obtain a new distance function, which is defined as follows:

$$\text{EMKL}(s, s') = \sum_h w_h \phi_h(s, s') \tag{6}$$

where $h > 1$ is the number of kernels to be combined and w_h is the weight of function ϕ_h, $\phi = \{\text{ERF}, \text{EPC}\}$.

4   Experiments

We evaluate the effectiveness of the EMKL method on the task of matching full-length interval-based temporal sequences on eight real-world datasets, namely *ASL-BU* [3], *ASL-BU2* [3], *Auslan2* [2], *Blocks* [2], *Context* [2], *Hepatitis* [1], *Pioneer* [2], *Skating* [2]. The weights of ERF and EPC are set to 2/3 and 1/3, respectively. We compare the 1-NN classification accuracy of EMKL with three state-of-the-art algorithms, namely the DTW-based [7], Artemis [7], and IBSM [8] methods. 1-NN classification accuracy is defined as follows:

Given database of e-sequences D and known class labels for every e-sequence, the effectiveness of a distance function can be assessed according to its accuracy when it is used for 1-NN classification. This accuracy is computed by making a 1-NN classification prediction for every $s \in D$ and then recording the fraction of correct predictions. To make a 1-NN classification prediction c' for an e-sequence s with class label c, first determine the matching set D' as the set of e-sequences in $D - s$ that have the minimum distance to s. Then let c' be the class label of the majority of the e-sequences in D', with ties broken arbitrarily. The prediction c' is correct if it is equal to c.

Table 1. Comparison of 1-NN classification accuracy of EMKL with baselines

Dataset	EMKL	IBSM	Artemis	DTW
ASL-BU	**89.50**	89.29	79.56	43.58
ASL-BU2	**82.75**	76.92	80.53	77.25
Auslan2	**37.5**	**37.5**	28.5	22
Blocks	**100**	**100**	99	87
Context	**97.50**	96.25	90	89
Hepatitis	77.11	**77.52**	72.09	74.03
Pioneer	**98.13**	95	97.5	93
Skating	94.72	**96.79**	84	77

Table 1 shows a comparison of EMKL with the competitors on the 1-NN classification accuracy for the datasets. We adopt the 1-NN classification results of the Artemis and DTW-based methods, as reported in Table 2 in [8] and those of IBSM reported in Table 5 in [9]. As shown, EMKL outperforms the Artemis and DTW-based methods on all datasets. Comparing to IBSM: EMKL wins on four datasets, ties on two datasets, and loses on two datasets.

5   Conclusion

We proposed three distance functions to match full-length event interval sequences. The ERF function measures the distance of e-sequences based on the relative frequency of the event intervals. The EPC function matches e-sequences based on the position codes of the event intervals, which explicitly represents the temporal order of event intervals and implicitly the temporal relations among them. Lastly, the EMKL function combines the ERF and EPC functions to measure three aspects that contribute to the similarity of e-sequences. The experimental evaluation of similarity matching, which was demonstrated by the 1-NN classification accuracy, suggests that EMKL is a better choice compared to the state-of-the-art methods. Overall, these results provide evidence that the EMKL method is an effective approach to the task of matching of full-length interval-based event sequences.

References

1. Patel, D., Hsu, W., Lee, M.L.: Mining relationships among interval-based events for classification. In: International Conference on Management of Data, SIGMOD 2008, pp. 393–404. ACM, New York (2008)
2. Mörchen, F., Fradkin, D.: Robust mining of time intervals with semi-interval partial order patterns. In: International Conference on Data Mining, SIAM, pp. 315–326 (2010)
3. Papapetrou, P., Kollios, G., Sclaroff, S., Gunopulos, D.: Mining frequent arrangements of temporal intervals. Knowl. Inf. Syst. **21**(2), 133 (2009)
4. Allen, J.F.: Maintaining knowledge about temporal intervals. Commun. ACM **26**(11), 832–843 (1983)
5. Mirbagheri, S.M., Hamilton, H.J.: FIBS: A Generic Framework for Classifying Interval-based Temporal Sequences. arXiv preprint arXiv:1912.09445 (2019)
6. Mirbagheri, S.M., Hamilton, H.J.: High Utility Interval-Based Sequences. arXiv preprint arXiv:1912.11165 (2019)
7. Kostakis, O., Papapetrou, P., Hollmén, J.: ARTEMIS: assessing the similarity of event-interval sequences. In: Gunopulos, D., Hofmann, T., Malerba, D., Vazirgiannis, M. (eds.) ECML PKDD 2011, Part II. LNCS (LNAI), vol. 6912, pp. 229–244. Springer, Heidelberg (2011). https://doi.org/10.1007/978-3-642-23783-6_15
8. Kotsifakos, A., Papapetrou, P., Athitsos, V.: IBSM: interval-based sequence matching. In: International Conference on Data Mining, SIAM, pp. 596–604 (2013)
9. Bornemann, L., Lecerf, J., Papapetrou, P.: STIFE: a framework for feature-based classification of sequences of temporal intervals. In: Calders, T., Ceci, M., Malerba, D. (eds.) DS 2016. LNCS (LNAI), vol. 9956, pp. 85–100. Springer, Cham (2016). https://doi.org/10.1007/978-3-319-46307-0_6

Classification of Rare Recipes Requires Linguistic Features as Special Ingredients

Elham Mohammadi[1,2,3], Nada Naji[1], Louis Marceau[1], Marc Queudot[1,3],
Eric Charton[1], Leila Kosseim[2], and Marie-Jean Meurs[3(✉)] (iD)

[1] Banque Nationale du Canada (BNC), Montreal, QC, Canada
{nada.naji,louis.marceau,eric.charton}@bnc.ca
[2] Concordia University, Montreal, QC, Canada
{elham.mohammadi,leila.kosseim}@concordia.ca
[3] Université du Québec à Montréal (UQAM), Montreal, QC, Canada
queudot.marc@courrier.uqam.ca, meurs.marie-jean@uqam.ca

Abstract. In this paper, we propose a joint model, composed of neural and linguistic sub-models, to address classification tasks in which the distribution of labels over samples is imbalanced. Different experiments are performed on tasks 1 and 2 of the DEFT 2013 shared task [10]. In one set of experiments, the joint model is used for both classification tasks, whereas the second set of experiments involves using the neural sub-model, independently. This allows us to measure the impact of using linguistic features in the joint model. The results for both tasks show that adding the linguistic sub-model improves classification performance on the rare classes. This improvement is more significant in the case of task 1, where state-of-the-art results are achieved in terms of micro and macro-average F1 scores.

Keywords: Text classification · Deep learning · Linguistic features

1 Introduction

Different techniques have been used to address the task of text classification throughout the years. In the era of statistical Natural Language Processing (NLP), machine learning approaches were used to automatically extract discriminative linguistic features that would be helpful for the specific task at hand. With the availability of larger corpora and the advent of deep learning, neural network architectures have become increasingly popular in performing different NLP tasks [1], including text classification. However, despite the advances made by deep learning and achieving state-of-the-art results in many tasks, not many studies have addressed the challenge of an imbalanced dataset, which is the case in many real-life scenarios and applications [13].

In this paper, we focus on 2 multi-class classification tasks with an imbalanced distribution of labels over data and measure the effectiveness of different methods in handling such a challenge. The tasks in question are the classification of

© Springer Nature Switzerland AG 2020
C. Goutte and X. Zhu (Eds.): Canadian AI 2020, LNAI 12109, pp. 426–437, 2020.
https://doi.org/10.1007/978-3-030-47358-7_44

cooking recipes into one of 4 difficulty levels (*Very Easy, Easy, Fairly Difficult,* and *Difficult*), and the classification of recipes based on meal type (*Starter, Main Dish,* or *Dessert*). The datasets, which are both imbalanced, are taken from the DEFT (Defi Fouille de Texte) 2013 shared task [10], tasks 1 and 2. To address the tasks, we experiment with two different architectures. The first architecture is a neural model which uses pretrained embeddings as input features. The second architecture is a joint model that is composed of neural and linguistic sub-models. We perform experiments using both architectures and 2 types of pretrained embeddings and measure the effectiveness of using a joint model in handling an imbalanced class distribution.

The rest of this paper is organized as follows: Sect. 2 goes over related previous work. Section 3 presents a statistics summary of the datasets that are used. In Sect. 4, the overall model architecture, the sub-models, and different utilized model configurations are presented. Section 5 includes the results achieved by different models and discusses their implications. Finally, Sect. 6 concludes this work.

2   Related Work

According to [13], 3 general approaches for handling imbalanced data in machine learning have been proposed:

1. Data-level approaches: These approaches involve changing the class distribution through under-sampling and over-sampling of training data to mitigate the class imbalance. However, both under and over-sampling can pose new challenges. Under-sampling decreases the number of samples, thus ignoring information that is available to further train the model. Over-sampling can lead to the creation of synthetic samples that are biased, since not all of the minority samples are included in the over-sampling process [18].
2. Algorithmic approaches: Instead of making modifications to the training data, algorithmic approaches adapt the process of learning to take into account the class imbalance. An example of this would be using class weights in the loss function to assign more penalty to a mistake made on a less frequent class compared to a more frequent one. The challenge in using these approaches is finding an optimal penalty matrix which can result in a better and less biased learning. Moreover, for extremely imbalanced data, such a method can make the classifier more prone to making mistakes on the more frequent classes, leading to a drop in overall performance [14].
3. Hybrid approaches: Data-level and algorithmic approaches can be combined for a better handling of an imbalanced distribution.

However, the mentioned approaches might not be helpful in the case of an extreme class imbalance. One possible avenue is extracting discriminative features, taking into account the imbalance present in the training samples [14]. Discriminative features which are extracted and used alongside distributed representations in a deep architecture have been shown to improve results in a variety of tasks, even when the main challenge is not class imbalance. To do non-factoid answer reranking, [3] use a recurrent architecture that encodes questions and answers, separately. A similarity matrix is calculated on the encoded pairs, followed by a Multi-Layer Perceptron (MLP) that performs the prediction. The results show that it is possible to improve the model by passing additional discourse features to the MLP, alongside features that are learnt through the recurrent architecture.

Researchers in [2] develop a model for Part-Of-Speech (POS) tagging that is made up of a bidirectional Long Short-Term Memory (BiLSTM) network, followed by a Conditional Random Field (CRF) layer that predicts the tags. They enrich this model with manually designed features at the embedding layer and achieve an improved performance, showing that combining manually designed and automatically learnt features can benefit such a task in the absence of a large annotated dataset.

For the classification of short texts, [19] employ a joint model which utilizes both implicit (pretrained word embeddings and character embeddings) and explicit (principal concepts in a text, extracted through a knowledge base) representations of texts. Feeding these representations to a multi-branch convolutional model, they achieve state-of-the-art results. Their results indicate that enriching the sample features through a knowledge base can result in a better classification.

In this work, we experiment with the addition of linguistic features to neural-based models and measure the difference in performance, with a focus on minority classes.

3   Datasets

The datasets that were used for our experiments have been taken from the DEFT 2013 shared task [10], task 1 and task 2. The dataset for task 1 consists of French cooking recipes that have been labelled with their respective level of difficulty on a 4 point scale that ranges from *Very Easy* to *Difficult*. As Table 1 shows, the distribution of labels in this dataset is very imbalanced, with more than 90% of the samples with either a *Very Easy* or *Easy* label, and a significantly smaller number of samples with a *Fairly Difficult* or *Difficult* label.

The dataset for task 2 consists of French cooking recipes that have been labelled with the recipe's meal type, *Starter, Main Dish*, or *Dessert*. Although the distribution of labels in this dataset is not as imbalanced as the dataset for

Table 1. Statistics of the train, development, and test datasets for task 1

Difficulty level	Train		Development		Test	
	# of Samples	Percentage	# of Samples	Percentage	# of Samples	Percentage
Very Easy	5569	50.2%	1393	50.2%	1132	49.0%
Easy	4601	41.5%	1151	41.5%	968	41.9%
Fairly Difficult	855	7.7%	213	7.7%	189	8.2%
Difficult	64	0.6%	16	0.6%	20	0.9%
Total	11089	100.0%	2773	100.0%	2309	100.0%

Table 2. Statistics of the train, development, and test datasets for task 2

Meal type	Train		Development		Test	
	# of Samples	Percentage	# of Samples	Percentage	# of Samples	Percentage
Starter	2599	23.4%	647	23.3%	562	24.4%
Main Dish	5167	46.6%	1280	46.1%	1084	47.0%
Dessert	3323	30.0%	846	30.5%	661	28.6%
Total	11089	100.0%	2773	100.0%	2307	100.0%

task 1, nearly half of the samples belong to the class *Main Dish*, causing the challenge of an imbalanced dataset in task 2, as well.

Finally, it should be noted that originally, for the DEFT 2013 shared task, the data for both tasks was released in the two stages of training and testing. For the experiments reported in this paper, 20% of the released training data was set aside for model validation (referred to as *Development* data in Tables 1 and 2).

4 Model Design

The joint model that we have developed for the classification of recipes is composed of two sub-models. The first sub-model is neural-based and the second sub-model utilizes linguistic features.

In this section, first, the architecture of the neural sub-model is presented. We then describe the linguistic features that were used to complement the extracted neural features and how the two parts of the model are combined for the final classification.

4.1 The Neural Sub-model

The Embedding Layer. The embedding layer is used to transform the concatenation of the preparation section (the main body) and the title of a recipe into dense vectors. In this work, two different types of pretrained transformer-based embeddings are used: the multilingual cased version of BERT embeddings [9], and CamemBERT embeddings [17] which have been trained only on French data using a BERT model. For both BERT and CamemBERT, only the features from the last layer of the models are extracted, resulting in a contextual dense representation of size 768 for each token.

It should be noted that all samples are limited to their first 100 tokens, and zero padding is used for the samples with less than 100 tokens.

The output of the embedding layer is then passed to either a recurrent or a convolutional architecture.

The Recurrent Architecture. For the hidden layer of the recurrent architecture, Gated Recurrent Units (GRUs) [6] were used, since GRUs are less prone to overfitting [7] because of having a smaller number of parameters (compared to

LSTMs [12]). A bidirectional GRU is used to process the embeddings consecutively in a forward and a backward pass. The output of the GRU layer is then passed to an attention layer which calculates its weighted average using Eq. 1:

$$Attention = \sum_{t=1}^{n} y_t \omega_t \qquad (1)$$

In the equation above, y_t stands for the output of the GRU layer at timestep t and ω_t refers to the weight assigned to y_t by the attention mechanism. The weight vector ω is calculated using the following process: First, a single N-to-1 fully connected layer is applied on the output of the hidden layer at each timestep (N being the size of the output representation at a timestep), resulting in a scalar for each timestep. The scalars for all timesteps are then concatenated and a softmax activation function is applied over them, producing the weight vector ω.

The Convolutional Architecture. In some experiments, a convolutional architecture is used instead of a recurrent one. In these experiments, first, a Convolutional Neural Network (CNN) [15] is used to process N-grams of input (N consecutive token representations) using convolution filters of size N. The hidden layer is followed by a pooling layer which is either average or max pooling or a combination of the two. Average pooling computes an average over the outputs of the hidden layer, while for max pooling, first, the output vectors of the hidden layer are passed through a Concatenated Rectified Linear Unit (CReLU) activation function.

4.2   The Linguistic Sub-model

The primary part of the linguistic sub-model is a feature extractor, which transforms each sample into a set of linguistic features proposed by [5]. The features extracted for each task are explained below.

Task 1 Features. For task 1, i.e. the classification of recipes based on their level of difficulty, the following features are used:

The number of tokens in the recipe title, the number of tokens in the recipe preparation part, the number of ingredients mentioned in the ingredient list of the recipe, the cost of the meal on a 3-point scale, the presence of 22 predefined words that have been identified to be discriminative, the presence of 48 predefined discriminative trigrams, and finally, the number of verbs in the recipe that belong to 3 different discriminative verb groups.

The extraction of these features results in a vector of size 77 for each recipe, which is used by the linguistic sub-model.

Task 2 Features. For this task, which is the classification of recipes based on meal type, the following features are used:

Similar to task 1, the number of tokens in the recipe title, the number of tokens in the preparation section, and the number of ingredients on the ingredient list are computed as the first 3 features. The 4th features is the cost associated with the meal on a 3-point scale. The remaining features consist of the presence or absence of 1231 ingredient names in the recipe, the presence of 48 discriminatve trigrams, and the number of verbs in the recipe belonging to each of the three predefined verb families. In the end, a feature vector of size 1286 is extracted for each recipe.

For a complete description of the selection process of the linguistic features, see [5]. The extracted feature vectors are then passed into a single-layer feedforward neural network, mapping each feature vector to a vector of the same size, resulting in output representations by the linguistic sub-model.

4.3   The Fusion Component

The fusion component first concatenates the output of the two sub-models, then applies a fully connected layer over the resulting vector, mapping them to vectors of size 4 in the case of task 1, and to vectors of size 3 in the case of task 2. The fully connected layer is followed by a softmax activation function that outputs the probabilities of the classes.

In order to measure the effect of the linguistic sub-model, some experiments involved utilizing only the neural sub-model. In those experiments, the fusion component was replaced with a fully connected layer, mapping the output of the attention/pooling layer to the number of classes, followed by a softmax activation function which produced the probability distribution over different classes.

4.4   Training

In order to train the models, a batch size of 32 was used. All models were trained for 20 epochs. The final model parameters were taken from the epoch that included the best micro score on the development dataset. Table 3 contains details regarding the hyperparameters of different models. Some specific important aspects of the training process are explained below.

Optimizer. AdamW [16] was used to optimize the training process. For all models in both tasks 1 and 2, an initial learning rate of 10^{-3} was used. This learning rate was adapted for CNN models to a rate of 10^{-4} after two epochs in task 1, and after five epochs in task 2.

Class Weights. In order to counter the effect of the imbalanced distribution of labels in both tasks, class weights were used in the utilized cross-entropy loss function. For experiments that utilized only the neural model, weights were automatically calculated, taking into account the proportion of the samples in each class over the number of all training samples for the task. In experiments that used the joint model, weights were manually set to 0.1, 0.1, 0.2, 0.6 (corresponding to the *Very Easy, Easy, Fairly Difficult,* and *Difficult* classes, respectively)

Table 3. Hyperparameters for each model, including neural models and joint models with both neural and linguistic sub-models. *#HL / #KH*: Number of hidden layers in recurrent models or kernel height in CNNs. *#HN / #K*: Number of hidden nodes in each recurrent layer or number of kernels in CNNs.

	Model	Task 1			Task 2		
		#HL / #KH	#HN / #K	Pooling	#HL / #KH	#HN / #K	Pooling
Neural	CNN-BERT	300, 200, 100, 100	1, 2, 3, 4	max	2	200	max
	GRU-BERT	1	64	attention	2	32	attention
	CNN-CamemBERT	2, 3	250, 50	max	2	200	max, average
	GRU-CamemBERT	2	32	attention	2	64	attention
Joint	CNN-BERT	2	250	max	2	200	max, average
	GRU-BERT	64	1	attention	2	32	attention
	CNN-CamemBERT	1	400	max, average	2	400	max, average
	GRU-CamemBERT	2	32	attention	2	32	attention

for task 1, and set to 0.6, 0.3, 0.1 (corresponding to classes *Starter*, *Main Dish*, and *Dessert*, respectively) for task 2.

Regularization. In order to regularize the network, the optimizer was used with a weight decay rate of 0.02. Moreover, a dropout layer with a probability of 0.2 was applied on the concatenation of the output of the two sub-models in the fusion component in joint models, and on the output of the pooling/attention layer in neural models.

Fine-tuning of BERT and CamemBERT Models. Since in the experiments which involved joint and neural models, the embedding layer was kept frozen, BERT and CamemBERT models were fine-tuned on the two tasks as additional experiments. The results achieved by the fine-tuned models are reported in Tables 4 and 6.

5   Results and Discussion

For both tasks, the results have been reported in terms of micro-average score, macro-average F1 score, macro-average precision and macro-average recall. The micro-average score stands for micro-average F1, precision, and recall, all 3 of which are equivalent in this work since evaluation is done on all classes.

The results of different experiments for task 1, alongside the results of DEFT 2013 teams achieving the best micro scores are shown in Table 4. Looking at the results achieved by the models that we have developed, it can be seen that in all cases, using a joint model has resulted in (often highly) superior performance in terms of micro and macro F1 scores on both development and test data, compared to a solely neural model. The improvement in results can be observed in terms of macro precision and macro recall, as well. This shows that the features

Table 4. Task 1 results.

	Model	Development				Test			
		Micro score	Macro F1	Macro P	Macro R	Micro score	Macro F1	Macro P	Macro R
Neural	CNN-BERT	61.5	39.3	41.1	37.6	58.8	37.7	39.4	36.1
	GRU-BERT	59.9	38.5	38.5	38.4	58.0	36.8	37.0	36.7
	Finetuned BERT	56.9	39.3	42.9	36.2	55.9	36.0	36.8	35.3
	CNN-CamemBERT	60.9	42.7	43.4	42.0	59.3	38.6	39.9	37.3
	GRU-CamemBERT	62.4	36.1	38.1	34.3	60.1	36.9	40.1	34.1
	Finetuned CamemBERT	61.2	37.3	38.5	36.2	59.3	37.6	38.9	36.4
Joint	CNN-BERT	64.5	49.1	60.0	41.6	62.0	47.3	59.3	39.3
	GRU-BERT	65.8	41.7	45.5	38.5	63.1	39.3	42.1	36.8
	CNN-CamemBERT	**66.4**	50.3	58.5	**44.2**	**63.8**	**50.0**	62.0	**42.0**
	GRU-CamemBERT	65.3	**51.1**	**68.5**	40.8	63.1	40.5	42.5	38.7
Deft 2013 Top Teams	First Team [5]	-	-	-	-	62.5	48.4	**68.2**	37.5
	Second Team [8]	-	-	-	-	61.2	45.1	52.4	39.5
	Third Team [4]	-	-	-	-	59.2	45.3	63.3	35.3

which are captured by a neural network can be complemented by linguistic features, resulting in a better classification.

A second observation is that the highest micro and macro scores (except for macro precision for which the highest score was achieved by [5]) belong to joint models that utilize pretrained CamemBERT embeddings in the neural submodel. This is to be expected since CamemBERT embeddings have been trained exclusively on French data, as opposed to multilingual BERT embeddings used by other models. Furthermore, Table 4 shows that the joint CNN-CamemBERT model outperformed the best baseline in terms of micro score and macro F1 and recall by the highest margin among all models, achieving state-of-the-art results in task 1.

Table 5 includes per-class results in terms of F1 score for task 1. On both development and test datasets, in 3 out of 4 classes, the best F1 score is achieved by joint models, specifically the ones that utilize CamemBERT as pretrained embeddings. Looking at the results by the 4 models that use CamemBERT embeddings, it can be seen that in all but one case (the 0% F1 score on the *Difficult* class by the GRU-CamemBERT models), adding linguistic features has improved the per-class performance, while this cannot be said about the models that use BERT, showing that the linguistic features have complemented CamemBERT embeddings more effectively than the BERT embeddings.

On the test set, the joint CNN-CamemBERT model achieved F1 scores higher than the best baseline model. This joint model also resulted in the highest F1 score of 25% on the *Difficult* class, which is the rarest among all classes. It should be noted that only 3 out of the 8 models achieved an F1 score higher than 0 on the *Difficult* class, 2 of which are joint models that utilize linguistic features. The per-class results show the effectiveness of linguistic features when the task involves a highly imbalanced dataset.

Table 6 shows the results that were achieved on task 2, by the models that we have developed alongside the models by the three top-performing teams in

Table 5. Task 1 per-class results, in terms of F1 score.

	Model	Development				Test			
		Very Easy	Easy	Fairly Difficult	Difficult	Very Easy	Easy	Fairly Difficult	Difficult
Neural	CNN-BERT	66.3	61.6	25.1	0.0	63.1	**59.7**	23.5	0.0
	GRU-BERT	68.0	56.0	**29.9**	0.0	65.8	55.2	26.1	0.0
	CNN-CamemBERT	71.0	51.9	22.6	19.5	69.9	51.2	19.2	9.8
	GRU-CamemBERT	71.1	56.7	7.3	0.0	68.7	55.0	12.6	0.0
Joint	CNN-BERT	72.1	60.8	24.9	21.1	70.0	58.1	22.5	17.4
	GRU-BERT	73.9	61.1	22.6	0.0	72.0	58.1	18.9	0.0
	CNN-CamemBERT	74.0	**61.8**	27.0	**27.3**	72.2	59.0	25.2	**25.0**
	GRU-CamemBERT	**74.4**	58.3	27.9	11.8	**72.5**	56.3	**29.4**	0.0
Deft 2013 Top Teams	First Team [5]	-	-	-	-	71.7	56.2	18.8	9.5
	Second Team [8]	-	-	-	-	69.2	57.0	26.1	16.0
	Third Team [4]	-	-	-	-	68.6	52.5	15.6	9.5

DEFT 2013. The first observation is that, among our models, the finetuned CamemBERT model achieved the best overall performance on the development dataset. However, this model is outperformed by the joint CNN-CamemBERT model, which was our best model in task 1, in the test phase. This shows that, in general, the joint CNN-CamemBERT model can generalize better to new samples. This model achieved the highest macro F1 score, alongside the top-performing team of DEFT 2013, and also the highest recall among all models, while it fell short of achieving the highest micro score by 0.3% and the highest macro precision by 0.4%.

Table 6 shows that all joint models outperform their neural counterparts in terms of micro and macro scores. However, unlike task 1, this improvement is not big enough to result in state-of-the-art results in terms of the micro-average score. One reason behind this performance can be the linguistic features that were used for task 2. It is possible that, compared to task 1, these features are not as representative of the classes. The higher sparsity of the linguistic feature matrix for task 2 could be another factor. Looking at the results of task 1 in Table 5, it can be observed that after the addition of the linguistic sub-model, when there is improvement, the amount of improvement is significantly higher in the case of rare classes. Therefore, it can be hypothesized that the strength of the proposed joint model is handling an imbalanced distribution of labels, resulting in a more significant improvement of results when the available dataset is more imbalanced. Knowing that the distribution of labels in task 1 is starkly more imbalanced than task 2, the joint model is more effective in the former than in the latter case.

Finally, Table 7 includes the per-class F1 scores achieved by different models on task 2. It also shows the per-class F1 scores achieved by the three top-performing teams in the DEFT 2013 shared task. Among the 8 models that we developed, the results show that the joint CNN-CamemBERT model achieves the highest F scores on all three classes on the test set. It also achieves state-of-the-art results on the class *Starter*, which is the rarest class in the dataset. This

Table 6. Task 2 results.

	Model	Development				Test			
		Micro score	Macro F1	Macro P	Macro R	Micro score	Macro F1	Macro P	Macro R
Neural	CNN-BERT	86.4	84.9	85.7	84.2	85.9	84.9	85.6	84.2
	GRU-BERT	84.4	83.2	83.2	83.2	84.8	84.0	84.0	83.9
	Finetuned BERT	86.3	85.8	85.1	86.5	86.4	86.2	85.6	86.8
	CNN-CamemBERT	87.6	86.8	86.5	**87.2**	88.1	87.6	87.6	87.7
	GRU-CamemBERT	86.5	85.6	85.5	85.6	87.1	86.5	86.6	86.4
	Finetuned CamemBERT	**88.2**	**87.1**	**87.3**	86.9	88.1	87.4	87.5	87.3
Joint	CNN-BERT	86.0	85.2	84.9	85.4	87.0	86.5	86.3	86.7
	GRU-BERT	85.0	84.2	83.9	84.6	85.5	85.0	84.8	85.2
	CNN-CamemBERT	87.5	86.8	86.4	87.1	88.6	**88.2**	88.0	**88.3**
	GRU-CamemBERT	86.9	86.1	85.8	86.5	87.8	87.3	87.2	87.4
Deft 2013 Top Teams	First Team [4]	-	-	-	-	**88.9**	88.2	**88.4**	88.1
	Second Team [5]	-	-	-	-	85.6	84.7	85.0	84.3
	Third Team [11]	-	-	-	-	84.9	84.1	84.2	84.1

Table 7. Task 2 per-class results, in terms of F1 score.

	Model	Development			Test		
		Starter	Main Dish	Dessert	Starter	Main Dish	Dessert
Neural	CNN-BERT	70.2	86.5	97.6	71.0	86.4	96.8
	GRU-BERT	67.9	83.8	97.8	70.5	85.0	96.4
	CNN-CamemBERT	**75.1**	**87.1**	98.2	77.1	88.0	97.7
	GRU-CamemBERT	72.8	86.4	97.4	75.4	87.8	96.4
Joint	CNN-BERT	72.0	85.6	97.7	75.0	87.2	97.3
	GRU-BERT	70.7	84.5	97.2	72.6	85.8	96.5
	CNN-CamemBERT	74.7	86.9	**98.6**	**78.1**	88.5	98.0
	GRU-CamemBERT	73.4	86.3	98.5	76.7	87.8	97.5
Deft 2013 Top Teams	First Team [4]	-	-	-	77.3	**88.8**	**98.6**
	Second Team [5]	-	-	-	70.3	85.6	97.9
	Third Team [11]	-	-	-	69.4	84.8	98.2

is in agreement with the hypothesis that the strength of the joint model is in the handling of rare classes.

6   Conclusion

In this paper, we proposed a joint model for the classification of imbalanced data. The model, composed of a neural and a linguistic sub-model, was utilized to address tasks 1 and 2 of the DEFT 2013 shared task [10], which involved the classification of French recipes based on difficulty level and meal type, using the datasets and the evaluation metrics specific to the two tasks. In order to measure the effect of the linguistic sub-model, experiments were performed using the joint model, while a second set of experiments involved using only the neural sub-model, independently. The results from these experiments show that, in both tasks, the joint models outperform their neural counterparts. In task 1, the joint model could achieve state-of-the-art results in terms of both micro and macro-average F1 scores, showing the effectiveness of this model in cases of highly imbalanced data.

Reproducibility

To ensure reproducibility and comparisons between systems, our source code is publicly released as an open source software in the following repository: https://github.com/cooking-classification/CAI2020.

The data could be obtained by contacting the DEFT 2013 shared task organizers (see https://deft.limsi.fr/2013/index.php?id=1&lang=en).

References

1. Amini, H., Farahnak, F., Kosseim, L.: Natural language processing: an overview. In: Frontiers in Pattern Recognition and Artificial Intelligence, vol. 5, Chap. 3, pp. 35–55. World Scientific, June 2019
2. Xuan Bach, N., Khuong Duy, T., Minh Phuong, T.: A POS tagging model for vietnamese social media text using BiLSTM-CRF with rich features. In: Nayak, A.C., Sharma, A. (eds.) PRICAI 2019, Part III. LNCS (LNAI), vol. 11672, pp. 206–219. Springer, Cham (2019). https://doi.org/10.1007/978-3-030-29894-4_16
3. Bogdanova, D., Foster, J., Dzendzik, D., Liu, Q.: If you can't beat them join them: handcrafted features complement neural nets for non-factoid answer reranking. In: Proceedings of the 15th Conference of the European Chapter of the Association for Computational Linguistics: Volume 1, Long Papers, pp. 121–131 (2017)
4. Bost, X., et al.: Systémes du LIA à DEFT 13. arXiv preprint arXiv:1702.06478 (2017)
5. Charton, E., Meurs, M.J., Jean-Louis, L., Gagnon, M.: Using collaborative tagging for text classification: from text classification to opinion mining. Informatics 1(1), 32–51 (2014)
6. Cho, K., et al.: Learning phrase representations using RNN encoder-decoder for statistical machine translation. In: Proceedings of the 2014 Conference on Empirical Methods in Natural Language Processing (EMNLP 2014), Doha, Qatar, pp. 1724–1734, October 2014
7. Chung, J., Gulcehre, C., Cho, K., Bengio, Y.: Empirical evaluation of gated recurrent neural networks on sequence modeling. In: NIPS 2014 Deep Learning and Representation Learning Workshop, Montreal, Canada, December 2014
8. Collin, O., Guerraz, A., Hiou, Y., Voisine, N.: Participation de orange labs à deft 2013. Actes du neuvième DÉfi Fouille de Textes, pp. 67–79 (2013)
9. Devlin, J., Chang, M.W., Lee, K., Toutanova, K.: BERT: pre-training of deep bidirectional transformers for language understanding. In: Proceedings of the 2019 Conference of the North American Chapter of the Association for Computational Linguistics: Human Language Technologies (ACL/HLT 2019), pp. 4171–4186. Minneapolis, Minnesota, June 2019
10. Grouin, C., Paroubek, P., Zweigenbaum, P.: DEFT2013 se met à table: présentation du défi et résultats. In: Actes de DEFT. TALN, Les Sables-d'Olonnes, France, 21 juin 2013

11. Hamon, T., Périnet, A., Grabar, N.: Efficacité combinée du flou et de l'exact des recettes de cuisine. Actes du neuvième DÉfi Fouille de Textes, p. 18 (2013)
12. Hochreiter, S., Schmidhuber, J.: Long short-term memory. Neural Comput. **9**(8), 1735–1780 (1997)
13. Johnson, J.M., Khoshgoftaar, T.M.: Survey on deep learning with class imbalance. J. Big Data **6**(1), 1–54 (2019). https://doi.org/10.1186/s40537-019-0192-5
14. Krawczyk, B.: Learning from imbalanced data: open challenges and future directions. Prog. Artif. Intell. **5**(4), 221–232 (2016). https://doi.org/10.1007/s13748-016-0094-0
15. LeCun, Y., Haffner, P., Bottou, L., Bengio, Y.: Object recognition with gradient-based learning. Shape, Contour and Grouping in Computer Vision. LNCS, vol. 1681, pp. 319–345. Springer, Heidelberg (1999). https://doi.org/10.1007/3-540-46805-6_19
16. Loshchilov, I., Hutter, F.: Decoupled weight decay regularization. In: Proceedings of the 3rd International Conference on Learning Representations (ICLR 2019), New Orleans, Louisiana, USA, May 2019
17. Martin, L., et al.: Camembert: A tasty French language model. arXiv preprint arXiv:1911.03894 (2019)
18. Oh, J.H., Hong, J.Y., Baek, J.G.: Oversampling method using outlier detectable generative adversarial network. Expert Syst. Appl. **133**, 1–8 (2019)
19. Wang, J., Wang, Z., Zhang, D., Yan, J.: Combining knowledge with deep convolutional neural networks for short text classification. In: International Joint Conference on Artificial Intelligence (IJCAI), pp. 2915–2921 (2017)

Happiness Analysis with Fisher Information of Dirichlet-Multinomial Mixture Model

Fatma Najar$^{(\boxtimes)}$ and Nizar Bouguila$^{(\boxtimes)}$

Concordia Institute for Information Systems Engineering (CIISE),
Concordia University, Montreal, QC, Canada
`f_najar@encs.concordia.ca, nizar.bouguila@concordia.ca`

Abstract. Emotion recognition requires robust feature representation and discriminative classification models. In this paper, we consider Fisher vectors for feature representation and Fisher scoring algorithm for learning the proposed model. We first propose a new Fisher scoring algorithm using an exact Fisher information matrix for the Dirichlet-multinomial (DM) mixture model. Subsequently, we present an exact derivation of the Fisher vectors for images representation and we analyze the intensity of happiness from EMOTIC database by applying the proposed framework. The obtained results prove the effectiveness and the robustness using Fisher vectors for emotion recognition.

Keywords: Emotion recognition · Fisher vectors · Exact Fisher information matrix · Dirichlet-multinomial mixture model · Fisher scoring

1 Introduction

We all live intending to be happy; which rises the interest for many online therapist applications in the interest of improving human mental health. The challenge here is how to measure happiness? Starting from text recognition which is based primarily on self-reported surveys to emotion recognition from images where researchers have recently succeeded to categorize the happiest moments. The majority of research works focused only on the six basic facial expressions ranging in *disgust, happiness, fear, anger, sad, surprise* using Action Units to encode the emotions. It has been shown in [4] that addressing only the face and the body pose does not give enough information to understand the emotional states of a person while adding scene context and more details recognize better the apparent emotional states. Alternatively, scientists employ the VAD emotional state model to express emotions through *Valence* which encodes the happiness and the pleasure, *Arousal* that measures the human agitation level, and *Dominance* to estimate the control level of such situation. Hence, using other visual features apart from the face, the social context that may constitues relevant information, plays a significant role to understand the emotional state of a person.

C. Goutte and X. Zhu (Eds.): Canadian AI 2020, LNAI 12109, pp. 438–444, 2020.
https://doi.org/10.1007/978-3-030-47358-7_45

In this paper, we model this combination of information in order to give more significant understanding of the real happiness state. We propose a new framework based on mixture of DM distributions and the use of the exact Fisher information matrix. We introduce a novel Fisher scoring learning algorithm and a different representation of images based on Fisher vectors.

2 Problem Statement

The DM has arisen to model count data in various contexts. In text modeling, Madsen and Elkan have proposed to solve the problem of burstiness [6] using the DM distribution. An important clue in every statistical model is the Fisher information matrix which measures the amount of information of a local parameter and it is widely practical for several tasks such as the calculation of the maximuum Likelihood Estimator (MLE) and for model selection. For the DM, Neerchal and Morel [8] approximated the Fisher information matrix and considered the same as the one of the Dirichlet distribution for large size clusters. However, as it is not always the case and clusters sizes are not all large, Paul et al. [9] presented the exact Fisher information matrix for the DM where they proved the difference between the exact and the approximate Fisher information. These facts motivated us to propose learning of DM mixture models based on the exact Fisher information matrix. A method that uses the Fisher scoring technique for computing the maximum likelihood estimates and considers the case where the sizes of the clusters are not all large.

3 The Mathematical Model

The DM distribution [7] is a combination between two distributions namely the multinomial distribution and the Dirichlet distribution. For more simplicity, another alternative representation for the DM was suggested by [3], and the Gamma function was replaced by rising polynomials to the discrete probability density function:

$$p(\overrightarrow{Y}|\overrightarrow{\phi},\vartheta) = \left(\frac{s}{\overrightarrow{Y}}\right)\frac{\prod_{d=1}^{D}\phi_d(\phi_d+\vartheta)\dots[\phi_d+(y_d-1)\vartheta]}{(1+\vartheta)\dots[1+(s-1)\vartheta]} \tag{1}$$

where $\overrightarrow{Y} = (y_1,\dots,y_D)$ is a D-dimensional count vector, $s = \sum_{d=1}^{D} y_d$, $\overrightarrow{\phi} = (\phi_1,\dots,\phi_D)$ is the proportion vector defined as $\phi_d = \frac{\theta_d}{|\theta|}$, $(d = 1,\dots,D)$, ϑ is the overdispersion parameter $\vartheta = \frac{1}{|\theta|}$ and $|\theta| = \sum_{d=1}^{D}\theta_d$ (θ is the parameter of the Dirichlet distribution)

Given N clusters where the i-th cluster ($i = 1,\dots,N$) is represented with a DM distribution, then a mixture of N DM distributions is given as follows

$$p(\overrightarrow{Y}|\boldsymbol{\Phi}) = \sum_{i=1}^{N} \pi_i \, \mathrm{DM}(\overrightarrow{Y}|\overrightarrow{\phi}_i,\vartheta_i) \tag{2}$$

where π_i is the mixing weight of the i-th cluster,

$\Phi = (\pi_1, \ldots, \pi_N, \phi_{11}, \ldots, \phi_{DN}, \vartheta_1, \ldots, \vartheta_N)$ is the set of all the parameters, and $DM(\overrightarrow{Y} | \overrightarrow{\phi}_i, \vartheta_i)$ is the DM distribution within the i-th cluster.

Suppose we have a set of W multimedia objects which could be text documents, images or videos, we represent each object by a vector of counts $\overrightarrow{Y}$ and the finite set of objects with $\mathcal{Y} = (\overrightarrow{Y}_1, \ldots, \overrightarrow{Y}_W)$, the complete log-likelihood apart from a constant is given by

$$\mathcal{L}(\mathcal{Y}|\Phi) = \sum_{j=1}^{W} \sum_{i=1}^{N} p(i|\overrightarrow{Y}_j) \Bigg\{ \log \pi_i + \sum_{d=1}^{D} \Big[\sum_{r=1}^{y_{jd}} \log(\phi_{id} + (r-1)\vartheta_i) \tag{3}$$
$$- \sum_{r=1}^{s} \log(1 + (r-1)\vartheta_i) \Big] \Bigg\}$$

where $p(i|\overrightarrow{Y}_j)$ is the posterior probability.

The Fisher scoring algorithm is based on the Fisher information matrix and the first derivate of the log-likelihood. The major principle is to iterate the following scoring equation

$$\Phi^{(t+1)} = \Phi^t + F(\Phi)^{-1} g(\Phi) \tag{4}$$

where $F(\Phi)$ is the Fisher information matrix which is a symmetric definite matrix where the elements of the exact matrix are determined based on the Beta-binomial probability function rather than the DM as following

$$F = \begin{bmatrix} E[-\frac{\partial^2 \mathcal{L}}{\partial \phi_{id}^2}] & E[-\frac{\partial^2 \mathcal{L}}{\partial \phi_{id} \partial \vartheta_i}] \\ E[-\frac{\partial^2 \mathcal{L}}{\partial \vartheta_i \partial \phi_{id}}] & E[-\frac{\partial^2 \mathcal{L}}{\partial \vartheta_i^2}] \end{bmatrix} \tag{5}$$

$$E[-\frac{\partial^2 \mathcal{L}}{\partial \phi_{id}^2}] = \sum_{j=1}^{W} p(i|\overrightarrow{Y}_j) \sum_{r=1}^{s} \frac{p(y_{jd} \geq r)}{(\phi_{id} + (r-1)\vartheta_i)^2} \tag{6}$$

$$E[-\frac{\partial^2 \mathcal{L}}{\partial \phi_{id} \partial \vartheta_i}] = \sum_{j=1}^{W} p(i|\overrightarrow{Y}_j) \sum_{r=1}^{s} (r-1) \frac{p(y_{jd} \geq r)}{(\phi_{id} + (r-1)\vartheta_i)^2} = E[-\frac{\partial^2 \mathcal{L}}{\partial \vartheta_i \partial \phi_{id}}] \tag{7}$$

$$E[-\frac{\partial^2 \mathcal{L}}{\partial \vartheta_i^2}] = \sum_{j=1}^{W} p(i|\overrightarrow{Y}_j) \sum_{r=1}^{s} (r-1)^2 \frac{p(y_{jd} \geq r)}{(\phi_{id} + (r-1)\vartheta_i)^2} - \frac{1}{(1 + (r-1)\vartheta_i)^2} \tag{8}$$

where $p(y_{jd} \geq r)$ has a Beta-binomial probability density function of the vector y_{jd} that should be bigger than the sum of the features [9] and $g(\Phi)$ is the score function calculated by the gradient of the log-likelihood with respect to the parameters

$$\frac{\partial \mathcal{L}}{\partial \phi_{id}} = \sum_{j=1}^{W} p(i|\overrightarrow{Y}_j) \sum_{r=1}^{y_{jd}} \frac{1}{\phi_{id} + (r-1)\vartheta_i} \tag{9}$$

$$\frac{\partial \mathcal{L}}{\partial \vartheta_i} = \sum_{j=1}^{W} p(i|\overrightarrow{Y}_j) \sum_{r=1}^{y_{jd}} \frac{r-1}{\phi_{id} + (r-1)\vartheta_i} - \sum_{r=1}^{s} \frac{r-1}{1 + (r-1)\vartheta_i} \tag{10}$$

We explain the different steps of the complete approach in the following algorithm.

Algorithm 1. Dirichlet-multinomial learning algorithm

Require: $\mathcal{Y}$ dataset, Number of components N;
1: Initialization using K-means and the Method of Moment as in [1]
2: **repeat**
3:     **for each** Component i **do**
4:         Estimate the posterior distribution $p(i|\overrightarrow{Y}_j)$ using Bayes rule
5:         Estimate the mixing weight components $\pi_i = \frac{1}{W} \sum_{j=1}^{W} p(i|\overrightarrow{Y}_j)$
6:         Update the parameters ϕ_{id} $(d = 1, \ldots, D)$ and ϑ_i using the Fisher scoring algorithm
7:     **end for**
8: **until** Convergence of Likelihood
Ensure: $\overrightarrow{\Phi}^*$

---

4 Experimental Evaluations

For feature representation, we use two different methods. We propose to use a new Fisher framework from the DM to represent the local features in images. Second, we consider the well-known representation technique namely the "Bag-of-visual-words" (BOV). Commonly, the Fisher information matrix is intractable for several models and Fisher vectors are approximated empirically as $F \approx \frac{1}{|Y|} \sum_y g(\boldsymbol{\Phi})g(\boldsymbol{\Phi})^T$. However, this approximation, for such cases, can be used only for the diagonal terms and leads to very poor approximation for a real world modeling perspective. For this reason, we derive, for the first time, exact Fisher vectors from the DM model.

Let $\overrightarrow{Y}_j$ be one image in the given database and $\overrightarrow{X} = (x_1, \ldots, x_T)$ the set of descriptors extracted from the images. We assume that those features are generated by a mixture of DMs: $p(\overrightarrow{X}) = \sum_{t=1}^{N} \pi_t \mathrm{DM}(\overrightarrow{X}|\overrightarrow{\phi}, \vartheta)$. The parameters of $\mathrm{DM}(\overrightarrow{X}|\overrightarrow{\phi}, \vartheta)$ are estimated through the proposed Fisher scoring algorithm (Algorithm 1) and based on those parameters, we derive the following Fisher vectors

$$\Psi(\boldsymbol{\Phi}) = F(\boldsymbol{\Phi})^{-\frac{1}{2}} g(\boldsymbol{\Phi}) \tag{11}$$

where F and g are the Fisher information matrix and score function, respectively, defined in the previous section.

For our experiments, we employed the EMOTIC database [5] which is a collection of 23,571 images of 34,320 annotated people containing 26 discrete categories and three continuous dimensions of the VAD Emotional State Model. This database covers a large-scale range of different emotional states, where we extracted a subset in the interest of analyzing the happiness intensity. We consider the following related emotional states: *affection, excitement, happiness, peace, pleasure, sympathy.* The considered categories represent a significant challenge as the emotions have similar scores and each image is annotated by five different annotates. Another challenge in this database is that more than 25% of people have their face partially occluded and not visible. For the sake of evaluating our results, we used the Average precision metric [2] which represents a trade-off between precision and recall for a set of top documents existing after each relevant document is retrieved. For both representations, we extract from each image the SIFT features from 16 × 16 patches on a regular grid every 8 pixels. For the BOV representation, we extracted also the MBLBP feature for the construction of the visual vocabulary. Following, we apply the proposed Fisher scoring technique on the entire features extracted. In the FV representation, for each image, we learn the parameters of the DM mixture model, we optimize the EM algorithm with Fisher scoring and we compute the exact and the approximated Fisher vectors. Then, for testing, we employ Support vectors machine technique. As a baseline comparison, we investigate the performance of the exact Fisher vectors image representation against the BOV representation and the related works. We found that Fisher vectors (exact and approximated) representation combined with SVM classification achieves high average precision with an increase of 30% and 10% respectively as shown in Table 1. We note also that in terms of accuracy rates, the proposed DM+FS+SIFT+EFV reaches 87.65% versus 40.55% using the same learning method but with the BOV representation DM+FS+SIFT+BOV, versus 34.36% and 37.04% for the BOV structure with the Expectation-Maximization inference method. This confirms our hypothesis about the Fisher scoring method and the Fisher vectors which are based on the exact calculation of the Fisher information matrix that takes into account the dependence between features especially the EMOTIC database is a such challenging dataset. Furthermore, our results using the DM for the both representation achieves better performance comparing to other methods reported in the literature [4]. This includes the convolutional neural network

Table 1. Evaluation results

Models	Accuracy	mAP (%)
CNN (B+I (SL_1)) [4]	–	40.83
CNN (B+ I (L_{Comb})) [5]	–	41.09
DM+EM+SIFT+BOV	34.36	50.84
DM+EM+MBLBP+BOV	37.04	54.50
DM+FS+SIFT+BOV	40.55	65.01
DM+FS+MBLBP+BOV	38.27	67.73
DM+FS+SIFT+AFV	86.42	50.04
DM+FS+SIFT+EFV	87.65	67.44

CNN (B+I (SL_1)) with 40.83% mean average precision (mAP) and CNN (B+ I (L_{Comb})) with 41.09% mAP. In Table 2, we analyze the happiness intensity by means of the average precision of each emotion class (affection, excitement, happiness, peace, pleasure, sympathy). The results demonstrate that exact Fisher vectors play a more important role than the approximated FV, the BOV in image representation, and also the CNN framework.

Table 2. Happiness intensity using BOV and Fisher vectors

	Affection	Excitement	Happiness	Peace	Pleasure	Sympathy
CNN (B+I (SL_1)) [4]	27.85	77.16	58.28	21.56	45.46	14.71
CNN (B+ I (L_{Comb})) [5]	26.01	78.51	55.21	22.94	48.65	15.25
DM+EM+SIFT+BOV	50.19	83.33	47.73	36.11	41.67	46.04
DM+EM+MBLBP+BOV	57.91	57.54	38.94	41.67	64.29	66.67
DM+FS+SIFT+BOV	25.69	72.75	54.67	100	69.84	67.06
DM+FS+MBLBP+BOV	56.71	74.56	36.07	95	91.67	52.42
DM+FS+SIFT+AFV	49.56	52.24	54.79	61.89	39.96	41.82
DM+FS+SIFT+EFV	74.93	36.28	80.15	47.56	90.93	74.81

5   Conclusion

In this paper, we propose a novel and efficient model to represent images using exact Fisher vectors. Based on the DM, we present the exact calculation of Fisher information matrix and we offer a new mixture model inference using the Fisher scoring algorithm to estimate the parameters of DM model. Compared to the BOV structure, the Fisher vectors offer better image representation as they encode statistics related to the DM rather than count of occurrences without semantic information. In addition, comparing the exact and the approximated calculation of FV, the exact one provides higher efficiency in terms of classification accuracy and average precision.

Acknowledgment. The completion of this research was made possible thanks to the Natural Sciences and Engineering Research Council of Canada (NSERC).

References

1. Bouguila, N., Ziou, D.: On fitting finite Dirichlet mixture using ECM and MML. In: Singh, S., Singh, M., Apte, C., Perner, P. (eds.) ICAPR 2005. LNCS, vol. 3686, pp. 172–182. Springer, Heidelberg (2005). https://doi.org/10.1007/11551188_19
2. Christopher, D.M., Prabhakar, R., Hinrich, S.: Introduction to information retrieval. Introd. Inf. Retriev. **151**(177), 5 (2008)
3. Haldane, J.B.: The fitting of binomial distributions. Ann. Eugenics **11**(1), 179–181 (1941)

4. Kosti, R., Alvarez, J., Recasens, A., Lapedriza, A.: Context based emotion recognition using emotic dataset. IEEE Trans. Pattern Anal. Mach. Intell. (2019)
5. Kosti, R., Alvarez, J.M., Recasens, A., Lapedriza, A.: Emotion recognition in context. In: Proceedings of the IEEE Conference on Computer Vision and Pattern Recognition, pp. 1667–1675 (2017)
6. Madsen, R.E., Kauchak, D., Elkan, C.: Modeling word burstiness using the Dirichlet distribution. In: Proceedings of the 22nd International Conference on Machine Learning, pp. 545–552. ACM (2005)
7. Mosimann, J.E.: On the compound multinomial distribution, the multivariate β-distribution, and correlations among proportions. Biometrika 49(1/2), 65–82 (1962)
8. Neerchal, N.K., Morel, J.G.: Large cluster results for two parametric multinomial extra variation models. J. Am. Stat. Assoc. 93(443), 1078–1087 (1998)
9. Paul, S.R., Balasooriya, U., Banerjee, T.: Fisher information matrix of the Dirichlet-multinomial distribution. Biom. J.: J. Math. Methods Biosci. 47(2), 230–236 (2005)

Personalized Multi-Faceted Trust Modeling in Social Networks

Alexandre Parmentier$^{(\boxtimes)}$ and Robin Cohen

Cheriton School of Computer Science, University of Waterloo, Waterloo, ON, Canada
{aparment,rcohen}@uwaterloo.ca

Abstract. In this work we develop a comprehensive approach for multi-faceted trust modeling (MFTM) and use it to model trustworthiness of peers in online social networks. We then show how this data-driven design supports clustering-based personalization of trust prediction, yielding better performance than competing solutions when applied to Yelp datasets for an item recommendation task. In all we demonstrate the promise of personalized trust modeling in social networks.

Keywords: Multiagent systems · Trust modeling · Social networks · Personalization

1 Introduction and Background

The aim of this research is to develop algorithms for presenting content in social networks to users, based on a modeling of peer relationships and trustworthiness. Predicting which reviews a user may benefit from the most on the Yelp business review site would help to reduce information overload and then allow the user to enjoy the most relevant services. We approach this problem through the lens of trust modeling: we predict which users are most likely to be trustworthy for other users. Specifically, we make use of multi-faceted trust modeling (MFTM) to approximate the trust formulation functions of users, then predict new "trust links" based on these approximations, essentially suggesting new potentially trustworthy users.

Our work is inspired by the works of Ardissono et al. and Mauro et al. [1,6], and of Fang et al. [3]. While each of these approaches is valuable, the Ardissono approach does not integrate data-driven learning of feature weights, while the Fang model offers a modest number of generic features. We build upon these proposals to offer a more comprehensive set of features, using a machine learning approach to weigh feature importance, and doing so in a way that supports personalization (through clustering) when applied to a content recommendation task. We then present results to demonstrate the value of predicting trust links using our methods, in comparison with competing approaches, running on a Yelp dataset.

© Springer Nature Switzerland AG 2020
C. Goutte and X. Zhu (Eds.): Canadian AI 2020, LNAI 12109, pp. 445–450, 2020.
https://doi.org/10.1007/978-3-030-47358-7_46

Multi-Faceted Trust Modeling (MFTM) is a technique for predicting trust-worthiness between users based on combinations of arbitrary suspected indicators of trustworthiness and reputation. By "trust indicator", we mean some quality of an agent which counts as evidence that they should/should not be trusted. For example, when choosing to hire a plumber, recommendations from friends, online reviews, and a quality web site with clearly indicated prices may all be relevant indicators of trustworthiness, which the judicious homeowner will combine into a single implicit score, effectively inducing a ranking among candidates.

Given this example, it is relatively straightforward to cast the trust prediction problem as a simplified binary classification task: for some agent requiring a partner for an interaction (truster), predict whether each of the other agents (trustee) is more likely to be trustworthy or untrustworthy. A dataset containing information describing user attributes and the presence of trust between some users can be used to train an off-the-shelf classifier to predict novel trust links. This summarizes the basic ideas underlying MFTM.

2   Dataset and Trust Indicators

We conduct experiments using the Yelp dataset[1]. This dataset contains approximately 1.6 million users and over 6 million reviews.

The main content produced by users on the site are reviews targeted at businesses. Following the norm in recommender systems literature, we refer to these businesses generically as "items". Each review contains a text portion and a final score in the range $[1, 5]$, with 5 being the most positive. The reviews generated by users can receive multiple types of feedback and endorsements from other users. For our experiments, we sampled 10000 users from the dataset.

We combined trust indicators proposed in previous works, adapting some to the specifics of the Yelp dataset. These trust indicators aim to quantify evidence of trustworthiness with respect to single users and between pairs of users. From [3] we implemented Benevolence, Competence and Integrity. Each of these indicators is quite generic, focusing on similarity between review scores produced by users. From [6] we implemented $elite_a$, $opLeader_a$ (reflecting number of fans), lup_a, vis_a, fb_a, and rel_{ab}. These indicators are relatively domain specific, being tuned towards the data that is available in the Yelp dataset. For instance, $elite_a$ is based on how many years user a has been deemed "elite" by the Yelp administrators. In addition to the indicators mentioned above, we computed some important modifications. For example, we adjusted $elite_a$, lup_a, $opLeader_a$ and vis_a to be sensitive to the number of years a user had been active on the site. In addition, we calculated the Jaccard distance between users with respect to items reviewed and categories of items reviewed.

Although space limits us from describing all these indicators, we define two indicators here. Benevolence between a and b is the Pearson Correlation

---

[1] https://www.yelp.com/dataset.

Coefficient (PCC) between the scores given to items by the two users [3], while rel_{ab} is the Jaccard Similarity between the friend groups of the two users [6].

3   Experiment Description

At a high level, we aim to examine the performance of (P)MFTM by contrasting multiple methods of predicting trust links in a dataset and evaluating the accuracy of these links by measuring their impact on the performance of a trust aware recommender system. The process is separated into three steps: clustering, trust link prediction, and recommender evaluation[2].

These 3 steps form a pipeline, with each step feeding results into the next. We evaluate solutions with and without prediction and personalization. In the case where personalization is not attempted, the Clustering step is skipped, with all users assigned to a single cluster. In our baseline experiment, which uses just the explicit friendship links in the Yelp dataset as a trust matrix, only the final step is performed.

Clustering: As an approach towards the personalization of trust link prediction (PMFTM), we clustered agents based on two distance measures present in the dataset: review score correlation and social circle Jaccard similarity.

We used a simple clustering method parameterized by a cluster size m. The first cluster is built by starting with a set containing only the most central point in the dataset (i.e. the point with the lowest mean distance to all other points). Points are added to the cluster in order of least mean distance to all points already in the cluster. For example, after choosing the first point, p_1, the next point added to the set, p_2, is the point p_i that minimizes $dist(p_1, p_i)$ (ties are broken randomly). The third point added is the point p_i that minimizes $\frac{1}{2}(dist(p_1, p_i) + dist(p_2, p_i))$. In general, the i'th point added to the cluster (for $i > 1$) is the point that minimizes $\frac{1}{i-1}\sum_{j=1}^{i-1} dist(p_j, p_i)$. This process continues until m points are in the set. The process then repeats to form the second cluster, except only those points not already assigned to a cluster are considered. The algorithm ends when all points are assigned to a cluster.

Trust Link Prediction: For each cluster of users (or for the totality of users, in the case where clustering was not being performed), a logistic regression classifier is learned to predict trust links between the users of that cluster and other users. Learning distinct classifiers for each cluster of users implements personalization of trust modeling. This solution has the capability to learn which factors are most important in predicting trust links for relatively small groups of users. However, in some cases the classifier learned for a cluster failed to converge to a minimally useful classifier (less than 70% accuracy on training data). In these

---

[2] The first two steps use all data and the last step reserves data for testing. While we acknowledge this approach may allow influence of test data on training, it was taken uniformly across the experiments compared in Sect. 4, allowing equal opportunity to all approaches.

cases, we substituted the predictions of the faulty classifier for a generic classifier trained on all users.

We predicted two type of trust links: an explicit friendship link (Yelp users can form up to 6000 mutual friendships each) and positive review PCC. Each is relevant to predicting who a user should trust for advice on Yelp: they naturally trust their friends, and they ought to trust users who have demonstrated similar preferences.

Recommendation Evaluation: We used the TrustMF [10] recommender system implementation distributed in LibRec [4] to evaluate the accuracy of predicted trust links. TrustMF is a collaborative filtering based matrix factorization recommender system which combines the traditional user-item matrix with a user-user social matrix in order to enhance recommendation. TrustMF is a convenient platform to evaluate the accuracy of trust link predictions produced by MFTM, as performance in the recommendation task ought to improve as the accuracy of the set of trust links provided to the algorithm improves.

In each of the experiments, a TrustMF model is trained on 80% of available ratings and all available trust links. The remaining 20% of rating data is reserved for measuring the accuracy of predictions. Each model was allowed to train for 200 iterations and used latent factor dimensionality equal to 10.

Prediction accuracy was evaluated on the basis of Mean Squared Error and Mean Absolute Error.

4   Experiments and Results

We ran a number of experiments in order to test the effects of applying different procedures to the trust prediction process.

- **Real Friends (Baseline):** This experiment uses the original friend links in the Yelp dataset. No prediction is performed.
- **FriendPredict (MFTM):** Based on a classifier trained on a balanced set of all friend links in the dataset, predict whether each pair of users are friends.
- **PCCPredict (MFTM):** Based on a classifier trained on a balanced set of PCC links in the dataset, predict whether each pair of users have a positive *review score* correlation.
- **Personalized Prediction (PMFTM):** We clustered users following the procedures outlined above, both on the basis of similar rating behaviour and social circle overlap. After clustering, we trained classifiers for each cluster of users to predict both similar review behaviour and friendship links. This resulted in four experiments: PCCCluster_PCCPredict, PCCCluster_FriendPredict, SocialCluster_PCCPRedict and SocialCluster_FriendPredict.

MAE and MSE results for recommendation experiments are illustrated in Figs. 1 and 2. For the sake of clarity, only the best results from each experiment class (Baseline, MFTM, PMFTM) are illustrated. These figures show how recommendation performance alters as the Social Regulation factor of the TrustMF system changes (this parameter controls the weight of social links in recomendation and is referred to as λ_T in [10]).

A full summary of results (best results are **bolded**, worst results are *italicized*) is presented in Table 1. All data points are the average of three experiments, run with the specified configuration and distinct random number generator seeds.

In summary, we show 1) noticeable improvements delivered by our

Table 1. Best results

Experiment	MAE	MSE
RealFriends	*0.871*	*1.352*
PCCPredict	0.864	1.318
FriendPredict	0.858	1.293
PCCCluster_FriendPredict	0.862	1.283
PCCCluster_PCCPredict	**0.857**	**1.267**
SocialCluster_FriendPredict	0.863	1.288
SocialCluster_PCCPredict	**0.857**	1.273

MFTM trust link predictors in comparison to using explicit trust links, in the context of TrustMF, 2) trust links predicted by PMFTM deliver improvements on both methods of 1).

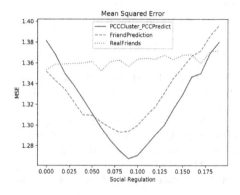

Fig. 1. MSE graph

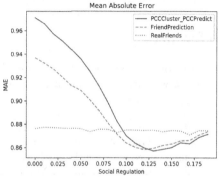

Fig. 2. MAE graph

5   Discussion and Conclusion

In this paper, we demonstrated the effectiveness of a personalized approach to multi-faceted trust modeling, while combining some of the best features from recent work in the field. We showed that the accuracy of a recommendation task can be improved by clustering similar users together and learning classifiers with respect to these clusters of users. Our best results were achieved by clustering users based on their preference similarity (i.e. PCC of submitted review scores).

With respect to social networks, our work aims to improve online experiences by supporting distinct presentation of content to differing users, achieved by reasoning about relationships with peers and the concept of trust. Our concern with trustworthiness of content relates well to companion efforts devoted to analyzing social networking posts in order to judge credibility [7] or to attempt to detect digital misinformation [2,11]. Some work on this topic ultimately focuses most on diffusion dynamics [5,9]: examining information spread is a distinct concern which may be interesting to explore as a basis for identifying clusters of users. Our work also relates to social networking research that examines the challenge of recommending content to users [8] through the networking relationships. In our case, we specifically examine friend relations and predictions of trust through rating behaviour.

While we find these results encouraging, many avenues for further experimentation remain open. Experiments with other datasets and an expanded set of metrics will offer further insight. We can imagine expanding our vision of the data analysis to include dynamic factors of agent behaviour (such as changes in values over time) and increasingly personalized predictions at the individual level, rather than on the basis of clusters of users. It is important to emphasize the potential of this research to address additional concerns with social networks. While we chose to demonstrate effectiveness in a recommendation task, the trust modeling we perform here, which identifies statistically likely trust links, has the potential to assist with companion issues in social networks such as moderation or detecting misinformation and hate speech.

References

1. Ardissono, L., Ferrero, M., Petrone, G., Segnan, M.: Enhancing collaborative filtering with friendship information. In: Proceedings of UMAP 2017, pp. 353–354. ACM, New York (2017)
2. Ciampaglia, G.L., Mantzarlis, A., Maus, G., Menczer, F.: Research challenges of digital misinformation: toward a trustworthy web. AI Mag. **39**(1), 65–74 (2018)
3. Fang, H., Guo, G., Zhang, J.: Multi-faceted trust and distrust prediction for recommender systems. Decis. Support Syst. **71**, 37–47 (2015)
4. Guo, G., Zhang, J., Sun, Z., Yorke-Smith, N.: LibRec: a java library for recommender systems. In: UMAP Workshops (2015)
5. Hui, P.M., Shao, C., Flammini, A., Menczer, F., Ciampaglia, G.L.: The Hoaxy misinformation and fact-checking diffusion network. In: Proceedings of the 12th International AAAI Conference on Web and Social Media (2018)
6. Mauro, N., Ardissono, L., Hu, Z.F.: Multi-faceted trust-based collaborative filtering. In: Proceedings UMAP 2019, pp. 216–224. ACM (2019)
7. Sardana, N., Cohen, R., Zhang, J., Chen, S.: A Bayesian multiagent trust model for social networks. IEEE Trans. Comput. Soc. Syst. **5**(4), 995–1008 (2018)
8. Tang, J., Aggarwal, C., Liu, H.: Recommendations in signed social networks. In: Proceedings of WWW 2016, International World Wide Web Conferences Steering Committee, Republic and Canton of Geneva, pp. 31–40, Switzerland (2016)
9. Tong, A., Du, D.Z., Wu, W.: On misinformation containment in online social networks. In: Advances in Neural Information Processing Systems, vol. 31, pp. 341–351. Curran Associates, Inc. (2018)
10. Yang, B., Lei, Y., Liu, D., Liu, J.: Social collaborative filtering by trust. Proc. IJCAI **2013**, 2747–2753 (2013)
11. Yang, S., Shu, K., Wang, S., Gu, R., Wu, F., Liu, H.: Unsupervised fake news detection on social media: a generative approach. In: Proceedings of AAAI 2019, February 2019

Mixing ICI and CSI Models for More Efficient Probabilistic Inference

Michael Roher and Yang Xiang[✉]

University of Guelph, Guelph, Canada
yxiang@uoguelph.ca

Abstract. Conditional probability tables (CPTs) in Bayesian Networks (BNs) have exponential space on the family size. Local models based on independence of causal influence (ICI) or context-specific independence (CSI) have been applied separately to improve the efficiency. We propose a framework to mix both local models in the same BN for improved efficiency. In particular, we show that ICI and CSI are orthogonal, and each is unable to express the other efficiently and accurately. We propose a formalism to encode both types of local models in the same BN, and to convert it into a homogenous representation to support exact inference. We report experimental evaluation where significant efficiency gain is obtained in exact inference.

Keywords: Bayesian networks · Probabilistic inference · Causal independence models · Context-specific independence models

1 Introduction

Discrete Bayesian networks (BNs) [6] exploit conditional independence among variables through directed acyclic graph (DAG) structures, and only quantify dependence of variables on their parents by conditional probability tables (CPTs). As tabular CPTs have exponential space, which extends to inference complexity, local models have been applied for further efficiency. Some exploit independence of causal influence (ICI), e.g., noisy-OR [6], noisy-MAX [4], DeMorgan [5], Non-Impeding Noisy-AND Tree (NIN-AND Tree or NAT) [11], and cancellation model [10]. Other local models exploit context-specific independence (CSI), e.g., CPT-trees [1], rule-based CSI [8], and algebraic decision diagrams [2].

These methods exploit ICI or CSI, but not both. Since ICI and CSI apply to individual families of variables in BNs, they can co-exist in an environment (see Sect. 7). In such cases, methods that exploit only one type of local models lose the opportunity afforded by also exploiting the other type.

We propose a framework that exploits both ICI and CSI for more efficient inference in BNs. When both exist, we apply NAT local models for ICI and CPT-tree local models for CSI, encoding both in the same BN. We convert each type

© Springer Nature Switzerland AG 2020
C. Goutte and X. Zhu (Eds.): Canadian AI 2020, LNAI 12109, pp. 451–463, 2020.
https://doi.org/10.1007/978-3-030-47358-7_47

of local models accordingly to obtain a homogeneous runtime representation for more efficient inference.

The remainder is organized as follows: Sect. 2 reviews background on NAT and CSI. We establish their orthogonality in Sect. 3. We analyze alternatives for mixing NAT and CSI in Sect. 4, and specifies our choice formalism. Section 5 formalizes CPT-tree transformation, and Sect. 6 combines it with NAT de-causalization [13] to obtain a homogeneous runtime representation. We report experimental results in Sect. 7.

2   Background

2.1   NAT Modelling of ICI

We review NAT modelling (see [11,13] for more details). A NAT model encodes dependency of an effect e on a set of uncertain causes $C = \{c_1, ..., c_n\}$, where $e \in \{e^0, ..., e^\nu\}$ ($\nu \geq 1$) and $c_i \in \{c_i^0, ..., c_i^{m_i}\}$ ($i = 1, ..., n$; $m_i \geq 1$). The effect and cause are inactive at e^0 and c_i^0, and are active at other values (may be written as e^+ or c_i^+) where higher indices may denote higher intensity. C and e form a family in BNs, where C is the parent set of e.

A causal event is a *success* or *failure* depending on if e is produced up to a certain value, is *single-* or *multi-causal* depending on the number of active causes, and is *simple* or *congregate* depending on the number of active effect values. A simple single-causal success is an event that cause c_i of value c_j^i ($j > 0$) renders e to occur at value e^k ($k > 0$), when other causes are inactive. Its probability is denoted $P(e^k \leftarrow c_i^j) = P(e^k | c_i^j, c_z^0 : \forall z \neq i, j > 0)$. A congregate multi-causal success is an event where a set of active causes $X = \{c_1, ..., c_q\}$ caused e to occur at e^k ($k > 0$) or higher intensity. Its probability is denoted $P(e \geq e^k \leftarrow c_1^{j_1}, ..., c_q^{j_q}) = P(e \geq e^k | c_1^{j_1}, ..., c_q^{j_q}, c_m^0 : c_m \in C \setminus X)$, where $j_i > 0$ for $i = 1, ..., q$, or $P(e \geq e^k \leftarrow \underline{x}^+)$ for simplicity.

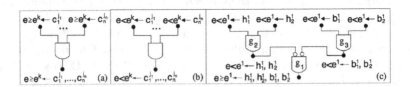

Fig. 1. (a) A direct NIN-AND gate. (b) A dual NIN-AND gate. (c) A NAT.

A NAT is composed of two types of NIN-AND gates, each over disjoint sets of causes $W_1, ..., W_q$. An input event of a *direct gate* (Fig. 1 (a)) is a causal success $e \geq e^k \leftarrow \underline{w_i}^+$, and the output event is $e \geq e^k \leftarrow \underline{w_1}^+, ..., \underline{w_q}^+$. An input of a *dual gate* (Fig. 1 (b)) is causal failure $e < e^k \leftarrow \underline{w_i}^+$, and the output event is $e < e^k \leftarrow \underline{w_1}^+, ..., \underline{w_q}^+$. Probability of output event is the product of input event probabilities.

Let e^k be an active effect value. $R = \{W_1, ..., W_m\}(m \geq 2)$ be a partition of a set $X \subseteq C$ of causes, $S \subset R$, and $Y = \cup_{W_i \in S} W_i$. Sets of causes in R *reinforce* each other relative to e^k, iff $\forall S\ P(e \geq e^k \leftarrow \underline{y}^+) \leq P(e \geq e^k \leftarrow \underline{x}^+)$. They *undermine* each other iff $\forall S\ P(e \geq e^k \leftarrow \underline{y}^+) > P(e \geq e^k \leftarrow \underline{x}^+)$.

A direct gate encodes undermining interactions and a dual gate encodes reinforcing interactions. They are combined in a NAT to express complex interactions among causes. Figure 1 (c) shows a NAT with 3 gates. Causes h_1 and h_2 reinforce each other, and so do b_1 and b_2. The two groups undermine each other.

A BN is *NAT-modelled* if the CPT of each variable of 2 or more parents is a NAT model. Its space is linear: $O(N \kappa n)$, where N is the number of variables, κ bounds variable domain sizes, and n bounds the number of parents per variable.

Common inference methods for BNs do not directly apply to NAT-modelled BNs. *Normalizing* NAT models to full tabular CPTs loses efficiency of NAT-modelling. Techniques to support efficient inference include *multiplicative factorization*, where NAT-modelled BNs are converted to equivalent, efficient Markov networks, and *de-causalization*, where they are converted to equivalent, efficient tabular BNs. For NAT-modelled BNs with high treewidth and low density (measured by percentage of links beyond being singly connected), two orders of magnitude speedup in inference has been demonstrated.

2.2 CPT-Tree Modelling of CSI

We review CSI (see [1] for more details). For a BN variable, a *context* is an assignment of values to some parents. For disjoint sets of variables X, Y, Z, and Cxt, X and Y are *contextually independent* given Z and context $Cxt = cxt$, denoted $I_c(X; Y | Z, Cxt = cxt)$, if $P(X | Z, cxt, Y) = P(X | Z, cxt)$ whenever $P(Z, cxt, Y) > 0$.

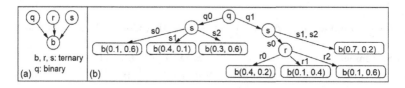

Fig. 2. (a) A BN family. (b) CPT-Tree for the family.

When CSI exists in a BN family, the CPT contain similar values. The BN family in Fig. 2 (a) admits $I_c(b; r | s, q = q_0)$ and $I_c(b; r | q = q_1, s \in \{s_1, s_2\})$. Its CPT has 14 parameters, though $P(b | q, r, s)$ generally has 36.

A CPT with CSI can be specified as CPT-tree (b)[1]. A CPT-tree for variable x and parents $\pi(x)$ is directed from the root. Each non-leaf is a variable in $\pi(x)$. Each path from the root to a leaf is a context, and the leaf specifies the

---

[1] The example generalizes CPT-trees in [1] slightly as explained in Sect. 5.

conditional probability distribution (CPD) of x, given the context. The CSI above are expressed by the left subtree and the rightmost branch, respectively.

We refer to BNs where some families are modelled by CPT-trees as *CPT-tree-modelled BNs*. Common inference methods for BNs do not directly apply to CPT-tree-modelled BNs. Techniques that support inference with CPT-tree-modelled BNs include network transformation and clustering [1], cutset conditioning [1], and variable elimination [7].

3   Orthogonality of NAT and CSI Models

A fundamental question that may undermine efforts to exploit a mixture of NAT and CSI is whether the local models are orthogonal. A negative answer renders the effort invalid, since one type of local models can be encoded by the other. For instance, alternative ICIs, noisy-OR, noisy-MAX, and DeMorgan, are all special NAT models. Below, we empirically answer the question positively.

First, we show that CSI generally cannot be exactly expressed as NAT models. A batch of 100 seed CPTs $P(x_0|x_1, x_2, x_3, x_4, x_5)$ are simulated, where variables have the same domain $\{1, 2, 3, 4, 5\}$.

Given a seed CPT P and a CSI, we generate a CSI CPT P^* as follows: For $I_c(x_0; x_1, x_2, x_3|x_4, x_5 = 5)$, P^* must satisfy $P^*(x_0|x_1, x_2, x_3, x_4, x_5 = 5) = P^*(x_0|x_1', x_2', x_3', x_4, x_5 = 5)$. For each distinct assignment (x_0, x_4), we arbitrarily assign (x_1, x_2, x_3), retrieve value $P(x_0|x_1, x_2, x_3, x_4, x_5 = 5)$ from P, and assign to every term $P^*(x_0|x_1', x_2', x_3', x_4, x_5 = 5)$.

We specified 3 alternative CSIs (Table 1). They allow different space reduction (2nd col.). Using the above method, we generated 3 CSI CPTs for each of 100 seed CPTs (a total of 400 source CPTs). We then compress each source CPT into NAT model, and evaluate the average Kullback-Leibler and Euclidean distances between the NAT and source CPTs (4th and 5th cols.).

Table 1. Summary of experiments on representing CSI CPTs as NAT models.

CSI Statement	# Src Para	# NAT Para	KL	ED
No CSI	12,500	80	0.738	0.219
$I_c(x_0; x_4, x_5|x_1 = 1, x_2 = 2, x_3 \in \{3, 4\})$	12,304	80	0.710	0.214
$I_c(x_0; x_2, x_3, x_4, x_5|x_1 = 1)$	10,004	80	0.692	0.210
$I_c(x_0; x_2, x_3, x_4, x_5|x_1 \in \{1, 2, 3, 4\})$	2,504	80	0.501	0.176

Table 1 reveals that source CPTs take much more space than resultant NAT models (>30 times). On the other hand, NAT models reduce space 30 to 150 times, but with errors. Though errors decrease as the numbers of source CPT parameters, NAT models generally cannot express CSI CPTs exactly.

Next, we show that NAT models generally cannot be suitably expressed as CSI models. Given a NAT CPT, if its probabilities can be grouped into tight clusters (small distance between member values), and the total number of such clusters is significantly less than the number of NAT parameters, then it pays to express the NAT CPT as a CSI model. A smaller number of clusters means space saving, and tight clusters mean small approximation errors. Based on this idea, given a NAT CPT and a distance bound δ (e.g., $\delta = 0.02$), we group values in the CPT into a set Ψ of clusters, such that the following conditions hold:

1. For each cluster $Q \in \Psi$ and each pair of values $p, q \in Q$, $|p - q| \leq \delta$.
2. For each two clusters $Q, R \in \Psi$, let $min_Q, min_R, max_Q, max_R$ be extreme values in Q and R, respectively. Either $max_Q < min_R$ or $max_R < min_Q$.
3. For clusters $Q, R \in \Psi$ where $max_Q < min_R$, we have $min_R - max_Q > \delta$.

Condition 1 bounds inner distance within each cluster. Condition 2 orders clusters by member values. Condition 3 bounds inter-cluster distance.

The number of clusters obtained is a lower-bound of the number of parameters when the NAT CPT is approximated by a CSI model. This is because values in the same cluster may refer to incompatible contexts, and cannot be encoded by the same CPT-tree leaf. We split such clusters as needed.

The clustering is applied to 100 generated NAT CPTs, each over a family of 5 parent variables. All variables are binary, with 32 parameters per CPT. Results with $\delta = 0.02$ are shown in Fig. 3. Each bar counts NAT CPTs that produced a particular number of clusters. As values in a cluster can be encoded by a single value with error $< \delta$, the number of clusters indicates the number of parameters needed if those NAT CPTs are encoded as CPT-trees. As is shown, all CPT-trees require at least 17 parameters, while the NAT CPT only needs 5.

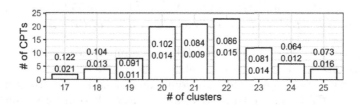

Fig. 3. Experiment results on representing NAT CPTs as CSI models.

CSI modelling errors are evaluated as follows: Compute the *centroid* of each cluster as the mean of its values, and use it as the CPT-tree parameter. The error to model a NAT CPT as CPT-tree is the Euclidean distance between the two CPTs. In Fig. 3, average modelling error for CPTs in each bar is at the top, with the standard deviation below it.

To summarize, it is generally not possible to encode CSI CPTs exactly as NAT models. When NAT CPTs are represented as CSI models, it generally not only introduces error, but also increases the number of parameters required. These evidences suggest that NAT models and CSI models are orthogonal.

4   Mixed NAT-CSI Bayesian Networks

ICI and CSI can each be exploited to improve space and inference efficiency in BNs. To our knowledge, no prior study considered inference on BNs that take advantage of both simultaneously. For that purpose, we resolve issues below:

First, we observe that ICI and CSI are applicable to individual families in BNs. Therefore, both can co-exist in an environment as well as in a BN: Some family of variables follow ICI local models and others follow CSI local models.

Second, suitable representation is needed for each type of local models. For ICI, we focus on NAT models for several reasons: They express both reinforcing and undermining interactions. They can mix such interactions recursively among cause subsets. They apply to multi-valued, ordinal, and nominal variables. They generalize other ICI models including noisy-OR, noisy-MAX, and DeMorgan, while having the same linear space.

For CSI, several formalisms are available. Rule bases and algebraic decision diagrams (ADDs) have been used for inference by variable elimination [2,7]. When used to answer multiple queries (over multiple unobserved variables), the compilation requires knowledge of evidence prior to inference. Loops in ADDs tend to increase treewidth of the resultant structure. As we aim at computing posteriors of all unobserved variables, with arbitrary evidence at inference time, we selected CPT-trees [1] to encode CSI.

We define the representation of choice as *mixed NAT-CSI Bayesian net* (MNCBN) (an example is given in Sect. 6):

Definition 1. *A MNCBN is a BN (V, D, P) over a set V of variables with dependency structure DAG D. The set P of CPTs is composed of one CPT per variable in V, partitioned into (TC, NM, CT), where TC is a set of tabular CPTs, NM is a set of NAT models, and CT is a set of CPT-trees.*

Third, neither NAT models nor CPT-trees support common BN inference algorithms directly. Each type of local models admits alternative processings before inference. NAT modelled BNs admit multiplicative factorization or de-causalization. CPT-tree-modelled BNs admit network transformation, cutset conditioning, or variable elimination. To prepare MNCBNs for inference, we have chosen to compile them into a homogeneous representation, by combining de-causalization for NAT models and network transformation for CPT-trees, as both convert local models into equivalent BN segments that tend to reduce treewidth. This choice assumes no prior knowledge on evidence and supports computing posteriors over all unobserved variables. In comparison, multiplicative factorization for NAT models and cutset conditioning with CPT-trees do not support a homogeneous representation. We demonstrate the choice in Sect. 6.

5   Formalizing CPT-Tree Transformation

We apply network (CPT-tree) transformation to convert CPT-trees in MNCBNs to BN segments. Although the idea is illustrated in [1] with a simple example over binary variables, to the best of our knowledge, no general algorithm has been formalized. We specify an algorithm suite, formalizing processing on multi-valued variables, set-valued CPT-tree branches, and multiplexer CPTs.

Let $dom(x)$ be the domain of variable x. A CPT-tree arc outgoing from node t may be labeled by a subset of $dom(t)$. A path from the root to a leaf, including such arcs, defines a set of contexts, e.g., the rightmost branch in Fig. 2 (b). CPT-trees with such set-valued arcs generalize those in [1], and allow more efficient CSI encoding.

Algorithm `SetDagSeg` takes a CPT-Tree T over variable x and parents $\pi(x)$, and builds a BN segment with auxiliary variables, all of which have domain $dom(x)$. Each node of T is at a level with the root at level 0, and each child of the root is at level 1. Transformation is driven by topology of T from level 0 onwards. For each node t in T, denote the path from root to t by $path(t)$. The output is a DAG G with a single leaf x, and G is constructed from x upwards.

Algorithm 1. $SetDagSeg(x, \pi(x), T)$

1   *initialize empty graph G with nodes $\{x\} \cup \pi(x)$;*
2   *denote the root of T by ρ and set $path(\rho) = \{\}$;*
3   *label x in G as $x_{path(\rho)}$;*
4   *for level $L = 0$ to max level in T,*
5   *for each node t in T at level L with $path(t)$;*
6   *find node v in G that is labelled $x_{path(t)}$ and add arc $t \to v$ in G;*
7   *if each child of t in T is leaf, continue;*
8   *denote partition of $dom(t)$ by arcs outgoing from t as $\{sd_1, ..., sd_m\}$;*
9   *for i=1 to m,*
10   *add node y to G with domain $dom(x)$ and label it $x_{\{path(t), t \in sd_i\}}$;*
11   *add arc $y \to v$ in G;*
12 *return G;*

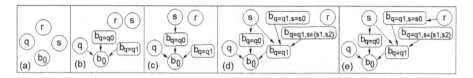

Fig. 4. Transformation of the CPT-Tree in Fig. 2.

For CPT-tree T in Fig. 2, `SetDagSeg` begins with G in Fig. 4 (a). For level $L = 0$ of T, q in T is processed with resultant G in (b). For $L = 1$, 1st instance of s in T is processed as in (c), followed by 2nd instance as in (d), where arcs outgoing from s partition $dom(s)$ into subdomains $\{s_0\}$ and $\{s_1, s_2\}$. For $L = 2$, r in T is processed to produce (e).

Node v in line 6 may not be processed by the *for* loop in line 9. If it is, multiple parents are added to v, and v is called a *multiplexer*. In Fig. 4 (e), $b_{\{\}}$ and $b_{q=q_1}$ are the only multiplexers.

Next, Algorithm `AssignCpt` assigns a CPT to each node in G except those in $\pi(x)$. They include $x_{\{\}}$ and nodes added by `SetDagSeg` line 10, divided as follows.

Type 1: They are added in line 10 and never processed after by line 6. Hence, they remain roots, e.g., $b_{q=q_1, s \in \{s_1, s_2\}}$. Type 2: They are processed in line 6 as v and by the *for* loop in line 9. They are multiplexers such as $b_{\{\}}$ and $b_{q=q_1}$.
Type 3: They are the remaining nodes that are processed in line 6 as v, passed test in line 7, and skipped the *for* loop in line 9, e.g., $b_{q=q_0}$ and $b_{q=q_1, s=s_0}$.

For Type 1 node v in G, traverse its path in T from root to a leaf, and assign its CPD to v. For $b_{q=q_1, s \in \{s_1, s_2\}}$ in Fig. 4 (e), follow path $(q = q_1, s \in \{s_1, s_2\})$ in Fig. 2 (b), and assign $P(b_{q=q_1, s \in \{s_1, s_2\}})$ *over* $(b_0, b_1, b_2) = (0.7, 0.2, 0.1)$.

For Type 3 node v, traverse its path in T to a node t. Since t passed test in line 7, each child of t is a leaf that specifies a CPD. Assemble them to form a CPT and assign to v. For $b_{q=q_0}$, follow path $(q = q_0)$ in Fig. 2 (b) to node s, retrieve CPDs, and assign CPT: $P(b_{q=q_0}|s_0) = (0.1, 0.6, 0.3), P(b_{q=q_0}|s_1) = ..., P(b_{q=q_0}|s_2) = ...$
Algorithm `AssignCpt` formalizes the above, where CPTs for Type 2 nodes are from `SetSwitchCpt` presented below.

Algorithm 2. *AssignCpt*$(x, \pi(x), T, G)$

```
1  for each node v in G,
2      if v ∈ π(x), continue;
3      if v is Type 1 with path(v),
4          traverse path(v) in T to leaf t;
5          retrieve CPD parameters at t and assign the CPD to v;
6      else if v is Type 3 with path(v),
7          traverse path(v) in T to node t;
8          for each child z of t in T, retrieve CPD parameters at z;
9          assemble the CPDs into CPT and assign to v;
10     else
11         denote the unique parent of v from π(x) by y;
12         call SetSwitchCpt(v, y, T, G) and assign the CPT returned to v;
13 return G;
```

Family of a Type 2 node v is created in `SetDagSeg` in line 6 (1st parent) and 11 (other parents). The 1st parent y is from $\pi(x)$, and is identified in line 11 of `AssignCpt`. Each other parent is auxiliary. CPT at v is to deterministically set v value according to an auxiliary parent. The right parent is decided by y value and path label of the parent. This is specified in Algorithm `SetSwitchCpt`.

Algorithm 3. $SetSwitchCpt(v, y, T, G)$

1 $initialize\ CPT\ P(v|y, u_1, ..., u_k)\ where\ \{y, u_1, ..., u_k\}\ is\ parent\ set\ of\ v\ in\ G;$
2 $for\ each\ assignment\ (v = v', y = y', u_1 = u'_1, ..., u_k = u'_k),$
3 $\quad find\ u_i\ in\ \{u_1, .., u_k\}\ whose\ path\ label = (path(v), y \in sd_i)\ and\ y' \in sd_i;$
4 $\quad if\ v' = u'_i,\ P(v'|y', u'_1, ..., u'_k) = 1;$
5 $\quad else\ P(v'|y', u'_1, ..., u'_k) = 0;$
6 $return\ P(v|y, u_1, ..., u_k);$

Consider CPT for $b_{q=q_1}$ with parent $s \in \pi(x)$ (Fig. 4 (e)). To set value $P(b_{q=q_1} = b_0|s = s_1, b_{q=q_1,s=s_0} = b_0, b_{q=q_1,s\in\{s_1,s_2\}} = b_1)$, variable $b_{q=q_1,s\in\{s_1,s_2\}}$ is selected since $s = s_1$ satisfies its path label. Value $b_{q=q_1,s\in\{s_1,s_2\}} = b_1$ is then compared with $b_{q=q_1} = b_0$. Since they do not match, it results in value 0 for the above probability.

6 Inference with Mixed NAT-CSI Bayesian Networks

We outline and demo a framework for exact inference with MNCBNs, that exploit both types of local models for improved efficiency.

First, the MNCBN is converted to standard BN: Each NAT model is de-causalized into an efficiency preserving BN segment. Each CPT-tree is transformed as described in Sect. 4. The result is a homogeneous, standard BN, encoding the same joint probability distribution (JPD), with a treewidth lower than that of MNCBN (if sparse). We refer to it as a *de-causalized and transformed BN* (DTBN). The DTBN supports exact inference with any common method.

Consider a MNCBN with the DAG in Fig. 5 (a). Family of g is a NAT model (b) and family of h is modelled by CPT-tree (c). All variables are ternary. Figure 6 shows DAG of the DTBN. The BN segment de-causalizing g family is dashed. In the dashed region, variables outside $\{g\} \cup \pi(g)$ are auxiliary. If all CPTs are tabular, the MNCBN has a total CPT size of 4506, where the largest CPT has size 2187. The DTBN has a total CPT size of 1267, where the largest CPT has size 243.

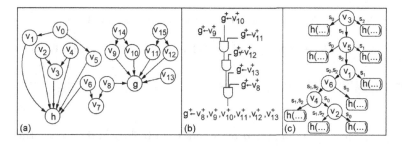

Fig. 5. (a) BN DAG. (b) NAT model for family of g. (c) CPT-tree over family of h.

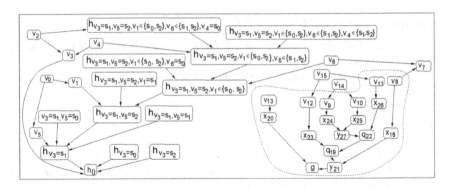

Fig. 6. The DTBN.

7   Experiments

Our experimental study aims to (i) confirm co-existence of NAT and CSI in real world BNs, (ii) evaluate computational gain by mixing NAT and CSI local models, and (iii) compare effectiveness of NAT and CSI models.

The 1st experiment confirms co-existence of NAT and CSI local models. NAT modelling has been applied to approximate 8 real world BNs from the *bnlearn* repository with reasonable inference errors [12]. Here, we test the 2 binary BNs from the 8, *Andes* and *Win95pts*, for CSI local models. We apply clustering in Sect. 3 to each CPT where the variable has 2 or more parents. Similar probabilities are grouped, subject to upper bound on inner-cluster distance and lower bound on inter-cluster distance. We use bound $\delta = 0.02$.

The results are shown in Table 2. The maximum number of parameters per CPT (4th col.) is 64 (6 parents) and 128 (7 parents), respectively. The maximum number of clusters found per CPT (5th col.) is 3 and 6, respectively. Hence, a significant amount of CSI exists in these CPTs. If modelled as CPT-trees, the Euclidean error for Win95pts is 0.041 (6th col.). The error for Andes is 0: For each cluster, all values are identical.

Table 2. Summary of results from clustering *Andes* and *Win95pts* BNs.

BN	#Node	#Fmly Proced	Max #Para/CPT	Max #Clus/CPT	Eu Dist
Andes	223	50	64	3	0
Win95pts	76	24	128	6	0.041

CSIs have also been identified by others from biological datasets, other BNs in *bnlearn*, and UCI datasets, e.g., [3,9]. These studies, the result above, and the NAT modelling study [12] suggest collectively co-existence of NAT and CSI local models in practice.

The 2nd experiment evaluates computational gain by mixing NAT and CSI models. We simulated MNCBNs each of 100 variables (binary or ternary), where 50% of families of 2 parents or more are NAT-modelled and the remaining 50% are CSI-modelled. The largest number of parents per node is 12, and each MNCBN has at least 2 such families. At least one such family is NAT-modelled, and at least one is CPT-tree modelled. Structural density of MNCBNs is controlled at $d = 5\%$ or 10% more links beyond being singly connected.

When each variable is unique in a CPT-tree, its transformation has no loop. Duplicated variables, e.g., s in Fig. 2, induce loops (see Fig. 4 (e)), and increases treewidth of the transformed structure. We control the number of variable duplications at $k = 0, 2, 4, 7, 10$. For each combination of (d, k), 30 BNs are generated. Hence, a total of 300 distinct MNCBNs are generated. Each MNCBN is converted into 4 standard BNs (encoding the same JPD) by methods D+N, D+T, N+N, and N+T, where D refers to De-causalizing NAT models, T refers to Transforming CPT-trees, and N referes to Normalizing NAT models and CPT-trees.

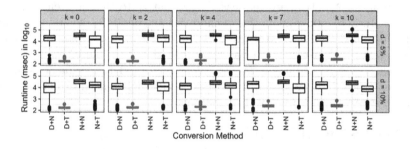

Fig. 7. Summary of inference runtimes

Each resultant BN is compiled for inference by lazy propagation (LP) (See Sect. 8 for rationale). Each BN has 10 inference runs, each with different observations over 10 randomly selected variables. Inference runs by BNs from the same MNCBN yielded the same posterior marginals (exact). Runtimes (2.9 GHz desktop) are shown in Fig. 7.

In all (d, k) combinations, N+N is the slowest. Both D+N and N+T are advantageous, even though they only exploit one type of local models. D+T is on average two orders of magnitude faster than alternatives, demonstrating clear computational advantage of exploiting both NAT and CSI in MNCBNs.

Relative performance of D+N and N+T is indiscernible in Fig. 7, partly due to existence of normalized CPTs. To evaluate relative gain from alternative local models, the 3rd experiment generated BNs in two steps: In the 1st step, only DAGs are generated with 200 variables each (binary or ternary). The largest number of parents per node is 12, and each DAG has at least 4 such families. We use $d = 5\%, 10\%$ and $k = 0, 2, 4, 7, 10$, a total of 10 combinations. For each combination, we simulated 30 DAGs, resulting in 300 distinct DAGs. In the 2nd step, a pair of BNs are created from each DAG: NM-BN where each family of 2 parents or more is NAT-modelled, and CM-BN where such families are modelled as CPT-trees. Hence, the pair share the DAG, and differ in JPDs.

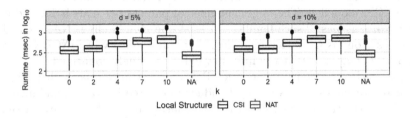

Fig. 8. Summary of inference runtimes by NM-BNs and CM-BNs.

We de-causalize NM-BNs and transform CM-BNs for lazy propagation. Ten inference runs are performed on each BN with random observations over 20 variables. The runtimes are shown in Fig. 8.

Runtimes of NM-BNs are the least, even relative to the most efficient CM-BNs ($k = 0$). Let m be domain size of child variable of CPT-tree family. Assuming single-valued CPT-tree arcs, each multiplexer has $m + 1$ parents. If CPT-tree has $k > 0$, transformation has loops. On the other hand, every node in de-causalized BN segment has at most 2 parents, and the segment is always loop-free. Hence, NAT modelled BNs are generally more efficient than CPT-tree modelled BNs, as confirmed by the experiment.

8   Conclusion and Remarks

The main contribution is a framework to mix ICI and CSI local models in BNs for more efficient inference. They are shown to be orthogonal, and hence neither subsumes the other. We have shown that NAT models and CPT-trees are suitable ways for mixing, and combining de-causalization and transformation enables a homogenous representation for exact inference. We report significant speedup in inference relative to exploitation of only one type of local models.

Two questions were received from peer review, to which we respond below. Due to space, we omit relevant references. (1) On why not adopt sum-product networks (SPNs): Although exact inference in BNs is NP-hard and that in SPNs is linear, when a general BN is compiled into a SPN, it incurs an exponential blow-up. Hence, SPNs are one way to explore special conditions in BNs, e.g., CSI, but not the only way, as this work shows. (2) On why not adopt simple propagation (SP): Published work on SP reported that SP is not always faster than LP. Other BN inference methods also exist. This work shows significant gains in inference efficiency by mixing ICI and CSI, and our performance comparison only necessitates identical inference method on BNs with and without mixing. There is no need for the fastest method, nor do we claim that LP is so.

Acknowledgement. Financial support from NSERC Discovery Grant to 2nd author is acknowledged.

References

1. Boutilier, C., Friedman, N., Goldszmidt, M., Koller, D.: Context-specific independence in Bayesian networks. In: Proceedings of the 12th Conference on Uncertainty in Artificial Intelligence, pp. 115–123 (1996)
2. Chavira, M., Darwiche, A.: Compiling Bayesian networks using variable elimination. In: Proceedings of the 20th IJCAI, pp. 2443–2449 (2007)
3. Friedman, N., Yakhini, Z.: On the sample complexity of learning Bayesian networks. In: Proceedings of the 12th Conference on UAI, pp. 274–282. Morgan Kaufmann (1996)
4. Henrion, M.: Some practical issues in constructing belief networks. In: Kanal, L., Levitt, T., Lemmer, J. (eds.) Uncertainty in Artificial Intelligence 3, pp. 161–173. Elsevier, New York (1989)
5. Maaskant, P., Druzdzel, M.: An independence of causal interactions model for opposing influences. In: Jaeger, M., Nielsen, T. (eds.) Proceedings of the 4th European Workshop on Probabilistic Graphical Models, Hirtshals, Denmark, pp. 185–192 (2008)
6. Pearl, J.: Probabilistic Reasoning in Intelligent Systems: Networks of Plausible Inference. Morgan Kaufmann, San Mateo (1988)
7. Poole, D.: Probabilistic partial evaluation: exploiting rule structure in probabilistic inference. In: Proceedings of the 15th IJCAI, pp. 1284–1291 (1997)
8. Poole, D., Mackworth, A., Goebel, R.: Computational Intelligence: A Logical Approach. Oxford University Press, New York (1998)
9. Talvitie, T., Eggeling, R., Koivisto, M.: Learning Bayesian networks with local structure, mixed variables, and exact algorithms. Int. J. Approximate Reasoning **115**, 69–95 (2019)
10. Woudenberg, S., van der Gaag, L., Rademaker, C.: An intercausal cancellation model for Bayesian-network engineering. Int. J. Approximate Reasoning **63**, 32–47 (2015)
11. Xiang, Y.: Non-impeding noisy-AND tree causal models over multi-valued variables. Int. J. Approximate Reasoning **53**(7), 988–1002 (2012)
12. Xiang, Y., Baird, B.: Compressing Bayesian networks: swarm-based descent, efficiency, and posterior accuracy. In: Bagheri, E., Cheung, J.C.K. (eds.) Canadian AI 2018. LNCS (LNAI), vol. 10832, pp. 3–16. Springer, Cham (2018). https://doi.org/10.1007/978-3-319-89656-4_1
13. Xiang, Y., Loker, D.: De-causalizing NAT-modeled Bayesian networks for inference efficiency. In: Bagheri, E., Cheung, J.C.K. (eds.) Canadian AI 2018. LNCS (LNAI), vol. 10832, pp. 17–30. Springer, Cham (2018). https://doi.org/10.1007/978-3-319-89656-4_2

RideSafe: Detecting Sexual Harassment in Rideshares

Shikhar Sakhuja[✉] and Robin Cohen[✉]

David R. Cheriton School of Computer Science, University of Waterloo,
Waterloo, Canada
{s2sakhuj,rcohen}@uwaterloo.ca

Abstract. Sexual harassment and abuse in rideshares is a growing problem. We propose a potential solution to this by using the voice recordings from the rideshare to detect cases of sexual harassment. Emotions such as fear, anger and disgust are most highly correlated to an individual being sexually harassed. Our solution aims to identify these emotions in a woman's voice as an indicator of sexual harassment. The Ryerson Audio-Visual Database of Emotional Speech and Song dataset was used and offered voice recordings from male and female actors speaking sentences in different emotions. We extract the Mel-Frequency Cepstral Coefficient (MFCC) of the recordings in the dataset and run it through Machine Learning methods such as CNN (Convolutional Neural Network), SVM (Support Vector Machines) and LSTM (Long-Short Term Memory). We achieved an F1-score of 95% with the CNN model on our dataset.

Keywords: Audio analysis · Emotion detection · Addressing harassment in rideshares

1 Introduction

The rideshare industry in the last 7 years has completely disrupted the long-standing taxi industry and consistently gen erates billions of dollars in revenue. The major players such as Uber reported a compounding growth of over 150% in 2016 [3]. What essentially started off as an American startup now has roots in countries around the world. The only things that have grown faster than the rideshare industry itself are the cases of sexual abuse and harassment against woman in these rideshares. Through this research, we aim to propose a technical solution for detecting cases of sexual harassment in the rideshare services through identifying emotions in voice recordings of the female rider such as fear, anger or disgust. The Ontario Human Rights Commission identifies sexual harassment as "engaging in a course of vexatious comment or conduct that is known or ought to be known to be unwelcome" [1]. This kind of behavior usually evokes feelings of fear, anger and/or disgust which gets reflected in the individual's response. We use this emotion expressed in the sound of the victim as a means

© Springer Nature Switzerland AG 2020
C. Goutte and X. Zhu (Eds.): Canadian AI 2020, LNAI 12109, pp. 464–469, 2020.
https://doi.org/10.1007/978-3-030-47358-7_48

to identify cases of harassment. Our approach proposes feature engineering that moves the audio recordings from the time-domain to the frequency domain. The frequency domain, especially, expressed by Mel-Frequency Cepstral Coefficients help capture pitch and emotion in the voice which we use as features to run our Machine Learning algorithms over.

2 Relevant Work

We explore work that has been done in Machine Learning for Speech Processing. Any Machine Learning problem has two primary steps to it – feature extraction and model selection. Feature extraction determines which features would be instrumental in getting the Machine Learning model to learn the space of the data and make reliable predictions. Feature extraction is very important in speech recognition since the raw data can be noisy, high-dimensional since it exists in time-domain, and very complex. Lithika et al. [5] famously proposed the idea of using Mel-Frequency Cepstral Coefficients (MFCC) as a means to model speech in order for Machine Learning algorithms to learn the features to predict emotion. Many people used this feature extraction method and used inherently linear classifiers such as Support Vector Machines (SVM) to detect emotions. Milton et al. [8] used SVMs on the Berlin Dataset [9] and achieved an accuracy of 68%. Lately, more and more researchers have shown the efficacy of using Recurrent Neural Networks (RNNs) and Convolutional Neural Networks (CNNs) on the MFCCs. Most famously, Lim et al. [6] used LSTMs, a special kind of RNN that deal with the issue of vanishing gradients, and achieved an F1 score of 78%. They also used 1D CNNs and achieved an accuracy of 86%. We primarily build upon these three methods, improve on their results and show how the application of emotion detection could deter sexual abuse in rideshares.

3 Dataset and Data Preprocessing

We use the RAVDESS dataset [7] that contains around 1440 files of voice actors speaking sentences in different emotions. The audio clips are 16 bit captured at 48 KHz in a .wav format. In particular, we have 12 male actors and 12 female actors who have 60 trials each to speak different sentences in 8 different emotions. These emotions include being sad, happy, neutral, calm, angry, disgusted, fearful and surprised. As a part of our research, we also interviewed 25 women from different nationalities to understand emotions that might be elicited in a situation where they were being sexually harassed by the rideshare driver. The most common emotions mentioned were anger, fear, and disgust. This gave us an incredible starting point since we could extract female voice that sounded angry, disgusted or fearful. We simply split the dataset into two parts, women who sounded angry, scared or disgusted and every other emotion and gender. This gave us a binary classification problem that we now attempt to solve.

Broadly speaking, the audio from the voice actors in the dataset can be broken down into two different kinds of features – time domain and frequency domain. The time domain usually represents the pressure that is captured by the microphone and is represented along with time as the X-axis. While this is helpful in many domains that use the time domain and need timestamps for application, it is relatively unhelpful in tasks such as Emotion Detection since we base our features on the pitch, rhythm and melody. For this reason, we move our audio to the frequency domain that give us more information about the aforementioned characteristics of the audio captured. In a nutshell, moving the audio to frequency domain allows us to separate the different sources of sound that comprise the audio clip. One of the most common ways of converting time to frequency domain is through a Fourier Transformation. It might be helpful to think of Fourier Transformation as a way of separating different audio sources from one unified audio clip, almost analogous to distinguishing all the different colors from a mixed color on a paint palette. Since Fourier Transformation simply gives us the different frequencies that make up the audio, we can go a step further and use Mel-Frequency Cepstral Coefficients (MFCCs) [10]. MFCC essentially model audio for computers to understand the same way humans perceive audio.

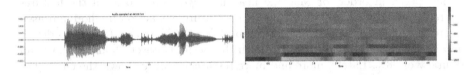

Fig. 1. Original audio signal   **Fig. 2.** MFCC of the audio signal

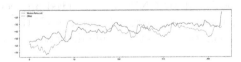

Fig. 3. Average MFCC bands of distressed women and non-distressed individuals

Pitch is one of the most important features of an audio signal and is calculated using the frequency of the signal. The following figures help visualize the difference between time domain audio-signal (see Fig. 1) and MFCCs (see Fig. 2). To visualize the statistical difference between our two classes – (1) a woman's voice eliciting disgust, anger, or fear; (2) men's and women's voices depicting other emotions – we can take the mean over the Mel Frequency bands over time (see Fig. 3) that we use as the input features to train our models. With the mean band values, we aim to train our models to differentiate between the positive and negative class.

4   Experiments and Analysis

Our input is the mean value over the MFCC bands (see Fig. 3) taken over time. This gives us an input size of 216 features that we can use to train our models

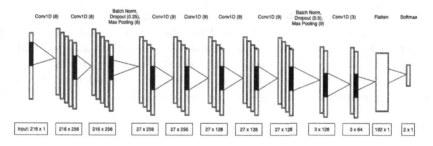

Fig. 4. CNN architecture

to recognize whether the input corresponds to the sound of a distressed woman or not. We divide our dataset – 720 negative samples and 192 positive samples – into a 70-15-15 split of training, testing and validation sets. We tune all our hyperparameters on the validation set. The results presented in Table 1 are from a test set of 228 samples.

Support Vector Machines: SVMs [2] are ideal for binary classification problems since they attempt to drop an optimal hyperplane between the features space of two classes. We train our SVM model with a Gaussian kernel, $C = 2$, and gamma = 0.05. We managed to achieve an F1 score of 91% on our test set with these configurations.

Convolutional Neural Networks: CNNs [11] allows vanilla NNs to learn different features in the feature space by focusing on having connections between localized regions of the input between the layers through the use of filters. We trained our model (see Fig. 4) and achieved an F1 score of 95% with our dataset.

Long-Short Term Memory: LSTMs [4] can deal with long term dependencies such as time series data and effectively model the feature space to make classifications. We used a stacked LSTM model with 3 LSTM layers each with 64 units stacked on top of each other. Further, we used a dropout of 0.2, 0.3 and 0.5 for each layer, respectively. With 50 epochs, we got an F1 score of 90%.

Table 1. Performance across different models

Algorithm	Precision	Recall	F1-Score	FPR	FNR
SVM	100%	84%	91%	0%	16%
LSTM	87%	92%	90%	43%	7%
1D-CNN	92%	98%	95%	25%	1%

All the results presented above are very competitive. However, we would hail 1D CNN as our best performer because it had the highest F1-Score and the lowest FNR (False Negative Rate). FNR corresponds the model classifying distressed emotion as a regular one. Our goal was to minimize the FNR since we shouldn't tolerate even a single instance of sexual harassment that might slip through our system. It is important to note here that we focus less on false positives. If the system flags a case of distressed female in rideshare, we can have the rideshare company either alert the driver/passenger to determine whether everything is okay or have manual curators listen in on the conversation to ensure the safety of the passengers.

5   Threats to Validity

Distress Might Not Always Mean Sexual Harassment: Most of the women we interviewed as a part of the research expressed that they would feel and sound scared, angry or disgusted if they were being sexually harassed. However, a woman could express these emotions without being sexually harassed or for another reason entirely. This would result in a higher FPR but still minimize the FNR. Low FNR is desirable because we can be certain that no potential case slips through the cracks. More oversight when a positive case is suspected may help to identify false labelling. And while manual curation of content may be of assistance, automated intelligent analysis of words and connotations of speech may also be of some help.

Privacy Violation: Bugging the rideshare with audio recorders might pose as a privacy violation for some. However, rideshares are an intrinsically public space and raising concern with having audio recorded in rideshares sounds similar to customers raising concerns about video cameras in stores and other public spaces. That being said, the rideshare users would be notified that the audio in the rideshare is going to be recorded. They will have to manually accept and consent to allow the audio recording to be uploaded to the cloud. Further, if having voice recording systems in rideshares can deter cases of sexual harassment, it seems like one of the most obvious cases of choosing the lesser evil. In our interviews, we did not come across any individual who raised concern against having their conversation with the rideshare driver uploaded to the cloud if it helped with the rampant issue of sexual harassment in rideshares and allowed women to feel safer.

6   Future Work and Conclusion

In this work, we judged instances of sexual harassment solely by assessing the emotion of a recorded conversation from a rideshare. This approach can be further improved by analyzing the content of the conversation in the rideshares, as a powerful tool in deterring sexual harassment. A dataset of rideshare conversations doesn't exist yet, but we could begin by looking at similar content on social media. It is also useful to expand our consideration of harassment cases and of the responses from victims. For instance, members of the trans and non-binary communities may be at even greater risk. And responses to trauma may well be less emotional, trying to prevent unsafe outcomes by aiming not to provoke the harasser. These realities may require new models for making use of audio transcripts towards addressing unwanted behaviour in rideshares.

The algorithms presented in this work can accurately classify a woman who sounds angry, disgusted or scared simply through the audio recording in a rideshare. Our CNN model enabled us to get a 1% FNR and an F1 score of 95% on our dataset. We hope that this work serves as motivation for researchers to collect a dataset of conversations from within rideshares. This would allow us to deploy this emotion recognition system and also train a model to analyze the content of the conversation to ensure safety in rideshares.

Acknowledgements. Thanks to Alice Kates for her valuable feedback and Yash Gupta for his initial suggestions on the topic.

References

1. Ontario Human Rights Commission: Identifying sexual harassment. http://www.ohrc.on.ca/en/policy-preventing-sexual-and-gender-based-harassment/2-identifying-sexual-harassment
2. Drucker, H., Burges, C.J., Kaufman, L., Smola, A.J., Vapnik, V.: Support vector regression machines. In: Advances in Neural Information Processing Systems, pp. 155–161 (1997)
3. Hensley, R.: Cracks in the ridesharing market-and how to fill them. https://www.mckinsey.com/industries/automotive-and-assembly/our-insights/cracks-in-the-ridesharing-market-and-how-to-fill-them
4. Hochreiter, S., Schmidhuber, J.: LSTM can solve hard long time lag problems. In: Advances in Neural Information Processing Systems, pp. 473–479 (1997)
5. Likitha, M., Gupta, S.R.R., Hasitha, K., Raju, A.U.: Speech based human emotion recognition using MFCC. In: International Conference on Wireless Communications, Signal Processing and Networking (WiSPNET), pp. 2257–2260. IEEE (2017)
6. Lim, W., Jang, D., Lee, T.: Speech emotion recognition using convolutional and recurrent neural networks. In: Asia-Pacific Signal and Information Processing Association Annual Summit and Conference (APSIPA), pp. 1–4. IEEE (2016)
7. Livingstone, S.R., Russo, F.A.: The Ryerson Audio-Visual Database of Emotional Speech and Song (RAVDESS): a dynamic, multimodal set of facial and vocal expressions in north American English. PloS one **13**(5), e0196391 (2018)
8. Milton, A., Roy, S.S., Selvi, S.T.: SVM scheme for speech emotion recognition using MFCC feature. Int. J. Comput. Appl. **69**(9), 34–39 (2013)
9. Pan, Y., Shen, P., Shen, L.: Speech emotion recognition using support vector machine. Int. J. Smart Home **6**(2), 101–108 (2012)
10. Practical-Cryptography-Website: MFCC generation algorihm. http://practicalcryptography.com/miscellaneous/machine-learning/guide-mel-frequency-cepstral-coefficients-mfccs/
11. Sermanet, P., Chintala, S., LeCun, Y.: Convolutional neural networks applied to house numbers digit classification. arXiv preprint arXiv:1204.3968 (2012)

Amalgamated Models for Detecting Duplicate Bug Reports

Sukhjit Singh Sehra[1]([✉]) [iD], Tamer Abdou[2] [iD], Ayşe Başar[2] [iD],
and Sumeet Kaur Sehra[3] [iD]

[1] Wilfrid Laurier University, Waterloo, Canada
ssehra@wlu.ca
[2] Ryesron University, Toronto, Canada
{tamer.abdou,ayse.bener}@ryerson.ca
[3] Guru Nanak Dev Engineering College, Punjab, India
sumeetksehra@gmail.com

Abstract. Automatic identification of duplicate bug reports is a critical research problem in the software repositories' mining area. The aim of this paper is to propose and compare amalgamated models for detecting duplicate bug reports using textual and non-textual information of bug reports. The algorithmic models viz. LDA, TF-IDF, GloVe, Word2Vec, and their amalgamation are used to rank bug reports according to their similarity with each other. The amalgamated score is generated by aggregating the ranks generated by models. The empirical evaluation has been performed on the open datasets from large open source software projects. The metrics used for evaluation are mean average precision (MAP), mean reciprocal rank (MRR) and recall rate. The experimental results show that amalgamated model (TF-IDF + Word2Vec + LDA) outperforms other amalgamated models for duplicate bug recommendations. It is also concluded that amalgamation of Word2Vec with TF-IDF models works better than TF-IDF with GloVe. The future scope of current work is to develop a python package that allows the user to select the individual models and their amalgamation with other models on a given dataset.

Keywords: TF-IDF · Word2Vec · GloVe · LDA

1 Introduction

Software bug reports can be represented as defects or errors' descriptions identified by software testers or users. These are generated due to the reporting of the same defect by many users. These duplicates cost futile effort in identification and handling. Developers, QA personnel and triagers consider duplicate bug reports as a concern. It is crucial to detect duplicate bug reports as it helps in reduced triaging efforts. The effort needed for identifying duplicate reports can be determined by the textual similarity between previous issues and new report [13]. Various approaches have been proposed to automate duplicate bug reports' detection. In early approaches, NLP [16], machine learning [1,10,20],

© Springer Nature Switzerland AG 2020
C. Goutte and X. Zhu (Eds.): Canadian AI 2020, LNAI 12109, pp. 470–482, 2020.
https://doi.org/10.1007/978-3-030-47358-7_49

information retrieval [19], topic analysis [1,9], deep learning [3], and combination of models [3,23] have been applied. Figure 1 shows the hierarchy of most widely used sparse and dense vector semantics [7].

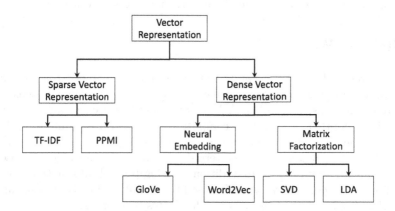

Fig. 1. Vector Representation in NLP

Our study has combined sparse and dense vector representation approaches to generate amalgamated models for duplicate bug reports' detection. The one or more models from LDA, TF-IDF, GloVe and Word2Vec are combined to create amalgamated similarity scores. The similarity score presents the duplicate (most similar) bug reports to bug triaging team. The proposed models takes into consideration textual information (description); and non-textual information (product and component) of the bug reports. TF-IDF signifies documents' relationships [16]; the distributional semantic models, Word2Vec and GloVe, use vectors that keep track of the contexts, e.g., co-occurring words.

LDA presents relationships between documents by transforming into a lower dimensional space. An amalgamated score is computed by merging the similarity scores from individual approaches. Thus, this score makes the basis for top k duplicate bug recommendations. The empirical evaluation has been performed on three datasets, namely, Apache, Eclipse, and KDE [17] with bug reports as discussed in Table 1. The effectiveness of the proposed approach is evaluated by three established performance metrics, viz. mean average precision (MAP), recall-rate@k, and mean reciprocal rank (MRR).

This study investigates and contributes into the following items:

- An empirical analysis of amalgamated models to rank duplicate bug reports.
- Effectiveness of amalgamation of models.
- Statistical significance and effect size of proposed models.

The paper has been divided into eight sections. The following section describes related work in detail. In third section, dataset and steps followed in pre-processing as given in [19] have been explained. The fourth section elaborates the methodology followed. The fifth section provides the insights into evaluation metrics. The sixth section discusses the results generated from the proposed models. The seventh section presents the threats to validity. In the final section, the paper is concluded and directions for future work are provided.

2   Related Work

Extensive research has been conducted in the area of detecting the duplicate bug reports automatically. Several methods have been developed focusing on these research areas [1,5,10,23]. A TF-IDF model has been proposed by modeling a bug report as a vector to compute textual features similarity [12]. An approach based on n-grams has been applied for duplicate detection [21]. All of the above methods are primary term-based methods and can diagnose the lexical duplicate bug reports. In addition to using textual information from the bug reports, the researchers have witnessed that additional features also support in the classification or identification of duplicates bug report.

The first study that combined the textual features and non-textual features derived from duplicate reports was presented by Jalbert and Weimer [6]. In year 2008, the execution traces were combined with textual information by Wang et al. [22]. In recent times, software engineering has witnessed the shift in the research focus towards the usage of Vector space models (VSMs). Word embedding is one of the most popular representation of document vocabulary. A method was proposed to use software dictionaries and word list to extract the implicit context of each issue report [1].

It has also been researched that Latent Dirichlet Allocation (LDA) provides great potential for detecting duplicate bug reports [5,9]. A combination of LDA and n-gram algorithm outperforms the state-of-the-art methods has been suggested Zou et al. [24]. Recently, deep learning technique for duplicate bug reports has been proposed by Budhiraja et al. [3]. Although in prior research many models have been developed and a recent trend has been witnessed to ensemble the various models. There exists no research which amalgamated the statistical, contextual, and semantic models to identify duplicate bug reports.

3   Dataset and Pre-processing

3.1   Dataset

A collection of bug reports that are publicly available for research purposes has been proposed by Sedat et al. [17]. The repository[1], presented three defect datasets extracted from Bugzilla in ".csv" format [17]. It contains the datasets for open source software projects: Apache, Eclipse, and KDE. The datasets contain information about approximately 914 thousands of defect reports over a

---

[1] https://zenodo.org/record/400614#.XaNPt-ZKh8x, last accessed: March 2020.

period of 18 years (1999–2017) to capture the inter-relationships among duplicate defects. Descriptive statistics are illustrated in Table 1. The dataset contains two categories of feature viz. textual and non-textual. The textual information is description given by the users about the bug i.e. "Short_desc". The non-textual information is presented by the features viz. "Product" and "Component", "Priority", "Bug severity", "Version", "Bug status", "current status" and "duplicate list". From these non-textual features "Product" and "Component" are used as filter, and "duplicate list" is used to create the ground truth for evaluation of the metrics.

Table 1. Dataset description

Project	Apache	Eclipse	KDE
# of reports	44,049	503,935	365,893
Distinct id	2,416	31,811	26,114
Min report opendate	2000-08-26	2001-02-07	1999-01-21
Max report opendate	2017-02-10	2017-02-07	2017-02-13
# of products	35	232	584
# of components	350	1486	2054

3.2 Pre-processing of Textual Features

Pre-processing and term-filtering were used to prepare the corpus from the textual features. In further processing steps, the sentences, words and characters identified in pre-processing were converted into tokens and corpus was prepared. The corpus preparation included conversion into lower case, word normalisation, elimination of punctuation characters, and lemmatization.

4 Methodology

The flowchart shown in Fig. 2 depicts the approach followed in this paper.

4.1 Latent Dirichlet Allocation

The bug reports textual information is the perfect example of the unstructured data as the content is written in natural language and LDA has emerged as efficient approach for pattern identification from unstructured data [18]. In this paper, LDA has been applied for querying the corpus data and identifying the latent patterns and the heuristic parameters proposed by Arun et al. [2] and Cao et al. [4] were used for deciding the topic count.

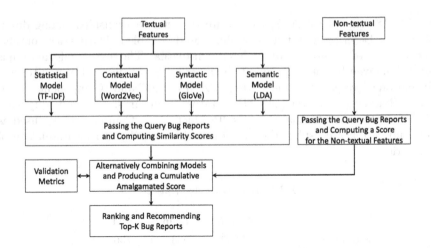

Fig. 2. Overall Methodology

4.2   Term Frequency-Inverse Document Frequency

The main idea behind the Term Frequency-Inverse Document Frequency (TF-IDF) is that the count of a term's occurrence in documents may be used to differentiate the documents. The weighted scheme for TF-IDF was adopted for representing one entity's significance relative to the other entities in the prepared corpus. The weight of an entity increases proportionally with a count of occurrences for a word in the document.

TF is document's local component measuring a normalized frequency of term occurrence; whereas the global component is represented by the inverse document frequency (IDF), i.e., $log[((1 + n_d)/(1 + dfi, j)] + 1$.

4.3   Word2Vec

Word2Vec is capable of capturing context of a word in a document, semantic and syntactic similarity, relation with other words, etc. Two variants of Word2Vec models namely, continuous bag-of-words (CBOW) and skip-gram are available. Both are capable to capture interactions between a centered word and its context words differently.

For a word vector $\hat{r}$ (predicted) and a word vector w_t (target), softmax function is applied to find the probability of the target word as given in Eq. 1.

$$P(w_t|\hat{r}) = \frac{exp(w_t, \hat{r})}{\sum_{w \in W} exp(w', \hat{r})} \tag{1}$$

Here W is the set of all target word vectors, where $exp(w_t, \hat{r})$ computes the compatibility of the target word w_t with the context $\hat{r}$. In this paper, gensim implementation of word2vec (skip-gram) pre-trained Google News corpus (3 billion running words) word vector model (3 million 300-dimension English word vectors) is used.

4.4   Global Vectors for Word Representation

Global Vectors for word representation (GloVe) is an unsupervised learning algorithm that combines the features of two model families, namely the global matrix factorization and local context window methods [8]. In this paper, GloVe used Google News pre-trained model to reduce the error between the dot product of (any two) word embedding vectors to the log of the co-occurrence probability. GloVe is based on matrix factorization on the word-context matrix. The model can be represented as in Eq. 2. In this, w and $\tilde{w}$ are word vectors.

$$F(w_i, w_j, \tilde{w}_k) = \frac{P_{ik}}{P_{jk}} \qquad (2)$$

Where i, j, and k are three words and the ratio P_{ik}/P_{jk} depends upon them.

4.5   Proposed Amalgamated Model

It has been identified that even the established similarity recommendation models such as *NextBug* [15] does not produce optimal and accurate results. Therefore, the current study created amalgamated models those merge one or more approaches viz. LDA, Word2Vec, GloVe and TF-IDF. The similarity scores vector (S_1, S_2, S_3, S_4) for k most similar bug reports is captured from individual approaches as shown in Fig. 2. Since the weights obtained for individual method have their own significance; therefore a heuristic ranking method is used to combine and create a universal rank all the results. The ranking approach assigns new weights to each element of the resultant similarity scores vector from the individual approach and assign it equal to the inverse of its position in the vector as in Eq. 3.

$$R_i = \frac{1}{Position_i} \qquad (3)$$

Once all ranks are obtained for each bug report and for each model selected, the amalgamated score is generated by summation of the ranks generated as given in Eq. 4, the ranks would be zero for left out models. It creates a vector of elements less than or equals to nk, where k is number of duplicate bug reports returned from each model and n is number of models being combined.

$$S = (R_1 + R_2 + R_3 + R_4) * PC \qquad (4)$$

Where S is amalgamated score (rank) of each returned bug report and R_1, R_2, R_3, and R_4 are the ranks returned from LDA, Word2Vec, GloVe, and TF-IDF, respectively as given in Eq. 3. Here PC is the product & component score and works as a filter. For instance, if two bug reports belong to same product and component then their similarity depend on the document similarity score. But if they belong to different product and component, then they are unlikely to be similar even if their document similarity score is high, thus made zero.

5   Evaluation Metrics

The evaluation metrics used to evaluate the one or more amalgamation of models are: recall-rate@k; mean average precision (MAP); and mean reciprocal rank (MRR). These metrics have been frequently used in recommendation systems to solve software engineering tasks [5, 6, 9, 15, 19].

5.1   Recall-Rate@k

Recall-rate is used to check the usefulness of top k recommendation. For a query bug q, it is defined as given in Eq. 5 as suggested by previous researchers [5, 19, 23].

$$RR(q) = \begin{cases} 1, & \text{if } if S(q) \cap R(q) \neq 0 \\ 0, & \text{otherwise} \end{cases} \tag{5}$$

Given a query bug q, $S(q)$ is ground truth and $R(q)$ represents the set of top-k recommendations from a recommendation system.

5.2   Mean Average Precision (MAP)

MAP is defined as the mean of the Average Precision ($AvgP$) values obtained for all the evaluation queries given in $MAP = \sum_{q=1}^{|Q|} \frac{AvgP(q)}{|Q|}$. In this equation, Q is number of queries in the test set.

5.3   Mean Reciprocal Rank (MRR)

Mean Reciprocal Rank (MRR) is calculated from the reciprocal rank values of queries. $MRR(q) = \sum_{i=1}^{|Q|} Reciprocal Rank(i)$ calculates the mean reciprocal rank and RR is calculated as in $Reciprocal Rank(q) = \frac{1}{index_q}$.

6   Results and Discussion

This section presents the results of the empirical evaluation. For evaluation of results, we used a Google Colab[2] machine with specifications as RAM: 24 GB Available; and Disk: 320 GB.

The amalgamated models compares the incoming query bug report against the already existing resolved bug report database and return the top-k duplicate bugs. The current research implements the algorithms in Python 3.5 and used "nltk", "sklearn", "gensim" [14] packages for model implementation. The default values of the parameters of the algorithms were used. The values of k has been taken as 1, 5, 10, 20, 30, and 50 to investigate the effectiveness of proposed approach.

---

[2] https://colab.research.google.com.

The current study has investigated the proposed models in comparison with the other combination of established approaches for duplicate bug report recommendation. For the empirical validation of the results, the developed models have been applied to open bug report data (discussed in Sect. 3) consisting of three datasets of bug reports. The datasets were divided into train and test data. In the open source software (OSS) bug repository datasets, one of the column contained the actual duplicate bug list i.e. if a bug report actually have duplicate bugs then the list is non-empty otherwise it is empty ('NA'). This list worked as ground truth to validate the evaluation parameters. All the bug reports with duplicate bug list are considered as test dataset for validation of the amalgamated models. The number of bug reports for test dataset for Apache, Eclipse, and KDE projects were 2,518, 34,316, and 30,377, respectively. The training dataset was used to convert the existing textual information into the vector representation for the models. The test data was used to detect the duplicate bug reports from the train dataset considered resolved. This helped to identify the duplicate bug reports and evaluate the models.

Table 2. Mean average precision of individual and amalgamated models using all dataset.

Models	Apache	Eclipse	KDE
TF-IDF	0.076	0.108	0.045
Word2Vec	0.115	0.171	0.132
GloVe	0.060	0.105	0.094
LDA	0.012	0.029	0.008
TF-IDF + LDA	0.149	0.127	0.082
TF-IDF + GloVE	0.138	0.128	0.098
TF-IDF + Word2Vec	0.144	0.173	0.126
TF-IDF + Word2Vec + LDA	0.161	0.166	0.158
TF-IDF + GloVe + LDA	0.163	0.123	0.130

6.1 Empirical Analysis

The empirical analysis of the proposed ensemble model has been performed on OSS datasets. The models take textual information from training dataset and create vocabulary to be used for finding the duplicates of test bug reports.

Apache Dataset. Apache dataset is smallest dataset of three datasets and contains 44,049 bug reports. These bug reports are generated for 35 products and 350 components. Figures 3(a) and 3(b) show that the amalgamation of models produces more effective results than the individual established approaches. Table 2 represents MAP values for the models. For the results, it is revealed that not all combinations produces good results.

Eclipse Dataset. The dataset of Eclipse contained 503,935 bug reports, and 31,811 distinct ids. It includes 232 products and 1486 components bug reports. Due to large dataset the random sampling of the full dataset was performed to select 10% of the dataset. The values of recall rate and MRR are presented in Figs. 3(c) and 3(d) respectively. The results obtained reveal that the amalgamated score has better value as compared to the scores obtained from individual approaches.

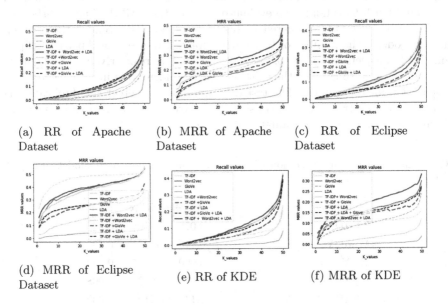

(a) RR of Apache Dataset

(b) MRR of Apache Dataset

(c) RR of Eclipse Dataset

(d) MRR of Eclipse Dataset

(e) RR of KDE

(f) MRR of KDE

Fig. 3. Performance of **(a)–(b)** Apache dataset, **(c)–(d)** Eclipse dataset, **(e)–(f)** KDE dataset

KDE Dataset. This dataset contains 365,893 bug reports of 584 products out of which 2054 were used. Due to large dataset the random sampling of the full dataset was performed to select 10% of the dataset. The evaluation metrics obtained from this dataset are depicted in Fig. 3(e) and 3(f).

6.2   Effectiveness of Amalgamation of Models

The results have demonstrated the superiority of the amalgamated models to identify the duplicate report as compared to individual approaches. Figure 3 shows the comparative performance of the proposed approach and the established approaches with parameter k varying for all the datasets. Further, it has been revealed that for two datasets Apache and KDE, the amalgamated model (TF-IDF + Word2Vec + LDA) produced the best results. Whereas for Ecilpse dataset a amalgamated model (TF-IDF + LDA) generated better than model (TF-IDF + Word2Vec + LDA). Another, conclusion from the results is that

Word2Vec individually is also very powerful to detect the duplicate reports. This study proposes the amalgamated model of TF-IDF + Word2Vec + LDA, that outperform other amalgamated models. It has also been concluded that Word2Vec and its combination produces better results as compared to GloVe.

6.3 Statistical Significance and Effect Size

To establish the obtained results of the proposed model, we performed the Wilcoxon signed-rank statistical test to compute the p-value, and measured the Cliff's Delta measure [11], and Spearman correlation. Table 3(a) depicts the interpretation of Cliff's Delta measure. By performing the Shapiro-Wilk test, the normality of the results was identified. Since it turned out to be non-Gaussian, non-parametric test Spearman correlation was applied to find out the relationship between the results of different approaches.

Table 3. Statistical significance and effect size

(a) Interpretation of Cliff's Delta Scores [11]	
Effect Size	Cliff's Delta (δ)
Negligible	$-1.00 \leq \delta < 0.147$
Small	$0.146 \leq \delta < 0.330$
Medium	$0.330 \leq \delta < 0.474$
Large	$0.474 \leq \delta \leq 1.00$

(b) p-value of Wilcoxon signed-rank test, Cliff's Delta and Spearman's correlation coefficient comparing the metrics of amalgamated (TF-IDF + Word2Vec + LDA) model with TF-IDF for Apache dataset

Metrics	Spearman's r	Cliff's Delta	p-value
Recall	0.99	0.4032	0.00051
MRR	0.99	0.8244	0.00043

Table 3(b) presents the p-value, Cliff's Delta measure and Spearman's correlation coefficient of amalgamated (TF-IDF + Word2Vec + LDA) model with TF-IDF in terms of two metrics for Apache dataset and KDE, respectively. The TF-IDF model has been compared with the amalgamated approach as TF-IDF has been presented as benchmark model in most of the previous studies. Table 3(b) presents that the results have a positive correlation, whereas there is a medium or large effect size, which means improvement is happening by amalgamation of models.

7 Threats to Validity

Internal Validity. The dataset repository contains the bug reports that contains dataset till the year 2017. The threat is that the size of textual information is small for each bug report. But, the current work applied the well-established methods of natural language processing to preparing the corpus from these large datasets. Therefore, we believe that there would not be significant threats to internal validity. While using LDA, a bias may have been introduced due to the choice of hyper-parameter values and the optimal number of topic solutions. However, to mitigate this, the selection of the optimal number of topic solutions was done by following a heuristic approach as suggested by Arun et al. [2] and Cao et al. [4].

External Validity. The generalization of results may be another limitation of this study. The similarity score was computed by following a number of steps and each of these steps has a significant impact on the results. However, verification of results is performed using open source datasets to achieve enough generalization.

8   Conclusion

The main contribution of this paper is an attempt to amalgamate the established natural language models for duplicate bug recommendation using bug textual information and non-textual information (product and component). The proposed amalgamated model combines the similarity scores from different models namely LDA, TF-IDF, Word2Vec, and GloVe. The empirical evaluation has been performed on the open datasets from three large open source software projects, namely, Apache, KDE and Eclipse. From the validation, it is evident that for Apache dataset the value of MAP rate increased from 0.076 to 0.163, which is better as compared to the other models. This holds true for all three datasets as shown in experimental results. Similarly, the values of MRR for amalgamated models is also high relative to the other individual models. Thus, it can be concluded that amalgamated approaches achieves better performance than individual approaches for duplicate bug recommendation. This study proposes the amalgamated model (TF-IDF + Word2Vec + LDA), that outperform other amalgamated models.

The future scope of current work is to develop a python package that allows the user to select the individual models and their amalgamation with other models on a given dataset. This would also allow the user to select combination of textual and non-textual features from dataset for duplicate bug detection.

References

1. Alipour, A., Hindle, A., Stroulia, E.: A contextual approach towards more accurate duplicate bug report detection. In: Proceedings of the 10th Working Conference on Mining Software Repositories, pp. 183–192. MSR 2013, IEEE Press, Piscataway (2013)
2. Arun, R., Suresh, V., Veni Madhavan, C.E., Narasimha Murthy, M.N.: On finding the natural number of topics with latent dirichlet allocation: some observations. In: Zaki, M.J., Yu, J.X., Ravindran, B., Pudi, V. (eds.) PAKDD 2010. LNCS (LNAI), vol. 6118, pp. 391–402. Springer, Heidelberg (2010). https://doi.org/10.1007/978-3-642-13657-3_43
3. Budhiraja, A., Dutta, K., Shrivastava, M., Reddy, R.: Towards word embeddings for improved duplicate bug report retrieval in software repositories. In: ICTIR 2018 Proceedings of the 2018 ACM SIGIR International Conference on Theory of Information Retrieval, pp. 167–170. ACM, Tianjin (2018)

4. Cao, J., Xia, T., Li, J., Zhang, Y., Tang, S.: A density-based method for adaptive LDA model selection. Neurocomputing **72**(7–9), 1775–1781 (2009)
5. Hindle, A., Onuczko, C.: Preventing duplicate bug reports by continuously querying bug reports. Empirical Softw. Eng. **24**(2), 902–936 (2018). https://doi.org/10.1007/s10664-018-9643-4
6. Jalbert, N., Weimer, W.: Automated duplicate detection for bug tracking systems. In: IEEE International Conference on Dependable Systems and Networks With FTCS and DCC (DSN), pp. 52–61 (2008)
7. Jurafsky, D., James H.M.: Vector semantics and embeddings. In: Speech and Language Processing: An Introduction to Natural Language Processing, Computational Linguistics, and Speech Recognition, pp. 94–122. Online, Stanford University, UK, third edn. (2019)
8. Kalajdziski, S., Ackovska, N. (eds.): ICT 2018. CCIS, vol. 940. Springer, Cham (2018). https://doi.org/10.1007/978-3-030-00825-3
9. Klein, N., Corley, C.S., Kraft, N.A.: New features for duplicate bug detection. In: Proceedings of the 11th Working Conference on Mining Software Repositories, pp. 324–327. ACM (2014)
10. Lazar, A., Ritchey, S., Sharif, B.: Improving the accuracy of duplicate bug report detection using textual similarity measures. In: Proceedings of the 11th Working Conference on Mining Software Repositories, pp. 308–311. MSR 2014, ACM, New York (2014)
11. Macbeth, G., Razumiejczyk, E., Ledesma, R.D.: Cliff's delta calculator: a nonparametric effect size program for two groups of observations. Universitas Psychologica **10**(2), 545–555 (2011)
12. Nagwani, N.K., Singh, P.: Weight similarity measurement model based, object oriented approach for bug databases mining to detect similar and duplicate bugs. In: Proceedings of the International Conference on Advances in Computing, Communication and Control, pp. 202–207 (2009)
13. Rakha, M.S., Shang, W., Hassan, A.E.: Studying the needed effort for identifying duplicate issues. Empirical Softw. Eng. **21**(5), 1960–1989 (2015). https://doi.org/10.1007/s10664-015-9404-6
14. Rehurek, R., Sojka, P.: Software framework for topic modelling with large corpora. In: Proceedings of the LREC 2010 Workshop on New Challenges for NLP Frameworks, pp. 45–50. ELRA, Valletta, Malta (2010)
15. Rocha, H., Valente, M.T., Marques-Neto, H., Murphy, G.C.: An empirical study on recommendations of similar bugs. In: 2016 IEEE 23rd International Conference on Software Analysis, Evolution, and Reengineering (SANER), vol. 1, pp. 46–56. IEEE (2016)
16. Runeson, P., Alexandersson, M., Nyholm, O.: Detection of duplicate defect reports using natural language processing. In: Proceedings of the 29th International Conference on Software Engineering, pp. 499–510. IEEE Computer Society (2007)
17. Sadat, M., Bener, A.B., Miranskyy, A.: Rediscovery datasets: connecting duplicate reports. In: IEEE International Working Conference on Mining Software Repositories, pp. 527–530 (2017)
18. Sehra, S., Singh, J., Rai, H.: Using latent semantic analysis to identify research trends in OpenStreetMap. ISPRS Int. J. Geo-Inf. **6**(7), 195 (2017)
19. Sun, C., Lo, D., Khoo, S.C., Jiang, J.: Towards more accurate retrieval of duplicate bug reports. In: Proceedings of the 26th International Conference on Automated Software Engineering, ASE 2011, pp. 253–262. IEEE Computer Society (2011)

20. Sun, C., Lo, D., Wang, X., Jiang, J., Khoo, S.C.: A discriminative model approach for accurate duplicate bug report retrieval. In: Proceedings of the 32nd ACM/IEEE International Conference on Software Engineering, vol. 1, pp. 45–54. ACM (2010)
21. Sureka, A., Jalote, P.: Detecting duplicate bug report using character N-gram-based features. In: Proceedings - Asia-Pacific Software Engineering Conference, APSEC, pp. 366–374 (2010)
22. Wang, X., Zhang, L., Xie, T., Anvik, J., Sun, J.: An approach to detecting duplicate bug reports using natural language and execution information. In: Proceedings of the 30th International Conference on Software Engineering, pp. 461–470. ACM (2008)
23. Yang, X., Lo, D., Xia, X., Bao, L., Sun, J.: Combining word embedding with information retrieval to recommend similar bug reports. In: Proceedings - International Symposium on Software Reliability Engineering, ISSRE, pp. 127–137. IEEE (2016)
24. Zou, J., Xu, L., Yang, M., Zhang, X., Zeng, J., Hirokawa, S.: Automated duplicate bug report detection using multi-factor analysis. IEICE Trans. Inf. Syst. **E99D**(7), 1762–1775 (2016)

Investigating Citation Linkage
as a Sentence Similarity Measurement
Task Using Deep Learning

Sudipta Singha Roy$^{(\boxtimes)}$ ⓘ, Robert E. Mercer ⓘ, and Felipe Urra

The University of Western Ontario, London, ON, Canada
ssinghar@uwo.ca, mercer@csd.uwo.ca

Abstract. Research publications reflect advancements in the corresponding research domain. In these research publications, scientists often use citations to bolster the presented research findings and portray the improvements that come with these findings, at the same time, to make the contents more understandable to the audience by navigating the flow of information. In the science domain, a citation refers to the document from where this information originates, but doesn't specify the text span that is actually being cited. This paper develops a framework which can create a linkage between the citing sentences from the ongoing research article and the related cited sentences from the corresponding referenced documents. Eventually, it will reduce the burden of the readers to go through all the sentences in the documents while acquiring the required background information. This citation linkage problem has been modelled as a semantic relatedness task where given a citing sentence the framework generates the sentence pairs with the citing sentence and each of the sentences from the reference document and then tries to determine which sentence pair is semantically similar and which pair is not. Construction of the citation linkage framework involves corpus creation and utilizing deep-learning models for semantic similarity measurement.

Keywords: Citation linkage · Textual semantic relatedness · Corpus creation

1 Introduction

Different types of written documents have different formats, writing patterns and serve different objectives. A research article can reflect research trends, a new invention, or perspectives to solve a problem in a particular domain. While writing a research document, the author discusses previous research that is either prominent to solve the same problem or has influenced the author's ideas presented in the ongoing research paper. This referencing to some other document while writing a research article is called a *citation* [10]. Thus, citations create links between various research articles. Usage of citations reduces the writing overload of authors as they don't have to write down the same thing.

© Springer Nature Switzerland AG 2020
C. Goutte and X. Zhu (Eds.): Canadian AI 2020, LNAI 12109, pp. 483–495, 2020.
https://doi.org/10.1007/978-3-030-47358-7_50

At the same time, it helps the readers to achieve some background knowledge over a topic which may be required to understand the ideas in the ongoing article.

In 1964 the idea of citation indexing was introduced [7]. As citation indexing became increasingly popular, more fine-grained analysis of citing were proposed. Garzone and Mercer [8] gave a method to identify the purpose of a citation in biochemistry and physics research articles. As well, modern applications such as multiple document summarization [25] and argumentation mining [24] use these citation links.

In the case of scientific research articles, a citation refers to the document from where the idea stated in the citing sentence originates. However, a citing sentence in an experimental biomedical research article typically refers to a small portion of the referenced document. If it were possible to pull out that specific text span from the reference document, it would be advantageous for applications such as those mentioned above. Additionally, it would reduce the burden of the readers having to read the complete document to find the piece that is being cited.

The objective of this paper is to establish a relationship between the citation sentence and its corresponding reference sentences from the cited paper. This task is called *citation linkage*. In this study, citation linkage is modelled as a textual semantic relatedness measurement task and the text span is delimited to a single sentence. Hence, this is why the citation linkage problem is formulated as a textual matching operation between a citation sentence and every sentence in the corresponding cited paper.

The major contributions of this work are creating a synthetic corpus for the citation linkage task containing more than sixty thousand sentence pairs from the biomedical domain, and building a framework to determine the appropriate cited sentences from a cited paper given a citation sentence. The rest of the paper is structured as follows: Sect. 2 elaborates on the citation linkage problem and Sect. 3 gives some related works on the citation linkage task as well as some deep-learning models for measuring textual semantic relatedness at the sentence level. Section 4 describes the data cleaning and pre-processing steps and how the corpus is built. A detailed description of the citation linkage framework is presented in Sect. 5. In Sect. 6 the results are provided and analysed, and a comparison with some previous work is given. We conclude the paper with a summary of the current work and present some directions for future study.

2   Citation Linkage

Citations establish a semantic link between the citing and the cited papers. A citation in a research paper indicates a portion of the reference paper called the *citation context* [10]. Usually, the citation context refers to some specific idea or topic introduced in the cited paper. The citation context contains few full sentences mentioning the method names, used appliances, or the findings and the hypothesis presented in the cited paper. However, citations don't specify which portion of the cited article is actually being referred to, rather a citation states

Table 1. Sample citation and corresponding target reference sentence [10]

Example 1	Citation sentence	Formalin fixation, the most widely used fixative in histopathology, has many advantages such as the ease of tissue handling, the possibility of long-term storage, an optimal histological quality and its availability in large quantities at low price [11]
	Target sentence	The advantages of formalin fixation are the ease of tissue handling, the possibility of long-term storage of wet material, and its low price [12]
Example 2	Citation sentence	Sample DNA is often damaged by exposure to formaldehyde and a potentially extremely acidic environment [26]
	Target sentence	However, DNA is relatively stable in mildly acidic solutions, but at around pH4 the beta glycosidic bond is in the purine bases are hydrolysed [5]
Example 3	Citation sentence	Different PCR buffer systems and/or different Taq poly-merases may yield different real time PCR results [11]
	Target sentence	A significant difference can be seen between the results from the different DNA polymerase-buffer systems [27]

only the name of the cited document. So, if a reader needs some background knowledge about that topic, he/she needs to read the whole referenced document. However, readers prefer to have precise information about the discoveries of the research work in a research article. For this they require concise background information which has influenced the work.

This paper investigates the ways to determine the referenced sentences from the cited paper given a citation sentence by means of measuring semantic relatedness. This problem is named as *citation linkage* and we have tried to solve it using different deep-learning methods. The citation context span may contain one or more than one sentences. However, for this work this span has been restricted to only one sentence and the citation linkage task has been designed as a semantic relatedness measurement task at the sentence level. This semantic relatedness task is actually a two class classification which operates on sentence pairs. Given a sentence pair it tries to predict whether they are semantically similar or not.

Table 1 shows a few examples of citation sentences and their corresponding reference sentences from the cited paper. Example 1 gives the sentence pair where the citation sentence is the paraphrase of the cited sentence and contains common words in a different order. In the second example, the citing sentence replaces the term "pH4" by "extremely acidic environment". To establish a linkage between these two terms a mapping is required between the scaling of pH to the acidic condition (which is currently beyond the methods used in this paper). In Example 3, the citing sentence interprets the information from the target sentence. From these examples it is clear that, for establishing linkage between citation and cited sentence, the proper mapping is essential in between sentences as well as words. As the final models used in the experiments for the textual semantic relatedness measurement are deep-learning techniques, proper word and sentence embedding techniques are required prior feeding the data to these models. Furthermore, while measuring semantic relatedness between two sentences, only few words rather than all the words in the sentences play the vital role. That's why the attention mechanism is influential to give proper importance to individual words.

3   Related Work

Interest in citations in scientific research literature has led to much work analyzing citations [7,8]. One type of analysis is: Given a citation sentence, citation analysis attempts to determine which section (i.e., abstract, introduction, methodology, result, and discussion) of the cited paper is being referred to by this sentence. However, this type of analysis is unable to specify the more precise citation span.

The ongoing CL-SciSumm Shared Task looks at three aspects of the citation linkage task: for each citation sentence ("citance") find the text span of the referenced paper that most accurately reflects the citance; for each cited text span, identify its discourse facet; and generate a summary of the referenced paper based on the text spans that are cited by multiple citances. The latter two subtasks are beyond the scope of this paper. The granularity of the text span can be a sentence fragment, a full sentence, or up to five consecutive sentences. We have focused on the single sentence text span in this paper. The CL-SciSumm Shared Task uses a corpus comprised of computational linguistics research papers. For the CL-SciSumm-17 shared task, Li et al. [15] applied a ruled-based approach using inverse document frequency and Jaccard similarity to determine the citation linkage between citing and cited sentence pairs. Later, they represented the sentences as the concatenation of word vectors in the sentence and calculated the cosine similarity between the sentence vectors [14]. In their work for the CL-SciSumm-19 shared task, they ran a convolutional neural network (CNN) over the sentence representations to generate better feature representations and then computed cosine similarity between the citing and cited sentences [16]. Other works, such as AbuRa'ed et al. [1] have also worked with the CL-SciSumm corpus.

Unfortunately, only a few works have approached this citation linkage task in the biomedical domain. In 2017, Houngbo and Mercer [9] developed a framework for the citation linkage task using a textual matching operation and built a small expert-annotated corpus comprised of sentence pairs from the biomedical domain. For the similarity measurement task, they used various traditional machine learning models. However, the accuracy they achieved was low.

Recently, deep learning based models are being used to generate the word embeddings which feed the neural network architectures described below. The models proposed by Bengio et al. [3] followed by Mikolov et al. [19,20] (Word2Vec) generate fixed-sized vector representations of words. However, these two models can't generate embeddings for the out of the vocabulary words. To solve this issue, Bojanowski et al. [4] proposed a model (fasttext) that generates the word embeddings from sub-word information. As the names of the very same chemical in the biomedical research articles come with different patterns (i.e., carbon dioxide is presented by CO_2, $C-O_2$, etc.), sub-word information is required to generate the embeddings for the chemical names. To cope with this issue, we have used fasttext for generating the word embeddings.

Currently, only one human annotated dataset is available for the citation linkage task in the biomedical domain [9]. It is too small to train the data thirsty deep learning models. To overcome this problem and to generate a synthetic dataset with more sentence pairs for the citation linkage task, different unsupervised methods for sentence embedding have been investigated. Kiros et al. [13] and Logeswaran and Lee [18] proposed unsupervised sentence embedding models. However, both of these models require pre-trained word embeddings as input and can't work with out of vocabulary words. To solve this issue, Pagliardini et al. [22] proposed a sentence embedding model (Sent2Vec) using compositional n-gram features. Here sentences are represented as fasttext CBOW generated pre-trained word vector's weighted mean. Because of its ability to work with unseen words, while generating synthetic data, we use Sent2Vec.

The inherent meaning of a sentence is typically provided by few words present in the sentence. To focus on portions of a sentence various attention mechanisms have been introduced in recent times. Badhanau et al. [2] introduced the idea of attention in the task of neural machine translation, Liu et al. [17] proposed a sentence encoding model for text entailment recognition task which they named "Inner Attention", Yang et al. [28] proposed a two-leveled hierarchical attention architecture to classify documents, and finally, Conneau et al. [6] utilized different attention mechanisms which have subsequently been named Infersent. They used multiple architectures like Bi-LSTM with max-pooling, hierarchical CNN, inner attention and hierarchical attention mechanisms. These models showed very prominent result over the semantic relatedness measurement task. In this paper, for the citation linkage task this Infersent [6] model is used.

4   Citation Linkage Corpus Creation

The only corpus available for the citation linkage task in the biomedical domain is from Houngbo and Mercer's work [9] which contains only method citations. Although in scientific research papers the citation span can be one or more sentences (or part of a sentence), for the above work, it is limited to a single sentence, so the citation linkage task involves sentence pairs. They had an expert annotate 3857 sentence pairs on a scale of 0 (not similar) and 1 (lowest similarity score) to 5 (highest similarity score). The dataset is highly imbalanced. Only 81 pairs are annotated with rating 4 or 5. Thus it becomes very hard to train deep-learning models with this dataset. At the same time, human annotation of a sufficient amount of data from biomedical research articles for this task is a costly process. To overcome this issue, we have developed a synthetic corpus of 68,898 sentence pairs over three biomedical topics: cell biology, biochemistry, and chemical biology. We are calling it a synthetic corpus, as it is developed not with any human annotation, but rather with an unsupervised sentence embedding technique called Sent2Vec. However, we have used the corpus built by Houngbo and Mercer [9] for the validation and test purposes with a few changes in the scoring factor. Scores 4 and 5 in this corpus are replaced with 1 and the remaining scores are replaced with 0.

Like other unsupervised sentence embedding models, Sent2Vec requires a lot of data for training. For this reason, 4,843,756 sentences from 28,310 research documents are collected. These documents, collected from BioMed Central, are from 90 biomedical domains like cell biology, bioinformatics, biochemistry, etc. As, these data are from different domains, they differ in format. Moreover, the same equation may have different variable names and symbols. For this reason all the equations are required to be replaced with the same symbol. The same replacement operation is needed for numbers as well. All of the equations are replaced with "$< equ >$" and all of the standalone numbers are replaced with "$< num >$". The sentences also contain a few unnecessary symbols in terms of their semantic representations which can be deleted. Furthermore, some sentences contain the citation numbers like "[xx]" which are unnecessary as well. These citation numbers are also deleted. Some confusions are found for the α letter. This symbol may appear in a equation to indicate in proportion to, or it may appear with a chemical name like "AUCO-α". For chemical names, this sign is replaced with the term "alpha". The symbol is kept as it is if it appears as a part of an equation. The reason to do this is its presence in a chemical name where only this symbol is different indicates a different chemical. Similar substitutions are done for other Greek letters. However, if a number appears as a part of a chemical name, it is preserved in its original form. In the next step, all the symbols with different representation formats are replaced with their corresponding common format representations. Table 2 shows the regex commands used for cleaning the data. Finally, all the characters are lower-cased.

After having cleaned the sentences, the next step is to train an unsupervised sentence embedding model. This step is important for the annotation of the sentence pairs. That is why Sent2Vec is trained with various parameter settings over the data. To generate the sentence pairs with citing and cited sentences, among 28,310 research articles, 112 were randomly selected from Biochemistry, Cell Biology and Chemical Biology. For these papers, 2289 papers which have cited them were manually collected. From these papers only the corresponding citation sentences are extracted. After cleaning the data, sentence pairs are generated, the first sentence in the pair being a sentence from the cited article and the second being the citing sentence such that there is one sentence pair for each sentence in the cited article. This step generates 475,807 sentence pairs.

Table 2. Regex for capturing different pattern throughout the data

Purpose	Regex command
Capture equations	[(({(\[\s]*([^\s(({(\[)))]%*+/:<=>\-\^|~¬±·×÷ΘΘ— ←→↔↦⇆⇔∀∇∈∉∈∂∏∑Ŧ∘∨∝∧∩∪∮⥲≤≥$∁¢∁⊆⊇∄⊆⊕ ⊙¬⊤⊥∪·\|⬚▷∘∘—⨿]+(\s*[(({(\[)]+\s*)+)?(∃\|∫\|log\|ln)?\s* ([^\s(({(\[)))]%\]]+\|(\d+[\.,])+\d+)?[)))]%\]]*[(({(\[\s]*[%*+/:< =>\^\|~¬±·×÷ΘΘ— ←→↔↦⇆⇔∀∇∈∉∈∂∏∑Ŧ∘∨∝∧∩∪∮⥲≤≥$∁¢∁⊆⊇∄⊆⊕ ⊙¬⊤⊥∪·|⬚▷∘∘—⨿]+\s*- ?[(({(\[\s]*(∃|∫|log|ln)?\s*([^\s(({(\[)))]%\]]+|(\d+[\.,])+\d+) \s*([(({(\[\s]*[^\s(({(\[)))]%*+/:<=>\^|~¬±·×÷ΘΘ— ←→↔↦⇆⇔∀∇∈∉∈∂∏∑Ŧ∘∨∝∧∩∪∮⥲≤≥$∁¢∁⊆⊇∄⊆⊕ ⊙¬⊤⊥∪·|⬚▷∘∘—⨿]+|(\d+[\.,])+\d+|-)?[\s)))]%\]]*)+
Capture numbers with no prior symbol	(^\d+|\K\d+)((\.\d+)|\^\d+|e[\-+]?\d+(\.\d+)?)?(?!|[\-\w])
Capture numbers with prior symbol	(?<!(\w|\d))[\-+]\d+((\.\d+)|\^\d+|e[\-+]?\d+(\.\d+)?)?
Capture citations	\[\d+([,\-]\d+)*\]
Capture chemical names ending with ∝ and β	-\s*∝|-\s* β

Individual sentences of each sentence pair are then fed to the already trained Sent2Vec model to get the vector representation. Cosine similarity is measured between the sentence vectors for each sentence pair. For different cutoff cosine similarity values, the performance is tested against a validation set which is a part of the human annotated data from Houngbo and Mercer's work [9]. Sentence pairs with cosine similarity values which are greater than or equal to the selected cutoff values are annotated with similarity value 1 (indicating the sentences in the pair are similar) and the remainder are annotated with 0 (indicating the sentences in the pair are not similar). However, among these 475,807 sentence pairs most of the pairs are annotated with zero value. If the models discussed below are trained with this data, they will likely be biased. For this reason, from these sentence pairs, 68,898 samples are chosen to balance the proportion of positive and negative samples. When choosing these samples, all of the sentence pairs annotated with similarity value 1 are kept. Then, for each citation sentence, n negative samples are chosen, where n is the number of positive samples found for that citation sentence. For the few citation sentences that have no corresponding positive cited sentences, five randomly chosen samples are inserted into the dataset. Thus, the dataset of 68,898 samples is created. This dataset contains 31,624 positively annotated sentence pairs, which is 45.89% of the samples.

5   Citation Linkage as a Semantic Similarity Measurement Task

Our goal is to link text in the cited document with a citing sentence. As we have restricted this text span to one sentence, we have posed this problem as a sentence level semantic similarity measurement task using various supervised deep-learned models.

To generate the word embeddings for the semantic similarity measurement models, fasttext [4] is trained using different parameter settings with the words from 28,310 research documents over 90 sub-domains of biomedicine as this model has the ability to generate word vectors for unseen words by utilizing the sub-word embeddings. The model with the best pearson and spearman values over the UMNSRS-Sim [23] dataset is chosen for the following experiments.

To solve the citation linkage task as a semantic relatedness measurement problem, the Infersent [6] model with various architectures is used. Infersent is a supervised sentence representation model which follows the siamese structure [21] in the core. Figure 1 gives an overview of the training process of the Infersent model for the citation linkage task. Two identical neural network architectures with identical parameter values are used as sentence encoders. One encoder is responsible to generate a sentence vector for the citing sentence (S_{citing}) while the other one is responsible for the cited sentence (S_{cited}). Then, a feature map is generated which accommodates the concatenation, absolute point-wise difference, and point-wise multiplication of these two sentence vectors. This feature map is then fed to the following dense and *softmax* layers to get the class prediction. To model the citation linkage task, the Infersent architecture is designed as a binary class classifier where class labels are 0 and 1. Here, 1 indicates that the citation sentence is citing the cited sentence where 0 indicates the opposite. In our experiments, we have tried four different models as the sentence encoder: Bi-LSTM with max-pooling, hierarchical CNN [29], inner attention [17] and hierarchical attention mechanism [28] over Bi-LSTM.

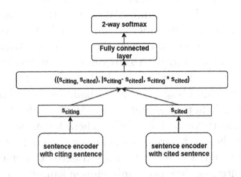

Fig. 1. Infersent training mechanism for the citation linkage task.

Finally, bootstrapping approaches are utilized for the citation linkage task. Bootstrapping is performed with the variants of the Infersent architectures. For this, the sentence pair dataset is separated into three portions. The annotation got after running Sent2Vec is kept as is for the first portion. Then with this data, the Infersent model is trained and validated against the human annotated validation data. The model with best validation accuracy is saved and used to annotate the second portion of the data. After annotating the second portion of the data, this data is added with the previously trained set and this combined data is used to train the Infersent model again. This time, the model is used to annotate the third portion of the data. After this third run is done, this annotated data is also added to the training set and Infersent is trained one more time with the latest trained dataset. The performance is then tested against the human annotated dataset. For a single bootstrapping, the same neural network architecture is used in all the trials, hence, four bootstrapping runs are made for the experiments.

6 Experimental Setup and Analysis of the Results

Citation linkage is posed here as a sentence level semantic relatedness problem. For this task, word embeddings, sentence embeddings, and supervised semantic similarity measurement models are used. To analyze the performance of the models this section is divided into two parts. Section 6.1 discusses our attempts to optimize the parameter settings of the different word embeddings, sentence embeddings, and Infersent architectures. The results of the models are discussed in detail in Sect. 6.2.

6.1 Network Parameters and Settings

For training fasttext, we have tried both the skip-gram and CBOW architectures. We have also varied the word embedding dimension, number of epochs, and window size. The n-grams at both character and word levels are kept static. Table 3 lists the different hyper-parameter configurations used for training and shows the selected model.

Table 3. Hyper-parameter settings for the training of the fasttext model.

Hyper-parameter	Ranges	Selected
Embedding dimension	200/300/600	300
Epochs	5/7/10	10
Window size	5/10/20	5
Maximum number of subwords	5	5
Learning rate	0.01/0.05/0.1	0.05
Architecture type	Skip-gram/CBOW	Skip-gram

Table 4. The hyper-parameter setting for Sent2Vec sentence embedding architecture

Hyper-parameter	Ranges	Selected
Embedding dimension	200/300/400/500/600/700/800	500
Epochs	5/7/10/15/20	10
Window size	10/20	20
Learning rate (LR)	0.01/0.05/0.1/0.2	0.2
Number of negative samples	10	10
Loss function	Negative sampling/ Hierarchical softmax/ Softmax	Negative sampling
Sampling threshold	0.0001	0.0001

Using the fasttext CBOW architecture, Sent2Vec generates the sentence vectors. For the training of the Sent2Vec architecture, various hyper-parameter values are tested. Table 4 shows the different values. The best sentence embedding hyper-parameter values are shown as well.

For the final Infersent architecture, the learning rate has been set to 0.1. Gradient clipping is used while training. For a decrease in validation set accuracy, the learning rate accuracy is divided by 5. The batch size we have tested with for all the architectures is 50 and the learning rate threshold is set to 0.0001. For the final Multi-Layer Perceptron, the hidden layer dimension is set to 512 and for the input of the LSTMs, 300 dimensional word vectors are used. For the training of the architectures, stochastic gradient descent is used. For the hierarchical CNN, four layers of convolution operations where each layer is followed by a max-pooling are used. Finally, the concatenation of each layer's output is used as the sentence representation. In case of both inner and hierarchical attention mechanism, four context vectors are used to focus on four different parts of the citing and cited sentences and finally, these feature maps are concatenated.

Finally, for the validation set, 800 samples from the human annotated corpus are chosen. As the human annotated dataset is highly imbalanced and only 81 positive samples are present, we randomly chose 20 positive samples to use in the validation set. The test set is separated with 3057 samples which contains 61 positive samples.

6.2   Performance Analysis

This section describes the results obtained throughout the citation linkage task experimentation. To analyze the performance of the word embedding models, pearson and spearman metrics are used. The best performance—pearson and spearman values are 0.519 and 0.403, respectively—is achieved for 300 dimensional word vector representations. For the training of Sent2Vec, different parameter settings are used and for each parameter setting, different cutoff values of cosine similarity between the vectors of the citing and the candidate cited sentence are tried. If the cosine similarity between any sentence pair goes above that cutoff point, that pair is annotated with value 1, else 0. Looking at MCC

Table 5. Performance analysis of different models for the citation linkage task. The models are: M1: Hierarchical ConvNet, M2: Bi-LSTM with max-pooling, M3: Bi-LSTM with inner attention, M4: Bi-LSTM with hierarchical attention. Furthermore, these four methods are also bootstrapped. The notation used is "Boot-" placed in front of each method name. The column headings: TP and FP: true and false positives, respectively; TN and FN: true and false negatives, respectively; P: Precision; R: Recall; F1: F1-score; TNR: True Negative Rate, (True Positive Rate not shown because it is the same as Recall); FPR and FNR: False Positive and Negative Rate, respectively; MCC: Matthews correlation coefficient; Acc.: Accuracy; and BACC: Balanced accuracy.

Model	TP	FP	TN	FN	P	R	F1	FPR	TNR	FNR	MCC	Acc. (in %)	BACC (in %)
M1	44	580	2416	17	0.07	0.72	0.13	0.19	0.81	0.27	0.18	80.5	76.38
M2	53	365	2631	8	0.13	0.87	0.22	0.12	0.88	0.13	0.30	87.81	87.35
M3	54	358	2638	7	0.13	0.89	0.22	0.12	0.88	0.11	0.31	88.07	88.28
M4	55	356	2640	6	0.13	0.90	0.23	0.12	0.88	0.09	0.32	88.17	89.14
Boot-M1	46	576	2420	15	0.07	0.75	0.13	0.19	0.81	0.25	0.20	80.69	78.09
Boot-M2	53	359	2637	8	0.13	0.87	0.22	0.12	0.88	0.13	0.31	88.02	87.45
Boot-M3	54	349	2647	7	0.13	0.89	0.23	0.12	0.88	0.11	0.32	88.38	88.43
Boot-M4	56	339	2657	5	0.14	0.92	0.25	0.11	0.89	0.08	0.34	88.75	90.24
Houngbo et al. [9]	34	995	2001	27	0.03	0.56	0.06	0.33	0.66	0.44	0.07	66.58	61.26
Li et al. [14]	39	779	2217	22	0.05	0.64	0.09	0.26	0.74	0.36	0.12	73.81	68.97

and BACC, which are the best metrics for imbalanced datasets, the best result is achieved for 500 dimensional sentence vectors with cutoff 0.57.

Table 5 shows the performance metrics found for different Infersent architectures. Among the four different Infersent architectures, the best result in terms of test set accuracy is found for Infersent with hierarchical attention mechanism and the worst performance is found for Hierarchical Convnet. For Hierarchical Convnet, not only is its accuracy is lower, but also it captures fewer true positives. For both the Bi-LSTM with hierarchical attention and bootstrapping with Bi-LSTM and hierarchical attention model, the true positive samples captured by the models are the same. Both of these models capture 56 true positive samples. However, the bootstrapping model captures fewer false positive samples compared to the other models. From the results with bootstrapping approaches, it is clear that the performance has improved in all cases, though the improvements are not significant. To compare the performance of the model with the existing models, we have trained those models with our synthetic corpus and then tested against the human annotated corpus. From Table 5 it is clearly visible that the models used here surpass the previous works' performances. In their own work, Houngbo and Mercer [9] reported that their approach captured only 48.5% positive samples out of 81 whereas if their model is trained with our synthetic dataset, it captures 55.73% positive samples, a significant improvement. This result gives proof of the fact that this synthetic dataset has been important for the citation linkage task in the biomedical domain.

7 Conclusions and Future Work

In this study, we have built a synthetic corpus for the citation linkage problem and we have shown that deep-learning models can achieve good performance to ascertain the cited sentences from a cited document given a citing sentence. However, in this work the citation linkage problem has been formulated as a sentence level semantic similarity measurement problem. In future, work can be done to map the citing sentence to a text span in the referenced document which is not limited to one sentence. Moreover, tree structured models can be applied to this task. So, for instance, chemical names containing multiple words can be treated as a single entity, which cannot be achieved bu the linear models used in this study.

References

1. AbuRa'ed, A., Chiruzzo, L., Saggion, H.: What sentence are you referring to and why? Identifying cited sentences in scientific literature. In: Proceedings of the International Conference Recent Advances in Natural Language Processing, RANLP 2017, pp. 9–17 (2017)
2. Bahdanau, D., Cho, K., Bengio, Y.: Neural machine translation by jointly learning to align and translate. arXiv preprint arXiv:1409.0473 (2014)
3. Bengio, Y., Ducharme, R., Vincent, P., Jauvin, C.: A neural probabilistic language model. J. Mach. Learn. Res. **3**, 1137–1155 (2003)
4. Bojanowski, P., Grave, E., Joulin, A., Mikolov, T.: Enriching word vectors with subword information. Trans. Assoc. Comput. Linguist. **5**, 135–146 (2017)
5. Bonin, S., Petrera, F., Niccolini, B., Stanta, G.: PCR analysis in archival postmortem tissues. Mol. Pathol. **56**(3), 184–186 (2003)
6. Conneau, A., Kiela, D., Schwenk, H., Barrault, L., Bordes, A.: Supervised learning of universal sentence representations from natural language inference data. arXiv preprint arXiv:1705.02364 (2017)
7. Garfield, E.: Citation analysis as a tool in journal evaluation. Science **178**(4060), 471–479 (1972)
8. Garzone, M., Mercer, R.E.: Towards an automated citation classifier. In: Hamilton, H.J. (ed.) AI 2000. LNCS (LNAI), vol. 1822, pp. 337–346. Springer, Heidelberg (2000). https://doi.org/10.1007/3-540-45486-1_28
9. Houngbo, H., Mercer, R.E.: Investigating citation linkage with machine learning. In: Mouhoub, M., Langlais, P. (eds.) AI 2017. LNCS (LNAI), vol. 10233, pp. 78–83. Springer, Cham (2017). https://doi.org/10.1007/978-3-319-57351-9_10
10. Houngbo, K.H.: Investigating citation linkage between research articles. Ph.D. thesis, The University of Western Ontario (2017)
11. Huijsmans, C.J., Damen, J., van der Linden, J.C., Savelkoul, P.H., Hermans, M.H.: Comparative analysis of four methods to extract DNA from paraffin-embedded tissues: effect on downstream molecular applications. BMC Res. Notes **3**(1), 239 (2010)
12. Kayser, K., Stute, H., Lübcke, J., Wazinski, U.: Rapid microwave fixation–a comparative morphometric study. Histochem. J. **20**(6–7), 347–352 (1988). https://doi.org/10.1007/BF01002728

13. Kiros, R., et al.: Skip-thought vectors. In: Advances in Neural Information Processing Systems, pp. 3294–3302 (2015)

14. Li, L., et al.: Computational linguistics literature and citations oriented citation linkage, classification and summarization. Int. J. Digit. Libr. **19**(2), 173–190 (2017). https://doi.org/10.1007/s00799-017-0219-5

15. Li, L., Zhang, Y., Mao, L., Chi, J., Chen, M., Huang, Z.: CIST@CLSciSumm-17: multiple features based citation linkage, classification and summarization. In: BIRNDL@ SIGIR (2), pp. 43–54 (2017)

16. Li, L., et al.: CIST@CLSciSumm-19: automatic scientific paper summarization with citances and facets. In: BIRNDL 2019 (2019)

17. Liu, Y., Sun, C., Lin, L., Wang, X.: Learning natural language inference using bidirectional LSTM model and inner-attention. arXiv preprint arXiv:1605.09090 (2016)

18. Logeswaran, L., Lee, H.: An efficient framework for learning sentence representations. arXiv preprint arXiv:1803.02893 (2018)

19. Mikolov, T., Chen, K., Corrado, G., Dean, J.: Efficient estimation of word representations in vector space. arXiv preprint arXiv:1301.3781 (2013)

20. Mikolov, T., Sutskever, I., Chen, K., Corrado, G.S., Dean, J.: Distributed representations of words and phrases and their compositionality. In: Advances in Neural Information Processing Systems, pp. 3111–3119 (2013)

21. Neculoiu, P., Versteegh, M., Rotaru, M.: Learning text similarity with Siamese recurrent networks. In: Proceedings of the 1st Workshop on Representation Learning for NLP, pp. 148–157 (2016)

22. Pagliardini, M., Gupta, P., Jaggi, M.: Unsupervised learning of sentence embeddings using compositional n-Gram features. In: Proceedings of the 2018 Conference of the North American Chapter of the Association for Computational Linguistics: Human Language Technologies, vol. 1 (Long Papers), pp. 528–540 (2018)

23. Pakhomov, S., McInnes, B., Adam, T., Liu, Y., Pedersen, T., Melton, G.B.: Semantic similarity and relatedness between clinical terms: an experimental study. In: AMIA Annual Symposium Proceedings 2010, pp. 572–576 (2010)

24. Palau, R.M., Moens, M.F.: Argumentation mining: the detection, classification and structure of arguments in text. In: Proceedings of the 12th International Conference on Artificial Intelligence and Law, pp. 98–107. ACM (2009)

25. Radev, D.R., Jing, H., Budzikowska, M.: Centroid-based summarization of multiple documents: Sentence extraction, utility-based evaluation, and user studies. In: NAACL-ANLP 2000 Workshop: Automatic Summarization (2000)

26. Wang, Y., et al.: High quality copy number and genotype data from FFPE samples using molecular inversion probe (MIP) microarrays. BMC Med. Genomics **2**(1), 8 (2009)

27. Wolffs, P., Grage, H., Hagberg, O., Rådström, P.: Impact of DNA polymerases and their buffer systems on quantitative real-time PCR. J. Clin. Microbiol. **42**(1), 408–411 (2004)

28. Yang, Z., Yang, D., Dyer, C., He, X., Smola, A., Hovy, E.: Hierarchical attention networks for document classification. In: Proceedings of the Conference of the North American Chapter of the Association for Computational Linguistics: Human Language Technologies, pp. 1480–1489 (2016)

29. Zhao, H., Lu, Z., Poupart, P.: Self-adaptive hierarchical sentence model. In: Twenty-Fourth International Joint Conference on Artificial Intelligence, pp. 4069–4076 (2015)

Improving Classification Using Topic Correlation and Expectation Propagation

Xavier Sumba$^{(\boxtimes)}$ and Nizar Bouguila

Concordia University, Montreal, QC, Canada
fsumbat@mail.concordia.ca, nizar.bouguila@concordia.ca

Abstract. Probabilistic topic models are broadly used to infer meaningful patterns of words over a mixture of latent topics that are commonly used for statistical analyses or as a proxy for supervised tasks. However, models such as Latent Dirichlet Allocation (LDA) assume independence between topic proportions due to the nature of the Dirichlet distribution; this effect is captured with other distributions such as the logistic normal distribution, resulting in a complex model. In this paper, we develop a probabilistic topic model using the generalized Dirichlet distribution (LGDA) in order to capture topic correlation while maintaining conjugacy. We make use of Expectation Propagation to approximate the posterior, resulting in a model that achieves more accurate inferences compared to variational inference. We evaluate the convergence of EP compared with the classical LDA by comparing the approximation to the marginal distribution. We show the obtained topics by LGDA and evaluate its predictive performance in two text classification tasks, outperforming the vanilla LDA.

Keywords: Topic modelling · Expectation propagation · Topic classification

1 Introduction

Topic models are among the best-known models to automatically organize documents. Especially, probabilistic topic models [3] have received great attention from the research community. They use statistical methods for uncovering topics from a collection of documents and are commonly used for annotating or organizing documents. Latent Dirichlet Allocation (LDA) [7] was proposed as an improvement of probabilistic Latent Semantic Analysis [16] and has become the most popular topic model since its introduction. Many variations have been introduced leading to applications [9] in a variety of domains. These applications have been possible due to the flexibility of the LDA model. LDA can be extended with other more complex models and adapted to a specific problem.

It is noticeable that the applicability of topic models are endless and due to digitalization, there is an exponential growth of information available online. Thus organizing and annotating those documents can be overwhelming and

© Springer Nature Switzerland AG 2020
C. Goutte and X. Zhu (Eds.): Canadian AI 2020, LNAI 12109, pp. 496–507, 2020.
https://doi.org/10.1007/978-3-030-47358-7_51

obtaining better topic models can substantially ease these tasks. For doing so, a lot of emphasis has been put in approximate inference since these models need to compute a posterior distribution which is intractable. Commonly sampling methods or deterministic approaches are used to deal with this intractable integral. For instance, Markov Chain Monte Carlo, a sampling method, is usually implemented using a Gibbs sampling algorithm [14,24]. Similarly, there are deterministic approaches such as Expectation Propagation (EP) and [20] and Variational Inference (VI) [4]. VI has been an active area of research having variations that are much faster and scale to great amounts of data by using stochastic optimization [15] or Autoencoding variational Bayes [23] that uses neural networks for approximating the posterior distribution.

In this work, we introduce a variation of LDA that models topic correlations leveraging the advantages of EP for approximating the posterior distribution. Topic correlations are important when, for example, a document about sports has content about soccer and athletics but lacks information about basketball. This correlation cannot be captured by LDA for the intrinsic nature of the Dirichlet distribution. However, the Generalized Dirichlet (GD) distribution is a generalization of the Dirichlet distribution that solves the limitations of its negative covariance matrix. It has been used successfully with count data [8], and apart from solving the restrictions of the Dirichlet distribution, maintains conjugacy in the LDA model. EP factorizes the joint distribution for later combining each factor with an approximation, and as a result, obtaining an overall approximation of the posterior distribution. This is appealing for models such as LDA since data partition allows EP to be distributed and scale to large datasets. In addition, EP has shown to obtain a better approximation than VI [19], which are biased.

The rest of this paper is organized as follows. First, Sect. 2 revisits the core methods upon our work is built on and related work in Sect. 3. Next, in Sect. 4, we outline the LGDA model, describe the expectation propagation approach, and derive a complete learning approach. Section 5 describes our experimental setup and evaluation of our proposed method. Finally, we conclude the paper in Sect. 6.

2 Background

This section gives groundwork of the main components upon our work is built on and introduces the notation used throughout this work.

2.1 Latent Dirichlet Allocation

LDA [7] is the most popular probabilistic topic model and since its introduction, it has become the most conventional and known unsupervised topic model for the discovery of latent topics. It can be described as a generative model, meaning that uses a probabilistic approach allowing to generate documents.

Following this generative process, each topic β_k is a distribution over a vocabulary V and a document has a mixture of topics $\beta = (\beta_1, \ldots, \beta_K)$, where K is the number of topics, which has to be known beforehand. All documents in the corpus share the topics β, but each document can express a topic in a different proportion θ_d. The generative process continues by drawing a word $w_{d,n}$ from topic $\beta_{z_{d,n}}$, where $z_{d,n}$ is the topic assignment for the word $w_{d,n}$. The topic assignment $z_{d,n}$ is drawn from a distribution over the document proportion θ_d.

Commonly, the document proportion is modeled with a Dirichlet distribution, and the topics and words with a Multinomial distribution. However, the evidence $p(w)$ of this model is intractable due to the coupling of θ and β. Thus, the posterior is frequently approximated with VI using the mean-field variational family, and by integrating out the latent variables, LDA is capable to infer the topic structure of a set of documents.

2.2   Expectation Propagation

EP [20] is a generalization of assumed density filtering (ADF), which is a one-pass sequential method. Unlike ADF, EP reuses data points to perform iterative refinements. In other words, EP handles partitioned data and combines partitions iteratively through message passing. In fact, EP is more computationally efficient than Markov Chain Monte Carlo [22] and has shown to be more accurate than VI [20,21].

Having the latent variable θ, EP approximates a target distribution $p(\theta \mid \mathcal{X})$, which is commonly the posterior, with a global approximation $q(\theta)$ that belongs to the exponential family. The target distribution must be factorizable such that the posterior can be split in V sites $p(\theta) \propto t_0(\theta) \prod_w^V t_w(\theta)$; the initial site t_0 is commonly represented with the prior distribution and the remaining t_w sites represent the contribution of each term to the likelihood. The approximating distribution must admit a similar factorization, i.e $q(\theta) \propto \prod_w^V \tilde{t}_w(\theta)$. Each approximating site has to be initialized and belongs to the exponential family. Consequently, each site is refined to create a cavity distribution by dividing the global approximation over the current approximate site $q^{\backslash w}(\theta) \propto q(\theta)/\tilde{t}_w(\theta)$.

Additionally, in order, to approximate each site, a new tilted distribution is introduced using the cavity distribution and the current site $q_w^*(\theta) \propto t_w(\theta)q^{\backslash w}(\theta)$. Subsequently, a new posterior is found by minimizing the Kullback Leibler divergence $D_{KL}(q_w^*(\theta) \parallel q^{new}(\theta))$ such that $\tilde{t}_w(\theta) \approx t_w(\theta)$. This minimization is equivalent to match the moments of those distributions [2,21]. Finally, the revised approximate site is updated by removing the remaining terms from the current approximation $\tilde{t}_w(\theta) \propto q^{new}(\theta)/q^{\backslash w}(\theta)$.

2.3   Generalized Dirichlet Distribution

A Dirichlet distribution can only capture negative correlations due to its negative covariance matrix. Additionally, when it is used as a prior, posing only one degree of freedom which hinders the ability to introduce variance information to each component of the random vector. Therefore the GD distribution [11,25] was introduced to alleviate these problems. It has positive parameters $\alpha = \alpha_1, \ldots, \alpha_K$ and $\kappa = \kappa_1, \ldots, \kappa_K$, and a random vector $\theta = \theta_1, \ldots, \theta_K$, where $\sum_k^K \theta_k \leq 1$ and $0 < \theta_k < 1$ for $k = 1, \ldots, K$. GD's PDF is illustrated in Eq. 1.

$$p(\theta \mid \alpha, \kappa) = \prod_k^K \frac{\Gamma(\alpha_k + \kappa_k)}{\Gamma(\alpha_k)\Gamma(\kappa_k)} \theta_k^{\alpha_k - 1} (1 - \sum_{j=1}^k \theta_j)^{\gamma_k}, \tag{1}$$

where $\gamma_k = \kappa_k - \alpha_{k+1} - \kappa_{k+1}$ for $k = 1, \ldots, K-1$ and $\gamma_K = \kappa_K - 1$; $\Gamma(\cdot)$ is the Gamma function. The mean and variance are shown in Eqs. 2 and 3 respectively.

$$\mu_k = \frac{\alpha_k}{\alpha_k + \kappa_k} \prod_{j=1}^{k-1} \frac{\kappa_j}{\alpha_j + \kappa_j} \tag{2}$$

$$Var(\theta_k) = \mu_k \left(\frac{\alpha_k + 1}{\alpha_k + \kappa_k + 1} \prod_{j=1}^{k-1} \frac{\kappa_j + 1}{\alpha_j + \kappa_j + 1} - \mu_k \right) \tag{3}$$

Additionally, Eq. 4 illustrates the covariance matrix, which has a more general structure. For instance, the Dirichlet distribution is just an special case of the GD distribution when $\kappa_k = \alpha_{k+1} + \kappa_{k+1}$.

$$Cov(\theta_m, \theta_n) = \mu_n \left(\frac{\alpha_m}{\alpha_m + \kappa_m + 1} \prod_{j=1}^{m-1} \frac{\kappa_j + 1}{\alpha_j + \kappa_j + 1} - \mu_m \right) \tag{4}$$

It is noteworthy that the GD distribution has K degrees of freedom which makes it more flexible and suitable for modeling correlated topics.

3 Related Work

The work in [19] proposes an inference alternative using Expectation Propagation (EP) for LDA model that does not bound the marginal probability as in [7] and leads to higher accuracy. However, in general, the LDA model is incapable of capturing topic correlation due to the limitation of the Dirichlet distribution for the document-topic probability. The Correlated Topic Model (CTM) [5] is proposed in order to capture a correlation of the topic proportions using a logistic normal distribution which results in a complicated model since the conjugacy with the Multinomial distribution is lost. Thus, [10] showed that the CTM can be modeled using a Generalized Dirichlet distribution (denominated GD-LDA or LGDA), maintaining conjugacy and leading to faster inference. Finally, the work of [1] and [17] propose inference alternatives to the LGDA model using collapsed variational bayes inference and variational bayes inference respectively.

4 Latent Generalized Dirichlet Allocation

This section provides an overview of the LGDA model and an approach to perform inference and estimation using expectation propagation.

4.1 Model

LGDA is a generative probabilistic model for count data. The generative process is similar to the vanilla LDA [7] with the difference that document-topic proportions $\boldsymbol{\theta}_d$ are drawn from a GD distribution.

1. Choose $\boldsymbol{\theta} \sim GenDir(\boldsymbol{\alpha}, \boldsymbol{\kappa})$
2. For each of the N words $\boldsymbol{w}_n$:
   (a) Choose a topic $\boldsymbol{z}_n \sim Multinomial(\boldsymbol{\theta})$
   (b) Choose a word $\boldsymbol{w}_n$ from $p(\boldsymbol{w}_n \mid \boldsymbol{z}_n, \boldsymbol{\beta})$

The model has the corpus level hyperparameters $\boldsymbol{\alpha}$ and $\boldsymbol{\kappa}$ for the prior GD distribution and $\boldsymbol{\beta}$ for the topics. Words are observed and represented with $\boldsymbol{w}$.

Given the hyperparameters, the joint distribution for a document of the model is given in Eq. 5.

$$p(\boldsymbol{\theta}, z, \boldsymbol{w} \mid \boldsymbol{\alpha}, \boldsymbol{\kappa}, \boldsymbol{\beta}) = p(\boldsymbol{\theta} \mid \boldsymbol{\alpha}, \boldsymbol{\kappa}) \prod_{n=1}^{N} p(z_n \mid \boldsymbol{\theta}) p(\boldsymbol{w}_n \mid z_n, \boldsymbol{\beta}) \qquad (5)$$

We can impose that each word among the documents belongs to a fixed vocabulary of size V. Then, because we assume there are K fixed topics in the corpus, and we are using a GD distribution prior, the word-topic probability matrix β is $K + 1 \times V$. Additionally, since we are dealing with probabilities, the topic proportions have to sum up to one $\sum_{k=1}^{K+1} \theta_k = 1$. It is evident that $\boldsymbol{\theta}$ is a different sample for each document, and as a result, each document exhibits a different topic proportion.

The topic assignment dictates which component to select from the topic mixture such that $p(z_n|\boldsymbol{\theta}) = \theta_{z_n}$. Similarly, a word topic probability is selected from β in a manner that $p(w_n \mid z_n, \boldsymbol{\beta}) = \beta_{z_n, w_n}$. Thus, we rewrite the joint distribution as a sum over the topic assignments z_n, obtaining Eq. 6.

$$p(\boldsymbol{\theta}, \boldsymbol{w} \mid \boldsymbol{\alpha}, \boldsymbol{\kappa}, \boldsymbol{\beta}) = p(\boldsymbol{\theta} \mid \boldsymbol{\alpha}, \boldsymbol{\kappa}) \prod_{n=1}^{N} \sum_{k=1}^{K+1} \theta_k \beta_{k, w_n} \qquad (6)$$

Each document has length N yet we can use a fixed vocabulary to represent the words over the collection of documents, and because of the ex-changeability assumption [7], the order of words is not relevant. Thus, the joint for a fixed vocabulary is represented in Eq. 7.

$$p(\boldsymbol{\theta}, \boldsymbol{w} \mid \boldsymbol{\alpha}, \boldsymbol{\kappa}, \boldsymbol{\beta}) = p(\boldsymbol{\theta} \mid \boldsymbol{\alpha}, \boldsymbol{\kappa}) \prod_{w=1}^{V} \left(\sum_{k}^{K+1} \theta_k \beta_{k, w} \right)^{n_w}, \qquad (7)$$

where n_w is the number of times that word w appears in the document.

Finally, the marginal probability of a document is obtained by integrating out the mixing topics $\boldsymbol{\theta}$ such that $p(\boldsymbol{w}) = \int p(\boldsymbol{\theta}, \boldsymbol{w}) d\boldsymbol{\theta}$. Now, it is more evident the coupling between $\boldsymbol{\theta}$ and β, which makes the posterior intractable [12]. Thus, in this work, we will make use of EP to approximate the posterior distribution. For instance, the probability of a collection of documents C is shown in Eq. 8.

$$p(C \mid \alpha, \kappa, \beta) = \prod_{d=1}^{D} \int p(\theta_d \mid \alpha, \kappa) \prod_{w=1}^{V} \left(\sum_{k=1}^{K+1} \theta_{d,k} \beta_{k,w} \right)^{n_{d,w}} d\theta_d \qquad (8)$$

4.2   Inference

As it is common in any Bayesian setting, the posterior distribution is defined by the hidden variables given the observed words $p(\theta, \mid w, \alpha, \kappa, \beta) \propto p(\theta, w \mid \alpha, \kappa, \beta)$. Hence, LGDA's evidence is intractable. Thus, we generate an approximation to $p(w)$ using EP since it has been shown that generates more accurate approximations [19,20]; unlike VI that tends to create biased approximations. Then, EP can provide an estimate for both the posterior and evidence, and sites can be defined as show in Eq. 9.

$$t_w(\theta) = \sum_{k=1}^{K+1} \theta_k \beta_{k,w} \qquad (9)$$

So, the posterior distribution can be factorized as shown in Eq. 10, where we use a GD distribution as prior.

$$p(\theta, \mid w, \alpha, \kappa, \beta) \propto p(\theta \mid \alpha, \kappa) \prod_{w=1}^{V} t_w(\theta)^{n_w} \qquad (10)$$

The approximate sites have a product form (Eq. 11). The parameter ϕ is a matrix $V \times K + 1$ and s_w is a normalization constant for the site w.

$$\tilde{t}_w(\theta) = s_w \prod_{k=1}^{K+1} \theta_k^{\phi_{w,k}} \qquad (11)$$

By making use of the approximate sites and the GD prior, an approximate posterior distribution can be calculated. Notice that because of conjugacy, we obtain an approximate GD distribution (Eq. 12)

$$q(\theta \mid \alpha', \kappa') \propto p(\theta \mid \alpha, \kappa) \prod_{w=1}^{V} \tilde{t}_w(\theta)^{n_w}, \qquad (12)$$

where $\gamma'_k = \kappa'_k - \alpha'_{k+1} - \kappa'_{k+1}$ for $k = 1, \ldots, K - 1$ and $\gamma'_K = \kappa'_K + \sum_{w=1}^{V} \phi_{w,K+1} n_w - 1$, and its parameters are shown in Eqs. 13 and 14 respectively.

$$\alpha'_k = \alpha_k + \sum_{w=1}^{V} \phi_{w,k} n_w \text{ for } k = 1, \ldots, K \qquad (13)$$

$$\kappa'_k = \kappa_k + \sum_{j=k+1}^{K+1} \sum_{w=1}^{V} \phi_{w,j} n_w \text{ for } k = 1, \ldots, K \tag{14}$$

In order to update the approximate site $\tilde{t}_w(\boldsymbol{\theta})$, a cavity distribution is introduced by removing it from the approximate posterior $q^{\backslash w}(\boldsymbol{\theta}) = q(\boldsymbol{\theta})/\tilde{t}_w(\boldsymbol{\theta})$. We obtain a cavity distribution that is another GD distribution with parameters $\boldsymbol{\alpha}^{\backslash w}$ and $\boldsymbol{\kappa}^{\backslash w}$ shown in Eqs. 15 and 16, where $\gamma_k^{\backslash w} = \kappa_k^{\backslash w} - \alpha_{k+1}^{\backslash w} - \kappa_{k+1}^{\backslash w}$ for $k = 1, \ldots, K - 1$ and $\gamma_K^{\backslash w} = \kappa_K^{\backslash w} + \sum_{w=1}^{V} \phi_{w,K+1} n_w - \phi_{w,K+1} - 1$.

$$\alpha_k^{\backslash w} = \alpha'_k - \phi_{w,k} \text{ for } k = 1, \ldots, K \tag{15}$$

$$\kappa_k^{\backslash w} = \kappa'_k - \sum_{j=k+1}^{K+1} \phi_{w,j} \text{ for } k = 1, \ldots, K \tag{16}$$

Next, the tilted posterior distribution can be obtained by using the site $t_w(\boldsymbol{\theta})$ and the cavity distribution such that

$$q_w^*(\boldsymbol{\theta}) = \frac{1}{z_w} t_w(\boldsymbol{\theta}) q^{\backslash w}(\boldsymbol{\theta}), \tag{17}$$

where the normalization constant $z_w(\boldsymbol{\alpha}^{\backslash w}, \boldsymbol{\kappa}^{\backslash w})$ is shown in Eq. 18.

$$z_w = \beta_{K+1,w} + \sum_{k=1}^{K} (\beta_{k,w} - \beta_{K+1,w}) \frac{\alpha_k^{\backslash w}}{\alpha_k^{\backslash w} + \kappa_k^{\backslash w}} \prod_{j=1}^{k-1} \frac{\kappa_j^{\backslash w}}{\alpha_j^{\backslash w} + \kappa_j^{\backslash w}} \tag{18}$$

Once found the tilted distribution, we proceed to match the moments with the approximate distribution in order to approximate the current site t_w with the approximate site $\tilde{t}_w$. Since moment matching is equivalent to minimizing the KL divergence, we obtain an optimal distribution $q^{new}(\boldsymbol{\theta})$ with parameters $\boldsymbol{\alpha}^{new}$ and $\boldsymbol{\kappa}^{new}$ that can be obtained from the system of equations shown in Eqs. 19 and 20. The values of the parameters can be obtained with fixed-point iteration method.

$$\Psi(\alpha_k^{new}) - \Psi(\alpha_k^{new} + \kappa_k^{new}) = \frac{1}{z_w} \frac{\partial z_w}{\partial \alpha_k^{\backslash w}} + \Psi(\alpha_k^{\backslash w}) - \Psi(\alpha_k^{\backslash w} + \kappa_k^{\backslash w}) \tag{19}$$

$$\Psi(\kappa_k^{new}) - \Psi(\alpha_k^{new} + \kappa_k^{new}) = \frac{1}{z_w} \frac{\partial z_w}{\partial \kappa_k^{\backslash w}} + \Psi(\kappa_k^{\backslash w}) - \Psi(\alpha_k^{\backslash w} + \kappa_k^{\backslash w}) \tag{20}$$

After matching the moments, the approximate site can be updated using the tilted distribution. In order to accomplish faster convergence and obtain a better representation of the global approximation, we use damping [13] with a step size μ. Notice when $\mu = 1$, no damping is applied. Hence, the factor updates are expressed in Eqs. 21 and 22.

$$s'_w = z_w \prod_{k=1}^{K} \frac{\Gamma(\alpha_k^{new} + \kappa_k^{new})}{\Gamma(\alpha_k^{new})\Gamma(\kappa_k^{new})} \frac{\Gamma(\alpha_k^{\backslash w})\Gamma(\kappa_k^{\backslash w})}{\Gamma(\alpha_k^{\backslash w} + \kappa_k^{\backslash w})} \tag{21}$$

$$\phi'_{w,k} = \mu(\alpha_k^{new} - \alpha_k^{\backslash w}) + (1 - \mu)\phi_{w,k}$$

$$\phi'_{w,K+1} = \frac{\mu}{2}\left(\kappa_K^{new} - \kappa_K^{\backslash w} + \phi_{w,K+1} - \sum_w \phi_{w,K+1}n_w\right) + (1 - \mu)\phi_{w,K+1} \tag{22}$$

Finally, we incorporate the contribution of the optimized site in the global approximate distribution $q^*(\boldsymbol{\theta}_d)$ by employing the cavity distribution and the optimal site; the updates are shown in Eq. 23.

$$\alpha_k'^{new} = \alpha_k' + n_w(\phi'_{w,k} - \phi_{w,k})$$

$$\kappa_k'^{new} = \kappa_k' + n_w\left(\sum_{j=k+1}^{K+1} \phi'_{w,j} - \phi_{w,j}\right) \tag{23}$$

4.3   Parameter Estimation

Finally, we obtain estimates of the model parameters by maximizing the ELBO with respect to α, κ, and β. Thus, we can write the ELBO as shown in Eq. 24.

$$\mathcal{L}(\boldsymbol{\alpha}, \boldsymbol{\kappa}, \boldsymbol{\beta}) = \sum_{d=1}^{D} \mathbb{E}_q\left[\log p(\boldsymbol{\theta}_d)\right] + \sum_{d=1}^{D} \mathbb{E}_q\left[\sum_{w=1}^{V} n_{d,w} \log(\sum_{k=1}^{K+1} \theta_{d,k}\beta_{k,w})\right] + C \tag{24}$$

Maximizing this expression with respect to α_k and κ_k lead us to the following system of equations (Eq. 25, which has no closed-form and can be approximated using Newton's method [18].

$$D\left[\Psi(\alpha_k + \kappa_k) - \Psi(\alpha_k)\right] = \sum_{d}^{D}\left[-\Psi(\alpha'_{d,k}) + \Psi(\alpha'_{d,k} + \kappa'_{d,k})\right]$$

$$D\left[\Psi(\alpha_k + \kappa_k) - \Psi(\kappa_k)\right] = \sum_{d}^{D}\left[-\Psi(\kappa'_{d,k}) + \Psi(\alpha'_{d,k} + \kappa'_{d,k})\right] \tag{25}$$

Next, we find the optimal topics by maximizing the ELBO w.r.t. $\beta_{k,w}$ (see Eq. 26) where we find an expectation that can be approximated using second-order Taylor expansion about $\mathbb{E}[\theta_d]$.

$$\beta_{k,w} \propto \sum_{d}^{D} n_{d,w}\mathbb{E}_q\left[\frac{\theta_{d,k}\beta_{k,w}}{\sum_{k=1}^{K+1} \theta_{d,k}\beta_{k,w}}\right] \tag{26}$$

5 Results

In this section, we test convergence by comparing the lower bounds and evaluate the LGDA model on a text classification task in order to evaluate the predictive performance due that correlation can lead to better predictive distributions.

Dataset. We use the Reuters-21578[1] corpus which is a collection of labeled newswire articles. The dataset consists of 21,578 documents, including documents without topics and typographical errors. Thus, we use the top-6 categories following the experiment performed by [1], resulting in approximately 9,000 documents. Table 1 summarizes the selected categories and number of documents per class. We preprocess the selected corpus by lowercasing words and removing punctuation. Next, words in third person are changed to first person and tenses are changed to present by using a standard lemmatizer. Stop words and words with less than three characters are filtered. Finally, we use a stemmer to reduce all the remaining words to its root form and tokenize to form the vocabulary.

Table 1. Classes and number of documents extracted from Reuters dataset

Category	Num. docs
acq	2369
crude	578
earn	3964
grain	582
interest	478
money-fx	717

Models. We compare the performance of LGDA-Expectation Propagation with LDA since it is the most commonly used topic model and has not only similar conjugacy properties but also a similar generative process. We use an implementation of LDA with variational Bayes inference[2].

Experiment Description. As noticed by [1], LGDA has a similar predictive power as LDA yet LGDA is better at discriminating related categories due that topics are correlated. Thus, we use train/test splits as specified in [1] and build two classifiers, a supervised LASSO regression with a Multinomial and Bernoulli distribution for multiclass and binary classification. We use the full dataset for the multiclass classifier which has a vocabulary size of $V = 10,123$ words, and similarly for the binary classifier, we use two related categories (i.e. *interest* and *money-fx*) resulting in a vocabulary size of $V = 4,233$ words. We use the number of topics K reported in [1].

---

[1] http://www.daviddlewis.com/resources/testcollections/reuters21578/.
[2] We use an implementation of LDA where no smoothing is applied [6].

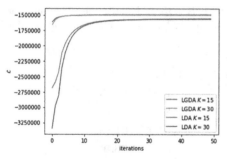

Fig. 1. Comparison of LGDA-EP and LDA in terms of evidence lower bound for $K = 15$ and $K = 30$ topics.

Table 2. Top five words on the full dataset with vocabulary size $10,123$ and $K = 15$ topics. Results of 5-fold cross-validation for aspect-specific sentiment extraction using different feature combinations.

LGDA-EP topics			
bank	dlrs	stock	say
market	billion	record	share
say	loss	april	company
billion	profit	dividend	dlrs
money	year	prior	offer

Topic Interpretability. We train LGDA-EP and LDA and evaluate the lower bounds using the full dataset with $K = 15$ and $K = 30$ topics as shown in Fig. 1. For EP, we initialize the approximate factors $\tilde{t}_w = 1$, and for LDA-VI, we initialize the variational parameters randomly. We can notice that LGDA-EP not only converged considerably faster but also reaches a better solution by looking at the approximate evidence. terms of ELBO.

We next look at the learned topics. Table 2 displays the 4 most used topics for LGDA-EP, as given by the average of the topic proportions θ_d. LGDA provide interpretable topics.

Topic Classification. We evaluate the predictive power of LGDA-EP and compare the obtained results with LGDA using variational Bayes inference (LGDA-VI) [1] and LDA [6]. We evaluate the models' performance in terms of accuracy. First, we build a binary classifier in order to evaluate the ability of LGDA to discriminate similar categories. We select the optimal number of topics as proposed by [1]. Table 3 illustrates the results of binary classification for the categories *money-fx* and *interest*. As expected LGDA is slightly better at discriminating similar categories obtaining 71% of accuracy.

Consequently, we build a classifier using the full-dataset, and as expected LGDA-EP provides similar or better predictive performance than the vanilla LDA as shown in Table 3. Figure 2 illustrates the confusion matrix for both LDA and LGDA with $K = 15$ topics. It is noticeable that LGDA is better not only at discriminating distinct categories but also similar categories which accounts for the accuracy's jump.

Table 3. Results for binary classification with K = 15 and multi-class classification with K = 15 and K = 30. Comparison using accuracy. VI: variational inference model; EP: expectation propagation.

Models	Accuracy		
	money-fx vs. interest	*All classes*	
	K = 15	K = 15	K = 30
LDA	69%	81%	78.8%
LGDA-VI [1]	70%	64.9%	64.8%
LGDA-EP	**71%**	**84%**	**78.9%**

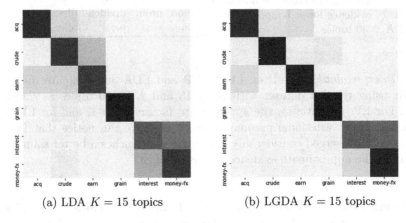

(a) LDA $K = 15$ topics   (b) LGDA $K = 15$ topics

Fig. 2. Confusion matrix

6   Conclusions

In this paper, we propose the use of Expectation Propagation (EP) for the Latent Generalized Dirichlet allocation model to learn a mixture of latent topics over documents and a vocabulary while maintaining topic correlation. We make use of EP in order to have accurate approximations since as opposed to variational inference, EP doesn't need to be bounded to create an approximation to the posterior. We additionally develop a method for parameter estimation. We evaluate topic interpretability by looking at the resulting topics and the predictive power of LGDA-EP showing the efficacy of the proposed method and showing superior results to the traditional LDA.

References

1. Bakhtiari, A.S., Bouguila, N.: A variational bayes model for count data learning and classification. Eng. Appl. Artif. Intell. **35**, 176–186 (2014)
2. Bishop, C.M.: Pattern Recognition and Machine Learning. Springer, Heidelberg (2006)

3. Blei, D.M.: Probabilistic topic models. Commun. ACM **55**(4), 77–84 (2012). https://doi.org/10.1145/2133806.2133826
4. Blei, D.M., Kucukelbir, A., McAuliffe, J.D.: Variational inference: a review for statisticians. J. Am. Statist. Assoc. **112**(518), 859–877 (2017)
5. Blei, D.M., Lafferty, J.D., et al.: A correlated topic model of science. Ann. Appl. Statist. **1**(1), 17–35 (2007)
6. Blei, D.M., Ng, A.Y., Jordan, M.I.: Latent Dirichlet allocation. In: Advances in Neural Information Processing Systems, pp. 601–608 (2002)
7. Blei, D.M., Ng, A.Y., Jordan, M.I.: Latent Dirichlet allocation. J. Mach. Learn. Res. **3**(Jan), 993–1022 (2003)
8. Bouguila, N.: Clustering of count data using generalized Dirichlet multinomial distributions. IEEE Trans. Knowl. Data Eng. **20**(4), 462–474 (2008)
9. Boyd-Graber, J., Hu, Y., Mimno, D., et al.: Applications of topic models. Found. Trends® Inf. Retrieval **11**(2–3), 143–296 (2017)
10. Caballero, K.L., Barajas, J., Akella, R.: The generalized Dirichlet distribution in enhanced topic detection. In: Proceedings of the 21st ACM International Conference on Information and Knowledge Management, pp. 773–782. ACM (2012)
11. Connor, R.J., Mosimann, J.E.: Concepts of independence for proportions with a generalization of the Dirichlet distribution. J. Am. Statist. Association. **64**(325), 194–206 (1969)
12. Dickey, J.M.: Multiple hypergeometric functions: probabilistic interpretations and statistical uses. J. Am. Statist. Assoc. **78**(383), 628–637 (1983)
13. Gelman, A., Vehtari, A., Jylänki, P., Robert, C., Chopin, N., Cunningham, J.P.: Expectation propagation as a way of life. arXiv preprint arXiv:1412.4869 157 (2014)
14. Griffiths, T.L., Steyvers, M.: Finding scientific topics. Proc. Natl. Acad. Sci. **101**(suppl 1), 5228–5235 (2004)
15. Hoffman, M.D., Blei, D.M., Wang, C., Paisley, J.: Stochastic variational inference. J. Mach. Learn. Res. **14**(1), 1303–1347 (2013)
16. Hofmann, T.: Unsupervised learning by probabilistic latent semantic analysis. Mach. Learn. **42**(1–2), 177–196 (2001)
17. Ihou, K.E., Bouguila, N.: Variational-based latent generalized Dirichlet allocation model in the collapsed space and applications. Neurocomputing **332**, 372–395 (2019)
18. Minka, T.: Estimating a Dirichlet distribution (2000)
19. Minka, T., Lafferty, J.: Expectation-propagation for the generative aspect model. In: Proceedings of the Eighteenth Conference on Uncertainty in Artificial Intelligence, pp. 352–359. Morgan Kaufmann Publishers Inc. (2002)
20. Minka, T.P.: Expectation propagation for approximate Bayesian inference. In: Proceedings of the Seventeenth Conference on Uncertainty in Artificial Intelligence, pp. 362–369. Morgan Kaufmann Publishers Inc. (2001)
21. Minka, T.P.: A family of algorithms for approximate Bayesian inference. Ph.D. thesis, Massachusetts Institute of Technology (2001)
22. Neal, R.M.: Probabilistic inference using Markov chain Monte Carlo methods (1993)
23. Srivastava, A., Sutton, C.: Autoencoding variational inference for topic models. arXiv preprint arXiv:1703.01488 (2017)
24. Steyvers, M., Griffiths, T.: Probabilistic topic models. Handb. Latent Semant. Anal. **427**(7), 424–440 (2007)
25. Wong, T.T.: Generalized dirichlet distribution in Bayesian analysis. Appl. Math. Comput. **97**(2–3), 165–181 (1998)

A Scheme for Generating a Dataset for Anomalous Activity Detection in IoT Networks

Imtiaz Ullah[✉] and Qusay H. Mahmoud

Department of Electrical, Computer and Software Engineering, Ontario Tech University, Oshawa, ON L1G 0C5, Canada
{imtiaz.ullah,qusay.mahmoud}@ontariotechu.net

Abstract. The exponential growth of the Internet of Things (IoT) devices provides a large attack surface for intruders to launch more destructive cyber-attacks. The intruder aimed to exhaust the target IoT network resources with malicious activity. New techniques and detection algorithms required a well-designed dataset for IoT networks. Firstly, we reviewed the weaknesses of various intrusion detection datasets. Secondly, we proposed a new dataset namely IoTID20 generated dataset from [1]. Thirdly we provide a significant set of features with their corresponding weights. Finally, we propose a new detection classification methodology using the generated dataset. Our proposed IoT botnet dataset will provide a reference point to identify anomalous activity across the IoT networks. The IoT Botnet dataset can be accessed from [2]. The new IoTID20 dataset will provide a foundation for the development of new intrusion detection techniques in IoT networks.

Keywords: Internet of Things · Intrusion detection · IDS · Dataset · DoS · Anomaly detection system · Flow-based intrusion detection · Infiltration · Cybersecurity

1 Introduction

Computer systems have become an important part of our daily life and IoT has recently gained tremendous attention in the IT industry because of its many benefits. The Internet of Things (IoT) incorporates physical objects from different domains and the Internet. IoT becomes an important technology to develop smart infrastructure and the adaption increased due to its analytics and the interconnectivity of machines and personal devices. A smart infrastructure would transform the way we manage critical services, the way we do business and the way we entertain ourselves. Smart grid, smart home, smart building, which are large-scale IoT applications are examples of smart infrastructure. The integration of IoT devices permits a smart infrastructure to achieve consistent and effective operations to reduce the operational cost significantly. Figure 1 shows a comparison of the world population and the number of IoT devices that need protection against the intruders. The exponential growth will make an IoT a smart object for the

© Springer Nature Switzerland AG 2020
C. Goutte and X. Zhu (Eds.): Canadian AI 2020, LNAI 12109, pp. 508–520, 2020.
https://doi.org/10.1007/978-3-030-47358-7_52

attackers to accomplish malicious activities and increase the attack surface of IoT networks. The effects of cyber-attacks become more destructive as a result many institutions experienced disruption of services, therefore, IoT devices required a sophisticated tool to identify malicious activity in the smart infrastructure. There are many Distributed Denial of Service (DDoS) launched via millions of IoT devices that slowdown or shutdown a dozen websites. A signature-based IDS can't detect novel attacks because the signature of each attack must be identified before a decision system can detect the attack. The complexity used by attackers and the increase in the zero-day attacks, an anomaly-based IDS considered well suited to the current environment.

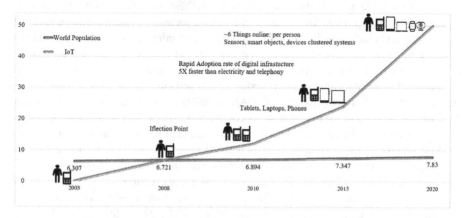

Fig. 1. Projecting the 'Things' behind the Internet of Things [3]

The IoT systems improve our life quality, make communication easier and increase data transfer and information sharing. Computer security is becoming more significant and essential due to the substantial expansion of IoT networks and increases the number of applications running in these IoT devices. A reliable and effective IoT service required a secure communication mechanism to improve the IoT sensing platforms. A security mechanism required to protect IoT devices to provide appropriate assurance for authentication, integrity, confidentiality, and non-repudiation.

Anomaly identification in big data becoming a challenging task in network security. Data mining and machine-learning methodologies play a significant role in evolving and developing anomaly-based intrusion detection systems. Data mining generally includes association rules learning, regression, classification, clustering, and visualization. Classification is the frequently used technique in supervised data mining practices. The objective of classification is to construct a model from categorized items to classify entities as correctly as possible. Modern intrusion detection algorithms and techniques evaluation required a new sophisticated dataset. In this paper, we developed a new dataset set adopted from [1] for detecting malicious activity in the IoT network.

The remainder of this paper is organized as follows. The related work is discussed in Sect. 2, followed by the testbed architecture for generating the IoTID20 dataset in Sect. 3. In Sect. 4 we present an analysis of different classification algorithms for intrusion detection using the generated dataset. Finally, in Sect. 5 we conclude the paper and offer ideas for future work.

2 Related Work

IoT provides a wide range of applications built on the merging smart objects and the Internet. The smart objects integrate the physical assets of an infrastructure to provide better utilization and effective management of resources. IoT brings many opportunities to the operations and services of many organizations. The main goal of the IoT is to provide a digital means for social needs in everyday transactions. The main challenges in IoT networks are Security; Data Management; Cost; Efficiency; Scalability; Availability; Application Development. Among these challenges, security is considered as one of the most prominent challenges in the development of smart infrastructure. An IDS monitors network traffic to identify malicious events or rules violations to generate an alert to a control station or implement a pre-emptive measure against a detected threat. The intrusion detection algorithm can be tested and evaluated via a dataset. The most common datasets for intrusion detection are the DARPA 98/99 [4, 5], which were developed at MIT Lincoln Lab via an emulated network environment. The DARPA 98 dataset contains seven days whereas the DARPA 99 contains five weeks of network traffic. However, the DARPA 98/99 datasets are widely used for network intrusion detection, but these datasets are frequently criticized because these datasets contain many redundant records. Lee and Stolfo [6] developed a framework to construct and extract features from DARPA 98/99 and named the new dataset KDD99. The KDD99 dataset is the most widely used dataset for intrusion detection. The KDD99 dataset contains TCP attributes but failed to give information about IP addresses. The KDD99 dataset contains 5 million of data instances and more than 20 different types of attacks. The dataset is publicly available with an explicit test subset.

Sperotto et al. [7] proposed a flow-based intrusion detection dataset for high-speed networks in 2008 at the University of Twente. This dataset was the first publicly available label dataset for flow-based intrusion detection. The dataset contains six days of network traffic collected from a honeypot server that offers SSH, Web and FTP services. Nearly all flows are malicious without the normal network flows because all data were collected from a honeypot. These high ratio malicious instances affect the learning algorithm to be biased in the direction of more frequent class and stop the detection algorithm from learning the least common class. On the other hand, these high ratio malicious instances in the testing set also cause the machine learning algorithm to produce biased evaluation results by a method which has better detection rates on the frequent records. Sangster et al. [8] established a combat competition to generate a labeled dataset for intrusion detection. The dataset contains four days of network traffic of the warfare competition. They comprehensively discussed the advantages and disadvantages of using warfare competition in developing a modern labeled dataset for intrusion detection. The dataset contains normal user actions as well as various types of attacks. The dataset is publicly available.

The ISCX dataset [9] was developed using a systematic approach to generate normal user behavior and malicious network traffics. The dataset contains seven days of network traffic for FTP, POP3, IMAP, SSH, SMTP, and HTTP protocols. A multistage attack scenario considered generating the malicious part of the dataset. The ISCX is a labeled dataset with realistic network traffic and diverse intrusion scenarios. The dataset is publicly available and can be downloaded from the website [10]. The NSL-KDD [11] dataset derived from the KDD99 dataset. The NSL-KDD removed redundant records from the KDD99 dataset. The training data of KDD99 contains 78% redundant instances while the testing data contains 75% redundant instances [11]. These redundant records produce biased results for frequent instances. The UNSW-NB15 dataset released by defense force academy University of New South Wales Australia [12]. The UNSW-NB15 dataset contains comprehensive modern normal network traffic as well as diverse intrusions scenario with a deep structured network traffic information. The UNSW-NB15 dataset comprises of realistic normal network traffic along with nine modern attack categories. The dataset consists of 49 features which are group into Flow, Basic, Content, Time, Additional Generated features, Connection and Labeled features. The UNSW-NB15 attack categories are Worm, Shellcode, Reconnaissance, Generic, Fizzers, Exploit, DoS, Backdoor and Analysis [12]. AWID dataset [13] focused on 802.11 networks and is publicly available at the website [14]. The AWID dataset contains 37 million packets and 156 features. The AWID dataset malicious traffic was created by implementing 16 detailed attacks against the Wi-Fi network. The AWID dataset labeled in two ways (4 classes and 16 classes) and split into a training and a test subset.

Ghorbani et al. [15] developed the CICIDS2017 dataset within an emulated environment at the Canadian Institute for Cybersecurity (CIC), University of New Brunswick Canada. The CICIDS2017 dataset consists of modern normal and malicious network traffic. The dataset consists of 80 network features and provides a reliable normal and malicious network flows. The data were collected for five days. ISCXFlow meter used to generate the CSV files of the dataset from Pcap files [16]. Normal user behavior and background network traffic generated via a B-Profile system. The CICIDS2017 dataset extracts normal and malicious behavior based on the SSH, FTP, HTTP, HTTPS, and email protocols. Distributed Denial of Service (DDoS) attack is a thoughtful threat to the computer networks that aimed to exhaust the network with abnormal network traffic. Ghorbani et al. [17] developed the CICDDOS2019 dataset with up-to-date normal and malicious DDOS network traffic. They used a B-Profile system to generate realistic normal background network traffic. The malicious part of the dataset consists of 12 DDoS attacks. The CICDDOS2019 dataset is publicly available at the website [10]. This dataset provides comprehensive metadata about IP addresses and malicious network traffic.

Moustafa et al. [18] developed an IoT botnet dataset via legitimate and emulated IoT networks. Smart fridge, Weather station, Motion-activated lights, Smart thermostat and Remotely activated garage door IoT services implemented using the Node-Red tool. A typical smart home configuration designed which contains five IoT devices operated locally and connected to the cloud infrastructure via a Node-Red tool to generate normal traffic. MQTT protocol used to transfer messages from IoT devices to the Cloud. The dataset is publicly available and consists of 49 features. IoT devices can be easily compromised than computer systems, which result in more IoT-based botnet attacks as compared to the computer systems. Meidan et al. [19] developed an IoT botnet dataset. The dataset was generated using nine commercial IoT devices and two IoT-based botnets BASHLITE and Mirai. The dataset contains 115 network features that provide a reliable normal and malicious network flow. The dataset consists of separate benign network traffic for each commercial device to ensure the normal network behavior for the training dataset.

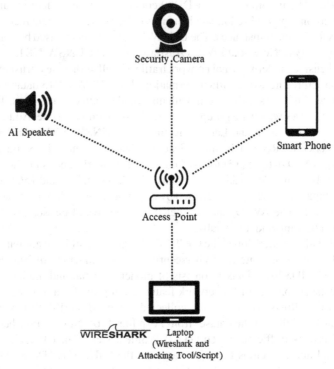

Fig. 2. IoTID20 dataset testbed environment [1]

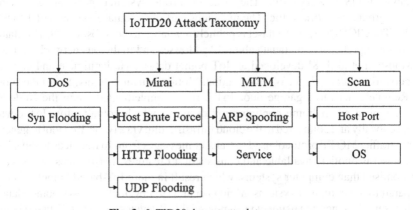

Fig. 3. IoTID20 dataset attack taxonomy

3   Testbed Architecture

The testbed for the IoTID20 dataset is a combination of IoT devices and interconnecting structures. A typical smart home environment was implemented which consists of smart home device SKT NGU and EZVIZ Wi-Fi camera to generate the IoTID20 dataset. Figure 2 shows the testbed environment for the IoTID20 dataset. These two IoT devices connected to a smart home Wi-Fi router. Other devices connected to the smart home router include laptops, tablets, smartphones. The SKT NGU and EZVIZ Wi-Fi camera are IoT victim devices and all other devices in the testbed are the attacking devices. The newly developed IoTID20 dataset adopted from the Pcap files available at the website [1]. Figure 3 shows the attack taxonomy of our proposed dataset.

We used CICflowmeter [16] application to extract features from Pcap files and generate a CSV format of the IoTID20 dataset. The next step is to label each instance of the IoTID20 dataset. The IoTID20 dataset consists of 80 network features and three label features. The label features are binary, category, and sub-category. Table 1 shows binary, category and sub-category labels of the IoTID20 dataset.

Table 1. Binary, category, and sub-category of IoTID20 Dataset

Binary	Category	Subcategory
Normal, Anomaly	Normal DoS, Mirai, MITM, Scan	Normal, Syn Flooding, Brute Force, HTTP Flooding, UDP Flooding ARP Spoofing Host Port, OS

Table 2. Normal and attacked instances in IoTID20 Dataset

Binary label distribution		Subcategory distribution	
Normal	40073	Type	Instances
Anomaly	585710	Normal	40073
		DoS	59391
Category label distribution		Mirai Ack Flooding	55124
Type	Instances	Mirai Brute force	121181
Normal	40073	Mirai HTTP Flooding	55818
DoS	59391	Mirai UDP Flooding	183554
Mirai	415677	MITM	35377
MITM	35377	Scan Host Port	22192
Scan	75265	Scan Port OS	53073

Table 3. IoTID20 dataset correlated features

Total features	Feature name
12	Active_Max, Bwd_IAT_Max, Bwd_Seg_Size_Avg, Fwd_IAT_Max, Fwd_Seg_Size_Avg, Idle_Max, PSH_Flag_Cnt, Pkt_Size_Avg, Subflow_Bwd_Byts, Subflow_Bwd_Pkts, Subflow_Fwd_Byts, Subflow_Fwd_Pkts

The most important benefits of the IoTID20 dataset; it replicates a modern trend of IoT network communication; it is among the few publicly available IoT intrusion detection dataset. The final version of the IoTID20 dataset consists of 83 network features and three label features. The IoTID20 intrusion detection dataset binary, category, and subcategory instances distribution are presented in Table 2.

The IoTID20 dataset analyzed and evaluated for features correlation, feature ranking and various machine-learning algorithms for classification. A preprocessing process of the IoTID20 dataset required because the data types and the format of some features are not suitable for machine learning algorithms. We used supervised machine learning algorithms and column normalization techniques to normalize and evaluate the IoTID20 dataset. The correlated features degrade the detection capability of a machine learning algorithm. Correlated features were removed from the IoTID20 dataset. A correlation coefficient of 0.70 was used to remove a list of correlated features. Table 3 shows a list of correlated features that removed from the IoTID20 dataset.

The features of the IoTID20 dataset were ranked using the Shapira-Wilk algorithm. The Shapira-Wilk algorithm measures the regularity of the distribution of occurrences with respect to the feature. Figure 4 shows feature ranking using the Shapiro-Wilk algorithm. More than 70% of the feature ranked with a value greater than 0.50. These high ranked features will improve the classification capability of detection algorithms and techniques. These high-rank features also support the feature selection process to improve the detection capability of the machine learning algorithms and to decrease the training time machine learning algorithms.

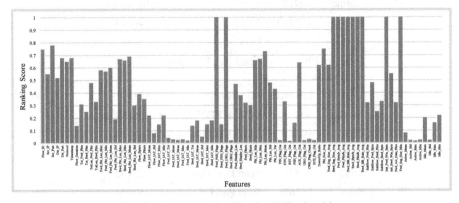

Fig. 4. Feature ranking Shapiro-Wilk algorithm

4   Analysis

Data mining and machine learning techniques play a significant role in developing an anomaly-based intrusion detection system. Data mining generally includes association rules learning, regression, classification, clustering, and visualization. Classification frequently used technique in supervised data mining practices. The objective of classification is to construct a model from categorized items to classify entities as correctly as possible. The accuracy of existing machine learning algorithms needs to be improved to detect new attacks because the attack patterns are changing every day. The IoTID20 dataset consists of 83 network features and 3 label features. There are many procedures to determine the performance of an estimator. Some of the most common measures are accuracy, precision, recall, F-measure.

$$\text{Accuracy} = \frac{TP + TN}{TP + TN + FP + FN} \tag{1}$$

$$\text{Precision} = \frac{TP}{TP + FP} \tag{2}$$

$$\text{Recall} = \frac{TP}{TP + FN} \tag{3}$$

$$F - \text{measure} = \frac{2\text{Precision} \cdot \text{Recall}}{\text{Precision} + \text{Recall}} \tag{4}$$

In machine learning, different cross-validation tests such as K-fold cross-validation, jackknife, and independent tests are used to evaluate the success rates of a classifier. Jackknife test is efficient and reliable but, the computational time of jackknife is an issue, especially using large datasets. Therefore, to minimize the running time, we used various K-fold cross-validation tests to evaluate the performance of the different classifiers. The IoTID20 dataset contains three labels for classifying normal network traffic and anomalous network traffic so we analyzed the dataset for binary, category and sub-category labels. Machine learning models developed through SVM, GaussianNB, LDA, Logic Regression, Decision Tree, Random Forest, and Ensemble classifiers.

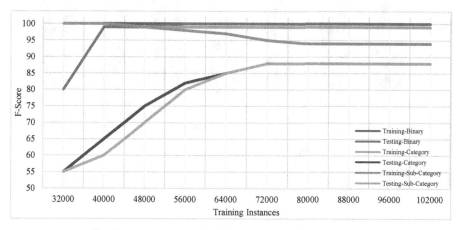

Fig. 5. Learning curve for label, category, and sub-category

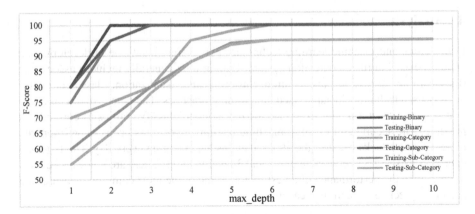

Fig. 6. Validation curve for label, category, and sub-category

The learning curve shows a relationship between the training and validation of an algorithm using various training samples. The learning curve determines how the algorithm can benefit by providing more data or the data provided enough for better performance of the algorithm. Figure 5 shows the F score learning curve for Binary, Category, and Sub-Category label classification using the decision tree algorithm. We used F-Score for the learning curve because the F-score is the harmonic mean of precision and recall. From the learning curve, it is concluded that a minimum of seventy thousand instances required to get better performance of decision tree for Binary, Category and Sub-Category labels classification. We also analyzed the learning curve of the IoTID20 dataset via GaussianNB, LDA, Logic Regression, Random Forest, SVM, and ensemble classifiers. After the investigation, it is determined that a minimum of seven thousand instances required to get a better classification score.

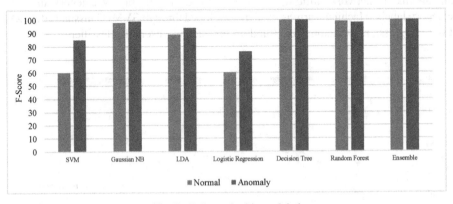

Fig. 7. F-Score for binary label

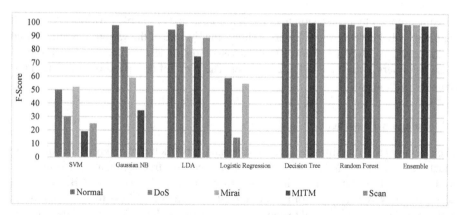

Fig. 8. F-Score for category label

A validation curve shows how effective a classifier on the data it is trained as well as how operative the classifier to the new input test data. A hyperparameter of a classifier essential to maximize the classification score. We used the max_depth of the decision tree with a depth value of 10 to produce the validation curve. The result of the validation curve for binary Label, Category, and Sub-Category classification are presented in Fig. 6. The classifier score converged after max_depth of 2 for binary label classification. The classifier score converged after max_depth of 6 for category and subcategory label classification. The classifier is not over fitted for the binary label and sub-category label classification but overfitted for category label classification.

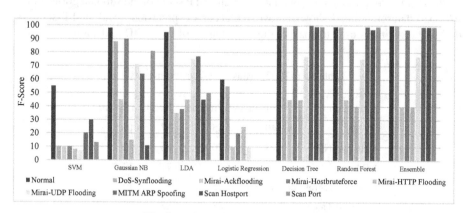

Fig. 9. F-Score for sub-category label

4.1 Binary Classification

The binary label classifies the dataset as normal network traffic or malicious network. SVM, Gaussian NB, LDA, and Logic regression poorly performed for binary label classification while the decision tree, random forest, and ensemble performed very well for binary label classification. Figure 7 shows the F-score of the various classifiers used in this paper. We used a 3, 5 and 10 fold cross-validation test to check the overfitting of these classifiers against the IoTID20 dataset. The result of the cross-fold validation test remains unchanged.

4.2 Category Classification

The category label classifies the dataset as normal network traffic or any of the following attack category DoS, Mirai, MITM or Scan. Decision tree estimator performs very well for all attack categories while logic regression, LDA, Gaussian NB, and SVM performance were very poor for most of the attack categories. Cross fold validation test with a K value of 3, 5, 10 used to check the overfitting of these classifiers. The outcome of the cross-fold validation assessment remains persistent. Figure 8 shows F-score for the category label classification.

4.3 Sub-category Classification

The sub-category label classifies the dataset into normal network traffic or any one of the categories as shown in Fig. 9. A better performance achieved by the decision tree classifier for the sub-categories but the classifier's logic regression, LDA, Gaussian NB, and SVM poorly performed for most of the attack sub-categories. Figure 9 shows the F-score for the sub-category label of the IoTID20 dataset. Some of these sub-category attacks were misclassified and we aimed to design and develop a machine learning model to improve the accuracy, precision, recall and F score of the sub-categories of the IoTID20 dataset. Table 4 shows an average accuracy, precision, recall and F-score of different algorithms used to analyze the subcategory of the IoTID20 dataset.

Table 4. IoTID20 dataset performance results

Algorithm	Accuracy	Precision	Recall	F Score
SVM	40	55	37	16
Gaussian NB	73	70	66	62
LDA	70	71	71	70
Logistic Regression	40	25	39	30
Decision Tree	88	88	88	88
Random Forest	84	85	84	84
Ensemble	87	87	87	87

5 Conclusion

The main contribution of this paper is a new IoTID20 dataset for anomalous activity detection in IoT networks and the evaluation of various machine learning algorithms through the IoTID20 datasets. The IoTID20 dataset contains various types of IoT attacks and families. We also reviewed various intrusion detection datasets and discussed weaknesses. Our proposed dataset includes 8 attack types to evaluate intrusion detection algorithms in IoT networks. Also, we provided a list of highly correlated and significant features for feature importance and feature ranking. We analyzed the dataset via the 7 most popular machine learning algorithms. According to the weighted average of Pr, Rc and F1 score, the highest score achieved by decision tree and ensemble algorithms. Our proposed dataset will provide a basis to develop a new methodology to detect malicious activity in IoT networks.

For future work, we plan to develop and evaluate a framework for anomalous activity detection models for IoT networks to improve accuracy.

References

1. Hyunjae, K., Kang Dong, H., Ahn Gyung, M., Lee Jeong, D., Yoo Kyung, H., Park Huy, K.: IoT Network intrusion dataset, 27 Sept 2019 (2019). http://dx.doi.org/10.21227/q70p-q449. Accessed 01 Jan 2020
2. Ullah, I., Mahmoud, Q.H.: IoT Network intrusion dataset (2020). https://sites.google.com/view/iot-network-intrusion-dataset. Accessed 18 Feb 2020
3. Press, G.: Internet Of Things by the numbers: what new surveys found (2018). https://www.forbes.com/sites/gilpress/2016/09/02/internet-of-things-by-the-numbers-what-new-surveys-found/#f439a4016a0e. Accessed 14 Jan 2020
4. Lippmann, R.P., et al.: Evaluating intrusion detection systems: the 1998 DARPA off-line intrusion detection evaluation. In: Proceedings of - DARPA Information Survivability Conference and Exposition DISCEX 2000, vol. 2, pp. 12–26 (2000)
5. Haines, J.W., Rossey, L.M., Lippmann, R.P., Cunningham, R.K.: Extending the DARPA off-line intrusion detection evaluations. In: Proceedings of - DARPA Information Survivability Conference and Exposition II, DISCEX 2001, vol. 1, pp. 35–45 (2001)
6. Lee, W., Stolfo, S.J.: A framework for constructing features and models for intrusion detection systems. ACM Trans. Inf. Syst. Secur. 3(4), 227–261 (2000)
7. Sperotto, A., Sadre, R., van Vliet, F., Pras, A.: A labeled data set for flow-based intrusion detection. In: Nunzi, G., Scoglio, C., Li, X. (eds.) IPOM 2009. LNCS, vol. 5843, pp. 39–50. Springer, Heidelberg (2009). https://doi.org/10.1007/978-3-642-04968-2_4
8. Sangster, B., et al.: Toward instrumenting network warfare competitions to generate labeled datasets. In: USENIX Security Workshop Cyber Security Experimentation Test, pp. 1–6 (2009)
9. Shiravi, A., Shiravi, H., Tavallaee, M., Ghorbani, A.A.: Toward developing a systematic approach to generate benchmark datasets for intrusion detection. Comput. Secur. 31(3), 357–374 (2012)
10. Ali Shiravi, A.A.G., Shiravi, H., Tavallaee, M.: Datasets (2012). https://www.unb.ca/cic/datasets/ids.html. Accessed 11 Jan 2020
11. Tavallaee, M., Bagheri, E., Lu, W., Ghorbani, A.A.: A detailed analysis of the KDD CUP 99 data set. In: IEEE Symposium Computational Intelligence Security Defence Applications CISDA 2009, pp. 1–6 (2009)

12. Moustafa, N., Slay, J.: UNSW-NB15: a comprehensive data set for network intrusion detection systems (UNSW-NB15 network data set). In: Proceedings of 2015 Military Communications and Information Systems Conference MilCIS 2015, pp. 1–6 (2015)
13. Kolias, C., Kambourakis, G., Stavrou, A., Gritzalis, S.: Intrusion detection in 802.11 networks: empirical evaluation of threats and a public dataset. IEEE Commun. Surv. Tutorials **18**(1), 184–208 (2016)
14. Kolias, C., Kambourakis, G., Stavrou, A., Gritzalis, S.: AWID a datasets focused on intrusion detection in 802.11 (2016). http://icsdweb.aegean.gr/awid/index.html. Accessed 10 Jan 2020
15. Sharafaldin, I., Lashkari, A.H., Ghorbani, A.A.: Toward generating a new intrusion detection dataset and intrusion traffic characterization. In: ICISSP 2018 – Proceedings of 4th International Conference Information System Security and Privacy, January 2018, pp. 108–116 (2018)
16. Lashkari, A.H., Gil, G.D., Mamun, M.S.I., Ghorbani, A.A.: Characterization of tor traffic using time based features. In: ICISSP 2017 – Proceedings of 3rd International Conference Information System Security and Privacy, January 2017, pp. 253–262 (2017)
17. Sharafaldin, I., Lashkari, A.H., Hakak, S., Ghorbani, A.A.: Developing realistic distributed denial of service (DDoS) attack dataset and taxonomy. In: Proceedings of International Carnahan Conference Security Technology, October 2019 (2019)
18. Koroniotis, N., Moustafa, N., Sitnikova, E., Turnbull, B.: Towards the development of realistic Botnet dataset in the Internet of Things for network forensic analytics: Bot-IoT dataset. Futur. Gener. Comput. Syst. **100**, 779–796 (2019)
19. Meidan, Y., et al.: N-BaIoT-Network-based detection of IoT bBotnet attacks using deep autoencoders. IEEE Pervasive Comput. **17**(3), 12–22 (2018)

Lexical Data Augmentation for Text Classification in Deep Learning

Rong Xiang[1]([✉]), Emmanuele Chersoni[1], Yunfei Long[2], Qin Lu[1], and Chu-Ren Huang[1]

[1] The Hong Kong Polytechnic University, 11 Yuk Choi Road,
Hung Hom, Hong Kong, China
xiangrong0302@gmail.com, emmanuelechersoni@gmail.com,
{qin.lu,churen.huang}@polyu.edu.hk
[2] School of Computer Science and Electronic Engineering, University of Essex,
Colchester, UK
yl20051@essex.ac.uk

Abstract. This paper presents our work on using **part-of-speech focused lexical substitution for data augmentation (PLSDA)** to enhance the prediction capabilities and the performance of deep learning models. This paper explains how PLSDA uses part-of-speech information to identify words and make use of different augmentation strategies to find semantically related substitutions to generate new instances for training. Evaluations of PLSDA is conducted on a variety of datasets across different text classification tasks. When PLSDA is applied to four deep learning models, results show that classifiers trained with PLSDA achieve 1.3% accuracy improvement on average.

Keywords: Data augmentation · Text classification · Lexical data augmentation · Deep learning

1 Introduction

Text classification aims to assign a set of pre-defined categorical labels to text. Typical classification applications include spam detection, topic modelling, sentiment analysis, fake news detection and etc. Deep learning methods, with more powerful data learning capability, have achieved significant improvements in text classification tasks. Recently proposed transformer-based methods such as BERT [1] and RoBERTa [3] have brought even more significant performance gains. However, more comprehensive learning models normally requires more training data. Yet, well-annotated training data is too expensive to get sufficient amount for any specific classification task, limiting the amount of tuning that can be done for a deep learning model. Data augmentation aims to use systematic ways to provide more training data for fine tuning.

Augmentation techniques have been used in some NLP studies such as machine translation, dialog systems, question answering as well as text classification. Lexical augmentation is a fundamental and efficient strategy in NLP augmentation studies [6,7] without changing syntactic structures. An early lexical

© Springer Nature Switzerland AG 2020
C. Goutte and X. Zhu (Eds.): Canadian AI 2020, LNAI 12109, pp. 521–527, 2020.
https://doi.org/10.1007/978-3-030-47358-7_53

augmentation method used a thesaurus to replace words with available synonyms [7]. WordNet [2] is another commonly used resource for synonym replacement [6]. In addition to using well-structured knowledge resources, interpolation by word embedding is also a feasible way to make use of semantically-close candidates for substitution [5]. Recent work proposed by Wei and Zou [6] extended word substitution by lexical insertion, deletion and swap methods for data augmentation. However, lexical insertion, deletion and swap process may infringe the semantic completeness and syntactic correctness.

In this paper, we conduct an in-depth study of data augmentation via lexical substitution to further improve the augmentation performance in text classification tasks. The proposed **part-of-speech focused lexical substitution for data augmentation** (PLSDA), as a lexical augmentation method, aims to create useful training data for natural language samples, and the substitution must consider both syntactic correctness as well as semantic closeness and diversity. More specifically, PLSDA first makes use of POS tags to determine words to be replaced for syntactic consistency. WordNet is then used to obtain synonyms for replacement with consideration of both similarity and diversity.

2   Design Principles of PLSDA

Lexical substitution refers to methods which create new instances from a given dataset by replacing a number of words in a text sampling with substitutes according to certain principles. POS focused Lexical Substitution Augmentation (PLSDA) consists of two main Parts: *Substitution Candidate Selection* and *Instance Generation*. For a given training sample *Substitution Candidate Selection* first follows its **syntactic consistency principle** and uses POS constraints to select candidate words for substitution. It then follows the **semantic consistency principle** to identify lexical units via semantic relatedness for each selected word to form a *Substitution Candidate Lists (SCLs)*. In the *Instance Generation*, whether a word is replaced or not is determined by sampling from Bernoulli distribution of *SCLs*, to form the final *Substitution Collection (SC)*. Lastly, substitutes in *SC* with respect to each position are used to generate augmented instances.

2.1   Substitution Candidate Selection

Let I denote a training instance with n words, $I = \{w_1, w_2, w_i, ..., w_n\}$. For each w_i, its POS tag t_{w_i}, can be readily obtained from available tools such as the Stanford NLP pipeline [4]. Replacement words for augmentation with the same POS tag, as the **principle of syntactic consistency** constraint, ensures that new text samples are syntactically identical to I. Candidates with the same POS for each w_i in I are obtained from WordNet. For example, a verb "chair" (a chairperson of an organization, meeting, or public event) will not be replaced with the noun "bench". In this work, substitutions are allowed only on certain word classes so that the newly created samples are likely to make sense. All w_i that satisfy the constraints are marked as replaceable.

Let SCL_{w_i} denote the substitution candidate list for each w_i with m synonyms. SCL_{w_i} is then obtained according to the following formula:

$$SCL_{w_i} = \{c^1_{w_i}, c^2_{w_i}, ...c^m_{w_i} \mid c^j_{w_i} \in Syn(w_i, j) \,\&\, t_{w_i} = t_{c^j_{w_i}}, \} \tag{1}$$

where $c^j_{w_i}$ is the j-th synonym for word w_i. $Syn(w_i, j)$ refers to the synonym set of w_i, where j is the membership subscript. Only w_i with at least one or more synonyms will be considered in Instance Generation ($m > 0$).

2.2   Instance Generation

To control the number of generated instances, Instance Generation selects appropriate candidates from the list of $SCLs$, each of has two values k and s, where k is the length of sentence I and s is the average number of substitutes, both can be determined for each given I. A sampling method is used to select a position i as a variable such that w_i is to be replaced. Bernoulli distribution $Ber(p_s)$ is applied to for every w_i having $SCLs$, where p_s as a probability is an algorithm parameter. For lack of any prior-knowledge, $p_s = 0.5$ can be used naively. The Bernoulli distribution below decides whether w_i with a non-empty SCL is selected as replacement points to forms the final SC.

$$P(w_i) = p_s^x(1 - p_s)^{1-x} \begin{cases} x = 1 \; w_i \; is \; selected, SC = SC \cup SCL_{w_i} \\ x = 0 \; w_i \; is \; not \; selected. \end{cases} \tag{2}$$

As there are typically multiple members for each SCL_{w_i}, two proposed strategies are investigated to select candidates from the average of s substitutes for each selected w_i. The first augmentation strategy is the **stochastic strategy**, which randomly picks a candidate from the words in SCL_{w_i} to avoid a rigorous selection algorithm. This random process samples from categorical distribution $Cat(p^1_{w_i}, p^2_{w_i}, ..., p^j_{w_i}, ..., p^m_{w_i})$, where $\sum p^j_{w_i} = 1$.

$$P(X = c^j_{w_i}) = p^j_{w_i}, \; j \in [1, m] \tag{3}$$

The second strategy is the **similarity-first strategy**, which makes use of similarity measures to pick candidates, exploiting similarity ranking. To use this strategy, candidates $\{c^1_{w_i}, c^2_{w_i}, ...c^m_{w_i}\}$ for a word w_i need be sorted according to their cosine similarity of word vectors. Augmented instances are picked according to their ranks.

3   Performance Evaluation

Eight benchmark datasets are used for NLP classification tasks: (1) **SST-2:** Stanford Sentiment Treebank dataset, (2) **Subj:** Subjectivity classification, (3) **MR:** movie review dataset, (4) **IMDB**, IMDB movie review dataset, (5) **Twitter** twitter sentiment classification dataset, (6) **AirRecord** airline customer service dataset, (7) **TREC:** question type identification dataset, and (8) **Liar:** fake news detection dataset. Four deep learning models are used in Performance evaluation including **LSTM, BiLSTM-AT, BERT** and **RoBERTa**.

Table 1. Accuracy of the models: the best is in bold and the second-best is underlined.

	SST-2	Subj	MR	IMDB	Twittter	AirRecord	TREC	Liar
LSTM	80.2	90.8	77.0	80.3	74.7	80.5	88.8	25.3
+EDA	80.9	91.3	77.6	81.2	75.7	81.2	89.3	26.0
+PLSDA	81.0	91.9	78.1	82.6	77.2	81.4	89.3	27.0
BiLSTM-AT	78.2	91.0	75.9	80.5	75.9	81.3	88.3	25.7
+EDA	78.9	91.5	76.6	81.8	76.9	81.9	88.9	26.3
+PLSDA	79.7	92.1	76.8	83.0	77.6	82.0	88.8	26.5
BERT	91.3	97.2	87.1	88.1	82.0	83.2	96.8	27.9
+EDA	92.0	97.4	88.0	88.9	82.7	83.9	97.5	28.2
+PLSDA	92.3	**98.4**	88.7	89.6	83.2	84.4	<u>97.6</u>	**29.0**
RoBERTa	93.0	97.3	90.3	89.1	83.3	84.3	96.5	27.2
+EDA	<u>93.7</u>	97.4	<u>90.7</u>	<u>90.0</u>	<u>84.1</u>	<u>85.5</u>	97.5	27.7
+PLSDA	**93.9**	<u>98.2</u>	**91.6**	**90.8**	**84.7**	**85.9**	**97.8**	<u>28.3</u>

3.1 Overall Performance

The performance of training with original training datasets, the current state-of-the-art augmentation method EDA, and PLSDA are presented in Table 1. Table 1 shows that BERT and RoBERTa, the recently proposed transformer models, significantly outperforms the previous deep learning models. BiLSTM-AT generally performs better than LSTM because BiLSTM-AT can obtain additional information from the reversed order and benefit from attention mechanism. Individual gains after training with PLSDA with respect to (w.r.t.) original training data range from 0.5% to 2.5%. Further calculation shows that the overall gain is 1.3% and 0.7% for PLSDA and EDA, respectively. This implies that lexicon substitution with appropriate syntactic constraint can further contribute to performance.

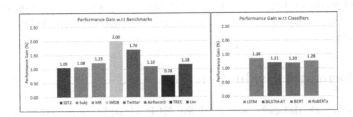

Fig. 1. Absolute performance gains (%) on average accuracy by PLSDA.

Figure 1 shows the absolute performance gains by using PLSDA. The left figure shows average performance gains w.r.t. datasets. The right figure shows average performance gains w.r.t. classifiers. Obviously, improvement on binary classification is more impressive than that on multi-class tasks. By observing different classification models, LSTM gains the largest improvement from PLSDA. Although BERT and RoBERTa are the state-of-the-art methods, they still obtain significant improvement through PLSDA.

3.2 Effectiveness of POS Types

The second experiment illustrates the effect of three different types of POS tags: Adjective/Adverb(A), Noun(N), Verb(V) and their combinations. The evaluation is conducted on BERT and RoBERTa. One dataset for each type of classification task is selected: Subj, IMDB, TREC and Liar.

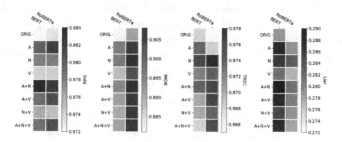

Fig. 2. Heatmaps of Lexicon POS; Accuracy bar is given besides each heatmap

Accuracy for each POS setting is shown as heatmaps in Fig. 2. Each model without PLSDA, denoted as ORIG (original), is reported in the first row as a reference. Generally, the performance of Adjective/Adverb and Noun replacement outperform Verb replacement. POS combinations A+N can be the best choice to get the best performance. A+N+V also results in a considerable accuracy although it is does not seem to be the best performed setting.

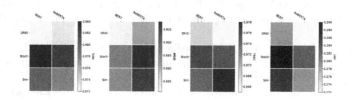

Fig. 3. Heatmaps of Sampling Strategy; Accuracy bar is given besides each heatmap

3.3 Sampling Strategy

The third experiment evaluates the two augmentation strategies. Evaluations are conducted for BERT and RoBERTa on Subj, IMDB, TREC and Liar. The combination of Adjective/Adverb and Noun is used.

Accuracy for the two augmentation strategies compared to their respective classifiers are shown as heatmaps in Fig. 3. This experiment gives a strong indication that even though both strategies are effective, stochastic substitution introduces more diversity in the augmentation and it is thus more appropriate for deep learning models.

4 Conclusion

In this paper, we present a part-of-speech focused lexical substitution approach for data augmentation, and investigate the effect of different lexical substitution strategies for eight text classification tasks. Performance evaluation shows that data augmentation improves the performance of deep learning models including state-of-the-art transformer-based models. Our investigation also found that nouns and adjectives/adverbs work better as replacement types even though their numbers of candidates are not necessarily large. Experimental results show that using stochastic sampling to find replacement outperform similarity-first strategy which indicates that augmentation by introducing diversity is better for training. In summary, data augmentation is as important in the deep learning age as it was during the conventional machine learning age.

Future work includes two directions. One is to investigate the performance of PLSDA on more publicly accessible datasets. The other direction is to explore the feasibility of PLSDA in other NLP tasks.

Acknowledgements. We acknowledge the research grants from Hong Kong Polytechnic University (PolyU RTVU) and GRF grant (CERG PolyU 15211/14E, PolyU 152006/16E).

References

1. Devlin, J., Chang, M.W., Lee, K., Toutanova, K.: Bert: pre-training of deep bidirectional transformers for language understanding. arXiv preprint arXiv:1810.04805 (2018)
2. Fellbaum, C.: WordNet. The Encyclopedia of Applied Linguistics (2012)
3. Liu, Y., et al.: Roberta: a robustly optimized Bert pretraining approach. arXiv preprint arXiv:1907.11692 (2019)
4. Toutanova, K., Klein, D., Manning, C.D., Singer, Y.: Feature-rich part-of-speech tagging with a cyclic dependency network. In: Proceedings of the 2003 Conference of the North American Chapter of the Association for Computational Linguistics on Human Language Technology, vol. 1, pp. 173–180. Association for computational Linguistics (2003)

5. Wang, W.Y., Yang, D.: That's so annoying!!!: a lexical and erame-semantic embedding based data augmentation approach to automatic categorization of annoying behaviors using# Petpeeve tweets. In: Proceedings of the 2015 Conference on Empirical Methods in Natural Language Processing, pp. 2557–2563 (2015)
6. Wei, J.W., Zou, K.: Eda: easy data augmentation techniques for boosting performance on text classification tasks. arXiv preprint arXiv:1901.11196 (2019)
7. Zhang, X., Zhao, J., LeCun, Y.: Character-level convolutional networks for text classification. In: Advances in Neural Information Processing Systems, pp. 649–657 (2015)

A Deeper Look at Bongard Problems

Xinyu Yun[(✉)], Tanner Bohn, and Charles Ling

Western University, London, ON N6A 3K7, Canada
{xyun,tbohn,charles.ling}@uwo.ca

Abstract. Machine learning, especially deep learning, has been successfully applied to a wide array of computer vision classification tasks in recent years. Infamous for requiring massive amounts of data to perform well at image classification problems, deep learning has so far been unable to solve Bongard problems (BPs), a set of abstract visual reasoning tasks invented in the 1960s. Each BP can be seen as a supervised learning task, with few training samples (6 for positive and 6 for negative), and often requiring highly abstract features to learn well. Automatically solving Bongard problems directly from images remains an ambitious goal, with very little machine learning literature on the topic. In this paper, we discuss several special properties of BPs as well as what it means to solve a BP. Making use of an expanded set of BP-like tasks to allow for a more careful evaluation of automated solvers, we develop and benchmark a deep learning based approach to solve these problems. To encourage work on this interesting problem, we also make freely available a dataset of over 200 BPs (https://github.com/XinyuYun/bongard-problems).

Keywords: Bongard problems · Convolutional neural networks · Feature extraction · Few-shot learning

1 Introduction

Despite recent successes in machine learning on many problems previously considered beyond the reach of artificial intelligence, tasks requiring divergent thinking, abstraction, and few-shot learning continue to be a challenge. While other tasks requiring one or more of these properties have seen recent attention and progress [9,12], Bongard problems (BP), which appear to require the solver to possess all three of these skills, continue to be largely unstudied. Created in the 1960s by Mikhail Bongard, these problems were designed to demonstrate the inadequacy of the standard pattern recognition tools of the day for achieving human-level visual cognition [1].

A typical Bongard problem consists of 12 tiles evenly divided into a left and a right class. To gauge the cognitive abilities of a test subject, the subject is shown the 12 tiles and then asked to provide a rule which distinguishes the tiles appearing on one side from the tiles on the other. For example, the intended rule for the second problem in Fig. 1 is 'clockwise spirals on the left, counterclockwise spirals on the right'.

© Springer Nature Switzerland AG 2020
C. Goutte and X. Zhu (Eds.): Canadian AI 2020, LNAI 12109, pp. 528–539, 2020.
https://doi.org/10.1007/978-3-030-47358-7_54

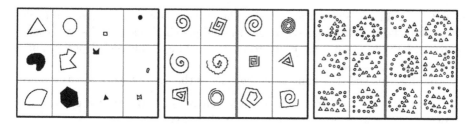

Fig. 1. Examples of easy, intermediate, and difficult Bongard problems.

As a classification task, BPs possess several properties which make it both interesting and difficult with respect to machine learning. A few of these properties are shared with other well-studied tasks, however, other properties also establish the Bongard problems as uniquely difficult[1].

Divergent Thinking. The three Bongard problems in Fig. 1, ranging from easy to difficult demonstrate the typical variation, both visually and in terms of solutions. Since there are a very large number of potential features to consider and many ways these features can be combined to define different rules, divergent thinking is required to perform well at Bongard problems. This property is also partially shared by Raven's Progressive Matrices (RPM) task [11], where deciding upon the tile which best completes the matrix requires considering many alternative hypotheses to find the one requiring the simplest justification. using a fixed set of visual features and sequence progression relations. There is considerably more diversity in the visual elements and rule types in Bongard problems.

Abstract Thinking. To solve second problem in Fig. 1, recognizing that the shapes have the characteristic of spiraling requires abstract thinking, because the property of spiraling is not physically present, but exists as non-trivial relationship between points on a curve. The patterns required to be identified to solve the problem often are not directly visible, but exists as a complex relationship between other abstract features. For example, finding the intended rule for the third problem in Fig. 1 likely requires observing that the individual shapes of a particular type should be grouped together to form the outlines of larger shapes.

Few-Shot Learning. To recognize that all objects on a given side share one potentially complex property among innumerable alternatives given only six samples per class requires few-shot learning. In contrast, datasets for image classification problems often have orders of magnitude more samples per class. This few-shot learning property is shared with both the popular Omniglot task (concerned with classification of hand-written characters) [9] and Raven's Progressive Matrices (matrix completion) [11].

For most of these properties, machine learning has had some success on associated problems. However, when multiple properties are present, as in the case

---

[1] A description of what does and does not make for a valid BP can be found here: http://www.foundalis.com/res/invalBP.html.

of BPs, learning to automatically solve the tasks becomes much more difficult. Due to this large performance gap and the unique challenges of BPs, we believe studying BPs is an efficient route towards reaching human-level performance across a variety of tasks.

Towards this end, the main contributions of the present work are as follows.

- We adapt a deep learning based approach to solve Bongard problems and overcome weaknesses in previous approaches (Sect. 3).
- We consider the set of properties which make BPs uniquely difficult and propose a set of metrics for automatic evaluation of BP solvers, which interprets BPs as few-shot classification tasks (Sect. 4).
- We evaluate our deep-learning based approaches on the BPs while examining the effects of pre-training and feature extraction methods (Sect. 5).

2   Related Work

Due to the difficulty of automatically solving BPs or the lack of awareness of them, few attempts at the task have been made.

Motivated by the appearance of Bongard problems in Godel, Escher, Bach [6], Hofstadter's own graduate student, Harry Foundalis, decided to approach the problem of automatically solving them in his dissertation [5]. Foundalis' approach consists of a cognitive architecture for visual pattern recognition called **Phaeaco**, which tries to solve BPs with the following process. First, working at the pixel level, Phaeaco attempts to explicitly extract the geometric primitives contained in each of the 12 tiles of a problem. Next, features shared among the tiles for each side are identified. This is repeated either until a rule is found or some stopping criterion is reached. The Phaeaco model is capable of finding solutions for up to 15 problems out of 200^2. Due to the non-deterministic nature of the program, the success rate of each of these problems varies dramatically, between 6% and 100%.

A more recent approach of solving Bongard problems is provided by [3]. Similar to Phaeaco, their pipeline begins with explicit extraction of visual features. Additionally, these visual features are then translated into a symbolic visual vocabulary. Candidate rules which split the 12 tiles are scored based on assigned prior probabilities of the grammar's production rules which produced the rule in such a way that shorter, less complex rules are preferred. Under this restriction, only 39 BPs are considered. The approach solves 35 of the 39 problems.

A recent approach utilizing deep learning to solve BPs was proposed in an intriguing blog post by [7][3]. While this approach does not entirely avoid manually defining the type of visual features that are important to consider, it comes close, and is the inspiration for the model we present in Sect. 3. Kharagorgiev's approach works roughly as follows: first, an image dataset of simple shapes is automatically constructed and used to train a convolutional neural network (CNN)

---

[2] Phaeaco results can be found here: http://www.foundalis.com/res/solvprog.htm.

[3] https://k10v.github.io/2018/02/25/Solving-Bongard-problems-with-deep-learning/.

as domain knowledge. Second, feature vectors are extracted and binarized with a manually chosen threshold for each of the 12 tiles with the CNN by taking globally-averaged feature maps, as proposed in [10]. Finally, finding a solution to a BP is then reduced to locating a feature where all tiles from each side have the same value, unique to that side. Of the 232 problems assembled by Foundalis[4], 47 problems are considered solved, 41 of which are correctly solved.

3   Our Models

Due to the uniqueness (both visually and in terms of solutions) of Bongard problems and the small size of the problem set compiled over the years (currently around 300), training a meta-learning model on a subset of the problems to try apply to new problems is difficult without overfitting to the specific rules types present in the training data. These properties make recent state-of-the-art approaches for few-shot classification problems [14], ill-suited for Bongard problems. In an attempt to overcome these hurdles, we apply transfer learning, a common deep learning based approach to learning with small data. The approach we take is to pre-train a convolutional neural network with synthetic images that contain visual features commonly present in BP tiles, then train a simple classifier on feature vectors for the 12 tiles produced by the CNN feature maps. Figure 2 provides a high level view of this process.

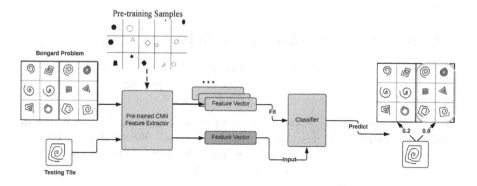

Fig. 2. Bongard problem solver pipeline.

Pre-training. Pre-training for image classification, as described in [4], popularized the insight that rather than learning to perform a new classification task from scratch, one can take advantage of knowledge coming from previously learned categories. By training a machine learning model to perform one task, it may implicitly discover features useful to learning to perform another similar

---

[4] The set of original BPs by Mikhail Bongard as well as those proposed by others can be found here: http://www.foundalis.com/res/bps/bpidx.htm.

task. Compared to past approaches to solving BPs which extracted visual and abstract features using hard-coded feature detectors and routines [3,5], we can influence what patterns a CNN discovers by simply augmenting the training task to require discovering those patterns, a much easier task than manually writing algorithms to detect those particular features. To ensure that the features we extract from the feature maps that are relevant to visual patterns presenting in BPs, and we pre-train the CNN on a related task: shape classification. Figure 2 shows some pre-training samples as well. In Sect. 5.2, we examine the effects of increasing variety of shapes on final BP solver performance.

Feature Extraction. To extract features for a given BP tile, we use global-averaged feature map activation, which computes the spatially averaged activation value for each kernel [10]. The magnitude of a globally-averaged value can be interpreted as the prevalence of a particular feature in the input image, with features in earlier layers often corresponding to simple visual features and later layers detecting features corresponding to more abstract concepts specific to the dataset and task [15]. In Sect. 5.2, we examine the effects of extracting features from layers of different depths in the pre-trained CNNs.

Classification. After calculating feature vectors for each tile in a BP, we train a classifier to distinguish between the two classes. While any classifier may be used, careful consideration should be made to influence the type of rule we want it to learn. In Sect. 4, we discuss the different types of solutions and rules to Bongard problems. In Sect. 5.2, we also observe the effects of the classifier on performance.

4   Evaluating Bongard Problem Solvers

To understand how to automatically evaluate a BP solver, it helps to understand what properties a solution may possess. In the present work, we consider proposed solutions to possess (or lack) the following properties.

Validity. We consider a proposed rule to be valid if it is able to correctly split (classify) the original 12 tiles, and invalid otherwise. We consider a **rule** to be a condition that can categorize tiles into left or right (either correctly or incorrectly), whereas a **solution** is a rule which is valid and can thus *correctly* categorize the 12 original tiles.

Robustness. We consider a solution to be robust if it is able to not only classify the original 12 tiles, but also additional ones which are classified left or right according to the intended rule, defined by the author of the problem.

Simplicity. An intuitive definition, although often impractical to use for evaluation, is that a simple solution takes few words to state. The opposite of a simple rule is a complex rule.

Figure 3 illustrates how valid rules (solutions) to a given problem may vary in robustness and simplicity. In Sect. 4.1 we discuss how to evaluate a solver with respect to validity, and in Sects. 4.2 and 4.3, we discuss evaluation with respect to robustness and simplicity.

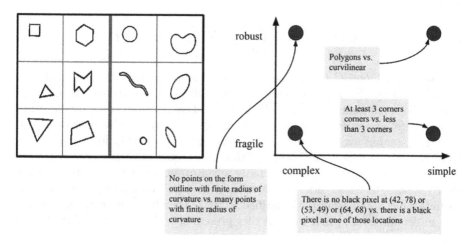

Fig. 3. Bongard #5 and various valid solutions (assuming each tile is 100 px by 100 px).

4.1 Measuring Validity

A model is said to produce a valid solution for a BP if the proposed rule correctly splits the 12 tiles into two groups. This corresponds to the evaluation method used by [5] and [3] (without the next step of subjective analysis). To condense the validity performance of a model into a single value, we average the validity scores across a set of Bongard problems:

$$validity = \frac{1}{\#\text{BPs}} \sum_{p \in BPs} pCC \tag{1}$$

where pCC is the set of all tiles in p correctly classified.

To accompany the validity score, we consider the average problem number where a valid solution is found. This allows us to observe whether our models have a bias, similar to humans, of solving more easy than difficult problems. This works due to the trend of problem difficulty increasing with problem number in the set of 200 BPs compiled by [5].

4.2 Measuring Robustness

Since the only way a solution can be robust is with respect to the intended solution, we use a functional definition of robustness. If a rule is able to correctly classify unseen samples from each class then it can be considered robust.

Here we define a subset $BPs(v)$ to represent Bongard Problems with valid solutions found by our model, and the set $\#BP(v)$ contains the total numbers of BPs for which valid rules are found. We average the robustness score based on BPs(v):

$$robustness = \frac{1}{\#BPs(v)} \sum_{p \in BPs(v)} newTilesCC \qquad (2)$$

where $newTilesCC$ if the fraction of new tiles for p correctly classified.

As noted by [3], Bongard problems are unlike usual classification problems in that the small number of examples for each class are often carefully chosen to have a single property in common while ruling out as many alternatives as possible. Leaving out even one or two tiles opens up to possibility for finding many non-intended solutions. Additionally, this interpretation of robustness ignores the case where a rule acts unexpectedly when presented with tiles that do not clearly belong to either side. If left vs right is circles vs. squares, what does it mean if a picture of a lamp is classified left? We therefore only consider robustness under the assumption that all tiles presented will belong to either the left or right.

4.3   Measuring Simplicity

Measuring the simplicity of a tile classification rule learned by an automated solver may be extremely difficult. This problem of interpreting how a deep learning model works is well studied with regards to image classification and often done with saliency maps, which show the parts of the input image which most influence the classification results [13, 15]. In the present work, we do not attempt to define a rule simplicity measure, however, in Sect. 5.3, we consider visualizing activation maps to gain insight into the types of rules discovered by out models.

5   Experiments and Results

In this section we analyze the performance and effects of hyperparameters of three variations of our problem solving model. The first model, **PT+SF**, uses pre-training and single feature classification (a decision tree of depth 1). Second is **PT+LR**, which also utilizes pre-training, but can propose rules combining many features using a logistic regression classifier.

First we discuss the experimental setup in Sect. 5.1, then we discuss observations made in 5.2, and in Sect. 5.3 we produce visualizations of the rules implicitly learned by a solver and examine their utility.

5.1   Setup

To observe the effect of feature abstraction on BP solver performance with as few confounding variables as possible, we use the same hyperparameters for each of the convolutional layers (architecture shown in Fig. 4):

- 64 kernels of size 3×3 with stride 1, ReLU non-linearity, and followed by 2×2 max-pooling with stride 2.

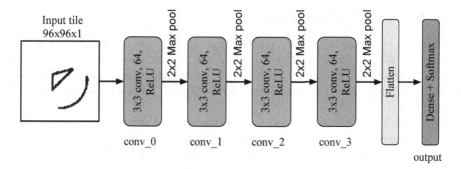

Fig. 4. Neural network architecture used. Four convolutional layers with the same hyperparameters are used.

- For the PT+SF and PT+LR models, the output of the last convolutional layer is mapped to shape class probabilities with a dense layer and softmax activation

The models are trained with categorical cross-entropy loss and the Adam optimizer [8] with the default hyperparameters defined by Keras [2].

We use 100,000 tiles (80/20 train/validation split) of size $96 \times 96 \times 1$. Table 1 contains the details of the five different pre-training data types of increasing complexity we experiment with and the number of training epochs we found to produce stable validation scores without overfitting. The final validation accuracy ranged from 100% for the easiest pre-training set to 93% for the most complex set.

Table 1. Pre-training data details.

Pre-training data type	Shape classes	# Shape classes	Training epochs
1	Single-segmented lines, dots, curves	3	3
2	#1, circles, ellipses	7	6
3	#2, 3-gons, equilateral 3-gons	11	20
4	#3, 2- and 3-segmented lines, 4-gons, equilateral 4-gons	17	20
5	#4, 5- and 6-gons, equilateral 5- and 6-gons	25	20

To evaluate the overall validity power of our models, we incrementally combine and keep all useful features from each convolutional layer, including the output with small size of shape classes that may carrying simple shape information to solve certain BPs, to obtain a consistent evaluation results. Sample layers would be like:

$$output, \ output + CL3, \ output + CL3 + CL2, ...$$

5.2 Results and Observations

At first, we evaluate the validity scores and average values of BP#(v) as defined in Sect. 4.1 for the 200 BPs. Then we manually created two additional test tiles for each problem (1 for each side) in order to estimate robustness based on the model's validity. All experiments results for PT+SF and PT+LR are averaged across 5 trials.

In Table 2 and Table 3 we see the validity and robustness for the proposed PT+SF and PT+LR models, containing the results with respect to the pre-training type and combined layers for feature extraction.

Table 2. Effects of pre-training and CNN combined layers used for feature extraction on PT+SF performance with 5 trials. CLi refers to the ith convolutional layer. The highest scores for each metric are bolded, and second and third highest underlined.

Metric	Layers	Pre-training type				
		1	2	3	4	5
Validity (avg BP#(v))	output	1.2% (89)	3.3% (50)	3% (21)	2.6% (49)	1.9% (66)
	output + CL3	14.3% (78)	18.6% (84)	19.5% (79)	22.1% (85)	19.2% (84)
	output + CL3 + CL2	18.8% (83)	22.1% (86)	25% (84)	27% (90)	24.7% (90)
	output + CL3 + CL2 + CL1	21.5% (87)	25.1% (87)	26.4% (85)	28.6% (90)	28.0% (91)
	output + CL3 + CL2 + CL1 + CL0	23.3% (90)	26.4% (88)	27.7% (88)	**30.2% (93)**	28.6% (91)
Robustness	output	**97.50%**	81.90%	64.94%	66.16%	84.16%
	output + CL3	61.18%	60.58%	62.92%	64.74%	66.30%
	output + CL3 + CL2	62.50%	62.36%	61.84%	62.66%	65.64%
	output + CL3 + CL2 + CL1	62.30%	63.30%	62.50%	64.58%	67.52%
	output + CL3 + CL2 + CL1 + CL0	63.86%	60.84%	63.36%	63.34%	66.26%

Table 3. Effects of pre-training type and CNN layers used for feature extraction on PT+LR performance. For the results shown, the logistic regression penalty is fixed to $l2$ and inverse regularization strength is chosen from $C = [1, 2, 4, 8, 16, 32, 64, 128]$. The highest scores for each metric are bolded, and second and third highest underlined.

Metric	Layers	Pre-training type				
		1	2	3	4	5
Validity (avg BP#(v))	output	1% (78)	2.7% (67)	1.8% (49)	2.5% (22)	3.7% (43)
	output + CL3	71.9% (96)	94.7% (96)	96.4% (97)	98.1% (98)	99.7% (99)
	output + CL3 + CL2	78.8% (99)	95.9% (97)	97.9% (98)	98.6% (98)	**99.8% (99)**
	output + CL3 + CL2 + CL1	79.2% (100)	95.9% (97)	98.1% (98)	98.7% (98)	**99.8% (99)**
	output + CL3 + CL2 + CL1 + CL0	78% (100)	95.9% (97)	98.1% (98)	98.7% (98)	**99.8% (99)**
Robustness	output	100.00%	84.28%	58.00%	74.32%	60.56%
	output + CL3	57.88%	54.64%	56.74%	54.74%	55.76%
	output + CL3 + CL2	56.08%	54.52%	56.68%	55.08%	55.92%
	output + CL3 + CL2 + CL1	56.40%	54.34%	56.86%	55.42%	56.72%
	output + CL3 + CL2 + CL1 + CL0	57.00%	54.34%	56.86%	55.22%	56.92%

Effects of Pre-training Complexity. For both PT+SF and PT+LR, increasing the variation of the shape set led to an improvement in both validity and robustness, with the effect being stronger when using the logistic regression classifier. The only exception is when the output class distributions are the only extracted features, in while case the robustness scores may be unusually high due to the smaller value of $BPs(v)$

Effects of Layers Combination. For both pre-trained models, it appears that including more convolutional layers produces better features when measuring validity, but robustness is less affected. This may be due to the deeper convolutional layers learning features specific to the shape classification task and thus less applicable to other tasks [15]. We can also observe that the PT+LR model can be seen as over-fitting when measuring validity (as indicated by the low corresponding robustness).

The output layer consistently performs poorly for both PT+SF and PT+LR in terms of validity, likely due to the small number of shape classes as listed in Table 1. This also suggests that just knowing what basic shapes are present in the image is helpful for solving only a small set of simple Bongard problems.

Effects of Classifier. In the PT+SF model, we used a decision tree with depth 1 to choose a single visual feature to serve as a rule for each Bongard problem. From the results, this very simple classifier has generally smaller validity scores compared with the PT+LR model, but is more robust. This observation matches the nature of BPs: they are often designed to be solved with only one abstract rule or feature as an intended solution. Thus, PT+SF may score higher in simplicity than PT+LR. The PT+LR model, using logistic regression, linearly combines many features. Not surprisingly, this more expressive classifier is capable of producing much higher validity.

Overall Performance. While direct performance comparisons should not be drawn to previous approaches due to differences in the types of rules automatically produced, our approaches are capable of finding valid solutions to a greater fraction of problems than previous approaches. Our PT+SF model finds valid solutions for up to 30.2% of the problems (~60/200) and correctly classifies two new test tiles for 66.3% of the problems(~38/60). The PT+LR achieve almost 100% validity, but at the cost of more complex solution rules. In contrast, [5] reports ~7.5% and the previous work most similar to ours and without further test set validation, [7], reported 18% of 232 problems solved (and 19% of the 200 problems we use).

5.3   Rule Visualization

In Fig. 5 we present 8 problems for which a valid solution was found by a PT+SF model which used pre-training set #4 and tile embeddings from the feature maps in the last convolutional layer. Highlighted areas indicate higher values in the activation map chosen by the BP solver for that problem. The intended rules are provided for each problem.

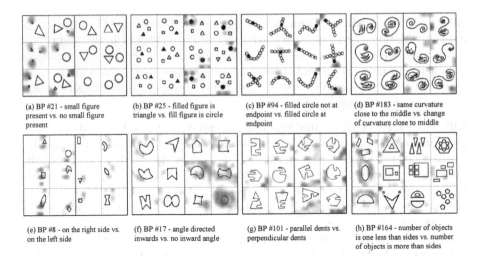

(a) BP #21 - small figure present vs. no small figure present

(b) BP #25 - filled figure is triangle vs. fill figure is circle

(c) BP #94 - filled circle not at endpoint vs. filled circle at endpoint

(d) BP #183 - same curvature close to the middle vs. change of curvature close to middle

(e) BP #8 - on the right side vs. on the left side

(f) BP #17 - angle directed inwards vs. no inward angle

(g) BP #101 - parallel dents vs. perpendicular dents

(h) BP #164 - number of objects is one less than sides vs. number of objects is more than sides

Fig. 5. Examples of interpretable (a to d) and non-interpretable (e to h) visualizations of valid rules found by the PT+SF model for Bongard problems.

Problems (a) to (d) in Fig. 5 have arguably interpretable rules. The intended rule for (a) is 'small shapes present on the left but not the right', and as expected, small figures are highlighted by the activation map of the automatically chosen filter. In both (b) and (c), the shapes clearly associated with the intended rules are highlighted. However, for (d), it appears that a valid, although non-intended solution was identified: there is more empty space around the corners on the right than on the left. The intended solution for this problem is 'same curvature close to the middle vs. change of curvature close to middle'. Problems (e) to (h) have also had valid solutions identified, but serve to demonstrate that a standard method of visualizing what a CNN has learned is frequently not well-suited for Bongard problems, as it is not always clear what part of the tiles should be highlighted to make the discovered rule more visible.

6   Conclusions

Bongard problems are a kind of visual puzzle which require skills central to human intelligence: divergent thinking, abstract thinking, and the ability to learn from little data. To solve these problems given raw images, we train a CNN to perform the related task of shape classification and use the globally-averaged feature maps to produce feature vectors for the tiles of a BP. We observed that increasing the shape variation of the pre-training data as well as extracting features from deeper convolutional layers tended to improve the quality of the extracted feature vectors, increasing the number of problems for which valid and robust solutions could be discovered.

The present work hints at many promising avenues. While the author of a problem may have a particular rule in mind, an automated solver may identify many valid solutions. Adding an active learning component to the Bongard problem requiring automated solvers to strategically test highly abstract hypotheses may be interesting. Developing a visualization technique capable of conveying the abstract rules learned by an automated solver is another task which may prove to be important.

References

1. Bongard, M.M.: The Problem of Recognition. Fizmatgiz, Moscow (1967)
2. Chollet, F.: Keras (2015). https://github.com/fchollet/keras
3. Depeweg, S., Rothkopf, C.A., Jäkel, F.: Solving Bongard problems with a visual language and pragmatic reasoning. arXiv preprint arXiv:1804.04452 (2018)
4. Fei-Fei, L., Fergus, R., Perona, P.: One-shot learning of object categories. IEEE Trans. Pattern Anal. Mach. Intell. **28**(4), 594–611 (2006). https://doi.org/10.1109/TPAMI.2006.79
5. Foundalis, H.: Phaeaco: a cognitive architecture inspired by Bongard's problems, Ph.D. thesis. Indiana University, Indiana, Bloomington (2006).
6. Hofstadter, D.R.: Gödel, Escher, Bach. Vintage Books, New York (1980)
7. Kharagorgiev, S.: Solving Bongard problems with deep learning, February 2018. https://k10v.github.io/2018/02/25/Solving-Bongard-problems-with-deep-learning/
8. Kingma, D.P., Ba, J.: Adam: a method for stochastic optimization. CoRR abs/1412.6980 (2014). http://arxiv.org/abs/1412.6980
9. Lake, B., Salakhutdinov, R., Gross, J., Tenenbaum, J.: One shot learning of simple visual concepts. In: Proceedings of the Annual Meeting of the Cognitive Science Society, vol. 33 (2011)
10. Lin, M., Chen, Q., Yan, S.: Network in network. arXiv preprint arXiv:1312.4400 (2013)
11. Raven, J.C., et al.: Raven's progressive matrices. Western Psychological Services, Los Angeles, CA (1938)
12. Santoro, A., Hill, F., Barrett, D., Morcos, A., Lillicrap, T.: Measuring abstract reasoning in neural networks. In: International Conference on Machine Learning, pp. 4477–4486 (2018)
13. Simonyan, K., Vedaldi, A., Zisserman, A.: Deep inside convolutional networks: visualising image classification models and saliency maps. arXiv preprint arXiv:1312.6034 (2013)
14. Vinyals, O., Blundell, C., Lillicrap, T., Wierstra, D., et al.: Measuring abstract reasoning in neural networks. In: Advances in Neural Information Processing Systems, pp. 3630–3638 (2016)
15. Zeiler, M.D., Fergus, R.: Visualizing and understanding convolutional networks. In: Fleet, D., Pajdla, T., Schiele, B., Tuytelaars, T. (eds.) ECCV 2014. LNCS, vol. 8689, pp. 818–833. Springer, Cham (2014). https://doi.org/10.1007/978-3-319-10590-1_53

Adversarial Models for Deterministic Finite Automata

Kaixuang Zhang[1(✉)], Qinglong Wang[2], and C. Lee Giles[1]

[1] Information Sciences and Technology, Pennsylvania State University,
State College, USA
{kuz22,clg20}@psu.edu
[2] School of Computer Science, McGill University, Montreal, Canada
qinglong.wang@mail.mcgill.ca

Abstract. We investigate a finer-grained understanding of the characteristics of particular deterministic finite automata (DFA). Specifically, we study and identify the transitions of a DFA that are more important for maintaining the correctness of the underlying regular language associated with this DFA. To estimate transition importance, we develop an approach that is similar to the approach widely used to expose the vulnerability of neural networks with the adversarial example problem. In our approach, we propose an adversarial model that reveals the sensitive transitions embedded in a DFA. In addition, we find for a DFA its critical patterns where a pattern is a substring that can be taken as the signature of this DFA. Our defined patterns can be implemented as synchronizing words, which represent the passages from different states to the absorbing state of a DFA. Finally, we validate our study through empirical evaluations by showing that our proposed algorithms can effectively identify important transitions and critical patterns. To our knowledge, this is some of the first work to explore adversarial models for DFAs and is important due to the wide use of DFAs in cyberphysical systems.

Keywords: Deterministic Finite Automata · Transition importance · Critical patterns · Adversarial model

1 Introduction

There has been a great deal of work on the computational power of deterministic finite automata (DFA, level 3 in the Chomsky hierarchy). Although DFA models can be significantly different in terms of the number of states, accepted strings, and complexity [12], the family of DFA models is usually studied as a whole for their computational power and compared with other formal computation models in the Chomsky hierarchy [8]. As such, our understanding of DFA remains at a relatively coarse-grained level. We believe it is still an open question regarding on how to differentiate different DFAs.

Here, we study individual DFAs for their fine-grained characteristics, including transition importance and critical patterns. Specifically, we examine the

C. Goutte and X. Zhu (Eds.): Canadian AI 2020, LNAI 12109, pp. 540–552, 2020.
https://doi.org/10.1007/978-3-030-47358-7_55

importance of transitions by relating this task with the adversarial example problem [10] often seen in deep learning. This problem describes the phenomenon where a model, which generalizes well on clean datasets, is strikingly vulnerable to adversarial samples crafted by slightly perturbing clean samples. Because string identification is an important problem in time series, speech, and other scenarios, this motivates research into understanding complicated learning models such as neural networks. One particular approach is to identify feature-level perturbations that significantly affect a learning model. Similar approaches have been used for examining the sample-level importance in building a learning model [9]. These studies use adversarial examples as a data-driven tool for probing the learning model's vulnerability, hence indirectly gaining an understanding of complicated learning models. In order to directly gain a better understanding of a DFA, we follow a similar approach but study the sensitivity of a DFA through model-level perturbations.

Next, we study critical patterns that can be used for identifying a specific DFA. Specifically, we formally define a critical pattern as a substring, which effectively identifies all strings accepted by a certain DFA. We show that for certain classes of DFA, we can identify these strings statistically by checking the existence of critical patterns embedded in their generated strings without exhaustively searching all possible strings or querying the underlying DFA [1,13]. We then develop an algorithm for finding the critical patterns of a DFA by transforming this task as a DFA synchronizing problem [6]. Last, we provide a theoretical approach for estimating the length of any existing perfect patterns and validate our analysis with empirical results.

We feel that our analysis on DFA models will help in research on the security of cyberphysical systems that are based on working DFAs, e.g., compilers, VLSI design, elevators, and ATMs. This could be especially for the case when the actual state machine is exposed to adversaries and be attacked. It is the intent of this work to open a discussion on these issues.

2 Transition Importance of a DFA

DFAs are one of the simplest automata in the Chomsky hierarchy of phrase structured grammars [4]. More formally, a DFA can be described by a five-tuple $A = \{\Sigma, S, I, F, T\}$, where Σ is the input alphabet (a finite, non-empty set of symbols), S denotes a finite and non-empty set of states, $I \in S$ represents the initial state while $F \subseteq S$ represents the set of accept states, and T is a set of deterministic transition rules. The transition rules of a certain DFA essentially describe how that DFA will process a string as it traverses its states. Throughout this paper all DFAs are complete minimal DFAs. Due to its deterministic nature, it is natural to assume different transitions are equally important for identifying a DFA. However, as will become clear from our analysis, this assumption does not generally hold. Here we illustrate this with the a DFA associated with the

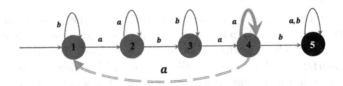

Fig. 1. Example for the Tomita-7 grammar. Red (Black) states are the accept (reject) states (Color figure online).

Tomita-7 grammar[1] shown in Fig. 1. Among all transitions, the cyclic transition (with input a) associated with state-4 is the most important one. This is because by substituting this transition by a transition to state-1 with the same input, we can add significantly more strings to the set of accepted strings.

2.1 Transition Importance Estimation as an Adversarial Model Problem

To estimate the importance of each transition and identify more important ones, we take an approach that is complementary to the approach used to identify sensitive features of a data sample viewed by a deep neural network (DNN) in the context of the adversarial example problem. As such, the transition importance estimated by our approach essentially reflects the sensitivity of a DFA with respect to a transition.

A typical formulation of the adversarial example problem is to maximize a loss function $\mathcal{L}$ with respect to a normal sample x_0 and a model f. Then finding an adversarial sample $\hat{x}$ is conducted by solving the following problem:

$$\hat{x} = \underset{|x-x_0|\leq\epsilon}{\arg\max}\, \mathcal{L}(x, f), \tag{1}$$

where ϵ denotes some predefined constraint on the scale of perturbation. Here we propose to transform the adversarial example problem (Eq. (1)) into the *adversarial model problem*, which considers model-level perturbations. Explicitly, given a model f_0 and a fixed set of string samples X, we try to solve the following problem:

$$\hat{f} = \underset{|f-f_0|\leq\epsilon}{\arg\max} \sum_{x\in X} \mathcal{L}(x, f). \tag{2}$$

Eq. (2) describes the problem of perturbing a target model in a constrained manner to cause maximal loss and provides an alternative view of the adversarial

---

[1] Tomita [11] defined the following grammars with a binary alphabet: (1) a^*, (2) $(ab)^*$, (3) an odd number of consecutive $'a's$ is always followed by an even number of consecutive $'b's$, (4) any binary string not containing "bbb" as a substring, (5) even number of bs and even number of $'a's$, (6) the difference between the numbers of $'b's$ and $'a's$ is a multiple of 3, (7) $b^*a^*b^*a^*$. These grammars have been widely used in grammatical inference.

example problem. Specifically, given an ideal mapping f from some functional space $\mathcal{F}$ for a certain learning task, and an arbitrarily small approximation error ϵ, the universal approximation theory [5] states that one can always find a candidate f' in some other functional space $\mathcal{F'}$ (generally taken as a subset of $\mathcal{F}$) satisfying $\|f - f'\| < \epsilon$. Given that DNNs already have very complicated architectures[2], we can only measure the difference between f and f' numerically, although these two functions might be quite different. Since these models built through analytical approaches may not necessarily have actions aligning with our intuition and expectation, we cannot easily, if not impossibly, establish a physical understanding of the gap between f' and f. Furthermore, in practice the gap between these two functions may be amplified by formulating the approximation problem as an optimization problem, and then applying various techniques to solve the latter [2]. These combined effects imply that the root cause of the adversarial example problem lies in both the ambiguity of the theoretical foundation for building a learning model and the imperfection in the practice of applying a learning model. Moreover, it is important to note that our transformation cannot be easily applied to complicated models like DNNs. The function represented by a DNN has too many parameters, including weights, neurons, layers, and all sorts of hyper-parameters. This results in an enormous perturbation space.

On the other hand, for a DFA, the perturbation space is significantly limited to only include its transitions and states. Furthermore, the perturbation of a state can be represented by a set of perturbations applied to the transitions associated with this state. Therefore, in the following, we only consider transition perturbations as they provide a more general description of the adversarial perturbations of a DFA. In addition, we only consider perturbations that make substitution operations on the transitions. This is because for a given DFA, inserting transitions is not allowed since this DFA is already complete and minimal. Also, removing a transition is equivalent to substituting this transition to the transition that connects the current state to an absorbing state, of which the outward transitions all loop back to itself. Our study of the adversarial DFA can be taken as a step in studying the adversarial phenomenon by restricting the underlying models to be physically interpretable and directly investigating the vulnerability of that model.

2.2   Transition Importance

The deterministic property of a DFA enables it to be naturally immune to adversarial examples. However, when the adversarial perturbation is applied to a DFA, it is possible to generate an adversarial DFA, which only differs from the original DFA by a limited number of transitions, yet recognizes a regular grammar that is dramatically different from the one associated with the original DFA. To quantitatively evaluate the difference between two sets of strings accepted by different DFAs, here we introduce the following metric:

---

[2] Recent research [7] on explaining DNNs have demonstrated the difficulty of analyzing and inspecting these powerful models.

Definition 1 (Intersection over Union (IoU)). *Given two arbitrary DFAs represented by A and $\hat{A}$, and their accepted sets of strings denoted by X and $\hat{X}$, respectively, then*

$$IoU(A, \hat{A}) = \frac{\left|X \cap \hat{X}\right|}{\left|X \cup \hat{X}\right|}. \tag{3}$$

It is easy to notice that the metric IoU is well-defined and lies between 0 and 1. One can apply the L'Hopital's rule to calculate it if both the numerator and the denominator approach infinity. By using the above definition of IoU, we express the adversarial model problem for a DFA as perturbing the transitions of a given DFA to reach a low IoU. Then we have the following theorem.

Theorem 1. *Given a DFA with alphabet $\Sigma = \{a_1, a_2\}$, we use A_1 and A_2 to denote its transition matrices associated with the first and second input symbol. Similarly, let $\hat{A}_1$ and $\hat{A}_2$ denote the transition matrices of perturbed DFA yielding*

$$IoU(A, \hat{A}) = \left(\frac{\sum_{n=1}^{\infty}(\mathbb{1} \otimes p)^{\mathrm{T}}(M_1 \otimes (A_1 + A_2) + M_2 \otimes (\hat{A}_1 + \hat{A}_2))^n(\mathbb{1} \otimes q)}{\sum_{n=1}^{\infty}(p \otimes p)^{\mathrm{T}}(A_1 \otimes \hat{A}_1 + A_2 \otimes \hat{A}_2)^n(q \otimes q)} - 1\right)^{-1}. \tag{4}$$

where $p \in \mathbb{B}^n$ is a one-hot encoding vector to represent the initial state, and $q \in \mathbb{B}^n$ denotes the set of accept states of a DFA with n states. We also have $\mathbb{1} = \begin{bmatrix} 1 \\ 1 \end{bmatrix}$, $M_1 = \begin{bmatrix} 1 & 0 \\ 0 & 0 \end{bmatrix}$, $M_2 = \begin{bmatrix} 0 & 0 \\ 0 & 1 \end{bmatrix}$, and $\otimes$ denotes the Kronecker product.

Due to space constraints, we only provide a sketch of this proof. For this we construct a new automaton that represents the union of two source DFAs. The initial state vector, accepting state vector, and the adjacency matrix of this constructed automaton are denoted as $\mathbb{1} \otimes p$, $\mathbb{1} \otimes q$, and $M_1 \otimes (A_1 + A_2) + M_2 \otimes (\hat{A}_1 + \hat{A}_2)$, respectively. Similarly, for the DFA that recognizes the intersection of two sets of strings accepted by two DFAs, we denote its initial state vector, accepting state vector, and the adjacency matrix as $p \otimes p$, $q \otimes q$, and $A_1 \otimes \hat{A}_1 + A_2 \otimes \hat{A}_2$, respectively. In order to compute the cardinality of the union set and the intersection set, we need to sum the number of strings for which the length varies from 1 to infinity. Now assume that there are two column vectors s_I and s_E, which represent the set of initial and ending states. Then the number of N-length strings that reach s_E from s_I is $s_I^{\mathrm{T}} P^N s_E$.

Theorem 1 provides directly an explicit formulation for computing our defined IoU. As such, the original adversarial model problem for the DFA can be transformed to an optimization problem. Furthermore, we require that this manipulation only allows one transition substitution to be applied to one of the transition matrices associated with different inputs. The allowed single transition substitution causes the Frobenius norm of the manipulated transition matrix to be changed by $\sqrt{2}$. This also avoids changes to the absorbing states of the source DFA (if they exist), so that any existing absorbing states will not be affected.

In addition, we require the set of accepted states remains the same. Therefore, we have the following optimization problem[3]:

$$\min_{\hat{A}_1,\hat{A}_2 \in \mathcal{T}} \frac{\sum_{n=1}^{\infty}(p \otimes p)^{\mathrm{T}}(A_1 \otimes \hat{A}_1 + A_2 \otimes \hat{A}_2)^n (q \otimes q)}{\sum_{n=1}^{\infty}(\mathbb{1} \otimes p)^{\mathrm{T}}(M_1 \otimes (A_1 + A_2) + M_2 \otimes (\hat{A}_1 + \hat{A}_2))^n(\mathbb{1} \otimes q)}$$

$$\text{s.t. } \& \left\|\hat{A}_1 - A_1\right\|_F^2 + \left\|\hat{A}_2 - A_2\right\|_F^2 = 2; \tag{5}$$

$$y(A_1 + A_2) = y(\hat{A}_1 + \hat{A}_2);$$

$$(A_1 + A_2)y^{\mathrm{T}} = (\hat{A}_1 + \hat{A}_2)y^{\mathrm{T}}.$$

where $y = 0$ when the source DFA does not have an absorbing state and $y = [0, 0, \cdots, 1]$. Otherwise, $\mathcal{T}$ denotes the set of transition matrices which contains exactly one 1 in each row, and $\|\cdot\|_F$ denotes the Frobenius norm.

In practice, it is possible that some additional constraints can be added to the above formulation. Specifically, here we require the perturbed DFA to remain strongly connected and no new absorbing states will be created. Since it is difficult to formulate these constraints in Eq. (5), we manually examine their violations in the obtained solutions. Note that these constraints can be easily checked by analyzing the spectrum of the perturbed transition matrix.

2.3 Evaluation of DFA Transition Importance

In the following, we use the Tomita grammars to demonstrate our estimation of the transition importance of DFAs. Since this is the first work on studying the adversarial scheme of formal computation models, our evaluation mainly focuses on examining the effectiveness of our proposed approach.

In the experiments, we select the Tomita-3/5/7 grammars as examples. These grammars are selected as they are representative of the exponential, proportional, and polynomial classes [12] of regular grammars with the binary alphabet, respectively (These classes are introduced in Sect. 3). Since it is impossible to sum up to infinity, for our evaluation we fix the maximum length N of binary strings to 20. Also, instead of solving the original objective which takes a quotient form, we apply the symmetry difference of two sets as an alternative. This choice is reasonable since the objective functions capture the same essence of minimizing the cardinality of the intersection and maximizing the cardinality of the union of two sets. Furthermore, as the original problem is formulated as a high-order integer programming problem, which is difficult to solve with existing solvers, we relax the constraints such that $\hat{A}_1$ and $\hat{A}_2$ are constrained as row-wise stochastic matrices as their entries. As such, we determine the final perturbation by selecting the one with the maximal value, which represents the maximal transition probability. We notice that our approximation may not yield the real optimal solution; however, as shown by the results in Table 1, it provides satisfying results in analyzing the transition importance.

---

[3] The constant number 1 is omitted for simplicity.

Table 1. Optimization results for the Tomita-3/5/7 grammars.

		IoU	
		Value from optimization	Value from randomization
The	3	1.48e-3	0.342
Tomita	5	0.152	0.289
Grammars	7	0.025	0.225

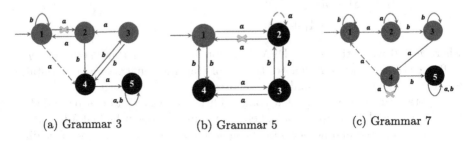

(a) Grammar 3   (b) Grammar 5   (c) Grammar 7

Fig. 2. Illustration of identified important transitions for example DFAs. The marked (with a yellow cross) and dashed lines demonstrate the most sensitive transitions of the original DFAs and the perturbed transitions, respectively. (Color figure online)

The effectiveness of our optimization approach when comparing it with a randomization approach is shown in Table 1. Specifically, for the randomization approach, we randomly select five legitimate perturbations (manually checked according to the constraints described above) and calculate and average the resulting IoU_{rand}. We then compare IoU_{rand} with the IoU_{opt} obtained by our optimization. It is clear that the results provided by the optimization approach are much more desirable. We also provide a visualization of the perturbations generated by our approach for each investigated grammar in Fig. 2.

3   Critical Patterns of DFA

3.1   Different Types of Critical Patterns

Here, we provide a relatively coarse-grained view, in contrast to what we described regarding transition importance, to investigate the characteristics of a DFA. Specifically, we identify critical patterns of a DFA, defined as:

Definition 2 (Absolute and relative patterns of a DFA). *Given the alphabet Σ of a DFA and a data space $X \subseteq \Sigma^*$, X is the union of two disjoint sets, i.e., $X = P \cup N$[4], and we define the following patterns:*

$$\text{Absolute pattern}: \hat{m} = \arg\max_{|m|=k} \left| Pr_{m \sim_f y}(y \in P) - Pr_{m \sim_f y}(y \in N) \right|. \quad (6)$$

---

[4] For a DFA, P (N) represents the space of strings accepted (rejected) by this DFA.

Relative pattern : $\hat{m} = \underset{|m|=k}{\arg\max} \left| Pr_{y \in P}(m \sim_f y) - Pr_{y \in N}(m \sim_f y) \right|,$ (7)

where y is a string in X and m $\sim_f$ y indicates that m is a factor (consecutive substring) of y.

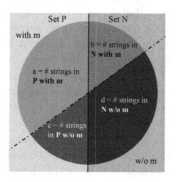

Fig. 3. An illustration of the difference between absolute and relative patterns.

Here we focus on the general case where all the strings follow the uniform distribution without using any particular prior knowledge. We illustrate the difference between the absolute and relative patterns with the example in Fig. 3 by splitting the entire data space X into four parts denoted as $\{a, b, c, d\}$. According to Eq. (6), an absolute pattern describes the substring m (has the length of k) such that, among all strings that contain m, it causes the largest discrepancy between the probabilities of a string that belongs to different disjoint sets. Thus, the absolute pattern differentiates strings in $\{a, b\}$, and the objective in Eq. (6) are equal to $\frac{|a-b|}{a+b}$. In contrast, a relative pattern is identified by considering the statistics of the entire data space, with the objective in Eq. (7) equal to $\left| \frac{a}{a+c} - \frac{b}{b+d} \right|$. Note that these two patterns are equivalent to each other under certain circumstances. For example, consider a DFA that rejects any binary string containing "*bbb*" as a substring[5]. In this case, both the absolute and relative patterns identify the factor "*bbb*". Here, we are concerned with the absolute pattern since it provides better insight as to the connection between identified patterns and the underlying DFA. In contrast a relative pattern mainly provides a conceptual understanding from a statistic perspective. Furthermore, we introduce the following definition:

Definition 3 (Perfect absolute pattern of a DFA).
   Let $\mathcal{A}_p = \{m \mid \max_m \left| Pr_{m \sim_f y}(y \in P) - Pr_{m \sim_f y}(y \in N) \right| = 1\}$, then the perfect absolute pattern is defined as:

$$\hat{m} = \underset{m \in \mathcal{A}_p}{\arg\min} |m|.$$ (8)

---

[5] The example DFA is associated with the Tomita-4 grammar.

A perfect absolute pattern describes a substring, which has minimal length among all absolute patterns and perfectly differentiates the strings from different disjoint sets. However, not all DFAs have perfect absolute patterns. Some DFAs, which have a cyclic property, contain recurrent or persistent states that contain both accepting and non-accepting states. This indicates that these DFAs can never determine the label of a string until they finish processing the entire string. These DFAs with a binary alphabet, as previously determined [12], belong to one of three classes, which are then categorized according to the complexity of different DFAs. Specifically, the complexity of a DFA is measured by its entropy value, which essentially reflects how balanced are the sets of strings accepted or rejected by a DFA. As such, a grammar with a higher complexity has a higher entropy value, hence recognizing more balanced string sets. Based on the entropy values of different DFAs, they can be categorized into three basic classes (1) polynomial class, where the number of accepted strings of a certain length is a polynomial function of the length; (2) exponential class, where the number of accepted strings of a certain length takes an exponential form of the length with the base value smaller than 2; (3) proportional class, where the number of accepted strings of a certain length is proportional to number of all binary strings with the same length. Interestingly, except for the proportional class, which contains DFAs with either 0 or 2 absorbing states, DFAs from other classes have exactly one absorbing state. See Wang et al. [12] for more details.

Upon inspection, it is just a random guess for identifying a string accepted by a DFA which has no absorbing state and by only checking its contained factors. Also, determining the pattern of a DFA, which contains two absorbing states, can be taken as performing a random guess twice. As such, we only focus in the following analysis on DFAs belonging to the polynomial and the exponential classes. Importantly, we find that identifying a perfect absolute pattern of a DFA is essentially analogous to designing a *synchronizing word* [6] for the absorbing state of a DFA. Therefore, instead of solving the optimization problem in Eq. (6), we propose a DFA synchronizing word approach and design a metric to evaluate the confidence for determining whether a certain string belongs to a particular class. We show that our metric is highly correlated with the probability in Eq. (6).

3.2 DFA Synchronizing Algorithm

Recall that the synchronizing word (or the reset sequence) is a substring that sends any state of this DFA to the same state. An absorbing state naturally fits this synchronizing scheme. As such, we can set the absorbing state as the state to be synchronized. And since all states will result in an absorbing state when applying the same substring to these states, the label of a string containing the substring can be determined definitely.

However, given a string of fixed length k, there is no guarantee that we can always reach an absorbing state. Thus, we design the following algorithm and metric to evaluate the efficiency of identifying an absolute pattern. More specifically, given a DFA with n states and a predefined length k, we then have its k-order transition matrix A_Σ^k by multiplying the transition matrix A_Σ by itself k times. We focus on the column associated with the absorbing state. This column represents the prefixes coming from all states to this absorbing state. We now choose the most frequent substring m appearing in this column. We denote that number of occurrences as $\hat{n}$ and determine m since an absolute pattern has confidence of $\hat{n}/n$. For the perfect absolute pattern, the confidence is 1, since the substring sends all other states to the absorbing state and it will appear in each entries of the column associated with the absorbing state of the k-order transition matrix. In experiments presented in the latter part of this section, we demonstrate the results of applying this algorithm.

Furthermore, given a DFA with one absorbing state, similar to the Černý's conjecture [3], we can estimate the length of a perfect absolute pattern associated with a DFA by providing a loose upper bound. That is, given a DFA with an absorbing state, we have the following theorem for estimating the minimal length of a synchronizing substring, which leads all states to the absorbing state.

Theorem 2. *The length of a perfect absolute pattern of a DFA with n states is at most $n(n-1)/2$.*

To obtain an upper bound of the length of a perfect absolute pattern, we need to consider the worst case. Specifically, the distance between each state and the absorbing state is $1, 2, \cdots, n-1$, respectively. In addition, at step t, synchronizing the nearest state is the optimal choice. Furthermore, after synchronizing the t-step nearest states, the distances between the rest $n-t$ states and the absorbing state range exactly from $t+1$ to $n-1$ during this iterative process. As such, to synchronize all states in the worst case, we have the length of a synchronizing substring at most $1 + 2 + \cdots + (n-1)$, which is equal to $n(n-1)/2$.

It is straightforward to check that the upper bound in Theorem 2 holds for any size of alphabet. As such, we conjecture that there exists a tighter upper bound, which depends on the number of states and the DFA alphabet size. Next, we provide some examples in order to further investigate the pattern length.

We demonstrate in Fig. 4a and b that when the number of states of a DFA is set to 3 or 4, we can construct a DFA for which the perfect absolute pattern meets the upper bound exactly. Specifically, for the DFAs shown in Fig. 4a and b, their associated patterns are *bab* and *babaab*. However, it is impossible to construct a 5-state DFA, for which the perfect absolute pattern has a length that reaches the upper bound in Theorem 2. More specifically, we have the following result:

Theorem 3. *The length of a absolute pattern of a 5-state DFA is at most 9.*

This theorem can be proved by using combinatoric and enumeration techniques, and is omitted due to space constraints. In Fig. 4c, we construct an example five-state DFA that has a pattern with a length of 9 in the worst case.

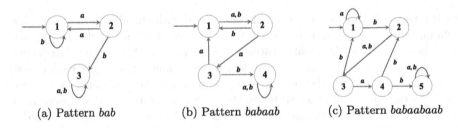

(a) Pattern *bab*   (b) Pattern *babaab*   (c) Pattern *babaabaab*

Fig. 4. DFA examples that illustrate Theorem 2.

Table 2. Transition matrices for example DFAs. $A_\sigma(s_1, s_2)$ represents a transition from state s_1 to s_2 via σ. We only provide the transitions matrices for DFA-3/4/5 and omit the matrices for DFA-1/2 (Tomita-4/7) due to space limit.

	Transition matrix
DFA 3	$A_a(1,4)$ $A_a(2,3)$ $A_a(3,4)$ $A_a(4,3)$ $A_a(5,4)$ $A_a(6,6)$
	$A_b(1,2)$ $A_b(2,3)$ $A_b(3,3)$ $A_b(4,5)$ $A_b(5,6)$ $A_b(6,6)$
DFA 4	$A_a(1,5)$ $A_a(2,4)$ $A_a(3,2)$ $A_a(4,1)$ $A_a(5,6)$ $A_a(6,1)$ $A_a(7,7)$
	$A_b(1,2)$ $A_b(2,3)$ $A_b(3,2)$ $A_b(4,3)$ $A_b(5,6)$ $A_b(6,7)$ $A_b(7,7)$
DFA 5	$A_a(1,2)$ $A_a(2,3)$ $A_a(3,4)$ $A_a(4,8)$ $A_a(5,3)$ $A_a(6,8)$ $A_a(7,4)$ $A_a(8,8)$
	$A_b(1,6)$ $A_b(2,1)$ $A_b(3,5)$ $A_b(4,7)$ $A_b(5,2)$ $A_b(6,7)$ $A_b(7,1)$ $A_b(8,8)$

Table 3. Patterns and their corresponding confidence for example DFAs. When several patterns have the same length, we randomly show only one of them.

Length	DFA 1		DFA 2			DFA 3		DFA 4		DFA 5	
	Pattern	Con.	Pattern	Con.	Prob.	Pattern	Con.	Pattern	Con.	Pattern	Con.
2	bb	2/3	ab	3/5	0.674	bb	1/2	ab	3/7	aa	5/8
3	bbb	1	bab	4/5	0.912	abb	1/2	abb	3/7	aaa	7/8
4			abab	1	1.0	abba	1/2	aabb	4/7	aaaa	1
5						baabb	2/3	aaabb	5/7		
6						bbaabb	1	aaaabb	5/7		
7								aaaabbb	5/7		
8								bbaaaabb	1		

3.3   Evaluation of DFA Pattern Identification

In the following experiments, we use the Tomita-4 and Tomita-7 grammars (indexed as DFA-1 and DFA-2), which are representative grammars for the exponential and the polynomial classes [12], respectively, and also use randomly generated other DFAs as shown in Table 2. For all DFAs, we set their starting state as state 1 and their absorbing states as the states with the largest indexes. By applying the algorithm we previously introduced, we obtain and demonstrate in Table 3 identified patterns for all evaluated DFAs.

We observe in Table 3 that our algorithm successfully identified the perfect absolute pattern for the Tomita-4 (DFA-1) grammar. We also find that the length of an identified perfect absolute pattern does not necessarily increase as the number of states increases. Moreover, we observe that the confidence for determining an absolute pattern for all DFAs is non-decreasing as the length of the identified pattern increases. To further understand the relationship between the confidence and the probability introduced in the definition of the absolute pattern, we design the following experiment and use DFA-2 as our demonstrative example. Specifically, we generate 1000 strings for each identified pattern with their lengths less than 15. We then calculate the frequency of the generated strings that appear in both the accepted and rejected sets, respectively. In particular, we use that frequency to approximate the probability by using the law of large numbers. We show in Table 3 that the probability difference and confidence have a positive correlation. Although we do not establish a theoretical relationship between the above mentioned two statistics, we empirically show that it is reasonable to replace the probability with the confidence. As such, we believe these results validate the effectiveness of our algorithm.

4   Conclusion

Here we have defined transition importance and critical patterns for DFAs, which we believe gives insight into understanding and identification of specific DFAs. Specifically, we transformed the widely-accepted adversarial sample scheme to an adversarial model scheme, which reveals the sensitivity of a model with respect to its components. For the case of a DFA, we focus on the components represented by its transitions. In addition, we have designed an effective synchronizing algorithm to find critical patterns of a DFA and studied the upper bound of the length of a perfect absolute pattern. Finally, we empirically validated our algorithms and the practicability of our metric with several grammars. Future work could focus on extending this work to understand more complex models and DFAs used in real applications.

References

1. Angluin, D.: Learning regular sets from queries and counterexamples. Inf. Comput. **75**(2), 87–106 (1987)
2. Bottou, L., Bousquet, O.: The tradeoffs of large scale learning. In: NIPS, pp. 161–168. Curran Associates Inc., (2007)
3. Černý, J.: Poznámka k homogénnym experimentom s konečnými automatmi (a note on homogeneous experiments with finite automata). Matematicko-fyzikálny časopis **14**(3), 208–216 (1964)
4. Chomsky, N.: Three models for the description of language. IRE Trans. Inf. Theory **2**(3), 113–124 (1956)
5. Cybenko, G.: Approximation by superpositions of a sigmoidal function. Math. Control Signals Syst. **2**(4), 303–314 (1989). https://doi.org/10.1007/BF02551274

6. Don, H., Zantema, H.: Finding DFAs with maximal shortest synchronizing word length. In: Drewes, F., Martín-Vide, C., Truthe, B. (eds.) LATA 2017. LNCS, vol. 10168, pp. 249–260. Springer, Cham (2017). https://doi.org/10.1007/978-3-319-53733-7_18

7. Guidotti, R., Monreale, A., Ruggieri, S., Turini, F., Giannotti, F., Pedreschi, D.: A survey of methods for explaining black box models. ACM Comput. Surv. **51**(5), 1–42 (2019)

8. Hopcroft, J.E., Motwani, R., Ullman, J.D.: Introduction to automata theory, languages, and computation. ACM Sigact News **32**(1), 60–65 (2001)

9. Koh, P.W., Liang, P.: Understanding black-box predictions via influence functions. In: Proceedings of the 34th International Conference on Machine Learning. ICML, pp. 1885–1894 (2017)

10. Serban, A.C., Poll, E.: Adversarial examples - a complete characterisation of the phenomenon. CoRR, abs/1810.01185 (2018)

11. Tomita, M.: Dynamic construction of finite-state automata from examples using hill-climbing. In: Proceedings of the Fourth Annual Conference of the Cognitive Science Society, pp. 105–108 (1982)

12. Wang, Q.: A comparative study of rule extraction for recurrent neural networks. arXiv preprint arXiv:1801.05420 (2018)

13. Weiss, G., Goldberg, Y., Yahav, E.: Extracting automata from recurrent neural networks using queries and counterexamples. In: Proceedings of Machine Learning Research. ICML, vol. 80, pp. 5244–5253. PMLR (2018)

Personalized Student Attribute Inference

Khalid Moustapha Askia and Marie-Jean Meurs[(⊠)][iD]

Université du Québec à Montréal (UQAM), Montreal, QC, Canada
moustapha_askia.khalid@courrier.uqam.ca, meurs.marie-jean@uqam.ca

Abstract. Accurately predicting their future performance can ensure students successful graduation, and help them save both time and money. However, achieving such predictions faces two challenges, mainly due to the diversity of students' background and the necessity of continuously tracking their evolving progress. The goal of this work is to create a system able to automatically detect students in difficulty, for instance predicting if they are likely to fail a course. We compare a naive approach widely used in the literature, which uses attributes available in the data set (like the grades), with a personalized approach we called *Personalized Student Attribute Inference* (PSAI). With our model, we create personalized attributes to capture the specific background of each student. Both approaches are compared using machine learning algorithms like decision trees, support vector machine or neural networks.

Keywords: Big data · Educational data mining · Knowledge tracing · Machine learning

1 Introduction

As all academic institutions aim to improve the quality of education, the success of their students is essential. To make university affordable and worthwhile, it is hence important to ensure that most of the students enrolled in a program succeed it and graduate on time. Therefore, early interventions for students who most likely will fail their courses can help them save both time and money. A possible solution towards this end is to build an automatic system that would successfully predict their future outcome. However, predicting students' performance is complex. The attributes frequently used by researchers are the Grade Point Average (GPA), internal assessment and students' demographic (gender, age etc). The issue with those attributes is that they tell nothing valuable about the student background and what s/he has been through. For that reason, predicting methods need to incorporate a way to capture students' background along with the historical student accomplishments (grades, credits obtained, GPA).

We developed a personalized model called *Personalized Student Attribute Inference* (PSAI), which creates personalized attributes to capture the specific background of each student. We compare our model with a naive approach, which uses directly the attributes available in the data set (like the grades, credits obtained, GPA, etc.). The next section explains our approach.

C. Goutte and X. Zhu (Eds.): Canadian AI 2020, LNAI 12109, pp. 553–557, 2020.
https://doi.org/10.1007/978-3-030-47358-7_56

Section 3 describes our experimental process and results, and finally, Sect. 4 discusses limitations and concludes.

2   Personalized Models

We focus on personalized models, which take into account as much as possible the specifics of each student profile, and therefore emphasizes the student's background. Grades, GPA, credits obtained, etc. are not sufficient since they cannot model a student's knowledge. For example, students SA and SB have both a GPA of 3.7 but SA took only easy courses (3 courses in total) and SB took the most difficult ones (5 courses in total). Both have the same GPA so we cannot automatically determine who is the most talented. It shows that static attributes (the ones that are recorded directly like the grades) do not actually provide much profile details about a student. Thus, for accurately predicting student performance, one should consider other attributes as for instance the difficulty of the courses. To do so, we grouped courses by similar level of difficulty then we assigned them a weight, increasing with difficulty. Also, we assigned a score to each student depending on the total number of courses s/he took, their difficulty and the grades s/he obtained.

2.1   Personalized Student Attribute Inference (PSAI)

For assigning weights to courses according to their difficulty, we take a scale which limits are the weights of "extremely" easy courses and those of courses "extremely" difficult. We experimentally assign weights as follows: "extremely" easy courses get a weight of 0.5 and "extremely" difficult courses get a weight of 2. An "extremely" difficult course is hence 4 times more difficult than an "extremely" easy course. By analyzing University marks system (where our data came from), for the "extremely" easy courses, we consider an average mark of 4.15 (between A (4.0) and A + (4.3)) and for the "extremely" difficult courses, an average of 1.15 (between D (1.0) and D + (1.3)).

Subsequently, in order to be able to assign a weight to a course according to the average mark obtained, we must consider a parametric function that will take this average mark as input and output the associated weight in accordance with the limits established above. The function must also be decreasing, *i.e.* if the input (the average mark) increases, the output (the weight) must necessarily decrease. Let $\beta \times \exp(-\alpha x)$ be an exponential function where β and α are parameters to be determined, and x is the average of the marks obtained by the students who took the course. To estimate the parameters β and α, we use the limits we fixed. Solving the following equations system:

$$\begin{cases} \beta \times \exp(-1.15\alpha) = 2 \\ \beta \times \exp(-4.15\alpha) = 0.5 \end{cases}$$

provides: $\alpha = \frac{\ln(4)}{3}$ and $\beta = 2\exp\left(\frac{1.15\ln(4)}{3}\right)$

Making use of this function that assigns a weight to a course according to its difficulty, we present in the following subsection our algorithm to create a personalized data set, which will be used to train machine learning algorithms and make performance predictions.

2.2   PSAI Algorithm

Algorithm 1. PSAI Algorithm for course A

Input: Prior information on courses (average mark in the course, marks obtained) took by students that took course A

1: $\alpha \leftarrow \frac{\ln(4)}{3}$

2: $\beta \leftarrow 2\exp\left(\frac{1.15\ln(4)}{3}\right)$

3: **For** each student that took course A:

4:      **For** each course i taken before course A:

5:           Let m_i be the average mark in the course i (according to all students that took this course)

6:           Let n_i be the mark of the current student in course i

7:           Compute the score of the student in the course i: $S_i = n_i \times \beta \times \exp(-\alpha m_i)$

8:      **End For**

9:      Compute the total score of the student: $S = $ mean of S_i

10: **End For**

11: Compute the weight of the course A: $W_A = \beta \times \exp(-\alpha m_A)$ where m_A is the average mark in course A

Output: A score for each student and the computed weight for course A

Algorithm 1 provides a dataset that will be used to train the prediction model. This dataset contains a score for each student and the weight of the course for which we want to predict the student performance. We also add as an attribute the overall success rate in the course.

3   Experiments and Results

For the sake of brevity, we present our results only for the following course (acronyms changed for non-disclosure reasons): ABC2222 : 6483 students with 5256 success and 1227 failures. The question asked to our models is the following: will a given student fail the course? The method of training and testing in our experiments is the cross validation [5]. We used the following machine learning algorithms : decision trees [6], K-Nearest Neighbors [3], Support Vector Machine (SVM) [2], random forest [1], an ensemble learning model (AdaBoost) [4] and neural network [7]. The evaluation metric is the F-measure (or F1-score) [8].

Tables 1 shows the results obtained with the different machine learning algorithms. We compare our results with those obtained using a direct (naive) method (the standard method) which only uses the attributes related to the course and the students present in the database. In our case these attributes are: admission base, citizenship, previous program, legal status, college program, age, gender, number of course credits obtained and GPA.

Table 1. Failure prediction in ABC2222 results

ABC2222		
Algorithm	Naive method	PSAI
	F-measure (%)	F-measure (%)
Neural network	32,87	68,47
Decision tree	44,37	66,37
Adaboost	45,70	**69,57**
k-NN	37,07	64,31
Random forest	**47,36**	58,30
SVM	27,81	62,75

4   Conclusion

PSAI outperforms standard methods. Despite the imbalanced nature of the data set including fewer failures than successes, our approach detects many of these failures. It proves the need to consider several "hidden" aspects including the difficulty of the courses taken. The main limitation of the approach is related to the data set itself. Some essential data are missing to improve the model. We do not have the information of students before their first registration at the university. Hence, our model can only be used from the second course taken by the student. In addition, as often when dealing with real data collected over a long period of time, a large part of the records is unusable because of missing data and errors/noises.

Reproducibility. This project is publicly released as an open source software in the following repository: https://gitlab.ikb.info.uqam.ca/khalid/psai

Acknowledgment. We acknowledge the support of the Natural Sciences and Engineering Research Council of Canada (NSERC), [MJM Canada NSERC Grant number 06487-2017]

References

1. Breiman, L.: Random forests. Mach. Learn. **45**(1), 5–32 (2001)
2. Cortes, C., Vapnik, V.: Support-vector networks. Mach. Learn. **20**(3), 273–297 (1995)

3. Cover, T., Hart, P.: Nearest neighbor pattern classification. IEEE Trans. Inf. Theory **13**(1), 21–27 (1967)
4. Dietterich, T.G.: Ensemble learning. In: The Handbook of Brain Theory and Neural Networks, vol. 2, pp. 110–125 (2002)
5. Kohavi, R., et al.: A study of cross-validation and bootstrap for accuracy estimation and model selection. In: IJCAI, Montreal, Canada, vol. 14, pp. 1137–1145 (1995)
6. Safavian, S.R., Landgrebe, D.: A survey of decision tree classifier methodology. IEEE Trans. Syst. Man Cybern. **21**(3), 660–674 (1991)
7. Sarle, W.S.: Neural networks and statistical models (1994)
8. Sasaki, Y., et al.: The truth of the F-measure. Teach. Tutor. Mater. **1**(5), 1–5 (2007)

Vehicle Traffic Estimation Using Weather and Calendar Data

Meetkumar Patel[(✉)]

Jodrey School of Computer Science, Acadia University, Wolfville, NS, Canada
meetpatel.p359@gmail.com

Abstract. Vehicular traffic is an important planning concern for commuters and businesses. However, there is no application available which can estimate traffic congestion and flow in the future up to five days. The problem is to develop an application that can predict vehicular traffic density and flow rate based on weather data, calendar data and special events data. This information would be valuable for commuters planning short or long- distance trips, and for transportation and infrastructure departments for better planning the maintenance of roads. The proposed research will combine image processing and machine learning methods.

Keywords: Traffic estimation · Machine learning · Image processing · Weather data · Calendar data

1 Introduction

People have access to the weather forecast and the current traffic situation, but they do not have a forecast for the traffic over the next couple of days taking into consideration of the forecasted weather. Future traffic information is important for if people can be informed about the traffic situation in advance, they can plan their travel accordingly. There is a need for an application that allows travelers to make better decisions about travel time and routes based on knowledge of forecasted weather and special events. Our challenge is to estimate traffic flow rate and density up to five days in advance for the Mackay Bridge in Halifax, Nova Scotia.

2 Background

There is a lot of research that has been done in the area of vehicle counting using video image processing and machine learning with good end results. Manchun et el. [1] presents a method to detect and count vehicle in complex scenes. Adaptive real-time block-wise background updating algorithm introduced in [2] produces highly accurate detection results. However, video data has been used in these methods for vehicle counting while in our case image data from NS webcam has variable interval of two to seven seconds between consecutive images

C. Goutte and X. Zhu (Eds.): Canadian AI 2020, LNAI 12109, pp. 558–561, 2020.
https://doi.org/10.1007/978-3-030-47358-7_57

and in addition to weather data, special events data also influence the prediction accuracy of traffic flow rate and density.

Prediction of traffic information can be affected by considering the impact of weather. Rainfall affects traffic volume and the inclusion of rainfall as a model input improves the prediction accuracy during rainfall events [3]. Jia et al. [4] concludes that the LSTM family can outperform the DBN in learning the time series characteristics of traffic data and the results highlights the importance of considering weather impacts and the considerable improvements attainable in prediction accuracy. Using deep learning, weather information has also used to improve the traffic flow prediction using data fusion at decision level. [5].

3   Approach

To predict traffic flow rate and density we need historical traffic data to train and test prediction models. This data will be extracted from Nova Scotia webcam [6] images. Parameters like camera position, image quality and frame rate affect the accuracy of the method. In the image data from NS webcam, varying capture intervals and the location of the camera, angled from the top left side of the bridge, makes this task more challenging.

Historical weather data will be obtained from Environment Canada while special event data, like Christmas or Natal Day, will be extracted from various web calendars. While weather data and calendar data are the primary factors affecting traffic density, time of day also plays a vital role as there is an increased number of cars on the road before and after office hours as compared to other times of the day.

3.1   Number of New Vehicles per Image Prediction

To estimate traffic flow rate and density, we need historical traffic information that contains flow rate and density per hour. To get this historical traffic data we are using image sequence data obtained from NS webcam over a 15-month period of the Mackay bridge in Halifax (Fig. 1(a)). Image processing and machine learning techniques are applied to the image sequence data to help estimate traffic count.

Before passing the images to the machine learning system they are preprocessed to make the learning task easier by identifying the moving objects and removing background from each image. The Background Subtraction algorithm is used to identify moving vehicles in an image sequence. The algorithm accepts a series of images as input and outputs masked images. Where vehicles are identified as a group of white pixels on an all black background. The image processing function called Closing is used to remove noise from the vehicles identified in the current frame. The closing operation is applied on the masked images to fill small holes inside the detected foreground vehicles.

By implementing element-wise multiplication between an original image and its closed transform, we get the actual vehicles without a background. To obtain the Region of Interest, unnecessary areas are cropped in each image. The NS Webcam images contain both inbound and outbound lanes of the bridge, but since the focus of this research is on the outbound lane of the bridge, we crop out the inbound lane from the image. After applying all these image processing techniques, the result is a relatively small image of three channels containing only vehicles (Fig. 1(e)).

The processed images are used to train a CNN-LSTM machine learning model as shown in Fig. 2. The goal is to train the model to predict the number of new vehicles in each frame. Each vehicle takes an average of 3 frames to pass the Region of Interest. Therefore, to predict the number of new vehicles in a given image, the model accepts a sequence of three consecutive frames (images) and the time in milliseconds since the last frame.

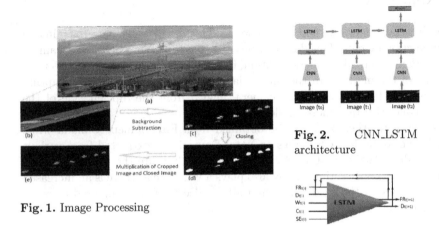

Fig. 2.   CNN_LSTM architecture

Fig. 1. Image Processing

Fig. 3. Auto-regression

3.2   Traffic Flow Rate and Density Prediction

An LSTM model will be trained to predict the traffic flow rate and density for the next hour using weather data, calendar data and special event data for each hour of every day over 2019. The model will be rolled forward to predict the traffic flow rate and density for each hour up to the next five days in the future.

For Data preparation, vehicular count must be converted into traffic flow rate and density on an hourly basis. To do this, we aggregate the predictions made by the "number of new vehicles" CNN-LSTM model for the given hour and apply the equations. To calculate the flow rate, $f = m/\Delta t$ where, m is the number of vehicles and Δt is the time period for which we are calculating traffic flow rate. To calculate the density, $d = n/\Delta x$ where n is the number of vehicles and Δx denotes an Area.

We are using weather data, calendar data and special event data formatted sequentially. To train a model to predict up to five days in advance, we will use our model to predict traffic for the next hour and then we will use that prediction as an auto-regressive value to predict the traffic flow rate and density for the next hour, up to five days into the future. Initially, the LSTM model will have one hidden layers. The results will be analyzed and the model will be modified to achieve the best accuracy.

Figure 3 illustrates the auto-regressive process, where $FR_t(i)$ and $FR_t(i+1)$ are the traffic flow rates for i^{th} hour and $i+1$ hour respectively, $D_t(i)$ and $D_t(i+1)$ are traffic densities for i^{th} hour and $i+1$ hour respectively, $W_t(i)$ is weather data, $C_t(i)$ is calendar data, and $SP_t(i)$ is special event data.

4 Empirical Studies Thus Far

We will use the first 12 months of data to train the models and the remaining 3 months to test the models. The mean absolute error (MAE) is used as the performance measure for both models. We have developed a CNN-LSTM model as described in Sect. 3.1, that produces MAE = 0.67 counts using a small dataset of only 150 images over only 6 min of time. By aggregating the vehicle counts the accuracy is 90.4%. Our goal is to develop a model that predicts traffic flow rate with a MAPE >85% and predicts traffic density >85%.

5 Conclusion

The result shows that the approach is promising with a small dataset. CNN-LSTM networks have the potential to solve the vehicular counting problem. The future work is to train a model with a large data set and then convert the results into traffic flow rate and density. These will be used as a label data for the model that will predict traffic flow rate and density based on weather data, calendar data and special event data.

References

1. Lei, M., Lefloch, D., Gouton, P., Madani, K.: A video-based real-time vehicle counting system using adaptive background method. In: IEEE International Conference on Signal Image Technology and Internet Based Systems, pp. 523–528. IEEE 2008
2. Liu, F., Zeng, Z., Jiang, R.: A video-based real-time adaptive vehicle-counting system for urban roads. PloS one **12**(11), e0186098 (2017)
3. Dunne, S., Ghosh, B.: Weather adaptive traffic prediction using neurowavelet models. IEEE Trans. Intell. Transp. Syst. **14**(1), 370–379 (2013)
4. Jia, Y., Wu, J., Xu, M.: Traffic flow prediction with rainfall impact using a deep learning method. J. Adv. Transp. **2017**, 10 (2017)
5. Koesdwiady, A., Soua, R., Karray, F.: Improving traffic flow prediction with weather information in connected cars: a deep learning approach. IEEE Trans. Veh. Technol. **65**(12), 9508–9517 (2016)
6. Nova Scotia webcams. https://www.novascotiawebcams.com/en/webcams/mackay-bridge/

Predicting Aggressive Responsive Behaviour Among People with Dementia

Maryam Tajeddin$^{(\boxtimes)}$ (iD)

Acadia University, Acadia Wolfville, NS B4P 2R6, Canada
152527t@acadiau.ca
https://cs.acadiau.ca

Abstract. Patients with dementia will have difficulty properly communicating life's challenges which can cause them to become agitated, resulting in verbal or physical aggression. Monitoring the risk of a resident harming themselves or others due to aggressive behaviour is a priority within a long-term care facility where dementia is present. Caregivers at long term care facilities record resident health and behaviour digitally either as structured data or unstructured text, providing an on-going log of each resident's patient history. We aim to use natural language processing (NLP) and machine learning (ML) techniques to develop models that can predict the probability of a resident exhibiting aggressive behaviours that may harm themselves or others within the next week.

Keywords: Natural language processing · Dementia · Machine learning · Aggressive behaviour

1 Introduction

Dementia is a general term for progressive diseases that cause a decline in memory, language, critical thinking and other reasoning aptitudes that influence an individual's capacity to perform regular activities. Difficulty in communicating makes dementia patients agitated and results in verbal or physical aggression. Caregivers at long term care facilities record resident health and behaviour digitally either as structured data or unstructured text, providing an on-going log of each resident's patient history. NLP techniques have been used to extract key features of persons' behaviour or social interactions from text such as emails, blogs, tweets, Facebook entries, etc. These features have been used to build ML models that can predict the persons' sentiments, propensity to purchase, and likely next action [1]. Similar work in this area requires additional data to be recorded from residents using sensors [3]. We propose that it is possible to develop similar ML models to predict the probability of a resident's aggressive behaviour based on structured and unstructured data that long term care facilities have in their computing system. Our objective is to train a deep LSTM-RNN

Supported by Shannex and Mitacs.

model to predict the probability of an aggressive act by a resident each day for the next seven days. Our approach will be split into six sections: 1. Structured and Unstructured data collection 2. Data cleaning 3. Feature Creation 4. Feature Consolidation 5. Development and Testing of deep RNN 6. Prediction and Visual Comparison.

1.1 Problem Statement

Patients with dementia will eventually experience a significant loss of cognitive function. As a result, they will have difficulty solving and communicating their physical problems and emotional pain appropriately. This is clinically described as the behavioural and psychological symptoms of dementia or responsive behaviours [8]. These can range from lower-risk behaviours such as restlessness and repetitiveness to higher risk behaviours such as agitation and verbal or physical aggression. Therefore, monitoring the risk of a resident harming themselves or others is always a priority within a long-term care facility where dementia is present. Caregivers are trained to recognize signs of degrading states of dementia and commons sources of stress for people with dementia but predicting when the aggressive behaviour is going to happen needs more investigation. We believe by using machine learning (ML) techniques we can predict the probability of a resident exhibiting aggressive behaviours.

2   Approach

At the long-term care facilities we are working with, care providers record resident behaviour using the following four EHR documentation tools: patients check-in demographic record, assessment reports (every 90 days), progress notes made routinely, and incident reports following a significant occurrence such as a violent outburst. Each of these are recorded digitally as structured data or unstructured text by the facilities' staff and provide an on-going log of each resident's patient history. When a violent situation occurs, often caregivers will review recent portions of the progress notes or assessment report to help determine the potential cause of such incidents. However, this commonly happens after the incident, to better understand why it occurred. Shannex, a family-owned and operated provider of long-term care in Nova Scotia, is exploring how to develop a system that predicts the probability of resident aggressive behaviours prior to an incident occurring.

A computerized clinical decision support (CDS) aims to aid decision making of health care providers and the public by providing easily accessible health-related information at the point and time it is needed [2]. NLP is instrumental in using free-text information to drive CDS, representing clinical knowledge and CDS interventions in standardized formats, and leveraging clinical narrative [5]. In principle, NLP could extract the facts needed to represent clinical knowledge and to develop many kinds of decision rules to be used within a long-term care facility. The following structured data and unstructured textual data for each patient will be collected: patients check-in demographic record, assessment reports progress notes made routinely, and incident reports and organized by date and time. Features will be created from the structured data using data engineering methods. Standard data engineering methods will be developed using Python and associated libraries such as Pandas. Features will be created from the unstructured textual data using NLP methods and data engineering methods. We will create an array of features using a bag-of-words approach that will include keywords (stop word removal, stemming), word letter capitalization, part-of-speech tagging, and n-grams of words using NLTK [4] and Stanford CoreNLP [7]. All features will be consolidated into a set of temporal records for each resident and include a date and time stamp as well as a label indicating the degree of aggression that has (or has not) occurred at that time. Assistance will be needed from Shannex personal in order to annotate records associated with aggressive behaviour. Our approach will be to create data covering from 2010 to 2019 up to 4010 residents on 15 Shannex facilities for developing and testing predictive models. Deep recurrent neural network (RNN) models will be developed to predict the probability of an aggressive act by a resident in the next 7 days given the resident's temporal record sequence to date. The model will be tested using an independent set of resident temporal records for residents who have and who have not harmed themselves or others at the facility. The success of the approach will be analyzed, and improvements will be made to increase the performance of the predictive model, in accordance with the success criteria.

2.1   Data Preparation

To protect patients' confidentiality from the academic research team, there are some privacy-preserving data sharing techniques for medical text data focusing on detection and removal of patient identifiers from the dataset. Anonymization or de-identification is one of the techniques to protect patients' privacy in sharing medical and healthcare data. In our approach, we use the Stanford CoreNLP system [7], an annotation-based NLP processing pipeline to tokenize and label the text. After anonymizing names, all features will be consolidated into a set of temporal records for each resident. Each patient's record consists of an incident report. In the incident reports, there is a category called "struck out" where all aggressive behaviours are recorded. Via NLP techniques, we will train our model to extract words such as punch, kick, swear, etc. indicating aggression and train our model to learn where and when an incident of aggression occurs. Each record will include a label indicating the degree of aggression that has (or has not) occurred at that time.

2.2   Recurrent Neural Network Model Development

Recurrent Neural Networks (RNNs), are good at processing sequence data using sequential memory for prediction [6]. Sequential memory is a mechanism that makes it easier for the human brain to recognize sequences over time. Sequence data comes in many forms and text is one such form. It can be broken up into a sequence of words to predict the possibility of an event such as a record of an aggressive act that would be recorded in an incident report. Models will be trained on residents who are not in the test set to ensure predictions generalize on unseen residents' records. Ultimately, to evaluate our approach we will use 5-fold cross-validation. F1 score will be used as our evaluation metric since binary accuracy may be affected by an unbalanced class presence.

3 Conclusion and Future Work

We aim to develop models using natural language processing (NLP) and machine learning (ML) techniques that can predict the probability of a resident exhibiting aggressive behaviours that may harm themselves or others within the next week. In the future, Shannex may use the predicted level of aggression, to electronically generate an alert for staff so that steps can be taken to reduce the probability of verbal or physically aggressive behaviour.

References

1. Belinkov, Y., Glass, J.: Analysis methods in neural language processing: a survey. Trans. Assoc. Comput. Linguist. **7**, 49–72 (2019)
2. Demner-Fushman, D., Chapman, W.W., McDonald, C.J.: What can natural language processing do for clinical decision support? J. Biomed. Inform. **42**(5), 760–772 (2009)
3. Khan, S.S., et al.: DAAD: a framework for detecting agitation and aggression in people living with Dementia using a novel multi-modal sensor network. In: 2017 IEEE International Conference on Data Mining Workshops (ICDMW), pp. 703–710 (November 2017). https://doi.org/10.1109/ICDMW.2017.98. ISSN:2375-9259
4. Loper, E., Bird, S.: NLTK: the natural language toolkit. In: arXiv preprint arXiv:cs/0205028 (2002)
5. Reyes-Ortiz, J.A., González-Beltrán, B.A., Gallardo-López, L.: Clinical decision support systems: a survey of NLP-based approaches from unstructured data. In: 26th International Workshop on Database and Expert Systems Applications (DEXA), pp. 163–167. IEEE (2015)
6. Shi, Z., Shi, M., Li, C.: The prediction of character based on recurrent neural network language model. In: 2017 IEEE/ACIS 16th International Conference on Computer and Information Science (ICIS), pp. 613–616. IEEE (2017)
7. Stanford CoreNLP - Natural Language Software - Stanford CoreNLP. https://stanfordnlp.github.io/CoreNLP/. Accessed 02 Aug 2020
8. Tible, O.: Phenomenological understanding of behavioural and psychological symptoms of Dementia: a clinical case presenting howls and shouting. Melancholic Type Phenomenol. Sep. J. Depress Anxiety **6**(290), 1044–2167 (2017)

Towards Analyzing the Sentiments in the Fields of Automobiles and Real-Estates with Specific Focus on Arabic Online Reviews

Ayman Yafoz[✉]

Department of Computer Sciene, University of Regina, Regina, SK, Canada
yafoz20a@uregina.ca

Abstract. The importance of performing sentiment analysis is noticed in various fields, such as politics, education, marketing, and so forth. However, in the Arabic language domain, there are limited studies that focus on analyzing the sentiments in the text in comparison to the English language. There is a lack of available annotated Arabic datasets covering specific domains (such as real-estates and automobiles) and containing data written in both modern standard Arabic (MSA) and the Gulf Cooperation Council (GCC) dialect. Furthermore, the limited and inadequate adoption of natural language processing and machine learning techniques is noteworthy in the current sentiment analysis contributions targeting the Arabic language. Therefore, the gap could be bridged by creating real-estates and automobiles datasets. Moreover, customizing, enhancing, and applying suitable natural language processing techniques and machine learning algorithms to analyze the sentiments in these datasets will also contribute to filling the current gap. Performing these steps will benefit the people interested in analyzing the sentiments related to real-estates and automobiles, and will add a new scope to the Arabic sentiment analysis field. Future researchers in this field could also be benefited by using the datasets that will be freely available. The aforementioned factors encouraged the researcher to conduct this research in order to fill the current gap in this area.

Keywords: Sentiment analysis · Arabic language · Automobiles · Real-estates · Machine learning

1 Problem Statement and Motivation

Sentiment analysis is a key research area in the field of applied linguistics. However, the sentiment analysis contributions addressing the Arabic language are not as mature as the contributions addressing the English language [5]. There is a lack of contributions specifically addressing sentiment analysis in automobile and real-estates online reviews in the Arabic language, particularly in the

© Springer Nature Switzerland AG 2020
C. Goutte and X. Zhu (Eds.): Canadian AI 2020, LNAI 12109, pp. 566–570, 2020.
https://doi.org/10.1007/978-3-030-47358-7_59

GCC dialect. To fill this gap, it is required to develop a special linguistic treatment and create dedicated datasets that work with specially adjusted machine learning classifiers to analyze the sentiments in the Arabic text. This challenge motivated the author to conduct this research.

2   Literature Review

The researchers in [1] proposed a system that applied a multi-way sentiment analysis approach to Arabic reviews. The dataset contains 63,257 reviews about books written in both colloquial and MSA. The highest accuracy result they found is 57.8% by the K-Nearest Neighbor for the classification approach that has four-levels hierarchical structure applied on five different partitions of the dataset. Another research conducted by the researchers in [4], targeted the classification of 54,716 Arabic tweets related to Egypt. The highest accuracy result they obtained is 69.10% by the SVM classifier.

3   Research Methodology

The novelty of this research lies in filling the gap of lacking a specific sentiment analyzer that works on datasets containing data written in both MSA and GCC dialect and are related to real-estates and automobiles. The proposed solution is composed of main stages, which are briefly described below.

3.1   Data Gathering and Annotation

The data related to automobile and real-estates is automatically gathered through specifically created web scrappers (developed using Python's BeautifulSoup library) from the three famous websites in GCC "Haraj for automobile data, Hawamer and Aqarcity for real-states data," and then filtered to leave data records containing sentiments (positive, negative or mixed). The automobiles dataset contains around 6,585 comments divided into three sentimental categories (positive, negative and mixed), and it is focusing on almost 29 topics related to automobiles. On the other hand, the real-estates dataset contains around 6,434 comments, and it is divided into three sentimental categories (positive, negative and mixed), and it focuses on almost 85 topics related to real-estates. For instance, the Arabic sentence "بسبب الغلاء الفاحش بالعقار الشباب و الشابات لم يستطيعوا الزواج" which means "Young men and women could not marry because of the high cost of real estate" represents negative records. Moreover, the Arabic sentence "مرتفعة شوي الباجيرو موتر ممتاز و قوي لكن أظن إن قطع غياره" which means "The Pajero is an excellent and a strong motor, but I think its spare parts are a little bit expensive" represents mixed records. Finally, the Arabic sentence "الألتيما ممتازة و سعرها جدا مناسب" which means "Altima is excellent and very affordable" represents positive records.

The data in both datasets is annotated by three annotators. Following that, the inter-rater agreement is computed through Fleiss Kappa to show the consistency and agreement level between annotators. For both datasets, the computed Fleiss Kappa degree for inter-rater agreement exceeded 82%, which reflects a degree of almost perfect agreement.

3.2   Data Preprocessing and Features Selection

The system is developed using Python programming language and libraries. The data is cleaned to ensure it does not contain irrelevant and noisy data and it is suitable for classification. A special treatment exclusively related to Arabic language text is followed in this stage to clean and normalize the data. This treatment includes addressing regular expressions and diacritics. Furthermore, a list containing around 683 Arabic stop words is prepared. These stop words are excluded from the text as they are considered redundant words lacking valuable semantics. In features selection step, the initial features are decreased and a subgroup retaining sufficient information for acquiring satisfactory performance outcomes is chosen. As part of features selection, both Stanford POS tagger and lemmatizer specifically developed to work on Arabic text are adopted. Moreover, Term Frequency-Inverse Document Frequency (TFIDF) and the N-Gram feature are also selected as features to reduce the dimensionality of the dataset [2].

3.3   Datasets Preparation

The datasets are divided into training and testing datasets. Before splitting the datasets, the datasets are randomly shuffled to redistribute and reorganize the data into suitable sections. The case of an unbalanced dataset is treated as well to avoid inadequate representations of minority classes in the training dataset. SMOTENC is adopted to perform oversampling on the minority classes. Following that, the datasets are partitioned into 70% for the training set and 30% for the testing set to ensure attaining adequate training and testing.

3.4   Data Processing and Visualization

A total of 22 machine learning classifiers are adopted to perform the classification process. The hyperparameter tuning process is performed to select the most optimum hyperparameters for those classifiers. The selected hyperparameters rendered the best scores when tested in the training stages. A random search is adopted to perform the hyperparameter optimization as it has proven to be more efficient than the manual and the grid search approaches [3]. Further, the word cloud is adopted to visually depict the most frequent and sentiment words in the datasets, and the ROC curve is chosen to plot the true and false-positive rates.

4   The Results

For the automobile dataset, both the Ridge and the Ridge CV classifiers outperformed all other adopted classifiers, scoring 82.6% in the accuracy results. In terms of the precision weighted averages, the Ridge, the Ridge CV, and the Logistic Regression classifiers topped the other classifiers by scoring 83%. In terms of the recall weighted average, both the Ridge and the Ridge CV classifiers exceeded the other classifiers by scoring 83%. Finally, in terms of the F1-score weighted averages, the Ridge, the Linear Support Vector, the Ridge CV and the Logistic Regression classifiers had the highest result of 82%. For the real estate dataset, the Ridge CV classifier outperformed all other classifiers by scoring 77.4% in the accuracy results. In terms of the precision weighted averages, the Ensemble Soft Vote and the Ridge CV classifiers outperformed all other classifiers by scoring 77%. In terms of the recall weighted average, the Ridge, the Ensemble Soft Vote, the Linear Support Vector, the Logistic Regression CV and the Ridge CV classifiers surpassed all other classifiers by scoring 77%. Finally, in terms of the F1-score weighted averages, the Ridge CV classifier outperformed the other classifiers by scoring 76%.

Finally, these results are competitive when compared with other contributions' results addressing the Arabic language; some of these contributions are shown in the literature review. This shows that the proposed system has an opportunity to be adopted and further enhanced in future contributions due to the satisfactory results.

5   Conclusion and Future Work

Notwithstanding the suggested improvements in the near future outlined below, the current outcomes of the proposed system in this research reflect a promising future. Nevertheless, in the near future, more experiments and studies could be conducted on how to enhance the results through improving the data cleaning process, addressing negated phrases, including a dictionary as a hybrid approach, and adopting advanced deep learning techniques. Moreover, the datasets could be enlarged to contain more positive, negative and mixed samples, which should improve the classification results.

References

1. Al-Ayyoub, M., Nuseir, A., Kanan, G., Al-Shalabi, R.: Hierarchical classifiers for multi-way sentiment analysis of Arabic reviews. Int. J. Adv. Comput. Sci. Appl. **7** (2016). https://doi.org/10.14569/IJACSA.2016.070269
2. Bachu, V., Anuradha, J.: A review of feature selection and its methods. Cybern. Inf. Technol. **19**, 3 (2019). https://doi.org/10.2478/cait-2019-0001
3. Bergstra, J., Bengio, Y.: Random search for hyper-parameter optimization. J. Mach. Learn. Res. **13**, 281–305 (2012). http://dblp.uni-trier.de/db/journals/jmlr/jmlr13.html#BergstraB12

4. Nabil, M., Aly, M., Atiya, A.: ASTD: Arabic sentiment tweets dataset, pp. 2515–2519 (January 2015). https://doi.org/10.18653/v1/D15-1299
5. Taboada, M., Brooke, J., Tofiloski, M., Voll, K., Stede, M.: Lexicon-based methods for sentiment analysis. Comput. Linguist. 267–307 (2011). https://doi.org/10.1162/COLI_a_00049

Author Index

Printed in the United States
By Bookmasters